NEAR RINGS, FUZZY IDEALS, AND GRAPH THEORY

BHAVANARI SATYANARAYANA
KUNCHAM SYAM PRASAD

CRC Press
Taylor & Francis Group
Boca Raton London New York

CRC Press is an imprint of the
Taylor & Francis Group, an **informa** business
A CHAPMAN & HALL BOOK

CRC Press
Taylor & Francis Group
6000 Broken Sound Parkway NW, Suite 300
Boca Raton, FL 33487-2742

First issued in paperback 2019

ISBN-13: 978-1-4398-7310-6 (hbk)
ISBN-13: 978-0-367-38004-5 (pbk)

Library of Congress Cataloging-in-Publication Data

Satyanarayana, Bhavanari.
  Near rings, fuzzy ideals, and graph theory / Bhavanari Satyanarayana and Kuncham Syam Prasad.
    pages cm
  Includes bibliographical references and index.
  ISBN 978-1-4398-7310-6 (hardback)
  1. Near-rings. 2. Fuzzy sets. 3. Graph theory. I. Prasad, Kuncham Syam. II. Title.

QA251.5.S28 2013
512'.46--dc23

2012043212

Visit the Taylor & Francis Web site at
http://www.taylorandfrancis.com

and the CRC Press Web site at
http://www.crcpress.com

Contents

Preface

Near rings, fuzzy ideals, and graph theory is a very fascinating course. Near ring theory has enormous applications in different subject areas such as digital computing, sequential mechanics, automata theory, graph theory, and combinatorics. The first step toward near rings was axiomatic research done by Dickson in 1905. He exhibited that there do exist "fields with only one distributive law." Near rings arise in a natural way. The set $M(G)$ of all mappings of a group $(G, +)$ into itself, with the usual addition and composition of mappings becomes a near ring. Another example is the set of all polynomials with addition and substitution.

This book provides the reader with a comprehensive idea about near ring theory with some links to fuzzy ideals and graph theory. It broadly covers three major topics: near rings, fuzzy ideals, and graph theory. Chapter 1 discusses all necessary fundamentals of algebraic systems. Chapters 2 and 3 cover the essentials of the fundamentals of near rings theory, appropriate examples, notations, and simple theorems. Chapter 4 covers the prime ideal concept in near rings. The rigorous approach of the dimension theory of N-groups, along with suitable illustrations, is presented in Chapter 5. Most of the chapters cover topics from the recent literature about certain notions and results such as prime ideals, essential ideals, uniform ideals, finite dimension, and primary and tertiary decompositions along with several characterizations. In Chapter 6, we make a brief study of matrix near rings with detailed proofs presented wherever necessary. The concept of gamma near ring, a generalization of the concepts of both gamma and near rings, is presented in Chapter 7 with suitable results. In Chapter 8, an introduction to fuzzy algebraic systems is presented. Particularly, the fuzzy ideals of near rings and gamma near rings are studied extensively. In Chapter 9, an attempt is made to discuss the concept of graph theory. A good presentation on some important concepts such as directed hypercubes, dimension, prime graphs, and graphs with respect to ideals in near rings completes the book.

This textbook can be used by students as an introductory course for MSc (pure/applied mathematics) and as a prescribed book for MPhil and PhD course work. In addition, one can select suitable parts from this textbook to frame the syllabus of the relevant papers for different courses according to one's needs.

Acknowledgments

The authors are pleased to express their thanks to the following for their constant encouragement in each of their endeavors:

Dr. Richard Wiegandt and Dr. Laszlo Marki of the Department of Algebra, A. Renyi Institute of Mathematics, Hungarian Academy of Sciences, Budapest, Hungary

Dr. H.H. Brungs, University of Alberta, Canada

Dr. P.V. Arunachalam, former vice chancellor of Dravidian University, Andhra Pradesh, India

Dr. D. Ramakotaiah, former vice chancellor of Acharya Nagarjuna University

Dr. A. Radha Krishna, former professor of mathematics, Kakatiya University, Andhra Pradesh, India

Dr. L. Radha Krishna, visiting professor of mathematics, Bangalore University, India

Dr. Vinod V. Thomas, Director, MIT, Manipal University

Dr. D. Srikanth Rao, Joint Director, MIT, Manipal University

Dr. Kedukodi Babushri Srinivas, MIT, Manipal University

Dr. Harikrishnan P.K., MIT, Manipal University

Dr. V.R.C. Murty, Department of Bio-Tech, MIT

Dr. M. Srinivasulu, MIT, Manipal University

(Dr.) Srikanth Prabhu, Department of CSE, MIT

Dr. Dasari Nagaraju, Manipal University, Jaipur

Dr. T.V. Pradeep Kumar, Acharya Nagarjuna University

Smt T. Madhavi Latha, PGT, APSWRS, Andhra Pradesh, India

Dr. P. Thrimurthy and Dr. Y. Venkateswara Reddy of Acharya Nagarjuna University

Dr. Satyanarayana places on record his deep sense of gratitude to his parents: Bhavanari Ramakotaiah (a teacher in an elementary school in the village of Madugula) and Bhavanari Anasuryamma, without whose constant encouragement and help it would not have been possible for him to pursue higher studies in mathematics. Also, he thanks his wife Bhavanari Jaya Lakshmi, and his children, Mallikharjuna, MTech, Satyasri, MBBS (China), and Satya Gnyana Sri (10+ student) for their constant patience with him and help in creating better output.

Dr. Prasad expresses his deep sense of gratitude and appreciation to his family members Koteswara Rao, Jayalakshmi, Anitha, and Bhanu Shashank, Narayana Rao Bodla (retired head master, Government High School), and other family members for their inspiration, without whose constant encouragement, it would not have been possible for him to pursue higher studies in mathematics.
Suggestions for the improvement of the book will be gratefully acknowledged.

Authors

Bhavanari Satyanarayana has been awarded the Council of Scientific and Industrial Research (CSIR) Junior Research Fellow (1980–1982), CSIR Senior Research Fellow (1982–1985), University Grants Commission Research Associateship (1985), and CSIR-POOL Officer (1988) awards. He has published about 66 research papers on algebra/fuzzy algebra/ graph theory in reputable national and international journals. He has co-authored/edited 39 books (undergraduate/postgraduate) including the book *Discrete Mathematics & Graph Theory* (Prentice  Hall India Limited), of which five books were published by VDM Verlag Dr. Müller, Germany. He was a selected scientist of the Hungarian Academy of Sciences, Budapest, and the University Grants Commission, New Delhi, in 2003. He worked with Prof. Richard Wiegandt at the A. Rényi Institute of Mathematics (Hungarian Academy of Sciences) during the period June–September, 2003. Subsequently, he became a selected senior scientist of the Hungarian Academy of Sciences, Budapest, and the Indian National Science Academy, New Delhi, for the period August–September, 2005. He was the principal investigator of three major research projects sponsored by the University Grants Commission, New Delhi. He is a fellow of the Andhra Pradesh Akademi of Sciences, and also an Andhra Pradesh Scientist awardee (2009).

Dr. Satyanarayana has supervised seven doctoral theses and ten MPhil degrees. He was a visiting professor at Walter Sisulu University, Umtata, South Africa (March–April, 2007). He has visited countries such as Austria (1988), Hong Kong (1990), South Africa (1997), Germany (2003), Hungary (2003, 2005), Taiwan (2005), Singapore (2005), Ukraine (2006), South Africa (2007), Thailand (2008), and Oman (2012) to deliver lectures and to participate in collaborative research work. He is the member secretary and managing editor of *Acharya Nagarjuna International Journal of Mathematics & Information Technology*. His recent international awards/positions include the Glory of India Award (Thailand, 2011), and fellow of ABI, USA (2012), Deputy Director General of the International Biographical Centre, England), Siksha Ratan Puraskar (awarded by the India International Friendship Society, New Delhi, in 2011), the Rajiv Gandhi Excellence Award (New Delhi, 2011), and the Bharat Vikas Ratan Award (New Delhi, 2011). Presently, he is a professor of mathematics in the Department of Mathematics, Acharya Nagarjuna University, Andhra Pradesh, India.

Kuncham Syam Prasad is a gold medalist, holding the first rank in MSc mathematics in 1994. He is a recipient of the Council of Scientific and Industrial Research (CSIR) Senior Research Fellowship. He obtained his MPhil (graph theory) in 1998 and PhD (algebra, near ring theory) in 2000 under the guidance of Dr. Bhavanari Satyanarayana (Andhra Pradesh Scientist awardee). He continued his post-doctoral research work on a senior research fellow-ship for about 11 months at Acharya Nagarjuna University, Andhra Pradesh, India.

He has published 21 research papers in reputed journals and has presented research papers in 14 national conferences and 10 international conferences held at Mobile (Alabama, USA), USA (1999), Germany (2003), Taiwan (2005), Ukraine (2006), Austria (2007), Thailand (2008), Indonesia (2009), Malaysia (2010), New Orleans, USA (2011), and Oman (2012). He also visited the Hungarian Academy of Sciences, Hungary, for joint research work with Dr. Bhavanari Satyanarayana (2003), and the National University of Singapore (2005) and the Center for Rings Theory and Applications at The Ohio University, USA (2011), for scientific discussions. He has coauthored nine books (undergraduate/postgraduate level), two of which have been published by *Discrete Mathematics and Graph Theory* (Prentice Hall India Ltd.) and *Finite Dimension in N-Groups and Fuzzy Ideals of Gamma Near Rings* (Monograph, VDM Verlag, Germany). He is also a recipient of the national level Best Research Paper Prize for the year 2000 by the Indian Mathematical Society for his research in algebra. He has received the Indian National Science Academy Visiting Fellowship Award (2004) for collaborative research work. He has supervised one doctoral degree and is a Life Member of eight academic organizations. Presently, he is working as an associate professor of mathematics at Manipal University, Karnataka, India.

Introduction

The book consists of nine chapters. Each chapter has been provided in a thorough manner with appropriate illustrations. The necessary prerequisites, such as notions of basic set theory, group theory, ring theory, the theory of vector spaces, the theory of modules, elementary graph theory, and an introduction to fuzzy sets with some algebraic structures, are provided in Chapter 1. Most of the symbols and notations used are initiated in this chapter so that the reader will be familiar with the rest. Chapter 2 provides the fundamental concepts of near rings, substructures, isomorphism theorems, and briefly introduces the theory of matrix near rings. Chapter 3 provides the notion of an *N*-group, related homomorphism theorems, chain conditions on ideals, and the direct and inverse systems. We also discuss results on the matrix near rings in this chapter. In Chapter 4, we present the concepts of prime and semiprime ideals; various characterizations of prime and semiprime ideals are also given. The concept of insertion of factors' property of near rings and *N*-groups is discussed in this chapter. A brief introduction to the finite spanning dimension has also been provided. In Chapter 5, the concept of the finite Goldie dimension is discussed. Several results on linearly independent elements in *N*-groups are presented. The concepts of primary and tertiary decomposition are provided. Chapter 6 contains the results of prime ideals in matrix near ring, and the finite Goldie dimensions in matrix near ring module are studied. The concept of the gamma near ring, a generalization of both the near ring and the gamma ring, is presented in Chapter 7. Substantial results on prime ideals and nilpotent ideals are presented. Further, introductory results on the modules over gamma near rings have been provided. Chapter 8 deals with the concept of fuzzy sets, which involves the study of uncertainty and vagueness. This concept is extended to algebraic systems such as near rings and gamma near rings. The fuzzy ideal structure, prime ideals, and homomorphism results are provided. Chapter 9 deals with the various concepts connected to graph theory. We provide the prime graph corresponding to a ring and the graph of a near ring with respect to an ideal. This covers some application domains of algebraic structures with graph theory.

1

Preliminaries

In this chapter, we present necessary fundamental definitions and results.

1.1 Set Theory

A set is considered as a primitive term, and is thus formally undefined, but we have an idea of what constitutes a set.

A **set** is a collection of objects in which we can say whether a given object is in the collection. We denote this by $a \in A$, read as "a belongs to A." The members of a set are called its **elements**.

If x is not an element of A, then we write $x \notin A$.

Suppose A and B are two sets. Then we say that A is a **subset** of B (written as $A \subseteq B$) if every element of A is also an element of B.

Two sets A and B are said to be **equal** (denoted by $A = B$) if A is a subset of B, and B is a subset of A.

If A and B are two sets, then the set $\{x \mid x \in A \text{ or } x \in B\}$ is denoted by $A \cup B$ and is called the **union** of A and B. The set $\{x \mid x \in A \text{ and } x \in B\}$ is denoted by $A \cap B$ and is called the **intersection** of A and B.

If A and B are two sets, then the set $\{x \in B \mid x \notin A\}$ is denoted by $B - A$ (or $B \backslash A$) and is called the **complement** of A in B. The set that contains no members is called the **empty set** and is denoted by $\varnothing$.

Definition 1.1.1

(i) If S and T are two sets, then the set $\{(s, t) \mid s \in S \text{ and } t \in T\}$ is called the **Cartesian product** of the sets S and T (here, the elements (s, t) are called ordered pairs, and the ordered pairs satisfy the property $(a, b) = (s, t)$ if and only if $a = s$ and $b = t$).
The Cartesian product of S and T is denoted by $S \times T$. Thus,

$$S \times T = \{(s, t) \mid s \in S \text{ and } t \in T\}.$$

Note that if S and T are two sets, then $S \times T$ and $T \times S$ may not be equal (an example is given in Example 1.1.2 (iv)).

(ii) If $S_1, S_2, \ldots, S_n$ are n sets, then the **Cartesian product** is defined as
$S_1 \times S_2 \times \ldots \times S_n = \{(s_1, s_2, \ldots, s_n) \mid s_i \in S_i \text{ for } 1 \le i \le n\}$.
Here, the elements of $S_1 \times S_2 \times \ldots \times S_n$ are called **ordered n-tuples**. The ordered n-tuples satisfy the condition

$$(s_1, s_2, \ldots, s_n) = (t_1, t_2, \ldots, t_n) \Leftrightarrow s_i = t_i, 1 \le i \le n.$$

Example 1.1.2

(i) {Rama, Sita, Lakshmana} is a set.
(ii) Suppose $A = \{a, b, c\}$ and $B = \{a, b, c, 3, 4\}$. Then A is a subset of B. If $X = \{c, b, a\}$, then $A = X$.
If $D = \{a, b, 2, 4\}$, then $A \cap D = \{a, b\}$ and $A \cup D = \{a, b, c, 2, 4\}$.
(iii) An empty set is a subset of every set. For any set Y, we have $\varnothing \subseteq Y$.
(iv) If $X = \{a, b\}$ and $Y = \{x, y\}$, then
$X \times Y = \{(a, x), (a, y), (b, x), (b, y)\}$ and $Y \times X = \{(x, a) (x, b), (y, a), (y, b)\}$.
Note that $X \times Y \ne Y \times X$.
(v) If $A = \{a, b\}$, $B = \{2\}$, $C = \{x\}$, then $A \times B \times C = \{(a, 2, x), (b, 2, x)\}$.

Example 1.1.3

List the elements of the set
$\{a/b \mid a$ and b are prime integers with $1 < a < 10$ and $3 < b < 9\}$.

Solution
We know that the prime numbers that are greater than 1 and less than 10 are 2, 3, 5, 7. Therefore, a may be 2 or 3 or 5 or 7; b may be 5 or 7. Therefore, the set is

$$\{2/5, 2/7, 3/5, 3/7, 5/5, 5/7, 7/5, 7/7\} = \{2/5, 3/5, 1, 7/5, 2/7, 3/7, 5/7\}.$$

Exercise 1.1.4

List the elements of $S \times T$, $T \times S$, where $T = \{a, b, c, d\}$ and $S = \{1, 2, 4\}$. Observe that the intersection of $S \times T$ and $T \times S$ is empty.

Definition 1.1.5

Let A_i be a collection of sets—one for each element i belongs to I, where I is some set (e.g., I may be the set of all positive integers).

We define $\bigcap_{i \in I} A_i = \{a \mid a \in A_i \text{ for all } i \in I\}$, and

$$\bigcap_{i \in I} A_i = \{a \mid a \in A_i \text{ for some } i \in I\}.$$

Note that if $\{A_i\}_{i \in I}$ is a collection of subsets of X, then $X - (\cup A_i) = \cap (X - A_i)$.

Example 1.1.6

Write $A_i = \{i, i + 1, i + 2, \ldots\}$ for each $i \in N$, the set of natural numbers. Then it is easy to observe that $\bigcup_{i \in N} A_i = N$ and $\bigcap_{i \in I} A_i = \emptyset$.

Definition 1.1.7

(i) Two sets A and B are said to be **disjoint** if $A \cap B = \emptyset$.
(ii) A collection $\{A_i\}_{i \in I}$ of sets is said to be **mutually disjoint** if $A_i \cap A_j = \emptyset$ for all $i \in I, j \in I$ such that $i \neq j$.

Example 1.1.8

(i) The sets $A = \{1, 2\}$, $B = \{a, b\}$, and $C = \{x, y, z\}$ are mutually disjoint.
(ii) If $B_i = \{2i, 2i + 1\}$ for all $i \in N$, then $\{B_i\}_{i \in N}$ is a collection of mutually disjoint sets.

Definition 1.1.9

Let A be a set. The power set of A is the set of all subsets of A. It is also denoted by $P(A)$.

Problem 1.1.10

If the set A has n elements, formulate a conjecture about the number of elements in $P(A)$.

Solution

Suppose A has n elements. Let m be an integer such that $0 \leq m \leq n$. We can select m elements from the given set A in ${}^nC_m$ ways. So, A contains ${}^nC_m$ distinct subsets containing m elements. Therefore the number of elements in $P(A) =$ number of subsets containing 0 number of elements

+ number of subsets containing only one element

+ ...

+ number of subsets containing n elements

$= {}^nC_0 + {}^nC_1 + {}^nC_2 + \ldots + {}^nC_n = 2^n.$

Definition 1.1.11

A **relation** R between the sets $A_1, A_2, \ldots, A_n$ is a subset of $A_1 \times A_2 \times \ldots \times A_n$. This relation R is called an **n-ary relation** (two-ary is called **binary**, three-ary is called **ternary**). In general, a **relation** means a binary relation on a set S (means a subset of $S \times S$). A relation R on S is said to be

   (i) **transitive** if $(a, b) \in R$, $(b, c) \in R$ implies $(a, c) \in R$;
  (ii) **reflexive** if $(a, a) \in R$ for all $a \in S$;
 (iii) **antisymmetric** if $(a, b) \in R$ and $(b, a) \in R \Rightarrow a = b$;
  (iv) **symmetric** if $(a, b) \in R$ implies $(b, a) \in R$;
   (v) **an equivalence relation** if it is reflexive, symmetric, and transitive.

If (a, b) is an element of the equivalence relation, then we write $a \sim b$ and we say that a and b are equivalent.

Let R be an equivalence relation on S and a an element of S. Then the set $[a] = \{s \in S \mid s \sim a\}$ is called the **equivalence class** of a (or the **equivalence class containing** a).

Example 1.1.12

   (i) If $A = \{a, b, c\}$ and $R = \{(a, a), (b, b), (c, c), (a, b), (b, a)\}$, then R is an equivalence relation on A and $[a] = \{a, b\}$, $[b] = \{a, b\}$, and $[c] = \{c\}$.
  (ii) Let Z be the set of all integers. Consider the set
       $X = \{\frac{a}{b} \mid a, b \in Z$ and $b \neq 0\}$. Define $\frac{a}{c} \sim \frac{b}{d}$ (or $\frac{a}{c} = \frac{b}{d}$) if $ad = bc$.
       Then this $\sim$ is an equivalence relation on X.
       An equivalence class of $\frac{a}{b}$ is called a **rational number**.
 (iii) Let $f: X \to Y$. Define a relation on X as follows: $x_1 \sim x_2 \Leftrightarrow f(x_1) = f(x_2)$. Show that this is an equivalence relation and describe the equivalence classes.
  (iv) On the set R of all real numbers, let us define a relation as follows:
       $x \sim y \Leftrightarrow x - y$ is an integer.
       Show that this is an equivalence relation and describe the equivalence classes.
   (v) For two sets X and Y, we say that X is **numerically equivalent** to Y, if there exists a bijection from X to Y. Take a set A and consider

$P(A) = \{X \mid X \subseteq A\}$. Then the relation "numerically equivalent" is an equivalence relation on $P(A)$.

(vi) Let S be the set of all the items for sale in a grocery store: we declare $a \sim b$ for $a, b \in S$, if the price of a equals to that of b. Clearly, $\sim$ is an equivalence relation. Note that in measuring this "generalized equality" on S, we ignore all the properties of the elements of S other than their prices. So, $a \sim b$ if they are equal as far as the price is concerned.

(vii) Let S be the set of all integers and $n > 1$ a fixed integer. The symbol "|" denotes divides. We define $a \sim b$ for $a, b \in S$, if $n \mid (a - b)$.

We verify that this is an equivalence relation. Since $n \mid 0$ and $0 = a - a$, we have $a \sim a$.

Because $n \mid (a - b)$ implies that $n \mid (b - a)$, it follows that $a \sim b$ implies that $b \sim a$.

Suppose $a \sim b$ and $b \sim c$, then $n \mid (a - b)$ and $n \mid (b - c)$; hence, $n \mid (a - b) + (b - c)$, that is, $n \mid (a - c)$. Therefore, $a \sim c$.

Definition 1.1.13

A **partition** of a set S is a set of subsets $\{S_i \mid \varphi \neq S_i \subseteq S$ and $i \in I$, where I is some index set$\}$ satisfying $\bigcup_{i \in I} S_i = S$ and $S_i \cap S_j = \varphi$ if $i \neq j$.

Example 1.1.14

(i) Write $A = \{1, 2, 3, 4, 5, a, b, c\}$, $S_1 = \{1, 2\}$, $S_2 = \{3\}$, $S_3 = \{4, 5, a\}$, and $S_4 = \{b, c\}$. Then S_1, S_2, S_3, S_4 form a partition for A.

(ii) Consider R, the set of all real numbers. The collection $\{(a, b) \mid a, b \in Z$ and $b = a + 1\}$ of subsets of R forms a partition for R.

Lemma 1.1.15 states that any two equivalence classes are either equal or disjoint.

Lemma 1.1.15

If R is an equivalence relation on S and $a, b \in S$, then either $[a] = [b]$ or $[a] \cap [b] = \varnothing$.

Proof

If $[a] \cap [b] = \varnothing$, then it is clear. Now suppose the intersection is nonempty. Let $x \in [a] \cap [b] \Rightarrow x \in [a]$ and $x \in [b] \Rightarrow x \sim a$ and $x \sim b \Rightarrow a \sim x$ and $x \sim b$ (since $\sim$ is symmetric) $\Rightarrow a \sim b$ (since $\sim$ is transitive). Now we show that $[a] = [b]$. For this, let $y \in [a] \Rightarrow y \sim a$. We already have $a \sim b$. Therefore, $y \sim b$ (by transitive property) $\Rightarrow b \sim y \Rightarrow y \in [b]$. Hence, $[a] \subseteq [b]$. Similarly, we get that $[b] \subseteq [a]$. Therefore, $[a] = [b]$.   ∎

Lemma 1.1.16

Let A be a set and $\sim$ an equivalence relation on A. Then, the set of all equivalence classes forms a partition for A.

Proof

The collection of all equivalence classes is $\{[a] \mid a \in A\}$. Since each $[a] \subseteq A$, it follows that $\bigcup_{a \in A}[a] \subseteq A$. Now, let $x \in A$. It is clear that $x \in [x] \subseteq \bigcup_{a \in A}[a]$. Therefore, it follows that $A = \bigcup_{a \in A}[a]$. By Lemma 1.1.15, we know that either $[a] = [b]$ or $[a] \cap [b] = \varnothing$ for any $a, b \in A$. Hence, the set of all equivalence classes forms a partition. ∎

Lemma 1.1.17

Let A be a set and $\{A_i \mid i \in I\}$ be a collection of nonempty subsets of A, which forms a partition for A. Then there exists an equivalence relation $\sim$ on A such that the equivalence classes are nothing but the sets of the partition.

Proof

Define the relation $\sim$ on A as $a, b \in A$, $a \sim b$ if there exists $i \in I$ such that $a, b \in A_i$. Now we show that this $\sim$ is an equivalence relation on A. To show $\sim$ is reflexive, take $a \in A$. Since $A = \cup A_i$, it follows that $a \in A_i$ for some $i \in I \Rightarrow a, a \in A_i \Rightarrow a \sim a$. Therefore, $\sim$ is a reflexive relation. To show $\sim$ is symmetric, take $a, b \in A$ such that $a \sim b$. This implies that there exists $i \in I$ such that $a, b \in A_i \Rightarrow b, a \in A_i \Rightarrow b \sim a$. Therefore, $\sim$ is symmetric. To show $\sim$ is transitive, let $a, b, c \in A$ such that $a \sim b$ and $b \sim c$. Now there exists $i, j \in I$ such that $a, b \in A_i$ and $b, c \in A_j$. So, $b \in A_i \cap A_j$. If $i \neq j$, then $b \in A_i \cap A_j = \varnothing$, which is a contradiction. Therefore, $i = j$ and $A_i = A_j$. Now $a, c \in A_i \Rightarrow a \sim c$. Therefore, $\sim$ is transitive. Hence, $\sim$ is an equivalence relation. To show the last part, take $i \in I$ and $x \in A_i$. Now we show that $[x] = A_i$. For this, let $y \in [x] \Rightarrow x \sim y \Rightarrow x, y \in A_i \Rightarrow y \in A_i$. Therefore, $[x] \subseteq A_i$. Let $z \in A_i$. Now, $x, z \in A_i \Rightarrow x \sim z \Rightarrow z \in [x]$. Therefore, $A_i \subseteq [x]$. Hence, $[x] = A_i$. So, every A_i is an equivalence class. Let $[b]$ be an equivalence class. Then $b \in A \Rightarrow$ there exists $j \in I$ such that $b \in A_j$. Now by the previous part, it follows that $[b] = A_j$. Therefore, the set of all equivalence classes is nothing but the sets of the given partition. ∎

Theorem 1.1.18

Let A be a set. If $\sim$ is an equivalence relation on A, then the set of all equivalence classes forms a partition for A. Conversely, if $\{A_i \mid i \in I\}$ is a partition on A, then there exists an equivalence relation on A whose equivalence classes are A_i, $i \in I$.

Proof

Combination of Lemmas 1.1.16 and 1.1.17. ∎

Definition 1.1.19

Let S and T be sets. A **function** f from S to T is a subset f of $S \times T$ such that

(i) for $s \in S$, there exists $t \in T$ with $(s, t) \in f$
(ii) $(s, u) \in f$ and $(s, t) \in f \Rightarrow t = u$

If $(s, t) \in f$, then we write $(s, f(s))$ or $f(s) = t$.
Here, t is called the **image** of s, and s is called the **preimage** of t.
The set S is called the **domain** of f, and T is called the **codomain**.
The set $\{f(s) \mid s \in S\}$ is a subset of T, and it is called the **image** of S under f (or the image of f). We denote the fact
"f is a function from S to T" by "$f: S \to T$."
$f: S \to T$ is said to be a **one–one function** (or **injective function**) if it satisfies the following condition: $f(s_1) = f(s_2) \Rightarrow s_1 = s_2$.
$f: S \to T$ is said to be an **onto function** (or **surjective function**) if it satisfies the following condition: $t \in T \Rightarrow$ there corresponds an element s in S such that $f(s) = t$.
A function is said to be a **bijection** if it is both one–one and onto.
Let $g: S \to T$ and $f: T \to U$. The **composition** of f and g is a function $f \circ g: S \to U$ defined by $(f \circ g)(s) = f(g(s))$ for all s in S.
A **binary operation** on a set S is a function from $S \times S$ into S.
An **n-ary operation** on a set S is a function from $S \times S \times \ldots \times S$ (n times) into S. A **unary operation** is a function from S into S. If f is a binary operation on S, then for any two elements a, b in S, the image of (a, b) under f is denoted by afb. A binary operation "$\cdot$" on S is said to be **associative** if $(s \cdot t) \cdot u = s \cdot (t \cdot u)$ for all elements s, t, u in S. A binary operation on S is said to be **commutative** if $s \cdot t = t \cdot s$ for all s, t in S.
A binary operation on S is said to have an **identity** if there exists an element e in S such that $e \cdot s = s = s \cdot e$ for all s in S.

Example 1.1.20

(i) Show that if $g: S \to T$ and $f: T \to U$ are one–one functions, then $f \circ g$ is also one–one.

Solution: Suppose that $(f \circ g)(s) = (f \circ g)(t)$ for $s, t \in S$. By the definition of composition of maps, we have $f(g(s)) = f(g(t))$. Since f is one–one, we get $g(s) = g(t)$. Since g is one–one, we get $s = t$. Therefore, $f \circ g$ is one–one.

(ii) If $g: S \to T$ and $f: T \to U$ are onto, then so is $f \circ g$.

Solution: Let $u \in U$. To show that $f \circ g$ is onto, we have to find an element s in S such that $(f \circ g)(s) = u$. Since f is onto, there exists t in T such that $f(t) = u$. Now since g is onto, there exists s in S such that $g(s) = t$. It is clear that $(f \circ g)(s) = f(g(s)) = f(t) = u$. Hence, $f \circ g$ is an onto function.

Definition 1.1.21

A function $f: S \to T$ is said to have an **inverse** if there exists a function g from T to S such that $(g \circ f)(s) = s$ for all s in S and $(f \circ g)(t) = t$ for all t in T. We call the function g the **inverse** of f. A function $f: S \to S$ is said to be an **identity function** if $f(s) = s$ for all s in S. The identity function on S is denoted by either I or I_S. The inverse of a function f, if it exists, is denoted by f^{-1}. Two functions $f: A \to B$ and $g: C \to D$ are said to be **equal** if $A = C$, $B = D$, and $f(a) = g(a)$ for all elements $a \in A = C$. If two functions f and g are equal, then we write $f = g$. The identity function is one–one and onto. A function g is inverse of f if and only if $f \circ g$ and $g \circ f$ are identity functions.

Problem 1.1.22

Prove that a function f has an inverse if and only if f is one–one and onto.

Solution

Suppose the inverse of $f: S \to T$ is $g: T \to S$.

By definition, $g \circ f(s) = s$ for all s in S and $f \circ g (t) = t$ for all t in T.

To show f is one–one, suppose $a, b \in S$ such that $f(a) = f(b)$.

By applying the function g on both sides, we get $g \circ f(a) = g \circ f(b)$.

Since $g \circ f$ is an identity, we get $a = g \circ f(a) = g \circ f(b) = b$.

Hence, f is one–one. To show f is onto, let t be an element of T. Write $x = g(t)$. Then $x \in S$ and $f(x) = f(g(t)) = f \circ g(t) = t$. Hence, f is onto.

Converse

Suppose f is one–one and onto.

Define $g: T \to S$ as $g(t) = s$, where $f(s) = t$. To verify that g is a function, suppose $g(t) = a$ and $g(t) = b$. Then $f(a) = t$ and $f(b) = t$.

So, $f(a) = f(b)$, which implies $a = b$ (since f is one–one). Therefore, g is a function.

For all s in S, it follows that $g \circ f(s) = g(f(s)) = g(t) = s$ (where $t = f(s)$).

Also, for all t in T, we have $f \circ g(t) = f(g(t)) = f(s) = t$. Hence, g is the inverse of f.

Note that the inverse of a function is unique. The identity function on a set is unique. An identity element in a set with respect to a binary operation is unique.

Notation 1.1.23

If S is a nonempty set, then we write
$$A(S) = \{f \mid f: S \to S \text{ is a bijection}\}.$$

Theorem 1.1.24

If $f, g, h \in A(S)$, then

   (i) $f \circ g \in A(S)$.
   (ii) $(f \circ g) \circ h = f \circ (g \circ h)$.
   (iii) There exists an element $I \in A(S)$ (the identity mapping on S) such that $f \circ I = f = I \circ f$.
   (iv) For $f \in A(S)$, there corresponds an element f^{-1} in $A(S)$ such that $f \circ f^{-1} = I = f^{-1} \circ f$.

Proof

   (i) $f, g \in A(S) \Rightarrow f, g$ are both one–one and onto
       $\Rightarrow f \circ g$ is both one–one and onto
       $\Rightarrow f \circ g$ is a bijection
       $\Rightarrow f \circ g \in A(S)$.
   (ii) This is a direct verification.
   (iii) We know that $I: S \to S$ defined by $I(s) = s$ for all $s \in S$,
       $(f \circ I)(s) = f(I(s)) = f(s)$, for all $s \in S \Rightarrow f \circ I = f$.
       $(I \circ f)(s) = I(f(s)) = f(s)$, for all $s \in S \Rightarrow I \circ f = f$.
   (iv) Define $f^{-1}: S \to S$ by $f^{-1}(y) = x$, if $f(x) = y$.
       Then f^{-1} is a function and $(f^{-1} \circ f)(x) = f^{-1}(f(x)) = f^{-1}(y) = x = I(x)$.
       So, $f^{-1} \circ f = I$. Similarly, $(f \circ f^{-1})(y) = f(f^{-1}(y)) = f(x) = y = I(y)$.
       This implies $f \circ f^{-1} = I$. Therefore, $f \circ f^{-1} = I = f^{-1} \circ f$. ∎

Problem 1.1.25

Let $|S|$ denotes the number of elements in S. If $|S| = n$, then show that $|A(S)| = n!$

Solution

Suppose $S = \{x_i \mid 1 \leq i \leq n\}$. If $f \in A(S)$, then f is a bijection.

To define $f: S \rightarrow S$, we have to define $f(x_i)$ as an element of S for each $1 \leq i \leq n$. To define $f(x_1)$, there are n possible ways (because $f(x_1) \in \{x_1, x_2, ..., x_n\}$). Since f is one–one, it follows that $f(x_1) \neq f(x_2)$.

So, after defining $f(x_1)$, there are $(n-1)$ ways to define $f(x_2)$, because $f(x_2) \in \{x_1, x_2, ..., x_n\} \setminus \{f(x_1)\}$. Thus, both $f(x_1)$ and $f(x_2)$ can be defined in $n(n-1)$ ways. Now, for $f(x_3)$, there are $(n-2)$ ways. Hence, $f(x_1), f(x_2), ..., f(x_n)$ can be defined in $n(n-1)(n-2) ... 2 \times 1 = n!$ ways. Therefore, $n!$ number of bijections can be defined from S to S. This means $|A(S)| = n!$

1.2 Group Theory

In this section, we study an algebraic object known as a "group," which serves as one of the fundamental building blocks for the development of "abstract algebra." In fact, group theory has several applications in many areas where symmetry occurs. Applications of groups can also be found in physics and chemistry. Some of the exciting applications of group theory have arisen in fields such as particle physics and binary codes.

Definition 1.2.1

For a nonempty set G, a **binary operation** on G is a mapping from $G \times G$ into G. In general, binary operations are normally denoted by *, ·, or o.

Definition 1.2.2

A nonempty set G together with a binary operation * is called a **group** if the algebraic system $(G, *)$ satisfies the following four axioms:

(i) Closure axiom: a, b are elements of G implies that $a * b$ is an element of G.

(ii) Associative axiom: $(a * b) * c = a * (b * c)$ for all elements a, b, c in G.

(iii) Identity axiom: There exists an element e in G such that $a * e = e * a = a$ for all elements a in G.

(iv) Inverse axiom: For any element $a \in G$, there corresponds an element $b \in G$ such that $a * b = e = b * a$.

Note 1.2.3

The element e of G (given in the identity axiom) is called an **identity element**. The element b (given in the inverse axiom) is called an **inverse** of a in G. Later, we prove that the identity element in a group is unique. The inverse of a given element is also unique.

Problem 1.2.4

Let S be a nonempty set. Let $A(S)$ be the set of all bijections on S. Suppose "o" is the composition of mappings. Then $(A(S), o)$ is a group.

Solution

(i) Closure axiom: Let f, g be in $A(S)$. Then f, g are bijections. Then $f o g$, the composition of two bijections, is also a bijection. Hence, $f o g$ is in $A(S)$.

(ii) Associative axiom: Let f, g, h be in $A(S)$. We know that for any mappings f, g, h, the condition of the associative law $f o (g o h) = (f o g) o h$ holds. Thus, associativity holds.

(iii) Identity axiom: We know that the identity mapping I on S, defined by $I(x) = x$ for all x in S, is a bijection on S. Therefore, I is in $A(S)$. Also, we know that $f o I = f = I o f$ for all mappings from S to S. Hence, I is an identity element in $A(S)$.

(iv) Inverse axiom: Let f be an element of $A(S)$. Then f is a bijection. Now the mapping g is defined by $g(x) = y$, whenever $f(y) = x$ is a bijection on S. Therefore, g is an element of $A(S)$. Also, it is easy to verify that $f o g = I = g o f$. Hence, g is the inverse of f in $A(S)$.

From these facts, we conclude that $(A(S), o)$ is a group.

Definition 1.2.5

Let $(G, *)$ be a group. Then $(G, *)$ is said to be a **commutative group** (or **Abelian group**) if it satisfies the commutative property: $a * b = b * a$, for all a, b in G.

TABLE 1.1

Multiplication Table

•	−1	+1
−1	1	−1
1	−1	1

Example 1.2.6

Consider the real numbers −1, +1. Take $G = \{−1, 1\}$. Then $(G, \cdot)$ is a commutative group with respect to the usual multiplication of numbers.

Verification

Closure axiom: From the multiplication table (Table 1.1) it is clear that $a \cdot b$ is in G for all a, b in G.

Associative axiom: We know that the real numbers satisfy the associative law. Since 1 and −1 are real numbers, this axiom holds good.

Identity axiom: From Table 1.1 it is clear that $1 \cdot a = a = a \cdot 1$ for all elements a in G. Hence, 1 is the identity element.

Inverse axiom: From Table 1.1 it is clear that 1 is the inverse of 1 and −1 is the inverse of −1.

Commutative law: From the table it is clear that $(−1) \cdot 1 = 1 \cdot (−1)$. Therefore, the commutative law holds well in $(G, \cdot)$. Hence $(G, \cdot)$ is a commutative group.

Definition 1.2.7

Let G be a group. If G contains only a finite number of elements, then G is called a **finite group**. If G contains an infinite number of elements, then G is called an **infinite group**. If G is a finite group, then the **order of G** is the number of elements in G. If G is an infinite group, then we say that the order of G is infinite. The order of G is denoted by $O(G)$.

Examples 1.2.8

(i) Let G be the set of all integers and + be the usual addition of numbers. Then $(G, +)$ is an Abelian group. Here, 0 is the additive identity and $−x$ is the additive inverse of x for any x in G. This $(G, +)$ is an infinite group, and so $O(G)$ is infinite.

(ii) Consider Q, the set of rational numbers, and R, the set of all real numbers. Clearly, these two are infinite Abelian groups with respect to the usual addition of numbers.

(iii) From Example 1.2.6, it is clear that the set G consisting of -1 and 1 is a group with respect to usual multiplication. This group is a finite group, and $O(G) = 2$.

Example 1.2.9

Let n be a positive integer. Write $G = \{a^i \mid i = 0, 1, ..., n\}$, where $a^0 = a^n = e$ for some fixed symbol a. Define an operation, denoted by "$\cdot$" on G, by

$$a^i \cdot a^j = a^{i+j} \text{ if } i + j < n$$
$$= a^{i+j-n} \text{ if } i + j \geq n.$$

Then $(G, \cdot)$ is a group of order n.
We call this group (given in Example 1.2.9) a **cyclic group** of order n.

Lemma 1.2.10

If G is a group, then
   (i) The identity element of G is unique.
  (ii) Every element in G has a unique inverse in G.
 (iii) For any a in G, we have $(a^{-1})^{-1} = a$.
 (iv) For all a, b in G, we have $(a \cdot b)^{-1} = b^{-1} \cdot a^{-1}$.

Proof

   (i) Let e, f be two identity elements in G. Since e is the identity, it follows that $e \cdot f = f$. Since f is the identity, it follows that $e \cdot f = e$. Therefore, $e = e \cdot f = f$. Hence, the identity element is unique.
  (ii) Let a be an element in G, and a_1, a_2 are two inverses of a in G. Now
$$a_1 = a_1 \cdot e \text{ (since } e \text{ is the identity)}$$
$$= a_1 \cdot (a \cdot a_2) \text{ (since } a_2 \text{ is the inverse of } a)$$
$$= (a_1 \cdot a) \cdot a_2 \text{ (by associative law)}$$
$$= e \cdot a_2 \text{ (since } a_1 \text{ is the inverse of } a)$$
$$= a_2.$$
Hence, the inverse of an element in G is unique.
 (iii) Let $a \in G$. Since $a \cdot a^{-1} = e = a^{-1} \cdot a$, it follows that a is the inverse of a^{-1}. Hence $(a^{-1})^{-1} = a$.
 (iv) Let a, b be elements in G. Consider $(b^{-1} \cdot a^{-1})(a \cdot b) = b^{-1} \cdot (a^{-1} \cdot a) \cdot b = b^{-1} \cdot e \cdot b = b^{-1} \cdot b = e$.

Similarly, $e = (a \cdot b) \cdot (b^{-1} \cdot a^{-1})$. This shows that $(a \cdot b)^{-1} = b^{-1} \cdot a^{-1}$. ∎

Definition 1.2.11

Let (G, o) be a group. A nonempty subset H of G is said to be a **subgroup** of G if H itself forms a group under the product in G.

The following lemma provides an equivalent condition for a subgroup.

Lemma 1.2.12

A nonempty subset H of a group G is a subgroup of G if and only if

(i) $a, b \in H$ implies $ab \in H$ and (ii) $a \in H$ implies $a^{-1} \in H$.

Proof

Suppose that H is a subgroup of G
$\Rightarrow H$ itself is a group under the product in G
$\Rightarrow$ (i) and (ii) hold.

Converse

Suppose H satisfies (i) and (ii). By (i), H satisfies the closure property. For any $a, b, c \in H$, we have $a, b, c \in G \Rightarrow a(bc) = (ab)c$. By (ii), for any $a \in H$, $a^{-1} \in H$. Now by (i), $e = aa^{-1} \in H$. Therefore $(H, \cdot)$ is a subgroup of $(G, \cdot)$. The proof is complete.   ■

Lemma 1.2.13

If H is a nonempty finite subset of a group G and H is closed under multiplication, then H is a subgroup of G.

Proof

Suppose H is a nonempty finite subset of a group G and H is closed under multiplication. Now we have to show that H is a subgroup of G. It is enough to show that $a \in H \Rightarrow a^{-1} \in H$ (by using Lemma 1.2.12). Since H is a nonempty set, there exists $a \in H$.

Now, $a, a \in H \Rightarrow a^2 \in H$. Similarly, $a^3 \in H, ..., a^m \in H, ...$ Therefore, $H \supseteq \{a, a^2, ...\}$. Since H is finite, it follows that there must be repetitions in $a, a^2, ...$ Therefore, there exist integers r, s with $r > s > 0$ such that $a^r = a^s \Rightarrow a^r \cdot a^{-s} = a^0$ $\Rightarrow a^{r-s} = e \Rightarrow e \in H$ (since $r - s > 0$ and $a \in H \Rightarrow a^{r-s} \in H$). Since $r - s - 1 \geq 0$, we have $a^{r-s-1} \in H$ and $a \cdot a^{r-s-1} = a^{r-s} = e \in H$.

Hence, a^{r-s-1} acts as the inverse of a in H. Hence, H is a subgroup.   ■

Example 1.2.14

Let S be a set. We know that $(A(S), \circ)$ is a group where "$\circ$" is the composition of mappings. Write $H(x_0) = \{f \in A(S) \mid f(x_0) = x_0\}$, where x_0 is a fixed point of S. Then $H(x_0)$ is a subgroup of $A(S)$.

Solution

Let $f, g \in H(x_0) \Rightarrow f(x_0) = x_0$ and $g(x_0) = x_0$. Now consider $(f \circ g)(x_0) = f(g(x_0)) = f(x_0) = x_0 \Rightarrow f \circ g \in H(x_0)$. Hence, the closure axiom holds well.

Also, $f(x_0) = x_0 \Rightarrow f^{-1}(x_0) = x_0$ (since f is a bijection, f^{-1} exists) $\Rightarrow f^{-1} \in H(x_0)$. Therefore, $H(x_0)$ is a subgroup of $A(S)$.

Definition 1.2.15

(i) Let G be a group and $a \in G$. Then $(a) = \{a^i \mid i = 0, \pm 1, \ldots\}$ is called the **cyclic subgroup** generated by the element $a \in G$.

(ii) Let G be a group. Then G is said to be a **cyclic group** if there exists an element $a \in G$ such that $G = (a)$.

Definition 1.2.16

If $\sim$ is an equivalence relation on S, then $[a]$, **the class of a**, is defined by $[a] = \{b \in S \mid b \sim a\}$.

Definition 1.2.17

Let G be a group and H be a subgroup of G, $a, b \in G$. We say that **a is congruent to b (mod H)**, written as $a \equiv b$ (mod H) if $ab^{-1} \in H$.

Lemma 1.2.18

The relation $a \equiv b$ (mod H) is an equivalence relation.

Proof

We have to show that

(i) $a \equiv a$ (mod H)

(ii) $a \equiv b$ (mod H) $\Rightarrow b \equiv a$ (mod H)

(iii) $a \equiv b$ (mod H), $b \equiv c$ (mod H) $\Rightarrow a \equiv c$ (mod H), for all $a, b, c \in G$.

(i) Since H is a subgroup of G, it follows that $aa^{-1} = e \in H$ for $a \in G \Rightarrow a \equiv a$ (mod H).

(ii) Suppose $a \equiv b$ (mod H)

$\Rightarrow ab^{-1} \in H \Rightarrow (ab^{-1})^{-1} \in H$ (since H is a subgroup of G)

$\Rightarrow (b^{-1})^{-1}a^{-1} \in H \Rightarrow ba^{-1} \in H \Rightarrow b \equiv a$ (mod H).

(iii) Suppose $a \equiv b$ (mod H), $b \equiv c$ (mod H)

$\Rightarrow ab^{-1} \in H, bc^{-1} \in H$

$\Rightarrow (ab^{-1})(bc^{-1}) \in H$ (since H is a subgroup of G)

$\Rightarrow a(b^{-1}b)c^{-1} \in H \Rightarrow aec^{-1} \in H \Rightarrow ac^{-1} \in H \Rightarrow a \equiv c$ (mod H).

Therefore, the relation $a \equiv b$ (mod H) satisfies (i) reflexive, (ii) symmetric, and (iii) transitive properties. Thus, the relation is an equivalence relation. ∎

Definition 1.2.19

If H is a subgroup of G and $a \in G$, then $Ha = \{ha \mid h \in H\}$ is called the **right coset** of H in G. The set $aH = \{ah \mid h \in H\}$ is called the **left coset**.

Lemma 1.2.20

For all $a \in G$, $Ha = \{x \in G \mid a \equiv x \pmod{H}\} = [a]$.

Proof

Consider $[a] = \{x \in G \mid a \equiv x \pmod{H}\}$, the equivalence class under the equivalence relation defined in Definition 1.2.17. Let $x \in Ha \Rightarrow x = ha$ for some $h \in H$. Since $h \in H$, it follows that $ax^{-1} = a(ha)^{-1} = aa^{-1}h^{-1} = eh^{-1} = h^{-1} \in H \Rightarrow ax^{-1} \in H \Rightarrow a \equiv x \pmod{H} \Rightarrow x \in [a]$.

Converse

Suppose $x \in [a] \Rightarrow a \equiv x \pmod{H} \Rightarrow ax^{-1} \in H$

$\Rightarrow (ax^{-1})^{-1} \in H \Rightarrow xa^{-1} \in H$

$\Rightarrow xa^{-1} = h$ for some $h \in H \Rightarrow x = ha \in Ha$.

Therefore $[a] = Ha$. ∎

Lemma 1.2.21

There is a one-to-one correspondence between any two right cosets of H in G.

Proof

Let H be a subgroup of G and Ha, Hb be two right cosets of H in G (for some $a, b \in G$). Define $\varphi: Ha \to Hb$ by $\varphi(ha) = hb$ for all $ha \in Ha$.

To show that φ is one–one, let $h_1a, h_2a \in Ha$ such that $\varphi(h_1a) = \varphi(h_2a) \Rightarrow h_1b = h_2b \Rightarrow h_1 = h_2$ (by the Cancellation Law) $\Rightarrow h_1a = h_2a$. Therefore, φ is one–one.

To show φ is onto, let $hb \in Hb \Rightarrow h \in H$. Now $ha \in Ha$ and $\varphi(ha) = hb$. Therefore, φ is onto. Thus, there is a one-to-one correspondence between any two right cosets of H. ∎

Note 1.2.22

Since $H = He$, it follows that H is also a right coset of H in G, and by Lemma 1.2.21, any right coset of H in G has $O(H)$ elements.

Theorem 1.2.23 (Lagrange's Theorem)

If G is a finite group and H is a subgroup of G, then $O(H)$ is a divisor of $O(G)$.

Proof

Let G be a finite group and H a subgroup of G with $O(G) = n$, $O(H) = m$ (note that since G is finite, H is also finite). We know that any two right cosets are either disjoint or identical. Now, suppose Ha_1, Ha_2, ..., Ha_k are only distinct right cosets of H in G.

This implies $G = Ha_1 \cup Ha_2 \cup ... \cup Ha_k$. Since every right coset has $O(H)$ elements, it follows that $O(G) = O(Ha_1) + O(Ha_2) + ... + O(Ha_k) = O(H) + O(H) + ... + O(H)$ (k times). This implies $O(G) = k \cdot O(H)$. This shows that $O(H)$ divides $O(G)$.

The converse of the Lagrange's theorem is not true, that is, the statement "If G is a finite group and $k \mid O(G)$, then there exists a subgroup H of G such that $O(H) = k$" is not true. ∎

Example 1.2.24

Consider the symmetric group S_4. We know that $S_4 = \{f : A \to A \mid f$ is a bijection and $A = \{1, 2, 3, 4\}\}$. Clearly, $|S_4| = 24$ ($= 4!$). Now, $A_4 =$ the set of all even permutations in S_4. Then $|A_4| = 12$. It can be verified that any six elements of A_4 cannot form a subgroup. Therefore, $6 \mid O(A_4)$ but A_4 contains no subgroup of order 6.

Definitions 1.2.25

  (i) If H is a subgroup of G, then the **index** of H in G is the number of distinct right cosets of H in G. It is denoted by $i(H)$.

  (ii) If G is a group and $a \in G$, then the **order** of a is defined as the least positive integer m such that $a^m = e$.

Note 1.2.26

If there is no positive integer n such that $a^n = e$, then a is said to be of **infinite order**.

Corollary 1.2.27

If G is a finite group and $a \in G$, then $O(a) \mid O(G)$.

Proof

Suppose G is a finite group and $a \in G$ and $O(G) = n$, $O(a) = m$. Let $H = \{a, a^2, \ldots, a^m = e\}$. Clearly, H is a subgroup of G. Now we have to show that $O(H) = m$. For this, suppose $O(H) < m \Rightarrow a^i = a^j$ for some $0 \le i, j \le m \Rightarrow a^i \cdot a^{-j} = a^j \cdot a^{-j}$ (if $j < i$) $\Rightarrow$ $a^{i-j} = a^0 = e$, where $0 < i - j < m$, which is a contradiction (since m is the least positive integer such that $a^m = e$).

    Therefore, $O(H) = m = O(a)$. Now, by the Lagranges theorem, we have $O(H) \mid O(G)$. This shows that $O(a) \mid O(G)$. ∎

Corollary 1.2.28

If G is a finite group and $a \in G$, then $a^{O(G)} = e$.

Proof

By Corollary 1.2.27, it follows that $O(a) \mid O(G) \Rightarrow$ there exists m such that $O(G) = m \cdot O(a)$. Now, $a^{O(G)} = a^{m \cdot O(a)} = (a^{O(a)})^m = e^m = e$.

    Let Z be the set of all integers and let $n > 1$ be a fixed integer. For the equivalence relation $a \equiv b \pmod{n}$ (a is congruent to b mod n), if $n \mid (a - b)$, the class of a (denoted by $[a]$) consists of all $a + nk$, where k runs through all the integers. We call this the **congruence class** of a. ∎

Theorem 1.2.29

Z_n forms a cyclic group under the addition $[a] + [b] = [a + b]$.

Proof

Consider $Z_n = \{[0], [1], \ldots, [n-1]\}$.

We define the operation $+$ in Z_n as $[a] + [b] = [a + b]$.

Suppose that $[a] = [a^1]$; then $n|(a - a^1)$. Also suppose that $[b] = [b^1]$; then $n|(b - b^1)$.

Hence, $n|((a - a^1) + (b - b^1)) \Rightarrow n|((a + b) - (a^1 + b^1))$.

Therefore, $(a + b) \equiv (a^1 + b^1) \pmod{n}$. Hence, $[a + b] = [a^1 + b^1]$.

This shows that the operation $+$ is well-defined on Z_n.

The element $[0]$ acts as the identity element and $[-a]$ acts as $-[a]$, the inverse of $[a]$. It can be verified that Z_n is a group under the operation $+$. Also, it is a cyclic group of order n generated by $[1]$. ∎

Note 1.2.30

(i) We define the multiplication on Z_n as $[a] \cdot [b] = [ab]$.

For instance, if $n = 9$, then $[2] \cdot [7] = [14] = [5]$ and $[3] \cdot [6] = [18] = [0]$. It can be verified that Z_n does not form a group under multiplication, since $[0] \cdot [a] = [0]$ for all a, and the unit element under multiplication is $[1]$, and $[0]$ cannot have a multiplicative inverse.

(ii) If $n = 6$, then $[2] \neq [0]$, $[3] \neq [0]$, yet $[2] \cdot [3] = [6] = [0]$; so, the nonzero elements (in general) need not form a group.

(iii) Let $U_n = \{[a] \in Z_n \mid (a, n) = 1\}$, noting that $(a, n) = 1$ if and only if $(b, n) = 1$ for $[a] = [b]$. If $(a, n) = 1$ and $(b, n) = 1$, then $(ab, n) = 1$. So, we have $[a] \cdot [b] = [ab]$.

Therefore, if $[a], [b] \in U_n$, then $[ab] \in U_n$. Therefore, U_n is closed under the operation product. The associativity condition can be easily verified (follows from the associative law of integers under the operation multiplication). $[1]$ is the identity element. Multiplication is commutative in U_n.

(iv) If $[a][b] = [a][c]$, where $[a] \in U_n$, then we have $[ab] = [ac]$, and so $[ab - ac] = [0]$. This shows that $n \mid a(b - c) = ab - ac$; but a is relatively prime to n. Now we have $n|(b - c)$, and so $[b] = [c]$. In other words, we have the cancellation property in U_n. Thus, U_n is a group.

(v) $|U_n| = $ the number of integers m with $1 \leq m < n$ such that $(m, n) = 1$.

Euler Function 1.2.31

The **Euler-φ-function** is defined for all integers n in the following way: $\varphi(1) = 1$, $\varphi(n) =$ the number of positive integers $<n$, and are relatively prime to n (if $n > 1$). Therefore, if $X_n = \{a \in Z \mid 0 < a < n \text{ and } (a, n) = 1\}$, then $\varphi(n) = O(X_n)$, if $n > 1$.

Notation 1.2.32

If H, K are subgroups of G, then $HK = \{x \in G \mid x = hk, h \in H, k \in K\}$. If $K \subseteq G$, then $K^{-1} = \{x^{-1} \mid x \in K\}$.

Note 1.2.33

H is a subgroup of $G \Rightarrow H^{-1} = H$.

Verification

Suppose H is a subgroup of G and $x \in H^{-1} \Leftrightarrow x^{-1} \in (H^{-1})^{-1} \Leftrightarrow x^{-1} \in H \Leftrightarrow (x^{-1})^{-1} \in H$ (since H is subgroup) $\Leftrightarrow x \in H$. Therefore, $H^{-1} = H$.

Note 1.2.34

(i) If G is a group and H is a subgroup, then $Ha = \{ha \mid h \in H\} = [a]$, the equivalence class containing a (under the equivalence relation $x \equiv y$ if and only if $xy^{-1} \in H$). Therefore, for any $a, b \in G$, it follows that either $Ha = Hb$ or $Ha \cap Hb = \varnothing$.

(ii) Let $\Lambda =$ the set of all left cosets of H in G, and
$P =$ the set of all right cosets of H in G.
Define $\varphi: P \to \Lambda$ by $\varphi(Ha) = a^{-1}H$. To verify that φ is one–one, let $\varphi(Ha) = \varphi(Hb) \Rightarrow a^{-1}H = b^{-1}H \Rightarrow (a^{-1})^{-1}b^{-1} \in H \Rightarrow ab^{-1} \in H \Rightarrow Ha = Hb$. Hence, φ is one–one. To verify that φ is onto, take a left coset $xH \in \Lambda$. Now, write $a = x^{-1}$; then $\varphi(Ha) = a^{-1}H = (x^{-1})^{-1}H = xH$. Therefore, φ is onto. Thus, there is a one-to-one correspondence between the set of all left cosets of H in G and the set of all right cosets of H in G.

Definition 1.2.35

A subgroup N of G is said to be a **normal subgroup** of G if for every $g \in G$ and $n \in N$, we have $gng^{-1} \in N$.

It is clear that a subgroup N is a normal subgroup of G, if and only if $gNg^{-1} \subseteq N$ for all $g \in G$.

Lemma 1.2.36

N is a normal subgroup of G, if and only if $gNg^{-1} = N$ for every $g \in G$.

Proof

Suppose N is a normal subgroup of G

$$\Rightarrow gNg^{-1} \subseteq N, \text{ for all } g \in G \tag{1.1}$$

Since $g \in G$, it follows that $g^{-1} \in G$.
Therefore, $g^{-1}N(g^{-1})^{-1} \subseteq N \Rightarrow g^{-1}Ng \subseteq N \Rightarrow g(g^{-1}Ng) \subseteq gN \Rightarrow Ng \subseteq gN \Rightarrow Ngg^{-1} \subseteq gNg^{-1}$

$$\Rightarrow N \subseteq gNg^{-1} \tag{1.2}$$

Therefore, from Equations 1.1 and 1.2, we have $N = gng^{-1}$.

Converse

Suppose $gng^{-1} = N$ for every g. Let $g \in G, n \in N$. Consider $gng^{-1} \in gNg^{-1} = N$. Therefore, $gNg^{-1} \subseteq N$. Hence, N is normal in G. ■

Lemma 1.2.37

The subgroup N of G is a normal subgroup of G, if and only if every left coset of N in G is a right coset of N in G.

Proof

Suppose N is a normal subgroup in $G \Rightarrow gNg^{-1} = N$ for all $g \in G \Rightarrow gNg^{-1}g = Ng \Rightarrow gN = Ng$ for all $g \in G$. Therefore, every left coset of N in G is a right coset of N in G.

Converse

Suppose that every left coset of N in G is a right coset of N in G. Then the left coset $xN = Ny$ for some $y \in G$. Since $x = xe \in xN = Ny$, it follows that $x \in Ny \Rightarrow x \in [y]$. Therefore, $Nx = [x] = [y] = Ny \Rightarrow xN = Nx \Rightarrow xNx^{-1} = Nxx^{-1} \Rightarrow N = xNx^{-1}$ for all $x \in G$. Therefore, N is a normal subgroup in G. ■

Lemma 1.2.38

A subgroup N of G is a normal subgroup of G if and only if the product of two right cosets of N in G is again a right coset of N in G.

Proof

Suppose that N is normal in G and $a, b \in G$. Consider $(Na)(Nb) = N(aN)b = N(Na)b$. Since N is normal, we have $Na = aN$ (by Lemma 1.2.37) $= NN(ab) = Nab$. Thus, the product of two right cosets of N in G is again a right coset of N in G.

Converse

Suppose that the product of two right cosets of N in G is again a right coset of N in G. We have to show that N is a normal subgroup in G. For this, take $x \in G$. Then $x^{-1} \in G$. Now Nx, Nx^{-1} are two right cosets of N in G. By converse hypothesis, $(Nx)(Nx^{-1})$ is also a right coset of N in G. Since $e = ex\, ex^{-1} \in NxNx^{-1}$ and $e \in Ne = N$, it follows that the product of the two right cosets, $Nx\, Nx^{-1}$ and Ne, have the common element e. Therefore, $NxNx^{-1} = Ne = N \Rightarrow n_1 xnx^{-1} \in N$ for some n_1 and for all $n \in N \Rightarrow xnx^{-1} \in n_1^{-1}N \subseteq N \Rightarrow xnx^{-1} \in N$ for all $x \in G$, and for all $n \in N$. Therefore, N is a normal subgroup of G. ∎

Theorem 1.2.39

If G is a group and N is a normal subgroup of G, then G/N is also a group. This group is called a quotient group or a factor group of G by N.

Proof

Let N be a normal subgroup of G. Consider $G/N = \{Na \mid a \in G\}$.

Closure property: Let $Na, Nb \in G/N \Rightarrow a, b \in G \Rightarrow ab \in G \Rightarrow (Na) \cdot (Nb) = Nab \in G/N$.

Associative property: Let $a, b, c \in G$. Then, $Na, Nb, Nc \in G/N$. Consider $Na(Nb \cdot Nc) = Na(Nbc) = Na(bc) = N(ab)c = NabNc = (Na \cdot Nb)Nc$.

Identity property: Let e be the identity in G. Then $Ne \in G/N$. Now, for any $Na \in G/N$, we have $(Na)(Ne) = Na = (Ne)(Na)$. This shows that $Ne \in G/N$ is the identity element.

Inverse property: Let $Na \in G/N$. Then $a \in G$, and hence $a^{-1} \in G$. Consider $Na^{-1} \in G/N$ and $(Na)(Na^{-1}) = Naa^{-1} = Ne = N(a^{-1}a) = (Na^{-1})(Na)$. This shows that Na^{-1} is the inverse of Na in G/N. Therefore, G/N is a group. ∎

Definition 1.2.40

A mapping $\varphi\colon G \to G^1$, where G, G^1 are groups, is said to be a **homomorphism** if for all $a, b \in G$, we have $\varphi(a \cdot b) = \varphi(a) \cdot \varphi(b)$.

Example 1.2.41

Let G be a group of real numbers under addition, and G^1 be the group of non-zero real numbers with ordinary multiplication. Define a mapping φ from $(G, +) \to (G^1, \cdot)$ by $\varphi(a) = 2^a$. Now consider $\varphi(a + b) = 2^{a+b} = 2^a \cdot 2^b = \varphi(a) \cdot \varphi(b)$. Therefore, φ is a homomorphism.

Lemma 1.2.42

Suppose G is a group and N is a normal subgroup of G. Define $\varphi\colon G \to G/N$ by $\varphi(x) = Nx$ for all $x \in G$. Then φ is an onto homomorphism. This is called the natural (or canonical) homomorphism or natural/canonical epimorphism. Also note that ker $\varphi = N$.

Proof

Let $x, y \in G$. Now $\varphi(xy) = Nxy = Nx \cdot Ny = \varphi(x) \cdot \varphi(y)$. To show φ is onto, take $Na \in G/N$. Then $a \in G$ and $\varphi(a) = Na$. Hence, φ is an onto homomorphism. $x \in \ker \varphi \Leftrightarrow \varphi(x) = Ne \Leftrightarrow Nx = Ne \Leftrightarrow xe^{-1} \in N \Leftrightarrow xe \in N \Leftrightarrow x \in N$. Therefore, ker $\varphi = N$. ∎

Definition 1.2.43

If φ is a homomorphism of G into G^1, then the **kernel of** φ (denoted by ker φ or k_φ) is defined by ker $\varphi = \{x \in G \mid \varphi(x) = e^1$, where e^1 is the identity in $G^1\}$.

Lemma 1.2.44

If φ is a homomorphism of G into G^1, then

(i) $\varphi(e) = e^1$, where e^1 is the identity element of G^1
(ii) $\varphi(x^{-1}) = [\varphi(x)]^{-1}$, for all x in G

Proof

(i) Let $x \in G \Rightarrow \varphi(x) \in G^1$. Since φ is a homomorphism, we have $\varphi(x) = \varphi(x) \cdot e^1$ and $\varphi(x) = \varphi(xe) = \varphi(x) \cdot \varphi(e)$. Therefore, $\varphi(x) \cdot e^1 = \varphi(x) \cdot \varphi(e) \Rightarrow e^1 = \varphi(e)$ (by cancellation laws).

(ii) By (i), $e^1 = \varphi(e) = \varphi(xx^{-1}) = \varphi(x) \cdot \varphi(x^{-1}) \Rightarrow \varphi(x^{-1})$ is the inverse of $\varphi(x)$. That is, $\varphi(x^{-1}) = [\varphi(x)]^{-1}$. This is true for all $x \in G$. ∎

Lemma 1.2.45

If φ is a homomorphism of G into G^1 with kernel K, then K is a normal subgroup of G.

Proof

First we show that $K \neq \varnothing$. Since $\varphi(e) = e^1$, where e^1 is the identity in G^1, it follows that $e \in \ker \varphi = K$. Therefore, $K \neq \varnothing$. Now we have to show that K is closed under multiplication and that every element in K has an inverse in K. Let $x, y \in K \Rightarrow \varphi(x) = e^1$ and $\varphi(y) = e^1 \Rightarrow \varphi(xy) = \varphi(x) \cdot \varphi(y)$ [since φ is a homomorphism] $= e^1 \cdot e^1 = e^1 \Rightarrow xy \in K$. Therefore, the closure axiom holds. Now, let $x \in K \Rightarrow \varphi(x) = e^1$. Therefore, $\varphi(x^{-1}) = [\varphi(x)]^{-1}$ (by Lemma 1.2.44) $= [e^1]^{-1} = e^1$. Therefore, $x^{-1} \in K$. Thus, every element in K has its inverse in K. Therefore, K is a subgroup of G.

Now we have to show that K is a normal subgroup of G. For this, take $g \in G$, $k \in K$. Consider $\varphi(gkg^{-1}) = \varphi(g)\varphi(k)\varphi(g^{-1}) = \varphi(g) \cdot e^1 \cdot \varphi(g^{-1})$ (since $k \in K$, we have $\varphi(k) = e^1) = \varphi(g)\varphi(g^{-1}) = \varphi(g)[\varphi(g)]^{-1}$ (by Lemma 1.2.44) $= e^1 \Rightarrow \varphi (gkg^{-1}) = e^1 \Rightarrow gkg^{-1} \in K$. Hence, K is a normal subgroup of G. ∎

Lemma 1.2.46

If φ is a homomorphism of G into G^1 with $\ker \varphi = K$, then the set of all inverse images of $g^1 \in G^1$ under φ in G is given by Kx, where x is any particular inverse image of g^1 in G (that is, $\varphi(x) = g^1$).

Proof

Let e, e^1 be the identities in G, G^1, respectively. Let $x \in G$ such that $\varphi(x) = g^1$. Write $\varphi^{-1}(g^1) = \{x \in G \mid \varphi(x) = g^1\}$. Now we have to prove that $\varphi^{-1}(g^1) = Kx$. Let $y \in Kx \Rightarrow y = kx$ for some $k \in K$. Then $\varphi(y) = \varphi(kx) = \varphi(k) \cdot \varphi(x) = e^1 . \varphi(x) = \varphi(x) = g^1 \Rightarrow \varphi(y) = g^1 \Rightarrow y \in \varphi^{-1} (g^1)$. Hence, $Kx \subseteq \varphi^{-1}(g^1)$.

Other Part

Let $z \in \varphi^{-1}(g^1) \Rightarrow \varphi(z) = g^1 = \varphi(x) \Rightarrow \varphi(z) = \varphi(x) \Rightarrow \varphi(z) [\varphi(x)]^{-1} = e^1 \Rightarrow \varphi(z) \cdot \varphi(x^{-1}) = e^1 \Rightarrow \varphi(zx^{-1}) = e^1 \Rightarrow zx^{-1} \in K \Rightarrow z \in Kx$. Therefore, $\varphi^{-1}(g^1) \subseteq Kx$. Hence, $Kx = \varphi^{-1}(g^1)$. ∎

Definition 1.2.47

A homomorphism φ from G onto G^1 is said to be an **isomorphism** if φ is a bijection. Two groups G, G^1 are said to be **isomorphic** if there is an isomorphism from G to G^1. In this case, we write $G \cong G^1$.

Corollary 1.2.48

An onto homomorphism $\varphi: G \to G^1$ with $K_\varphi = K = \ker \varphi$ is an isomorphism if and only if $K = \{e\}$.

Proof

Suppose φ is an onto homomorphism from G to G^1. Assume that φ is an isomorphism. Let $x \in K \Rightarrow \varphi(x) = e^1$, where e^1 is an identity in $G^1 \Rightarrow \varphi(x) = \varphi(e) \Rightarrow x = e$ (since φ is one–one). Therefore, $K = \{e\}$.

Converse

Suppose $K = \{e\}$. Now we show that φ is one–one. Let $a, b \in G$ such that $\varphi(a) = \varphi(b) \Rightarrow \varphi(a)[\varphi(b)]^{-1} = e^1 \Rightarrow \varphi(a) \cdot \varphi(b^{-1}) = e^1 \Rightarrow \varphi(ab^{-1}) = e^1 \Rightarrow ab^{-1} \in K = \{e\} \Rightarrow ab^{-1} = e \Rightarrow ab^{-1}b = eb \Rightarrow a = b$. Therefore, φ is one–one. The proof is completed. ∎

Theorem 1.2.49 (Fundamental Theorem of Homomorphisms)

Let φ be a homomorphism of G onto G^1 with kernel K. Then $G/K \cong G^1$.

Proof

Since φ is an onto homomorphism from G to G^1, we have $\varphi(G) = G^1$. That is, G^1 is the homomorphic image of φ. Define $f: G/K \to G^1$ by $f(Ka) = \varphi(a)$ for all $Ka \in G/K$.

To show that f is well-defined, let $Ka = Kb$ (for $a, b \in G$) $\Rightarrow ab^{-1} \in K \Rightarrow \varphi(ab^{-1}) = e^1 \Rightarrow \varphi(a) \cdot [\varphi(b)]^{-1} = e^1 \Rightarrow \varphi(a) = \varphi(b) \Rightarrow f(Ka) = f(Kb)$.

To show f is one–one, suppose $f(Ka) = f(Kb) \Rightarrow \varphi(a) = \varphi(b) \Rightarrow \varphi(a) \cdot [\varphi(b)]^{-1} = e^1$
$\Rightarrow \varphi(a) \cdot \varphi(b^{-1}) = e^1 \Rightarrow \varphi(ab^{-1}) = e^1 \Rightarrow ab^{-1} \in K \Rightarrow Ka = Kb$.
Therefore, f is one–one.

To show that f is onto, let $y \in G^1$. Since $\varphi: G \to G^1$ is onto, it follows that there exists $x \in G$ such that $\varphi(x) = y$. Since $x \in G$, it follows that $Kx \in G/K$. Now $f(Kx) = \varphi(x) = y$. Therefore, f is onto.

To show that f is a homomorphism, let $Ka, Kb \in G/K$. Consider $f(Ka \cdot Kb) = f(Kab) = \varphi(ab) = \varphi(a) \cdot \varphi(b)$ (since φ is a homomorphism) $= f(Ka) \cdot f(Kb)$. Therefore, f is a homomorphism.

Hence, $f: G/K \to G^1$ is an isomorphism. ∎

Theorem 1.2.50 (Correspondence Theorem)

Let φ be a homomorphism of G onto G with kernel K. Let H be a subgroup of G. Define $H = \{x \in G \mid \varphi(x) \in H\}$.
Then (i) H is a subgroup of G and $K \subseteq H$.
(ii) If H is normal in G, then H is normal in G.

Proof

Suppose φ is a homomorphism of G onto G, and $K = \ker \varphi$. Consider H as in the statement.

(i) To show that H is a subgroup, let $x, y \in H$. This implies that $\varphi(x)$, $\varphi(y) \in H \Rightarrow \varphi(x) \cdot \varphi(y) \in H$ (by the closure property of H) $\Rightarrow \varphi(xy) \in H$ (since φ is a homomorphism) $\Rightarrow xy \in H$ (by the definition of H). Hence, the closure property holds well in H. Now, let $x \in H$. This implies that $\varphi(x) \in H \Rightarrow [\varphi(x)]^{-1} \in H \Rightarrow \varphi(x^{-1}) \in H \Rightarrow x^{-1} \in H$. Therefore, H is a subgroup of G. Now, let $x \in K \Rightarrow \varphi(x) = e$ (since $K = \ker \varphi$) $\Rightarrow \varphi(x) = e \in H \Rightarrow x \in H$. Therefore, $K \subseteq H$.

(ii) Suppose H is normal in G. Let $h \in H, g \in G$. Then $\varphi(h) \in H$ and $\varphi(g) \in G$ $\Rightarrow \varphi(g)\varphi(h)\varphi(g)^{-1} \in H$ (since H is a normal subgroup) $\Rightarrow \varphi(g)\varphi(h)\varphi(g^{-1}) \in H$ $\Rightarrow \varphi(ghg^{-1}) \in H \Rightarrow ghg^{-1} \in H$. This shows that H is normal in G. ∎

Theorem 1.2.51

Let φ be a homomorphism of G onto G with kernel K, and let N be a normal subgroup of G, $N = \{x \in G \mid \varphi(x) \in N\}$. Then $G/N \cong (G/K)/(N/K)$.

Definition 1.2.52

Let G be a group. An isomorphism from G to G is called an **automorphism** of G.

Example 1.2.53

For any set S, the set of all bijections on S forms a group with respect to the composition of mappings.

Example 1.2.54

If G is a group, then $A(G)$ = the set of all automorphisms of G is a group with respect to the composition of mappings.

Solution

Consider the set $A(G) = \{f\colon G \to G \mid f$ is an automorphism$\}$. Now we have to show that $A(G)$ is a group under the operation, the composition of mappings.

Closure property: Let $f, g \in A(G) \Rightarrow f\colon G \to G$, $g\colon G \to G$ be isomorphisms. Then $f \circ g \in A(G)$. [**Verification:** $(f \circ g)(ab) = f(g(ab)) = f(g(a) \cdot g(b))$ (since g is a homomorphism) $= f(g(a)) \cdot f(g(b))$ (since f is a homomorphism) $= (f \circ g)(a) \cdot (f \circ g)$ (b). Therefore, $f \circ g$ is a homomorphism.]

Associative property: Let $f, g, h \in A(G)$. We know that functions satisfy the associative law with respect to the composition of mappings. Therefore $(f \circ g) \circ h = f \circ (g \circ h)$.

Existence of identity: Consider the identity mapping $I\colon G \to G$ defined by $I(x) = x$ for all $x \in G$. We know that this is a one–one and onto mapping. Also, $I(xy) = xy = I(x) \cdot I(y)$. That is, I is a homomorphism. Therefore, I is an automorphism. Thus, $I \in A(G)$. Now for any $f \in A(G)$, we have $(f \circ I)(x) = f(I(x)) = f(x)$. Therefore, $f \circ I = f$. Similarly, $I \circ f = f$. Hence, $f \circ I = I \circ f = f$. This shows that I acts as the identity in $A(G)$.

Existence of inverse: Let $f \in A(G)$. Since $f\colon G \to G$ is a bijection, $f^{-1}\colon G \to G$ by $f^{-1}(x) = y$ if $f(y) = x$ is also a bijection. Now we show that f^{-1} is a homomorphism. Let $a, b \in G$. Write $a^1 = f(a)$, $b^1 = f(b)$. Consider $f^{-1}(a^1b^1) = f^{-1}(f(a) \cdot f(b)) = f^{-1}(f(ab)) = ab = f^{-1}(a^1) \cdot f^{-1}(b^1)$ (since $a = f^{-1}(a^1)$ and $b = f^{-1}(b^1)$). Therefore, f^{-1} is a homomorphism. Hence, f^{-1} is an automorphism. So, $f^{-1} \in A(G)$. Now we know that $f^{-1} \circ f = I = f \circ f^{-1}$. Therefore, f^{-1} is the inverse of f in $A(G)$. This shows that $A(G)$ is a group with respect to the operation composition of mappings.

1.3 Ring Theory

The concept of a "group" was introduced in the previous section. A group is an algebraic system with a single binary operation satisfying certain axioms. We know that addition and multiplication are two binary operations on such number systems as integers, rational numbers, real numbers, and complex numbers. With respect to addition, all these systems satisfy the axioms of a group. Also, the systems of the set of nonzero rational numbers, the set of nonzero real numbers, and the set of all nonzero complex numbers form a group under the multiplication binary operation. Several of the properties of numbers depend simultaneously on both the addition and multiplication operations. The addition and multiplication operations, on number systems are interrelated, and lead mathematicians to study algebraic systems with two binary operations. A ring is an algebraic system with addition and multiplication binary operations. In this section, we introduce the concept of a ring and study some fundamental concepts of ring theory.

Definition 1.3.1

A nonempty set R is said to be a **ring** (or an **associative ring**) if there exist two operations: + (called addition) and · (called multiplication) on R, satisfying the following three conditions:

(i) $(R, +)$ is an Abelian group
(ii) $(R, \cdot)$ is a semigroup, and
(iii) $a(b + c) = ab + ac$ and $(a + b)c = ac + bc$ for all $a, b, c \in R$

Definition 1.3.2

Let $(R, +, \cdot)$ be a ring.

(i)   If there exists an element $1 \in R$ such that $a \cdot 1 = 1 \cdot a = a$ for every $a \in R$, then we say that R is a **ring with identity** (or unit) element. The element "1" is called as the identity element of the ring R.
(ii) If the ring R satisfies the condition $a \cdot b = b \cdot a$ for all $a, b \in R$, then we say that R is a **commutative** ring.

Examples 1.3.3

(i) $(Z, +, \cdot)$ is a commutative ring with identity.
(ii) $(2Z, +, \cdot)$ is a commutative ring without identity.

(iii) $(Q, +, \cdot)$ is a commutative ring with identity.

(iv) $(Z_n, +, \cdot)$ is a commutative ring with identity.

Definition 1.3.4

(i) If R is a commutative ring, then $0 \neq a \in R$ is said to be a **zero divisor** if there exists $0 \neq b \in R$ such that $ab = 0$.

(ii) A commutative ring is said to be an **integral domain** if it has no zero divisors.

(iii) A ring R is said to be a **division ring** if $(R^*, \cdot)$ is a group (where $R^* = R - \{0\}$).

(iv) A division ring is said to be a **field** if it is commutative.

We may define the concept "field" as follows:

Definition 1.3.5

An algebraic system $(F, +, \cdot)$, where F is a nonempty set and $+, \cdot$ are two binary operations, is said to be a **field** if it satisfies the following three properties:

(i) $(F, +)$ is an Abelian group

(ii) $(F^*, \cdot)$ is a commutative group, where $F^* = F - \{0\}$, and

(iii) $a(b + c) = ab + ac$ and $(a + b)c = ac + bc$ for all a, b, c in F

Example 1.3.6

Let R be the set of all real valued continuous functions on the closed unit interval $[0, 1]$. For $f, g \in R$ and $x \in [0, 1]$, define $(f + g)(x) = f(x) + g(x)$ and $(f \cdot g)(x) = f(x) \cdot g(x)$.

(i) Define $h: [0, 1] \to R$ by $h(x) = \frac{1}{2} - x$ if $0 \leq x \leq \frac{1}{2}$ and $h(x) = 0$ if $\frac{1}{2} \leq x < 1$. Then h is continuous and so $h \in R$.

Now, we verify that h does not have its inverse. Let us suppose $h^1 \in R$ such that $hh^1 = e$. Then $(hh^1)(1/2) = e(1/2) = 1 \Rightarrow h(1/2) \cdot h^1(1/2) = 1 \Rightarrow 0 \cdot h^1(1/2) = 1 \Rightarrow 0 = 1$, a contradiction. Hence, h does not have its inverse. Thus, the ring R cannot be a field.

(ii) Define $f: [0, 1] \to R$ by $f(x) = 0$ if $0 \leq x \leq \frac{1}{2}$ and $f(x) = x - \frac{1}{2}$ if $\frac{1}{2} \leq x < 1$. Then f is continuous, and so $f \in R$.

Since $h \neq 0$, $f \neq 0$, and $hf = 0$, it follows that the ring R is not an integral domain.

(iii) If $d \in R$ for all $d(x) \neq 0$ for all $x \in [0, 1]$, the function d^1 is defined by $d^1(x) = 1/d(x)$ for all x with $0 \leq x \leq 1$ a continuous mapping, and $d \cdot d^1 = e$, then, d^1 is the inverse of d.

Conversely, if $d \in R$ has an inverse, then $d \cdot (d^{-1}) = e \Rightarrow d(x) \cdot d^{-1}(x) = 1$ for all $x \in [0, 1] \Rightarrow d^{-1}(x) = 1/d(x)$ for all $x \in [0, 1]$.

From this discussion, we can conclude that $d \in R$ has an inverse if and only if $d(x) \neq 0$ for all $0 \leq x \leq 1$, and the function defined by $d^1(x) = 1/d(x)$ for all $x \in [0, 1]$ is continuous and the inverse of the continuous function d.

(iv) From (i), it is clear that there exist nonzero elements in R that do not have a multiplicative inverse. Hence, this ring R cannot be a division ring.

Example 1.3.7

(i) Let F be the field of real numbers, and let R be the set of all formal square arrays $\begin{pmatrix} a & b \\ c & d \end{pmatrix}$, where a, b, c, d are real numbers.

We define

$$\begin{pmatrix} a_1 & b_1 \\ c_1 & d_1 \end{pmatrix} + \begin{pmatrix} a_2 & b_2 \\ c_2 & d_2 \end{pmatrix} = \begin{pmatrix} a_1 + a_2 & b_1 + b_2 \\ c_1 + c_2 & d_1 + d_2 \end{pmatrix}.$$

It is easy to verify that R forms an Abelian group under addition with $\begin{pmatrix} 0 & 0 \\ 0 & 0 \end{pmatrix}$ acting as the zero element with respect to the addition operation + defined.

The element $\begin{pmatrix} -a & -b \\ -c & -d \end{pmatrix}$ is the additive inverse of $\begin{pmatrix} a & b \\ c & d \end{pmatrix}$.

We define the operation multiplication on R by

$$\begin{pmatrix} a & b \\ c & d \end{pmatrix} \begin{pmatrix} r & s \\ t & u \end{pmatrix} = \begin{pmatrix} ar + bt & as + bu \\ cr + dt & cs + du \end{pmatrix}.$$

The element $\begin{pmatrix} 1 & 0 \\ 0 & 1 \end{pmatrix}$ acts as a multiplicative identity element.

With respect to the addition and multiplication operations defined earlier, R becomes a ring.

Since $\begin{pmatrix} 1 & 0 \\ 0 & 0 \end{pmatrix} \begin{pmatrix} 0 & 0 \\ 1 & 0 \end{pmatrix} = \begin{pmatrix} 0 & 0 \\ 0 & 0 \end{pmatrix}$, it follows that R is not an integral domain.

Since $\begin{pmatrix} 1 & 0 \\ 0 & 0 \end{pmatrix} \cdot \begin{pmatrix} 0 & 0 \\ 1 & 0 \end{pmatrix} = \begin{pmatrix} 0 & 0 \\ 0 & 0 \end{pmatrix} \neq \begin{pmatrix} 0 & 0 \\ 1 & 0 \end{pmatrix} = \begin{pmatrix} 0 & 0 \\ 1 & 0 \end{pmatrix} \begin{pmatrix} 1 & 0 \\ 0 & 0 \end{pmatrix}$, it fol-

lows that R is not a commutative ring.

(ii) **(Field of Real Quaternions, invented by William Rowan Hamilton (1805–1865))**: Let F be the field of real numbers. Consider the set of all formal symbols $\alpha_0 + \alpha_1 i + \alpha_2 j + \alpha_3 k$, where $\alpha_0, \alpha_1, \alpha_2, \alpha_3 \in F$. Equality and addition of these symbols are defined as follows:

$\alpha_0 + \alpha_1 i + \alpha_2 j + \alpha_3 k = \beta_0 + \beta_1 i + \beta_2 j + \beta_3 k$, if and only if $\alpha_0 = \beta_0$, $\alpha_1 = \beta_1$, $\alpha_2 = \beta_2$, and $\alpha_3 = \beta_3$, and $(\alpha_0 + \alpha_1 i + \alpha_2 j + \alpha_3 k) + (\beta_0 + \beta_1 i + \beta_2 j + \beta_3 k) = (\alpha_0 + \beta_0) + (\alpha_1 + \beta_1)i + (\alpha_2 + \beta_2)j + (\alpha_3 + \beta_3)k$.

We now define multiplication of vectors. [When Hamilton discovered the concept of quaternions on October 16, 1843, he used his penknife to carve out the basic rules of multiplication on Brougham Bridge in Dublin.]

Multiplication of vectors.

The product is based on the conditions:

$$i^2 = j^2 = k^2 = -1, \ ij = k, \ jk = i, \ ki = j \text{ and } ji = -k, \ kj = -i, \ ik = -j.$$

If we go round the circle clockwise, the product of any two successive ones is the next one, and going round counterclockwise, we get the negatives. We can write out the product of any two quaternions as follows:
$(\alpha_0 + \alpha_1 i + \alpha_2 j + \alpha_3 k)(\beta_0 + \beta_1 i + \beta_2 j + \beta_3 k) = \Upsilon_0 + \Upsilon_1 i + \Upsilon_2 j + \Upsilon_3 k$, where

$$\left.\begin{aligned}
\Upsilon_0 &= \alpha_0 \beta_0 - \alpha_1 \beta_1 - \alpha_2 \beta_2 - \alpha_3 \beta_3 \\
\Upsilon_1 &= \alpha_0 \beta_1 + \alpha_1 \beta_0 + \alpha_2 \beta_3 - \alpha_3 \beta_2 \\
\Upsilon_3 &= \alpha_0 \beta_0 + \alpha_1 \beta_2 - \alpha_2 \beta_1 + \alpha_3 \beta_0
\end{aligned}\right\} \Upsilon_2 = \alpha_0 \beta_2 - \alpha_1 \beta_3 + \alpha_2 \beta_0 + \alpha_3 \beta_1$$

$$\tag{1.3}$$

If some α_i is 0 in $x = \alpha_0 + \alpha_1 i + \alpha_2 j + \alpha_3 k$, we shall omit it in expressing x; thus,
$0 + 0i + 0j + 0k$ will be written as 0,
$1 + 0i + 0j + 0k$ as 1,
$0 + 3i + 4j + 0k$ as $3i + 4j$.
It can be observed that
$$(\alpha_0 + \alpha_1 i + \alpha_2 j + \alpha_3 k)(\alpha_0 - \alpha_1 i - \alpha_2 j - \alpha_3 k) = \alpha_0^2 + \alpha_1^2 + \alpha_2^2 + \alpha_3^2 \tag{1.4}$$
This has a very important consequence.
Suppose that $x = \alpha_0 + \alpha_1 i + \alpha_2 j + \alpha_3 k \neq 0$ (implies that some α is nonzero). Since the α's are real, $\beta = \alpha_0^2 + \alpha_1^2 + \alpha_2^2 + \alpha_3^2 \neq 0$.

Thus $(\alpha_0 + \alpha_1 i + \alpha_2 j + \alpha_3 k)\left(\dfrac{\alpha_0}{\beta} - \dfrac{\alpha_1}{\beta} i - \dfrac{\alpha_2}{\beta} j - \dfrac{\alpha_3}{\beta} k\right) = 1.$

We observed that if $x \neq 0$, then x has an inverse. This shows that the set of all real quaternions forms a noncommutative division ring.

Example 1.3.8

Let R be a ring and M be the set of all $n \times n$ matrices over R (that is, the entries of the matrices are elements from the ring R). Then M becomes a ring with respect to the usual addition and multiplication of matrices. This ring M is called the ring of $n \times n$ matrices over R, and it is denoted by $M_n(R)$.

Definition 1.3.9

A ring R is said to be a **Boolean ring** if $x^2 = x$ for all $x \in R$.

Note 1.3.10 (The Pigeonhole Principle)

If a objects are distributed over m places and if $a > m$, then some place receives at least two objects.

Theorem 1.3.11

A finite integral domain is a field.

Proof

It is clear that in the case of an integral domain, we have $ab = 0 \Rightarrow a = 0$ or $b = 0$. Now it is enough to show that every nonzero element has a multiplicative inverse. Let D be an integral domain. Now we show that

(i) there exists $1 \in D$ such that $a\,1 = a$ for all $a \in D$, and
(ii) $0 \neq a \in D \Rightarrow$ there exists $b \in D$ such that $ab = 1$.

Let $D = \{x_1, x_2, \ldots, x_n\}$ and $0 \neq a \in D$. Now, $x_1 a, x_2 a, \ldots, x_n a$ are all distinct (for $x_i a = x_j a \Rightarrow (x_i - x_j)a = 0 \Rightarrow x_i - x_j = 0 \Rightarrow x_i = x_j$ (since $a \neq 0$)). Therefore, $D = \{x_1 a, x_2 a, \ldots, x_n a\} \Rightarrow a = x_k a$ for some $1 \leq k \leq a$ (since $a \in D$). Since D is commutative, it follows that $x_k a = a = a x_k$. We show that x_k is the identity element. For this, let $y \in D$; then $y = x_i a$ for some i. Now consider $y x_k = (x_i a)x_k = x_i(a x_k) = x_i a = y$. Thus, $y x_k = y$ for all $y \in D$. Therefore, x_k is the identity element. Now, for $x_k \in D = \{x_1 a, x_2 a, ..., x_n a\} \Rightarrow x_k = x_j a$ for some $1 \leq j \leq a$. Therefore, x_j is the multiplicative inverse of a. Hence, D is a field. ∎

Problem 1.3.12

If p is a prime number, then J_p (that is, Z_p), the ring of integers modulo p, is a field.

Solution

We know that J_p contains exactly 0, 1, 2, ... $(p-1)$. Let $a, b \in J_p$ such that $ab \equiv 0$ (mod p) $\Rightarrow ab - 0$ is divisible by $p \Rightarrow p$ divides $ab \Rightarrow p$ divides a or p divides b (since p is prime) $\Rightarrow a - 0$ is divisible by p or $b - 0$ is divisible by $p \Rightarrow a \equiv 0$ (mod p) or $b \equiv 0$ (mod p). Therefore, Jp is an integral domain, and by Theorem 1.3.11, J_p is a field.

Definition 1.3.13

(i) An integral domain D is said to be of **characteristic 0** if the relation $ma = 0$, where $0 \neq a \in D$ and m is an integer, can hold only if $m = 0$.

(ii) An integral domain D is said to be of **finite characteristic** if there exists a positive integer m such that $ma = 0$ for all $a \in D$. If D is of finite characteristic, then we define the characteristic of D to be the smallest positive integer p such that $pa = 0$ for all $a \in D$. The characteristic of R is denoted by Cha. R.

Examples 1.3.14

(i) $(Z, +, \cdot)$ is an integral domain with characteristic 0.

(ii) $(Z_p, +, \cdot)$ is an integral domain with characteristic p because for every $a \in Z_p$, we have $pa = a + a + ... + a$ (p times) $= ap = a \cdot 0 = 0$ (since $p = 0$ in Z_p).

(iii) $(Z_6, +, \cdot)$ is a commutative ring, but not an integral domain (since $2 \cdot 3 = 6 = 0$, but $2 \neq 0 \neq 3$).

Definition 1.3.15

Let $(R, +, \cdot)$, $(R^1, +, \cdot)$ be two rings. A mapping $\varphi : R \to R^1$ is said to be a **homomorphism** (*or* a **ring homomorphism**) if it satisfies the following two conditions:

(i) $\varphi(a + b) = \varphi(a) + \varphi(b)$, and
(ii) $\varphi(ab) = \varphi(a) \varphi(b)$, for all $a, b \in R$.

Problem 1.3.16

If $\varphi: R \to R^1$ is a homomorphism, then (i) $\varphi(0) = 0$ and (ii) $\varphi(-a) = -\varphi(a)$ for all $a \in R$.

Solution

(i) Consider $0 + \varphi(0) = \varphi(0) = \varphi(0 + 0) = \varphi(0) + \varphi(0) \Rightarrow \varphi(0) = 0$.
(ii) $0 = \varphi(0) = \varphi(a + (-a)) = \varphi(a) + \varphi(-a) \Rightarrow \varphi(-a) = -\varphi(a)$.

Note 1.3.17

Let $\varphi: R \to R^1$ be a nonzero homomorphism.

(i) If R is an integral domain, $\varphi(1) = 1$.
(ii) If φ is onto, then $\varphi(1) = 1$ (here R, R^1 are rings with identity).

Verification

(i) Since φ is a nonzero mapping, it follows that $\varphi(R) \neq 0$. This implies that there exists $x \in R$ such that $\varphi(x) \neq 0$. Now, $1 \cdot \varphi(x) = \varphi(x) = \varphi(1 \cdot x) = \varphi(1) \cdot \varphi(x) \Rightarrow [1 - \varphi(1)]\varphi(x) = 0 \Rightarrow 1 - \varphi(1) = 0$ (since $\varphi(x) \neq 0$ and R^1 is an integral domain) $\Rightarrow \varphi(1) = 1$.
(ii) First, we show that $\varphi(1)$ is the identity in R^1. For this, let $y \in R^1$. Since φ is onto, it follows that there exists $x \in R$ such that $\varphi(x) = y$. Now, $\varphi(1) \cdot y = y = \varphi(1) \cdot \varphi(x) = \varphi(1 \cdot x) = \varphi(x) = y$. Therefore, $\varphi(1) \cdot y = y$ for all $y \in R^1$. Similarly, $y \cdot \varphi(1) = y$, for all $y \in R^1$. This shows that $\varphi(1)$ is the identity in R^1. Since the identity element is unique, it follows that $\varphi(1) = 1$.

Note 1.3.18

For any two rings R and R^1, if we define $\varphi: R \to R^1$ by $\varphi(x) = 0$ for all $x \in R$, then φ is a homomorphism. This homomorphism is called the zero homomorphism.

Definition 1.3.19

(i) Let $\varphi: R \to R^1$ be a homomorphism. Then the set $\{x \in R \mid \varphi(x) = 0\}$ is called the **kernel** of the homomorphism φ, and it is denoted by ker φ or $I(\varphi)$.

(ii) A homomorphism $\varphi\colon R \to R^1$ is said to be an **isomorphism** if φ is both one–one and onto.

(iii) Two rings R and R^1 are said to be **isomorphic** if there exists an iso-morphism $\varphi\colon R \to R^1$.

Problem 1.3.20

Let $\varphi\colon R \to R^1$ be a homomorphism. Then

(i) ker $\varphi = \{0\} \Leftrightarrow \varphi$ is one–one, and

(ii) If φ is onto, then φ is an isomorphism if and only if ker $\varphi = \{0\}$.

Solution

(i) Suppose ker $\varphi = \{0\}$. To see if φ is one–one, suppose $x, y \in R$ such that $\varphi(x) = \varphi(y)$. This implies $\varphi(x) - \varphi(y) = 0 \Rightarrow \varphi(x) + \varphi(-y) = 0 \Rightarrow \varphi(x - y) = 0 \Rightarrow x - y \in$ ker $\varphi = \{0\} \Rightarrow x - y = 0 \Rightarrow x = y$. Therefore, φ is one–one.

Converse

Suppose φ is one–one. Since $\varphi(0) = 0$, it follows that $0 \in$ ker $\varphi \Rightarrow \{0\} \subseteq$ ker φ. Now let $y \in$ ker $\varphi \Rightarrow \varphi(y) = 0 = \varphi(0)$ (we know that $\varphi(0) = 0$) $\Rightarrow y = 0$ (since φ is one–one $\Rightarrow$ ker $\varphi \subseteq \{0\}$. Therefore, ker $\varphi = \{0\}$.

(ii) Suppose φ is an isomorphism $\Rightarrow \varphi$ is a bijection $\Rightarrow \varphi$ is one–one $\Rightarrow$ ker $\varphi = \{0\}$ (by (i)). **Converse:** Suppose ker $\varphi = \{0\} \Rightarrow \varphi$ is one–one (by (i)). Since φ is onto, it follows that φ is a bijection. Hence, φ is an isomorphism.

Definition 1.3.21

A nonempty set I of a ring R is said to be

(i) a **left ideal** of R if I is a subgroup of $(R, +)$ and $ra \in I$ for every $r \in R$, $a \in I$

(ii) a **right ideal** of R if I is a subgroup of $(R, +)$ and $ar \in I$ for every $r \in R$, $a \in I$

(iii) an **ideal** (or two-sided ideal) of R if I is both left ideal and right ideal

Problem 1.3.22

If $\varphi\colon R \to R^1$ is a homomorphism, then ker φ is an ideal of R.

Solution

To verify that ker φ is an ideal of R, we have to show that (i) ker φ is a subgroup of $(R, +)$ and (ii) $a \in$ ker φ, $r \in R \Rightarrow ar, ra \in$ ker φ.

(i) Let $a, b \in$ ker $\varphi \Rightarrow \varphi(a) = 0 = \varphi(b)$. Now, $\varphi(a - b) = \varphi(a) - \varphi(b) = 0 - 0 = 0$. This implies $a - b \in$ ker φ. Hence, ker φ is a subgroup of $(R, +)$.

(ii) Now let $a \in$ ker φ. Then $\varphi(a) = 0$. Therefore, $\varphi(ar) = \varphi(a) \cdot \varphi(r) = 0 \cdot \varphi(r) = 0$, and so $ar \in$ ker φ. Similarly, we can show that $\varphi(ra) = 0$, and so $ra \in$ ker φ. Therefore, ker φ is an ideal of R.

Lemma 1.3.23

Let $f: R \rightarrow R^1$ be a ring homomorphism.

(i) If J is an ideal of R, then $f(J) = \{f(x) \mid x \in J\}$ is an ideal of R^1.

(ii) If J_1, J_2 are ideals of R such that $f(J_1)$ and $f(J_2)$ are equal, and ker $f \subseteq J_2$, then $J_1 \subseteq J_2$.

(iii) If W^1 is an ideal of R^1, then $f^{-1}(W^1) = \{x \in R \mid f(x) \in W^1\}$ is an ideal of R.

Remark 1.3.24

Let R be a ring and I an ideal of R.

(i) Define a relation $\sim$ on R as $a \sim b$ if and only if $a - b \in I$ for $a, b \in R$. Now we verify that this relation is an equivalence relation. Since $a - a = 0 \in I$, it follows that $a \sim a$. This shows that the relation is reflexive. Suppose $a \sim b \Rightarrow a - b \in I \Rightarrow b - a \in I \Rightarrow b \sim a$. This shows that the relation is symmetric. Suppose $a \sim b$, $b \sim c \Rightarrow a - b \in I$, $b - c \in I \Rightarrow a - c = (a - b) + (b - c) \in I \Rightarrow a \sim c$. This shows that the relation is transitive. Since this relation is reflexive, symmetric, and transitive, it follows that the relation $\sim$ is an equivalence relation on R.

(ii) Write $a + I = \{a + x \mid x \in I\}$ for $a \in R$. Now $a + I$ is the equivalence class containing a. We say that $a + I$ is a coset of I.

(iii) Write $R/I = \{a + I \mid a \in R\}$, the set of all equivalence classes (or) the set of all cosets.

(iv) Define $+$ and $\cdot$ on R/I as $(a + I) + (b + I) = (a + b) + I$ and $(a + I) \cdot (b + I) = (a \cdot b) + I$. Then $(R/I, +, \cdot)$ becomes a ring. In this ring, $0 + I$ is the additive identity, and $(-a) + I$ is the additive inverse of $a + I$. This ring R/I is called the **quotient ring** of R modulo the ideal I.

Now we have the following conclusion:

Theorem 1.3.25

If R is a ring and I is an ideal of R, then R/I is also a ring.

Problem 1.3.26

Let R be a ring, I an ideal of R, and R/I the quotient ring. If we define $\varphi: R \to R/I$ as $\varphi(a) = a + I$, then φ is an epimorphism (that is, an onto homomorphism). The verification is left as an exercise for the reader.

Definition 1.3.27

(i) Consider $\varphi: R \to R/I$ defined in Problem 1.3.26. This φ is called the **canonical epimorphism** from R onto R/I.
(ii) The ring $(R/I, +, \cdot)$ is called the **quotient ring** of the ring R by an ideal I.

Note 1.3.28

Consider $\varphi: R \to R/I$ as it was defined in Problem 1.3.26. Then $x \in \ker \varphi$ if and only if $\varphi(x) = 0 = 0 + I \Leftrightarrow x + I = 0 + I \Leftrightarrow [x] = [0] \Leftrightarrow x \sim 0 \Leftrightarrow x - 0 \in I \Leftrightarrow x \in I$. Therefore, $\ker \varphi = I$.

Lemma 1.3.29

Let $f: R \to R^1$ be a ring homomorphism.

(i) If J is an ideal of R, then $f(J) = \{f(x) \mid x \in J\}$ is an ideal of R^1.
(ii) If J_1, J_2 are ideals of R such that $f(J_1) = \{f(x) \mid x \in J_1\}$ and $f(J_2) = \{f(x) \mid x \in J_2\}$ are equal, and $\ker f \subseteq J_2$, then $J_1 \subseteq J_2$.
(iii) If W^1 is an ideal of R^1, then $f^{-1}(W^1) = \{x \in R \mid f(x) \in W^1\}$ is an ideal of R.

Theorem 1.3.30

Let R, R^1 be two rings, and $f: R \to R^1$ is an epimorphism such that $I = \ker f$. Then

(i) **(First Homomorphism Theorem)** $(R/\ker f) = (R/I)$ is isomorphic to R^1.
(ii) **(Correspondence Theorem)** There exists a one-to-one correspondence between the set of ideals of R^1 and the set of ideals of R containing I.

(iii) (**Third Homomorphism Theorem**) This correspondence can be achieved by associating with an ideal W^1 of R^1, the ideal W of R defined by $W = \{x \in R \mid f(x) \in W^1\}$. With W so defined, R/W is isomorphic to R^1/W^1.

Theorem 1.3.31 (Second Homomorphism Theorem)

Let A be a subring of a ring R and I an ideal of R. Then $A + I = \{a + I \mid a \in A, I \in I\}$ is a subring of R, I is an ideal of $A + I$, and $(A + I)/I \cong A/(A \cap I)$.

Note 1.3.32

Let $f: R \to R^*$ be a ring homomorphism. Write $R^1 = \mathrm{Im}\, f$. We know that R^1 is a subring of R^*. So R^1 is a ring in its own right. Clearly, $f: R \to \mathrm{Im}\, f = R^1$ is the ring homomorphism. Also, it is onto. Therefore, by Theorem 1.3.31, $R/\ker f \cong R^1 = \mathrm{Im}\, f$. So, for any homomorphism f we get $R/\ker f \cong \mathrm{Im}\, f$.

Problem 1.3.33

If R is a ring with identity and $\varphi: R \to R^1$ is a ring epimorphism, then show that $\varphi(1)$ is the identity in R^1.

Solution

Let $y \in R^1$. We have to show that $\varphi(1) \cdot y = y = y \cdot \varphi(1)$. Since φ is onto, it follows that there exists $x \in R$ such that $\varphi(x) = y$. Consider $\varphi(1) \cdot y = \varphi(1) \cdot \varphi(x) = \varphi(1 \cdot x) = \varphi(x) = y$. Also, $y \cdot \varphi(1) = \varphi(x) \cdot \varphi(1) = \varphi(x \cdot 1) = \varphi(x) = y$. Therefore, $\varphi(1)$ is the identity in R^1.

Notation 1.3.34

(i) Let $X \subseteq R$. The intersection of all ideals of R containing X is called the ideal generated by X. It is denoted by $<X>$. If $X = \{a\}$, then we write $<a>$ for $<X>$. For an element $a \in R$, the ideal $<a>$ is called the **principal ideal** generated by a.

(ii) It is easy to verify that for $a \in R$, the following is true:

$$<a> = \left\{ ra + as + \left(\sum_{i=1}^{k} r_i a s_i \right) + na \mid r_i, s_i \in R, n, k \in \mathbb{Z} \right\}.$$

Definition 1.3.35

Let R be a ring and $S \subseteq R$. If S is a ring in its own rights, then S is called a **subring** of R (equivalently, if $(S, +)$ is a subgroup of $(R, +)$ and $ab \in S$ for all a, $b \in S$, then S is called a **subring** of R).

Example 1.3.36

Let $R = \left\{ \begin{pmatrix} a & b \\ 0 & a \end{pmatrix} \mid a, b \in \mathbb{R} \right\}$; R is a subring of the 2×2 matrices over the set of real numbers.

Let $I = \left\{ \begin{pmatrix} 0 & b \\ 0 & 0 \end{pmatrix} \mid b \in \mathbb{R} \right\}$, where $\mathbb{R}$ is the set of real numbers;

(i) It is easy to verify that I is an additive subgroup of R. Now the following two steps show that I is an ideal:

$$\begin{pmatrix} x & y \\ 0 & x \end{pmatrix}\begin{pmatrix} 0 & b \\ 0 & 0 \end{pmatrix} = \begin{pmatrix} 0 & xb \\ 0 & 0 \end{pmatrix} \in I.$$

$$\begin{pmatrix} 0 & b \\ 0 & 0 \end{pmatrix}\begin{pmatrix} x & y \\ 0 & x \end{pmatrix} = \begin{pmatrix} 0 & bx \\ 0 & 0 \end{pmatrix} \in I.$$

(ii) Take $\begin{pmatrix} a & b \\ 0 & a \end{pmatrix} \in R$.

Now $\begin{pmatrix} a & b \\ 0 & a \end{pmatrix} = \begin{pmatrix} a & 0 \\ 0 & a \end{pmatrix} + \begin{pmatrix} 0 & b \\ 0 & 0 \end{pmatrix}$, so that

$$\begin{pmatrix} a & b \\ 0 & a \end{pmatrix} + I = \left(\begin{pmatrix} a & 0 \\ 0 & a \end{pmatrix} + \begin{pmatrix} 0 & b \\ 0 & 0 \end{pmatrix}\right) + I = \begin{pmatrix} a & 0 \\ 0 & a \end{pmatrix} + I, \text{ since } \begin{pmatrix} 0 & b \\ 0 & 0 \end{pmatrix} \text{ is in } I.$$

Thus, all the cosets of I in R may be represented in the form $\begin{pmatrix} a & 0 \\ 0 & a \end{pmatrix} + I$.

If we map this element onto a, that is, if we define $\psi\left(\begin{pmatrix} a & 0 \\ 0 & a \end{pmatrix} + I\right) = a$, then it is easy to verify that ψ is an isomorphism onto the field of real numbers. Therefore, $R/I \cong \mathbb{R}$.

(iii) Define $\varphi: R \rightarrow \mathbb{R}$ by $\varphi \begin{pmatrix} a & b \\ 0 & a \end{pmatrix} = a$. We claim that φ is a homomorphism.

For the given $\begin{pmatrix} a & b \\ 0 & a \end{pmatrix}, \begin{pmatrix} c & d \\ 0 & c \end{pmatrix}$, we have $\varphi \begin{pmatrix} a & b \\ 0 & a \end{pmatrix} = a, \varphi \begin{pmatrix} c & d \\ 0 & c \end{pmatrix} = c,$

$\begin{pmatrix} a & b \\ 0 & a \end{pmatrix} + \begin{pmatrix} c & d \\ 0 & c \end{pmatrix} = \begin{pmatrix} a+c & b+d \\ 0 & a+c \end{pmatrix}.$

Also, $\begin{pmatrix} a & b \\ 0 & a \end{pmatrix} \cdot \begin{pmatrix} c & d \\ 0 & c \end{pmatrix} = \begin{pmatrix} ac & ad+bc \\ 0 & ac \end{pmatrix}$; and hence

$\varphi \left(\begin{pmatrix} a & b \\ 0 & a \end{pmatrix} + \begin{pmatrix} c & d \\ 0 & c \end{pmatrix} \right) = \varphi \begin{pmatrix} a+c & b+d \\ 0 & a+c \end{pmatrix} = a+c = \varphi \begin{pmatrix} a & b \\ 0 & a \end{pmatrix} + \varphi \begin{pmatrix} c & d \\ 0 & c \end{pmatrix},$ and

$\varphi \left(\begin{pmatrix} a & b \\ 0 & a \end{pmatrix} \cdot \begin{pmatrix} c & d \\ 0 & c \end{pmatrix} \right) = \varphi \begin{pmatrix} ac & ad+bc \\ 0 & ac \end{pmatrix} = ac = \varphi \begin{pmatrix} a & b \\ 0 & a \end{pmatrix} \varphi \begin{pmatrix} c & d \\ 0 & c \end{pmatrix}.$

Hence, φ is a homomorphism of R onto $\mathbb{R}$.

(iv) If $\begin{pmatrix} a & b \\ 0 & a \end{pmatrix} \in \ker \varphi$, then $\varphi \begin{pmatrix} a & b \\ 0 & u \end{pmatrix} = a,$ and also $\varphi \begin{pmatrix} a & b \\ 0 & a \end{pmatrix} = 0,$ since

$\begin{pmatrix} a & b \\ 0 & a \end{pmatrix} \in \ker \varphi.$ Thus, $a = 0$ and so $\begin{pmatrix} a & b \\ 0 & a \end{pmatrix} \in I.$ From this it follows

that $I = \ker \varphi.$ So $R/I \cong$ image of $\varphi = \mathbb{R}.$

Example 1.3.37

Let R be any commutative ring with 1. If $a \in R$, let $(a) = \{xa \mid x \in R\}$, then (a) is an ideal of R.

Thus, $u \pm v = xa \pm ya = (x \pm y)a \in (a)$. Also, if $u \in (a)$ and $r \in R$, then $u = xa$; hence, $ru = r(xa) = (rx)a$, and so is in (a). Thus, (a) is an ideal of R. Note that if R is not commutative, then (a) need not be an ideal; but it is certainly a left ideal of R.

Theorem 1.3.38

Let R be a commutative ring with a unit element whose only ideals are (0) and R itself. Then R is a field.

Example 1.3.39

If U, V are ideals of R, then the two sets $U + V = \{a + b \mid a \in U, b \in V\}$ and $UV = \{\sum_{i=1}^{n} a_i b_i \mid a_i \in U, b_i \in V \text{ and } n \text{ is a positive integer and } 1 \leq i \leq n\}$ are ideals of R.

Example 1.3.40

If A is a right ideal and B is a left ideal of R, then $A \cap B$ need not be a left (or right) ideal of R. We observe this by a suitable example.

Let $R^1 = M_2(R) =$ the ring of all 2×2 matrices over the set of real numbers $\mathbb{R}$.

Write $A = \left\{ \begin{bmatrix} a & b \\ 0 & 0 \end{bmatrix} \mid a, b \in R \right\}$ and $B = \left\{ \begin{bmatrix} c & 0 \\ d & 0 \end{bmatrix} \mid c, d \in R \right\}$.

Then, A is a right ideal of R and B is a left ideal of R.

Now $A \cap B = \left\{ \begin{bmatrix} a & 0 \\ 0 & 0 \end{bmatrix} \mid a \in R \right\}$.

Let $a \in R$, $a \neq 0$. Since $\begin{bmatrix} 3 & 4 \\ 1 & 0 \end{bmatrix} \cdot \begin{bmatrix} a & 0 \\ 0 & 0 \end{bmatrix} = \begin{bmatrix} 3a & 0 \\ a & 0 \end{bmatrix} \notin A \cap B$, we conclude that $A \cap B$ cannot be a left ideal.

Since $\begin{bmatrix} a & 0 \\ 0 & 0 \end{bmatrix} \cdot \begin{bmatrix} 3 & 4 \\ 1 & 0 \end{bmatrix} = \begin{bmatrix} 3a & 4a \\ a & 0 \end{bmatrix} \notin A \cap B$, we conclude that $A \cap B$ cannot be a right ideal. Therefore, $A \cap B$ is neither a left ideal nor a right ideal of R.

Definition 1.3.41

(i) An ideal I of R is said to be a **maximal ideal** if it is maximal among the set of all proper ideals of R.

(ii) A ring R is called **simple** if it has exactly two ideals (that is, (0) is the maximal ideal of R).

(iii) A ring R is said to be **primitive** if (0) is the largest among the ideals of R contained i n a maximal right ideal.

Definition 1.3.42

(i) A left (respectively, right) ideal I of R is said to be a **monogenic ideal** if there exists $0 \neq a \in I$ such that $Ra = I$ (respectively, $aR = I$).

(ii) A ring R is a **subdirect product** of a family of rings $\{S_i \mid i \in I\}$ if there is a monomorphism $k: R \to S = \prod_{i \in I}^{S_i}$ such that $\pi_i \circ k$ is an epimorphism for all $I \in I$, where $\pi_i: S \to S_i$ is the canonical epimorphism.

For the necessary definitions and results that are not mentioned here, we refer to the books written by Herstein (1964), Hungerford (1974), and Lambek (1966).

In this chapter, we use the term "ring" to mean an associative ring (not necessarily commutative).

Now we present the definition of a prime ideal and some related results. We start with the following definition.

Definition 1.3.43

(i) A proper ideal P of R is called **prime** if for any two ideals A and B, $AB \subseteq P \Rightarrow A \subseteq P$ or $B \subseteq P$.
(ii) A proper ideal S of R is said to be **semiprime** if $I^2 \subseteq S$ and I is an ideal of R imply $I \subseteq S$.
(iii) A ring R is said to be a **prime ring** if the zero ideal is a prime ideal.

Theorem 1.3.44 (Proposition 2, P27 of Lambek, 1966)

The proper ideal P of a commutative ring R is prime if and only if for all elements a and b, $ab \in P$ implies $a \in P$ or $b \in P$.

Theorem 1.3.45 (Proposition 4, P28 of Lambek, 1966)

The ideal M of a commutative ring R is prime if and only if R/M is an integral domain.

Theorem 1.3.46 (Exercise 17, P134 of Hungerford, 1974)

Let I be an ideal of R and $f: R \to R/I$, the canonical epimorphism. An ideal P of R containing I is a prime ideal in R if and only if $f(P)$ is a prime ideal in R/I.

Definition 1.3.47

Let R be a commutative ring.

(i) The intersection of all prime ideals of R is called the **prime radical** of R. This prime radical is denoted by "rad R."

(ii) An element $r \in R$ is called **nilpotent** if $r^n = 0$ for some natural number n.

Definition 1.3.48

A commutative ring R is called **semiprime** if its prime radical is 0.

Example 1.3.49

Consider $\mathbb{Z}$, the ring of integers:

$$\text{rad } \mathbb{Z} = \bigcap_{P \text{ is a prime ideal}} P \quad \subseteq (0) \text{ (since (0) is a prime ideal)}.$$

Therefore, rad $\mathbb{Z} = 0$. Hence, $\mathbb{Z}$ is semiprime.

Verification

Suppose $0 \neq m \in \text{rad } \mathbb{Z} = \cap p\mathbb{Z}$.

This means that $m \in p\mathbb{Z}$ for all prime p, and so $p|m$ for all primes.

Since there exist an infinite number of prime numbers, there exists a prime number q such that $m < q$.

Now q is a prime number, and q cannot divide m, which is a contradiction.

Theorem 1.3.50 (Proposition 8, p. 29 of Lambek, 1966)

In a commutative ring, we have rad $R = \{r \in R \mid r \text{ is nilpotent}\}$.

Definition 1.3.51

(i) An element $a \in R$ is said to be **strongly nilpotent** provided every sequence $a_0, a_1, \ldots, a_n \ldots$ with the property $a_0 = a$, $a_{n+1} \in a_n R a_n$ for all positive integers n is ultimately zero.

(ii) An ideal I of R is said to be **nilpotent** if $I^n = 0$ for some positive integer n.

(iii) An ideal I of R is called a **nil ideal** if every element of I is nilpotent.

Note 1.3.52

(i) Every strongly nilpotent element is nilpotent. Nilpotent elements need not be strongly nilpotent (for this, observe Example 1.3.53).
(ii) If R is commutative, then the concepts of nilpotent and strongly nilpotent are the same.

Example 1.3.53

Consider the ring of all 3×3 matrices over the ring of integers.

$$\text{Write } a_0 = \begin{pmatrix} 0 & 1 & 1 \\ 0 & 0 & 1 \\ 0 & 0 & 0 \end{pmatrix} \text{ and } x = \begin{pmatrix} 1 & 1 & 1 \\ 1 & 1 & 1 \\ 1 & 1 & 1 \end{pmatrix}.$$

Define $a_1 = a_0 x a_0, \ldots, a_n = a_{n-1} x a_{n-1}, \ldots$ Now $a_n \neq 0$ for all $n \geq 0$, and hence a_0 is not strongly nilpotent. But $a_0^3 = 0$. This shows that a_0 is nilpotent.

Hence, a_0 is a nilpotent element that is not strongly nilpotent.

Theorem 1.3.54 (Proposition 1, p. 56 of Lambek, 1966)

The prime radical of R is the set of all strongly nilpotent elements.

Theorem 1.3.50 states that in the case of a commutative ring R, the prime radical is equal to the set of all nilpotent elements of R. Note 1.3.52, together with Example 1.3.53, clarifies that the prime radical of an associative ring need not be equal to the set of all nilpotent elements of R.

Result 1.3.55 (P70, Lambek, 1966)

Let R be a ring and I an ideal of R. Then R is nil if and only if I and R/I are nil.

The concept of a Γ-ring was introduced by Nobusawa (1964). Later, this concept was generalized by Barnes (1966).

Definition 1.3.56 (Nobusawa, 1964)

Let M be an additive group whose elements are denoted by $a, b, c, \ldots$, and Γ another additive group whose elements are $\alpha, \beta, \gamma, \ldots$ Suppose that $a\alpha b$ is defined to be an element of M and that $\alpha a \beta$ is defined to be an element of Γ

for every a, b, α, and β. If the products satisfy the following three conditions for every $a, b, c \in M$, $\alpha, \beta \in \Gamma$.

(i) $(a + b)\alpha c = a\alpha c + b\alpha c$,
  $a(\alpha + \beta)b = a\alpha b + a\beta b$,
  $a\alpha(b + c) = a\alpha b + a\alpha c$.

(ii) $(a\alpha b)\beta c = a\alpha(b\beta c) = a(\alpha b\beta)c$.

(iii) If $a\alpha b = 0$ for all a and b in M, then $\alpha = 0$,

then M is called a Γ-ring.

Definition 1.3.57 (Barnes, 1966)

Let M and Γ be additive Abelian groups. M is said to be a Γ-ring if there exists a mapping $M \times \Gamma \times M \to M$ (the image of (a, α, b) is denoted by $a\alpha b$) satisfying the following conditions:

(i) $(a + b)\alpha c = a\alpha c + b\alpha c$
  $a(\alpha + \beta)b = a\alpha b + a\beta b$
  $a\alpha(b + c) = a\alpha b + a\alpha c$.

(ii) $(a\alpha b)\beta c = a\alpha(b\beta c)$ for all $a, b, c \in M$ and $\alpha, \beta \in \Gamma$.

Most authors have studied Γ-rings in the sense of Barnes. So, in this book, we include the results related to the concept: Γ-ring in the sense of Barnes.

Henceforth, all Γ-rings considered are Γ-rings in the sense of Barnes. A natural example of a Γ-ring can be constructed in the following way.

Example 1.3.58

From the following observation, we may conclude that every ring is a Γ-ring.

Let R be a ring. Write $M = R$, $\Gamma = R$. Now we verify that M is a Γ-ring.

Take $a, b, c \in M$, $\alpha, \beta, \gamma \in \Gamma$.

$a\alpha b$ is the product of a, α, b in R. So, $a\alpha b \in R = M$.

Using the distributive laws in R, it is easy to verify that

$(a + b)\alpha c = a\alpha c + b\alpha c$,
$a(\alpha + \beta)b = a\alpha b + a\beta b$,
$a\alpha(b + c) = a\alpha b + a\alpha c$.

Since the product in R is associative, it follows that $(a\alpha b)\beta c = a\alpha(b\beta c)$. Hence, M is a Γ-ring.

1.4 Vector Spaces

We recall the "addition" and "scalar multiplication" defined on the space of solutions of a homogeneous system of linear equations. We define a vector space to be a set on which similar operations are defined. More precisely, we define the concept "vector space" as follows.

Definition 1.4.1

Let $(V, +)$ be an Abelian group and F a field. Then V is said to be a **vector space** over the field F if there exists a mapping $F \times V \to V$ (the image of (α, v) is denoted by $\alpha \cdot v$ or αv) satisfying the following conditions:

(i) $\alpha(v + w) = \alpha v + \alpha w$
(ii) $(\alpha + \beta) v = \alpha v + \beta v$
(iii) $\alpha(\beta v) = (\alpha\beta)v$ and
(iv) $1 \cdot v = v$

for all $\alpha, \beta \in F$ and $v, w \in V$ (here 1 is the identity element of F with respect to multiplication).

Note 1.4.2

We use F to denote a field. The elements of F are called scalars, and the elements of V are called vectors.

Remark 1.4.3

Let $(V, +)$ be a vector space over F. Let $\alpha \in F$. Define $f\colon V \to V$ by $f(v) = \alpha v$ for all $v \in V$. Then

(i) f is a group homomorphism (or group endomorphism)
    (for this, consider $f(v_1 + v_2) = \alpha(v_1 + v_2) = \alpha v_1 + \alpha v_2 = f(v_1) + f(v_2)$).
(ii) If $\alpha \neq 0$, then f is an isomorphism.

Verification

To show f is one–one, suppose $f(v_1) = f(v_2)$. Then $\alpha v_1 = \alpha v_2$. This implies $\alpha^{-1}\alpha v_1 = \alpha^{-1}\alpha v_2 \Rightarrow v_1 = v_2$. To see if f is onto, let $w \in V$. Write $v = \alpha^{-1}w$. Then $f(v) = \alpha v = \alpha\alpha^{-1}w = w$.

Example 1.4.4

(i) Let K be a field and F a subfield of K. Write $V = K$. For any $\alpha \in F$ and $v \in V = K$, we have the product $\alpha v \in V = K$ (since K is a field and $\alpha \in F \subseteq K, v \in K \Rightarrow \alpha v \in K$). Now $V = K$ is a vector space over F. Therefore, every field is a vector space over its subfields.

(ii) Let R be the field of real numbers.
$R^n = \{(x_1, x_2, ..., x_n) \mid x_i \in R, 1 \le i \le n\}$. Define the scalar product of $r \in R$ and $(x_1, x_2, ..., x_n) \in R^n$ as $r(x_1, x_2, ..., x_n) = (rx_1, rx_2, ..., rx_n)$. Then R^n is a vector space over R. This R^n is called the Euclidean space of dimension n.

(iii) Let F be a field. Write $V = F^n = \{(x_1, x_2, ..., x_n) \mid x_i \in F, 1 \le i \le n\}$. Define the scalar product of $\alpha \in F$ and $(x_1, x_2, ..., x_n) \in F^n$ as $\alpha(x_1, x_2, ..., x_n) = (\alpha x_1, \alpha x_2, ..., \alpha x_n)$. Then, F^n is a vector space of dimension n.

(iv) Let F be a field. Consider $F[x]$, the ring of polynomials over F.

Let $\deg f(x)$ denotes the degree of the polynomial $f(x)$, which is the highest of the degree of all its terms. Write $V_n = \{f(x) \mid f(x) \in F[x]$ and $deg(f(x)) \le n\}$. Then $(V_n, +)$ is a group where "+" is the addition of polynomials.

(i) $f(x), g(x) \in V_n \Rightarrow deg(f(x)) \le n$ and $deg(g(x)) \le n \Rightarrow deg(f(x) + g(x)) = \max\{deg(f(x)), deg(g(x))\} \le n$. Hence, $f(x) + g(x) \in V_n$. This shows that "+" is closed on V_n.

(ii) The associative law is true since it is true in $F[x]$.

(iii) $0(x) = 0 + 0x + ... + 0x^n$ is of $deg \le n$, and also $0(x) \in V_n$ is the identity element of V_n.

(iv) $-f(x)$ is the inverse of $f(x)$ for any $f(x) \in V_n$.

(v) We know that "+" is commutative in $F[x]$.

Hence $(V_n, +)$ is an Abelian group.

Now for any $\alpha \in F$ and $f(x) = a_0 + a_1 x + ... + a_n x^n \in V_n$, we define scalar multiplication as follows: $\alpha(f(x)) = \alpha a_0 + \alpha a_1 x + ... + \alpha a_n x^n$.

Then $deg[\alpha(f(x))] \le n$, and hence $\alpha(f(x)) \in V_n$.

Now it is easy to verify that V_n is a vector space over F.

Definition 1.4.5

Let V be a vector space over F and W be an additive subgroup of V. Then W is called a **subspace** of V if W is a vector space over F under the same scalar multiplication.

Theorem 1.4.6

Let V be a vector space over F and $\varnothing \neq W \subseteq V$. Then the following two conditions are equivalent.

(i) W is a subspace of V.
(ii) $\alpha, \beta \in F$ and $w_1, w_2 \in W \Rightarrow \alpha w_1 + \beta w_2 \in W$.

Proof

(i) $\Rightarrow$ (ii): $\alpha \in F, w_1 \in W \Rightarrow \alpha w_1 \in W; \beta \in F, w_2 \in W \Rightarrow \beta w_2 \in W$. Since αw_1, $\beta w_2 \in W$ and W is a subspace, we have $\alpha w_1 + \beta w_2 \in W$.

(ii) $\Rightarrow$ (i): Let $w_1, w_2 \in W$. Since $1 \in F$, it follows that $w_1 + w_2 = 1 \cdot w_1 + 1 \cdot w_2 = \alpha w_1 + \beta w_2 \in W$ (here $\alpha = 1$ and $\beta = 1$). Therefore, $+$ is a closure operation on W. Since $W \subseteq V$, the associative law holds. $x \in W \Rightarrow x, x \in W \Rightarrow 0 = 1 \cdot x + (-1)x = \alpha x + \beta x \in W$ (here $\alpha = 1$, $\beta = -1$). Therefore, the additive identity 0 is in W.

Let $x \in W$. Then $-x = 0 \cdot x + (-1)x = \alpha x + \beta x \in W$ (here $\alpha = 0$, $\beta = -1$).

Therefore, for any element $x \in W$, we have the additive inverse $-x \in W$. We verified that $(W, +)$ is a group. Since $W \subseteq V$ and V is Abelian, it follows that $(W, +)$ is also Abelian. It is easy to verify the other conditions of vector space. Hence, W is a subspace of V. ∎

Note 1.4.7

For a subset W of vector space V, to show that W is a subspace of V, it is enough to verify that $\alpha w_1 + \beta w_2 \in W$ for all $\alpha, \beta \in F$ and $w_1, w_2 \in W$.

Definition 1.4.8

If U and V are vector spaces over F, then the mapping $T: U \to V$ is said to be a **homomorphism** if

(i) $T(u_1 + u_2) = Tu_1 + Tu_2$ and (ii) $T(\alpha u_1) = \alpha T(u_1)$ for all $u_1, u_2 \in V$ and $\alpha \in F$.

If a homomorphism T is one–one, then we say that it is a one–one **homomorphism** (or **monomorphism**). If T is onto, then we say that it is an **onto homomorphism** (or **epimorphism**). If it is a bijection, then we say that it is an **isomorphism**. If $U = V$, then a homomorphism $T: U \to V$ is called an **endomorphism**. If $T: U \to U$ is an isomorphism, then it is called an **automorphism**.

Notation 1.4.9

Let U and V be two vector spaces over F. Then we write Hom $(U, V) = \{f\colon U \to V \mid f$ is a vector space homomorphism$\}$. Also, Hom $(U, F) = \{f\colon U \to F \mid f$ is a vector space homomorphism$\}$.

Lemma 1.4.10

If V is a vector space over F, then we have the following four conditions:

(i) $\alpha 0 = 0$ for all $\alpha \in F$.
(ii) $0 \cdot v = 0$ for all $v \in V$.
(iii) $(-\alpha) v = -(\alpha v)$ for all $\alpha \in F$ and $v \in V$.
(iv) If $0 \neq v \in V$, $\alpha \in F$, then $\alpha v = 0 \Rightarrow \alpha = 0$.

Proof

(i) $0 + \alpha 0 = \alpha 0 = \alpha(0 + 0) = \alpha 0 + \alpha 0 \Rightarrow 0 = \alpha 0$ (by right cancellation law).
(ii) $0 + 0v = 0v = (0 + 0)v = 0v + 0v \Rightarrow 0 = 0v$ (by right cancellation law).
(iii) $0 = 0v = [(-\alpha) + (\alpha)]v = -\alpha v + \alpha v \Rightarrow -\alpha v$ is the additive inverse of αv
$\Rightarrow (-\alpha v) = -(\alpha v)$.
(iv) Suppose $0 \neq v \in V$, $\alpha v = 0$. If $\alpha \neq 0$, then $\alpha^{-1} \in F$. Now $\alpha v = 0 \Rightarrow \alpha^{-1}(\alpha v) = \alpha^{-1}(0) = 0 \Rightarrow v = 0$, a contradiction. Hence, $\alpha = 0$. ∎

Construction of Quotient Space 1.4.11

Let V be a vector space and W be a subspace of V. Define a relation $\sim$ on V as $a \sim b$ if and only if $a - b \in W$. Clearly, this $\sim$ is an equivalence relation. We denote the equivalence class of $a \in V$ by $a + W$. Write $V/W = \{a + W \mid a \in V\}$. Define "+" on V/W as $(a + W) + (b + W) = (a + b) + W$. Since V is an Abelian group, it follows that $(V/W, +)$ is also an Abelian group (called as the quotient group of V by W). Now, to get a vector space structure, let us define the scalar product between $\alpha \in F$ and $a + W \in V/W$ as $\alpha(a + W) = \alpha a + W$. With respect to these operations $(V/W, +)$ becomes a vector space over F. This vector space is called the **quotient space of** V by W.

Note 1.4.12

From the previous discussion, we conclude the following: If V is a vector space over F and if W is a subspace of V, then V/W is a vector space over F, where the operations are defined as follows:

(i) $(v_1 + W) + (v_2 + W) = (v_1 + v_2) + W$ and
(ii) $\alpha(v_1 + W) = \alpha v_1 + W$
for $v_1 + W, v_2 + W \in V/W$ and $\alpha \in F$.

Theorem 1.4.13

If T is a homomorphism of U onto V with kernel W, then V is isomorphic to U/W. Conversely, if U is a vector space and W is a subspace of U, then there exists a homomorphism of U onto U/W.

Proof

Given that $T: U \rightarrow V$ is an onto homomorphism with ker $T = W$. To show that $V \cong U/W$, define $f: U/W \rightarrow V$ by $f(a + W) = T(a)$ for all $a + W \in U/W$. Now $a + W = b + W \Leftrightarrow a - b \in W = \ker T \Leftrightarrow T(a - b) = 0 \Leftrightarrow Ta = Tb \Leftrightarrow f(a + W) = f(b + W)$.

Hence, f is well-defined and one–one.

To show f is onto, let $v \in V$. Since T is onto, there exists $a \in U$ such that $Ta = v$. Therefore, $f(a + W) = Ta = v$. Hence, f is onto.

To show f is a homomorphism, consider $f[(a + W) + (b + W)] = f[(a + b) + W] = T(a + b) = Ta + Tb = f(a + W) + f(b + W)$. Also, $f[\alpha(a + W)] = f[\alpha a + W] = T(\alpha a) = \alpha(Ta) = \alpha(f(a + W))$. Hence, f is a vector space homomorphism.

Converse

Suppose U is a vector space and W is a subspace of U. Define $\varphi: U \rightarrow U/W$ by $\varphi(a) = a + W$ for all $a \in V$. Then it is easy to verify that φ is a vector space homomorphism. Also, for any $a + W \in U/W$, we have $\varphi(a) = a + W$. This shows that φ is onto. Hence, φ is an onto homomorphism. ∎

Definition 1.4.14

Let V be a vector space over F and let U_i, $1 \le i \le n$ be subspaces of V. Then V is said to be the **internal direct sum** of U_i, $1 \le i \le n$, if every element $v \in V$ can be written in one and only one way as $v = u_1 + u_2 + \ldots + u_n$ for $u_i \in U_i$, $1 \le i \le n$.

Definition 1.4.15

Suppose V_i, $1 \le i \le n$ are vector spaces over a field F. Write $V = \{(v_1, v_2, \ldots, v_n) \mid v_i \in V_i, 1 \le i \le n\}$. Define "+" on V componentwise (that is, $(v_1, v_2, \ldots, v_n) + (w_1, w_2, \ldots, w_n) = (v_1 + w_1, v_2 + w_2, \ldots, v_n + w_n)$). Then $(V, +)$ is an Abelian group. In this group, $0 = (0, 0, \ldots, 0)$ is the additive identity and $(-v_1, -v_2, \ldots, -v_n)$ is the additive inverse of $(v_1, v_2, \ldots, v_n)$. To prove a vector space structure for V, for $\alpha \in F$ and $(v_1, v_2, \ldots, v_n) \in V$, define $\alpha(v_1, v_2, \ldots, v_n) = (\alpha v_1, \alpha v_2, \ldots, \alpha v_n)$. Then, it is easy to verify that V is a vector space over the field F. This vector space V is called the **external direct sum** of the vector spaces V_i, $1 \le i \le n$, and we write $V = V_1 \oplus V_2 \oplus \ldots \oplus V_n$ or $V = \Sigma_{1=1}^n V_i$.

Theorem 1.4.16

If V is the internal direct sum of U_i, $1 \le i \le n$, then V is isomorphic to the external direct sum of U_i, $1 \le i \le n$.

Definition 1.4.17

Suppose V is a vector space over F, $v_i \in V$ and $\alpha_i \in F$ for $1 \le i \le n$. Then $\alpha_1 v_1 + \alpha_2 v_2 + \ldots + \alpha_n v_n$ is called the **linear combination** (over F) of $v_1, v_2, \ldots, v_n$.

Definition 1.4.18

Let V be a vector space and $S \subseteq V$. We write $L(S) = \{\alpha_1 v_1 + \alpha_2 v_2 + \ldots + \alpha_n v_n \mid n \in N, v_i \in S$ and $\alpha_i \in F$ for $1 \le i \le n\}$ = the set of all linear combinations of finite number of elements of S. This $L(S)$ is called the **linear span of** S.

Note 1.4.19

If $v \in S$, then $1 \cdot v$ is a linear combination, and hence $v \in L(S)$. So, $S \subseteq L(S)$.

Lemma 1.4.20 (5.20 of Satyanarayana and Prasad, 2009)

If S is any subset of a vector space V, then $L(S)$ is a subspace of V.

Lemma 1.4.21 (5.21 of Satyanarayana and Prasad, 2009)

If S and T are subsets of a vector space V, then

(i) $S \subseteq T \Rightarrow L(S) \subseteq L(T)$.
(ii) $L(S \cup T) = L(S) + L(T)$.
(iii) $L(L(S)) = L(S)$.

Definition 1.4.22

(i) The vector space V is said to be **finite-dimensional** (over F) if there is a finite subset S in V such that $L(S) = V$.

(ii) If V is a vector space and $v_i \in V$ for $1 \le i \le n$, then we say that v_i, $1 \le i \le n$ are **linearly dependent** over F if there exist elements $a_i \in F$, $1 \le i \le n$, not all of them equal to zero, such that $a_1 v_1 + a_2 v_2 + \ldots + a_n v_n = 0$.

(iii) If the vectors v_i, $1 \le i \le n$ are not linearly dependent over F, then they are said to be **linearly independent** over F.

Examples 1.4.23

Consider $\mathbb{R}$, the set of reals, and write $V = \mathbb{R}^2$; then

(i) V is a vector space over the field $\mathbb{R}$.

(ii) Consider $S = \{(1, 0), (1, 1), (0, 1)\} \subseteq \mathbb{R}^2$. Then $L(S) = \{\alpha_1(1, 0) + \alpha_2(1, 1) + \alpha_3(0, 1) \mid \alpha_i \in \mathbb{R}, 1 \le i \le 3\} = \{(\alpha_1 + \alpha_2, \alpha_2 + \alpha_3) \mid \alpha_i \in \mathbb{R}, 1 \le i \le 3\} \subseteq \mathbb{R}^2$. If $(x, y) \in \mathbb{R}^2$ then write $\alpha_1 = x$, $\alpha_2 = 0$, $\alpha_3 = y$. It is clear that $(x, y) = (\alpha_1 + \alpha_2, \alpha_2 + \alpha_3) \in L(S)$. Hence, $L(S) = \mathbb{R}^2 = V$. This shows that $\mathbb{R}^2$ is a finite-dimensional vector space.

(iii) Write $v_1 = (1, 0)$, $v_2 = (2, 2)$, $v_3 = (0, 1)$, $v_4 = (3, 3)$. Then $\alpha_1 v_1 + \alpha_2 v_2 + \alpha_3 v_3 + \alpha_4 v_4 = 0$, where $\alpha_1 = 2$, $\alpha_2 = -1$, $\alpha_3 = 2$, $\alpha_4 = 0$. Thus, there exist scalars $\alpha_1, \alpha_2, \alpha_3, \alpha_4$, not all of them equal to zero, such that $\alpha_1 v_1 + \alpha_2 v_2 + \alpha_3 v_3 + \alpha_4 v_4 = 0$. Hence, $\{v_i \mid 1 \le i \le 4\}$ is a linearly dependent set.

(iv) Suppose $v_1 = (1, 0)$, $v_2 = (0, 1)$. Suppose $\alpha_1 v_1 + \alpha_2 v_2 = 0$ for some $\alpha_1, \alpha_2 \in \mathbb{R}$. Then $\alpha_1(1, 0) + \alpha_2(0, 1) = 0 = (0, 0) \Rightarrow (\alpha_1, \alpha_2) = (0, 0) \Rightarrow \alpha_1 = 0 = \alpha_2$. Hence, v_1, v_2 are linearly independent.

Lemma 1.4.24 (5.24 of Satyanarayana and Prasad, 2009)

Let V be a vector space over F. If $v_1, v_2, \ldots, v_n \in V$ are linearly independent, then every element in their linear span has a unique representation in the form $\lambda_1 v_1 + \lambda_2 v_2 + \ldots + \lambda_n v_n$ with $\lambda_i \in F$, $1 \le i \le n$.

Lemma 1.4.25 (5.25 of Satyanarayana and Prasad, 2009)

Let $S = \{v_i \mid 1 \le i \le n\}$ be a subset of vector space V. If v_j is a linear combination of its preceding ones, then $L(\{v_1, v_2, \ldots, v_{j-1}, v_{j+1}, \ldots, v_n\}) = L(S)$.

Theorem 1.4.26 (5.26 of Satyanarayana and Prasad, 2009)

If $v_i \in V$, $1 \leq i \leq n$, then either they are linearly independent or some v_k is a linear combination of the preceding ones $v_1, v_2, ..., v_{k-1}$.

Definition 1.4.27 (5.29 of Satyanarayana and Prasad, 2009)

A subset S of a vector space V is called a **basis** of V if S consists of linearly independent elements (that is, any finite number of elements in S is linearly independent) and $V = L(S)$.

If S contains finite number of elements, then V is a **finite-dimensional vector space**. If S contains an infinite number of elements, then V is called an **infinite-dimensional vector space**.

Corollary 1.4.28 (5.31 of Satyanarayana and Prasad, 2009)

If V is a finite-dimensional vector space and if $u_1, u_2, ..., u_m$ spans V, then some subset of $\{u_1, u_2, ..., u_m\}$ forms a basis of V (equivalently, every finite-dimensional vector space has a basis $\{v_i \mid 1 \leq i \leq n\}$).

Lemma 1.4.29 (5.32 of Satyanarayana and Prasad, 2009)

If $v_1, v_2, ..., v_n$ is a basis of V over F, and if $w_1, w_2, ..., w_m$ in V are linearly independent over F, then $m \leq n$.

Corollary 1.4.30 (5.33 of Satyanarayana and Prasad, 2009)

If V is a finite-dimensional vector space over F, then any two bases of V have that same number of elements.

Result 1.4.31 (5.34 of Satyanarayana and Prasad, 2009)

Let V and W be two vector spaces over F. Suppose $h: V \rightarrow W$ is an isomorphism and $\{v_i \mid i \in I\}$ is a basis of V. Then $\{h(v_i) \mid i \in I\}$ is a basis for W.

Definition 1.4.32

   (i) If V is a finite-dimensional vector space, then the integer n in Result 1.4.31 is called the **dimension of** V **over** F. Equivalently, the condition (ii).

  (ii) If V is a finite-dimensional vector space and $S = \{v_i \mid 1 \leq i \leq n\}$ is a linearly independent set such that $L(S) = V$, then n is called the **dimension of** V. In this case, we write $n = \dim(V)$.

 (iii) If V is not a finite-dimensional vector space, then we say that the dimension of V is infinite. In this case we say that V is an **infinite-dimensional vector space**.

Corollary 1.4.33

Any two finite-dimensional vector spaces over F of the same dimension are isomorphic.

(Sketch of the proof: Let V and W be two vector spaces with $\dim(V) = \dim(W) = n$. Suppose $\{v_1, v_2, \ldots, v_n\}$ and $\{w_1, w_2, \ldots, w_n\}$ are bases for V and W, respectively. Define $f: V \to W$ by $f(\alpha_1 v_1 + \alpha_2 v_2 + \ldots + \alpha_n v_n) = \alpha_1 w_1 + \alpha_2 w_2 + \ldots + \alpha_n w_n$. Then it is easy to verify that f is a one–one, onto vector space homomorphism.)

Lemma 1.4.34 (5.38 of Satyanarayana and Prasad, 2009)

If V is a finite-dimensional vector space over F, and if $u_1, u_2, \ldots, u_m \in V$ are linearly independent, then we can find vectors $u_{m+1}, u_{m+2}, \ldots, u_{m+r}$ in V such that $\{u_i \mid 1 \leq i \leq (m + r)\}$ is a basis of V.

Lemma 1.4.35 (5.39 of Satyanarayana and Prasad, 2009)

If V is a finite-dimensional vector space and if W is a subspace of V, then the following conditions are true:

   (i) W is a finite-dimensional vector space.

  (ii) $\dim(W) \leq \dim(V)$.

 (iii) $\dim(V/W) = \dim(V) - \dim(W)$.

1.5 Module Theory

The concept of a "module" over a ring is a generalization of a "vector space" over a field. The definition is given as follows.

Definition 1.5.1

Let M be an additive Abelian group and R a ring. We say that M is a **left R-module** if there exists a mapping from $R \times M$ to M (we denote the image of (r, m) by rm) satisfying the following conditions:

(i) $r(m_1 + m_2) = rm_1 + rm_2$
(ii) $(r_1 + r_2)m = r_1m + r_2m$
(iii) $r_1(r_2m) = (r_1r_2)m$ (iv) $1 \cdot m = m$ if $1 \in R$ for all $r, r_1, r_2 \in R, m, m_1, m_2 \in M$.

If M is a left R-module, then we denote this by $_RM$.

If we take the elements of the ring R into the right-hand side, then we call M a right R-module.

In this case, we consider the mapping $M \times R$ to M (the image of (m, r) is denoted by mr).

It is clear that if R is a division ring, then a left R-module is called a **vector space** over the division ring R.

Theorem 1.5.2

Any additive Abelian group is a module over the ring of integers $\mathbb{Z}$.

Proof

Let $(A, +)$ be an additive Abelian group and $\mathbb{Z}$ be the ring of integers. Define a mapping "." from $\mathbb{Z} \times A$ to A as follows:

Let $k \in \mathbb{Z}$ and $a \in A$. If k is a positive integer, then define
$$ka = a + a + \ldots + a \ (k \text{ times}).$$

If $k = 0$, then define $ka = 0$, where 0 is the additive identity in A.

If k is a negative integer, then we can write $k = -p$, where p is a positive integer.

Now define $ka = (-p)a = -(pa)$, where $-(pa)$ is the additive inverse of pa.

First, we show $k(a_1 + a_2) = ka_1 + ka_2$ for all $k \in \mathbb{Z}$, and $a_1, a_2 \in A$.

If $k > 0$, then $k(a_1 + a_2) = (a_1 + a_2) + \ldots + (a_1 + a_2) \ (k \text{ times})$
$= (a_1 + a_1 + \ldots + a_1) \ (k \text{ times}) + (a_2 + a_2 + \ldots + a_2) \ (k \text{ times})$
$= ka_1 + ka_2$. If $k = 0$, then
$0(a_1 + ka_2) = 0 = 0 + 0 = 0a_1 + 0a_2 = k(a_1 + a_2) = ka_1 + ka_2$.

If $k < 0$, then $k = -p$, where p is a positive integer.

Now $k(a_1 + a_2) = -p(a_1 + a_2)$

$$= -(p(a_1 + a_2)) = -(pa_1 + pa_2)$$
$$= -pa_1 - pa_2 = (-pa_1) + (-pa_2)$$
$$= (-p)a_1 + (-p)a_2 = ka_1 + ka_2.$$

Therefore, $k(a_1 + a_2) = ka_1 + ka_2$ for all $k \in \mathbb{Z}, a_1, a_2 \in A$.

Similarly, we can show that $(k_1 + k_2)a = k_1a + k_2a$.

$(k_1k_2)a = k_1(k_2a)$, $1 \cdot a = a$ for all $k_1, k_2 \in \mathbb{Z}, a \in A$. Therefore, A is a left $\mathbb{Z}$-module. So, an additive Abelian group A is a module over the ring of integers $\mathbb{Z}$. ∎

Example 1.5.3

   (i) Every ring R is a module over itself.
      [**Verification:** Let R be a ring and $M = R$. Then we have the scalar multiplication $f: R \times M \to M$ defined by $f(r, m) = rm$. With respect to this scalar multiplication $(M, +)$ becomes a module over R.]
   (ii) Every Abelian group is a module over $\mathbb{Z}$.
  (iii) Every vector space over a field F is a module over the ring F.

Note 1.5.4

Let M be a left R-module. Then

   (i) $0 \cdot m = 0$ for all $m \in M$
  (ii) $r \cdot 0 = 0$ for all $r \in R$
 (iii) $r(-m) = (-r)m = -rm$ for all $r \in R$ and $m \in M$.

Verification

   (i) $0 + 0 \cdot m = 0 \cdot m = (0 + 0)m = 0 \cdot m + 0 \cdot m$. Therefore, $0 = 0m$.
  (ii) $0 + r0 = r0 = r(0 + 0) = r0 + r0 \Rightarrow 0 = r0$.
 (iii) $0 = 0m = (r + (-r))m = rm + (-r)m \Rightarrow -(rm) = (-r)m$.
      Similarly, we can prove the other part.

Throughout the book, we use the term "module" to mean "left module," unless otherwise stated.

Example 1.5.5

   (i) Every ring R is a left R-module.
  (ii) Every left ideal M of a ring R is a left R-module.

Example 1.5.6

   (i) Let M be the set of all $m \times n$ matrices over the ring R. Then M is a left R-module.

(ii) Let M be an additive Abelian group and R be the ring of all endomorphisms of M. Then M is a left R-module.

(iii) Let M, N be two R-modules. Then the Cartesian product $M \times N$ is also an R-module by defining
$(x_1, y_1) + (x_2, y_2) = (x_1 + x_2, y_1 + y_2)$ and $r(x, y) = (rx, ry)$ for all $r \in R$, x_1, $x_2 \in M$, y_1, $y_2 \in N$.

(iv) The polynomial ring $R[x]$ (that is, the ring of all polynomials in one indeterminate x) over a given ring R is an R-module.

Problem 1.5.7

Let M be an additive Abelian group. Then show that there is only one way of making M a $\mathbb{Z}$-module.

Proof

We know that an additive Abelian group M is a $\mathbb{Z}$-module by defining $k \cdot a$ as in Theorem 1.5.2.

Suppose M is a $\mathbb{Z}$-module with respect to the mapping $f : \mathbb{Z} \times M \to M$. Now we show that $f(k, a) = k \cdot a$ for all $k \in \mathbb{Z}$, $a \in M$.

Since M is a $\mathbb{Z}$-module with respect to f, we have

(i) $f(k, a_1 + a_2) = f(k, a_1) + f(k, a_2)$
(ii) $f(k_1 + k_2, a) = f(k_1, a) + f(k_2, a)$
(iii) $f(k_1 k_2, a) = f(k_1, f(k_2, a))$
(iv) $f(1, a) = a$ for all $a, a_1, a_2 \in M$, $k, k_1, k_2 \in \mathbb{Z}$.

If $k > 0$, $f(k, a) = f((\underbrace{1 + 1 + \ldots + 1}_{k \text{ times}}), a)$

$$= f(1, a) + f(1, a) + \ldots + f(1, a) \, (k \text{ times})$$

$$= a + a + \ldots + a \, (k \text{ times}) = ka.$$

If $k = 0$, then consider

$$f(1, a) = f(1 + 0, a) = f(1, a) + f(0, a)$$

$$\Rightarrow f(1, a) + 0 = f(1, a) + f(0, a)$$

$$\Rightarrow 0 = f(0, a) \Rightarrow 0 \cdot a = f(0, a).$$

Therefore, $f(k, a) = ka$ if $k = 0$.

If $k < 0$, then $k = -p$, where p is a positive integer.

Now $f(1, a) + f(-1, a) = f(1 + -1, a) = f(0, a) = 0$
$$\Rightarrow f(-1, a) = -f(1, a) = -a.$$

Therefore, $f(k, a) = f(-p, a) = f((-1) + (-1) + \ldots + (-1), a)$

$$= f(-1, a) + f(-1, a) + \ldots + f(-1, a)$$

$$= ((-a) + (-a) + \ldots + (-a))$$

$$= p(-a) = -pa = (-p)a = ka.$$

Therefore, $f(k, a) = ka$ for all $k \in \mathbb{Z}, a \in M$.

So there is only one way of making an additive Abelian group into a $\mathbb{Z}$-module.   ∎

Definition 1.5.8

A nonempty subset N of an R-module M is called an **R-submodule** (or **submodule**) of M if

   (i) $a - b \in N$ for all $a, b \in N$ and
   (ii) $ra \in N$ for all $r \in R, a \in N$

Example 1.5.9

   (i) Every left ideal of a ring R is an R-submodule of the left R-module R.
   (ii) If M is an R-module and $x \in M$, then the set $Rx = \{rx \mid r \in R\}$ is an R-submodule of M.
   (iii) Consider $\mathbb{Z}$ (the ring of integers) as a $\mathbb{Z}$-module. Then every $\mathbb{Z}$-submodule of $\mathbb{Z}$ is of the form $n\mathbb{Z}$, where $n \in \mathbb{Z}$.

Note 1.5.10

Let R be a ring and $I \subseteq R$. Then I is a submodule of $_R R$ if and only if I is a left ideal of R.

Theorem 1.5.11

If $\{N_i\}_{i \in \Delta}$ is a family of R-submodules of an R-module M, then $\bigcap_{i \in \Delta} N_i$ is also an R-submodule of M.

Proof

Write $N = \bigcap\limits_{i \in \Delta} N_i$.

We have to show that N is an R-submodule of M.

(i) Let $a, b \in N$. This implies that $a, b \in N_i$ for all $i \in \Delta$. Since each N_i is a submodule, it follows that $a - b \in N_i$ for all $i \in \Delta$. This implies that $a - b \in \bigcap\limits_{i \in \Delta} N_i = N$.

(ii) Let $a \in N$ and $r \in R$. This implies that $a \in N_i$ and $r \in R$ for all $i \in \Delta$. Since each N_i is a submodule, it follows that $ra \in N_i$ for all $i \in \Delta$. This means $ra \in \bigcap\limits_{i \in \Delta} N_i = N$.

Therefore, N is an R-submodule of M. ∎

Remark 1.5.12

(i) The union of two submodules of M need not be a submodule of M.
   [**Example:** Let $\mathbb{Z}$ be the ring of integers. Then we can consider $\mathbb{Z}$ as a module over $\mathbb{Z}$.
   Write $A = 2\mathbb{Z}$, $B = 3\mathbb{Z}$, and $C = A \cup B$.
   Now, A and B are two submodules of the module $\mathbb{Z}$. Since $2 \in A \subseteq C$, $3 \in B \subseteq C$, and $1 = 3 - 2 \notin C$, it follows that C cannot be a submodule of the module $\mathbb{Z}$.]

(ii) If A and B are two submodules of M, then $A + B = \{a + b \mid a \in A, b \in B\}$ is also a submodule of M.

Theorem 1.5.13

If M is an R-module and $x \in M$, then the set $K = \{rx + nx \mid r \in R, n \in \mathbb{Z}\}$ is the smallest R-submodule of M containing x. Moreover, if R has the unit element, then $K = Rx$.

Proof

Part 1: In this part, we show that K is an R-submodule of M.
Let $k_1, k_2 \in K$. Then $k_1 = r_1x + n_1x$, $k_2 = r_2x + n_2x$, where $r_1, r_2 \in R$ and $n_1, n_2 \in \mathbb{Z}$.
Now $k_1 - k_2 = (r_1x + n_1x) - (r_2x + n_2x)$
$$= r_1x - r_2x + n_1x - n_2x$$
$$= (r_1 - r_2)x + (n_1 - n_2)x \in K.$$

Let $a \in R$ and $rx + nx \in K$.

Now we have to show that $a(rx + nx) \in K$.

If $n > 0$, $a(rx + nx) = a(rx) + a(nx)$

$$= (ar)x + a(x + \ldots + x) \, (n \text{ times})$$

$$= (ar)x + (ax + ax + \ldots + ax) \, (n \text{ times})$$

$$= (ar + \underbrace{a + \ldots + a}_{n \text{ times}})x$$

$$= vx, \text{ where } v = ar + \underbrace{a + \ldots + a}_{n \text{ times}} \in R$$

$$= vx + 0x, \text{ where } v \in R, 0 \in \mathbb{Z}$$

$$\in K.$$

If $n = 0$, then $a(rx + nx) = a(rx + nx) = a(rx + 0) = arx \in K$.

If $n < 0$, then $n = -p$, where p is a positive integer.

Now, $a(rx + nx) = a(rx + (-p)x)$

$$= a(rx) + a((-p)x)$$

$$= (ar)x + a((-x) + (-x) + \ldots + (-x))$$

$$= (ar)x + a(-x) + a(-x) + \ldots + a(-x) \, (p \text{ times})$$

$$= (ar)x + (-a)x + (-a)x + \ldots + (-a)x$$

$$= (ar + \underbrace{(-a) + (-a) + \ldots + (-a)}_{p \text{ times}})x$$

$$= ux \text{ (for some } u \in R)$$

$$\in K.$$

So, in all cases, $a(rx + nx) \in K$ for all $a \in R$, $rx + nx \in K$.

Therefore, K is an R-submodule of M.

Clearly, $x = 0x + 1x \in K$. Therefore, K is an R-submodule of M containing x.

Part 2: Let L be an R-submodule of M containing x.

Now $x \in L$. Since L is an R-submodule, we have $rx \in L$ for all $r \in R$ and $nx \in L$ for all $n \in \mathbb{Z}$. Now, $rx + nx \in L$ for all $r \in R, n \in \mathbb{Z}$.

So, $K \subseteq L$. Therefore, K is the smallest R-submodule of M containing x.

Part 3: Suppose R has the unity element e.

Now we show that $K = Rx$.

Since $x \in K$, we have $Rx \subseteq K$.

Let $rx + nx \in K$.

For $n > 0$, $rx + nx = rx + n(ex) = rx + (ex + ex + \ldots + ex) \, (n \text{ times}) = (r + e + \ldots + e)x = ux \in Rx$ (for some $u \in R$).

Similarly, if $n \leq 0$ we can show that $rx + nx \in Rx$.

Therefore, $K \subseteq Rx$. Hence, $K = Rx$. ∎

Definition 1.5.14

Let M be an R-module and S be a nonempty subset of M. Then the intersection of all R-submodules of M containing S is called the **R-submodule** (or **submodule**) **generated** by S. The submodule generated by the set S is denoted by either $<S>$ or (S). If $S = \{a\}$, then $<S>$ is denoted by $<a>$.

Note 1.5.15

(i) The submodule generated by S is the smallest R-submodule of M containing S.

(ii) Let M be an R-module and $x \in M$. Then the submodule generated by x is given by $(x) = \{rx + nx \mid r \in R, n \in \mathbb{Z}\}$.

If R has the unity element, then the submodule generated by x is Rx.

Definition 1.5.16

An R-module M is called a **finitely generated module** if $M = <\{x_1, x_2, ..., x_n\}>$, where $x_i \in M$, $1 \le i \le n$. These elements $x_1, x_2, ..., x_n$ are called the **generators** of M.

Remark 1.5.17

The set of generators of a module need not be unique.

[To justify this, let S be the set of all polynomials in x of degree $\le n$, over a field F.

Then S is a vector space over F with the usual addition and multiplication. It is easy to see that S is generated by the set $\{1, x, ..., x^n\}$.

Let $f(x) \in S$.

Then $f(x) = a_0 + a_1 x + ... + a_n x^n$ (since $\{1, x, ... x^n\}$ is a generating set)

$= a_0 - a_1 + a_1 + a_1 x + ... + a_n x^n$

$= (a_0 - a_1) + a_1(1 + x) + a_2 x^2 + ... + a_n x^n$.

So, we can also write a general element $f(x)$ as a linear combination of the elements $1, (1 + x), ..., x^n$.

Therefore, S is also generated by $\{1, 1 + x, x^2, ..., x^n\}$.]

Definition 1.5.18

An R-module M is called a **cyclic module** if $M = <x>$ for some $x \in M$.

Theorem 1.5.19

If an R-module M is generated by a set $\{x_1, x_2, \ldots, x_n\}$ and $1 \in R$, then $M = \{r_1 x_1 + r_2 x_2 + \ldots + r_n x_n \mid ri \in R\}$, and the right-hand side is symbolically denoted by $\Sigma_{i=1}^{n} R x_i$.

Proof

Given that M is generated by the set $\{x_1, x_2, \ldots, x_n\}$.

Then M is the smallest R-submodule of M containing $x_1, x_2, \ldots, x_n$.

Write $\Sigma_{i=1}^{n} R x_i = \{r_1 x + r_2 x + \ldots + r_n x_n \mid r_i \in R\}$. Now we show that $M = \Sigma_{i=1}^{n} R x_i$. Since M is an R-module and $x_1, x_2, \ldots, x_n \in M$, it is clear that $\Sigma_{i=1}^{n} R x_i = \{r_1 x + r_2 x + \ldots + r_n x_n \mid r_i \in R\} \subseteq M$. Now we show that $\Sigma_{i=1}^{n} R x_i$ is an R-submodule of M containing $x_1, x_2, \ldots, x_n$.

Let $a, b \in \Sigma_{i=1}^{n} R x_i$. Then $a = r_1 x_1 + r_2 x_2 + \ldots + r_n x_n$ and $b = s_1 x_1 + s_2 x_2 + \ldots + s_n x_n$ for $r_i, s_i \in R$, $i = 1, 2, \ldots n$. Now,

$a - b = (r_1 x_1 + r_2 x_2 + \ldots + r_n x_n) - (s_1 x_1 + s_2 x_2 + \ldots + s_n x_n) = r_1 x_1 + r_2 x_2 + \ldots + r_n x_n - s_1 x_1 - s_2 x_2 - \ldots - s_n x_n = (r_1 - s_1) x_1 + (r_2 - s_2) x_2 + \ldots + (r_n - s_n) x_n \in \Sigma_{i=1}^{n} R x_i$ and

$$ra = r(r_1 x_1 + r_2 x_2 + \ldots + r_n x_n) = rr_1 x_1 + rr_2 x_2 + \ldots + rr_n x_n$$

$$= (rr_1) x_1 + (rr_2) x_2 + \ldots + (rr_n) x_n \in \Sigma_{i=1}^{n} R x_i \text{ for all } r \in R.$$

Therefore, $\Sigma_{i=1}^{n} R x_i$ is an R-submodule of M. Moreover,

$$x_i = 0 x_1 + 0 x_2 + \ldots + 1 x_i + \ldots + 0 x_n \in \Sigma_{i=1}^{n} R x_i, \text{ for all } i = 1, 2, \ldots, n.$$

Therefore, $\Sigma_{i=1}^{n} R x_i$ is an R-submodule of M containing $x_1, x_2, \ldots, x_n$. But M is the smallest R-submodule of M containing $x_1, x_2, \ldots, x_n$. So, $M \subseteq \Sigma_{i=1}^{n} R x_i$. Hence, $M = \{r_1 x_1 + r_2 x_2 + \ldots + r_n x_n \mid n \in R\}$. ∎

Definition 1.5.20

If an element $m \in M$ can be expressed as $m = a_1 x_1 + a_2 x_2 + \ldots + a_n x_n$, where $a_i \in R$ and $x_i \in M$ for all $i = 1, 2, \ldots, n$, then we say that m is a **linear combination of elements** $x_1, x_2, \ldots, x_n$ over R.

Definition 1.5.21

Let $\{N_i\}$, $1 \leq i \leq k$ be a family of R-submodules of a module M. Then the submodule generated by $\bigcup_{i=1}^{k} N_i$, that is, the smallest submodule of M containing $N_1, N_2, \ldots, N_k$, is called the **sum of submodules** N_i, $1 \leq i \leq k$, and is denoted by $\Sigma_{i=1}^{k} N_i$.

Definition 1.5.22

Let $\{N_i\}_{i \in \Delta}$ be an arbitrary family of submodules of an R-module M. Then the submodule generated by $\bigcup_{i \in \Delta} N_i$ is called the **sum of submodules** $\{N_i\}_{i \in \Delta}$, and it is denoted by $\Sigma_{i \in \Delta} N_i$.

Theorem 1.5.23

Let $\{N_i\}$, $1 \leq i \leq k$ be a family of R-submodules of an R-module M. Then

$$\sum_{i=1}^{k} N_i = \{x_1 + x_2 + \ldots + x_k \mid x_i \in N_i \text{ for all } i\}.$$

Proof

Write $S = \{x_1 + x_2 + \ldots + x_k \mid x_i \in N_i \text{ for all } i\}$.

Now we show that S is the smallest R-submodule of M containing $N_1, N_2, \ldots, N_k$. Let $a, b \in S$. Then

$$a = x_1 + x_2 + \ldots + x_k \text{ and } b = y_1 + y_2 + \ldots + y_k$$
$$\text{for } x_i, y_i \in N_i, 1 \leq i \leq k.$$
$$\text{Now } a - b = (x_1 + x_2 + \ldots + x_k) - (y_1 + y_2 + \ldots + y_k)$$
$$= x_1 + x_2 + \ldots + x_k - y_1 - y_2 - \ldots - y_k$$
$$= (x_1 - y_1) + (x_2 - y_2) + \ldots + (x_k - y_k) \in S.$$

Let $r \in R$ and $a \in S$. Then $ra = r(x_1 + x_2 + \ldots + x_k) = rx_1 + rx_2 + \ldots + rx_k \in S$ (since $rx_i \in N_i$).

Therefore, S is an R-submodule of M. Clearly, S contains $N_1, N_2, \ldots, N_k$.

Let K be any R-submodule of M containing $N_1, N_2, \ldots, N_k$. Then K contains all elements of the form $x_1 + x_2 + \ldots + x_k$, where $x_i \in N_i \Rightarrow K$ contains S, that is, $S \subseteq K$.

Therefore, S is the smallest R-submodule of M containing $N_1, N_2, \ldots, N_k$. But, by definition, $\Sigma_{i=1}^{k} N_i$ is the smallest R-submodule of M containing $N_1, N_2, \ldots, N_k$. This implies $\Sigma_{i=1}^{k} N_i = S$. ∎

Note 1.5.24

Let $\{N_i\}_{i\in\Delta}$ be an arbitrary family of submodules of an R-module M. Then $\Sigma_{i\in\Delta} N_i = \left\{ \underset{\text{finite}}{\Sigma}\, x_i \mid x_i \in N_i \right\}$, where $\underset{\text{finite}}{\Sigma}\, x_i$ stands for the sum of the finite number of elements from the R-submodules N_i, $i \in \Delta$.

Definition 1.5.25

The sum $\Sigma_{i\in\Delta} N_i$ of a family $\{N_i\}_{i\in\Delta}$ of R-submodules of an R-module M is said to be **direct** if each element x of $\Sigma_{i\in\Delta} N_i$ can be uniquely written as

$$x = \sum_{i\in\Delta} x_i, \; x_i \in N_i,$$

and $x_i = 0$ for almost all i in Δ.

Note 1.5.26

(i)   The direct sum of submodules $\{N_i\}_{i\in\Delta}$ is denoted by $\oplus\Sigma_{i\in\Delta} N_i$.

(ii) If D is a finite set, that is, if $D = \{1, 2, ..., k\}$, then the direct sum of submodules $\{N_i\}$ $1 \le i \le k$ is denoted by $\oplus\Sigma_{i=1}^{k} N_i (\text{or})\, N_1 \oplus ... \oplus N_k$.

(iii) If $\oplus\Sigma_{i\in\Delta} N_i$ is a direct sum, then each N_i is called a **direct summand** of $\oplus\Sigma_{i\in\Delta} N_i$.

Theorem 1.5.27

Let $\{N_i\}_{i\in\Delta}$ be a family of R-submodules of an R-module M. Then the following conditions are equivalent:

(i) $\displaystyle\sum_{i\in\Delta} N_i$ is a direct sum.

(ii) $0 = \displaystyle\sum_{i} x_i \in \sum_{i\in\Delta} N_i \Rightarrow x_i = 0$ for all $i \in \Delta$.

(iii) $N_i \cap \left(\displaystyle\sum_{\substack{j\in\Delta \\ j\neq i}} N_j \right) = (0), i \in \Delta.$

Proof

(i) $\Rightarrow$ (ii): Assume (i), that is, $\sum_{i \in \Delta} N_i$, is a direct sum. Then every element of $\sum_{i \in \Delta} N_i$ has a unique representation. Let $0 = \sum_i x_i \in \sum_{i \in \Delta} N_i$. We know that $0 = \sum_i 0$.

Therefore, by our assumption, $x_i = 0$ for all i. This proves (ii).

(ii) $\Rightarrow$ (iii):
Assume (ii), that is, $0 = \sum_i x_i \in \sum_{i \in \Delta} N_i \Rightarrow x_i = 0$ for all $i \in \Delta$.

Let $x \in N_i \cap \sum_{\substack{j \in \Delta \\ j \neq i}} N_j$. Then $x \in N_i$ and $x \in \left(\sum_{\substack{j \in \Delta \\ j \neq i}} N_j \right) \Rightarrow x \in N_i$, and $x = \sum_{\substack{j \in \Delta \\ j \neq i}} a_j$,

where $a_j \in N_j$ and all but a finite number of a_j's are 0. This implies

$$0 = \sum_{\substack{j \in \Delta \\ j \neq i}} a_j + (-x) \in \sum_{i \in \Delta} N_i \Rightarrow x = 0 \text{ (by our assumption)}.$$

Therefore, $N_i \cap \left(\sum_{\substack{j \in \Delta \\ j \neq i}} N_j \right) = (0)$, $i \in \Delta$. This proves (iii).

(iii) $\Rightarrow$ (i): Assume (iii), that is, $N_i \cap \left(\sum_{\substack{j \in \Delta \\ j \neq i}} N_j \right) = (0)$, $i \in \Delta$. Take $x \in \sum_{i \in \Delta} N_i$.

Suppose $x = \sum_{i \in \Delta} a_i$, $x = \sum_{i \in \Delta} b_i$, $a_i, b_i \in N_i$, and all but a finite number of a_i, b_i are 0. Then $0 = \sum_{i \in \Delta} (a_i - b_i)$

$$\Rightarrow a_i - b_i = -\sum_{\substack{j \in \Delta \\ j \neq i}} (a_j - b_j) \in N_i \cap \left(\sum_{\substack{j \in \Delta \\ j \neq i}} N_j \right) = (0)$$

$$\Rightarrow a_i - b_i = 0 \text{ for all } i \in \Delta \Rightarrow a_i = b_i \text{ for all } i \in \Delta.$$

Therefore, every element of $\sum_{i \in \Delta} N_i$ has a unique representation. This implies $\sum_{i \in \Delta} N_i$ is a direct sum. This proves (i). ∎

Definition 1.5.28

Let M, N be any two R-modules. A mapping $f: M \to N$ is said to be an **R-homomorphism** or **homomorphism** or **R-linear mapping** or **linear mapping** if

(i) $f(x + y) = f(x) + f(y)$; and
(ii) $f(rx) = r \cdot f(x)$ for all $x, y \in M$, $r \in R$.

Moreover, if f is one–one, then the homomorphism f is said to be a **monomorphism**. If f is onto, then the homomorphism f is said to be an **epimorphism**. If f is a bijection, then the homomorphism f is said to be an **isomorphism**.

Note 1.5.29

The set of all R-homomorphisms of an R-module M into an R-module N is denoted by $\mathrm{Hom}_R(M, N)$, and it is easy to verify that $\mathrm{Hom}_R(M, N)$ is an additive Abelian group in which the additive identity is the zero mapping. If $f \in \mathrm{Hom}_R(M, N)$, then the mapping $(-f)$ defined by $(-f)(x) = -(f(x))$ is the additive inverse of f.

Example 1.5.30

(i) Let M be an R-module. Define a mapping $f: M \to M$ as $f(x) = x$ for all $x \in M$. Then f is a module homomorphism and this homomorphism is called the **identity endomorphism** of M.
(ii) Let M be an R-module. Define a mapping $f: M \to M$ as $f(x) = 0$ for all $x \in M$. Then f is an R-homomorphism and this homomorphism is called the **zero endomorphism** of M.
(iii) Let M be an R-module and $a \in R$. Define
$f: M \to N$ as $f(x) = ax$ for all $x \in M$. Then f is an R-homomorphism.
(iv) Let R be a ring and n be a fixed positive integer. Then the set R^n is an R-module. Define $\pi_i: R^n \to R$ as $\pi_i(x_1, x_2, \ldots, x_n) = x_i$ for all $(x_1, x_2, \ldots, x_n) \in R^n$. Then π_i is an R-homomorphism, and it is called the **projection of R^n onto the ith component** or the **canonical homomorphism** for $i = 1$ to n.

Note 1.5.31

Let $f: M \to N$ be an R-homomorphism. Then

(i) $f(0) = 0$

(ii) $f(-x) = -f(x)$

(iii) $f(x - y) = f(x) - f(y)$ for all $x, y \in M$.

Definition 1.5.32

Let $f: M \to N$ be an R-homomorphism. Then the set $\{x \in M \mid f(x) = 0\}$ is called the **kernel** of f and is denoted by ker f (or ker(f)). It is easy to verify that ker f is an R-submodule of M.

Definition 1.5.33

Let $f: M \to N$ be an R-homomorphism. Then the set $\{f(x) \mid x \in M\}$ is called the **image** of f, and it is denoted by Im(f). It is easy to verify that Im(f) is an R-submodule of N.

Note 1.5.34

Suppose $f: M \to N$ and $g: N \to K$ are module homomorphisms. Then

(i) $g \circ f: M \to K$ is a module homomorphism.

(ii) ker $f = (0)$, if and only if f is one–one.

(iii) If f is an isomorphism, then $f^{-1}: N \to M$ is also an isomorphism.

(iv) If f and g are monomorphisms, then so is $g \circ f$.

(v) If f and g are epimorphisms, then so is $g \circ f$.

(vi) If $g \circ f$ is a monomorphism, then so is f.

(vii) If $g \circ f$ is an epimorphism, then so is g.

Note 1.5.35

Let N be an R-submodule of an R-module M. For any $a_1, a_2 \in M$, define $a_1 \equiv a_2 \pmod{N}$ if $a_1 - a_2 \in N$; then "$\equiv$" is an equivalence relation on M.

For any $a \in M$, let $\bar{a}$ be the equivalence class containing a. Then

$\bar{a} = \{b \in M \mid b = \pmod{N}\} = \{a + x \mid x \in N\} = a + N$.

Write $M/N = \{\bar{a} \mid a \in M\} = \{a + N \mid a \in M\}$.

For any $a + N, b + N \in M/N$, and for any $r \in R$, define addition and scalar multiplication on M/N as follows:

$(a + N) + (b + N) = (a + b) + N$ and $r(a + N) = ra + N$.

Then M/N becomes an R-module. Here $0 + N$ is the identity in M/N.

For $x + N \in M/N$, the element $(-x) + N$ is the additive inverse of $x + N$.

The module M/N is called **a quotient module** or a **factor module of** M **modulo** N.

Define $f: M \to M/N$ as $f(x) = x + N$ for all $x \in M$. Then f is an epimorphism, and it is called the **canonical epimorphism** of M onto M/N.

Theorem 1.5.36

The submodules of the quotient module M/N are of the form U/N, where U is a submodule of M containing N.

Proof

Let $f: M \to M/N$ be a canonical epimorphism. Let X be an R-submodule of M/N. Write $U = \{m \in M \mid f(m) \in X\}$. Let $a, b \in U \Rightarrow f(a), f(b) \in X$. Now $f(a - b) = f(a) - f(b) \in X \Rightarrow a - b \in U$. For any $r \in R, a \in U$, we have $f(ra) = rf(a) \in X \Rightarrow ra \in U$.

Therefore, U is an R-submodule of M.

Let $a \in N$. Now $f(a) = a + N = N = \bar{0} \in X \Rightarrow a \in U$, and so $N \subseteq U$.

Therefore, U is an R-submodule of M containing N.

Clearly, $f(U) \subseteq X$. Let $x \in X$. Since f is onto, there exists $y \in M$ such that $f(y) = x$.

Now $f(y) = x \in X$, where $y \in M \Rightarrow y \in U$. So, $x = f(y), y \in U \Rightarrow x \in f(U)$.

Therefore, $X \subseteq f(U)$. Hence, $X = f(U) = U/N$.

So, any R-submodule of M/N is of the form U/N for some R-submodule U of M containing N. ∎

Theorem 1.5.37 (Fundamental Theorem of Module Homomorphisms)

Let f be an R-homomorphism of an R-module M into an R-module N. Then $M/\ker f \cong f(M)$.

Proof

For convenience, we write $K = \ker f$. Define $\varphi: M/K \to f(M)$ by $\varphi(x + K) = f(x)$ for all $x + K \in M/K$. Let $x_1 + K, x_2 + K \in M/K$.

Now, $x_1 + K = x_2 + K$ if and only if $x_1 - x_2 \in K$ if and only if $f(x_1 - x_2) = 0$ if and only if $f(x_1) - f(x_2) = 0$ if and only if $\varphi(x_1 + K) = \varphi(x_2 + K)$. Therefore, φ is well-defined and one–one.

Let $x^1 \in f(M)$. Then $x^1 = f(m)$ for some $m \in M$.

Now $m + K \in M/K$ and $\varphi(m + K) = f(m) = x^1$.

Therefore, for each $x^1 \in f(M)$, there exists $m + K \in M/K$ such that $\varphi(m + K) = x_1$, and so φ is onto. Now we show that φ is an R-homomorphism.

For any $x + K, y \in K \in M/K$, we have

$\varphi((x + K) + (y + K)) = \varphi((x + y) + K) = f(x + y) = f(x) + f(y) = \varphi(x + K) + \varphi(y + K)$.

For any $x + K \in M/K$ and $r \in R$, we have

$\varphi(r(x + K)) = \varphi(rx + K) = f(rx) = r \cdot f(x) = r \, \varphi(x + K)$.

Therefore, φ is an R-homomorphism.

Hence, $\varphi \colon M/K \to f(M)$ is an isomorphism. The proof is complete. $\blacksquare$

Theorem 1.5.38

Let A and B be R-submodules of R-modules of M and N, respectively. Then

$$\frac{M \times N}{A \times B} \cong \frac{M}{A} \times \frac{N}{B}.$$

Proof

Define $\varphi \colon M \times N \to \frac{M}{A} \times \frac{N}{B}$ by $\varphi(m, n) = (m + A, n + B)$ for all $m \in M, n \in N$. Now we show that φ is an epimorphism.

Let $(m_1, n_1), (m_2, n_2) \in M \times N$ be such that $(m_1, n_1) = (m_2, n_2)$

$\Rightarrow m_1 = m_2, n_1 = n_2$

$\Rightarrow m_1 + A = m_2 + A, n_1 + B = n_2 + B$

$\Rightarrow (m_1 + A, n_1 + B) = (m_2 + A, n_2 + B)$

$\Rightarrow \varphi(m_1, n_1) = \varphi(m_2, n_2)$.

Therefore, φ is well-defined.

For any $(m_1, n_1), (m_2, n_2) \in M \times N$,

$\varphi((m_1, n_1) + (m_2, n_2)) = \varphi(m_1 + m_2, n_1 + n_2)$

$= ((m_1 + m_2) + A, (n_1 + n_2) + B))$

$= ((m_1 + A) + (m_2 + A), (n_1 + B) + (n_2 + B))$

$= \varphi(m_1, n_1) + \varphi(m_2, n_2)$

For any $r \in R$, and for any $(m, n) \in M \times N$,

$\varphi(r(m, n)) = \varphi(rm, rn)$

$$= (rm + A, rm + B)$$

$$= (r(m + A), r(m + B))$$

$$= r(m + A, m + B) = r(\varphi(m, n)).$$

Therefore, φ is an R-homomorphism.

Let $x \in \frac{M}{A} \times \frac{N}{B}$.

Then $x = (m + A, n + B)$ for some $m \in M$, $n \in N$.

Now $(m, n) \in M \times N$ and $\varphi(m, n) = (m + A, n + B) = x$.

Therefore, φ is onto.

So, $\varphi: M \times N \to \frac{M}{A} \times \frac{N}{B}$ is an epimorphism.

Therefore, by the fundamental theorem of homomorphisms,

$\frac{M \times N}{\ker \varphi} \cong \varphi(M \times N) \Rightarrow \frac{M \times N}{\ker \varphi} \cong \frac{M}{A} \times \frac{N}{B}$.

Now we show that $\ker \varphi = A \times B$.

$\ker \varphi = \{(m, n) \in M \times N \mid \varphi(m, n) = 0\}$

$$= \{(m, n) \mid (m + A, n + B) = (A, B)\}$$

$$= \{(m, n) \mid m + A = A, n + B = B\}$$

$$= \{(m, n) \mid m \in A, n \in B\} = A \times B.$$

Therefore, $\frac{M \times N}{A \times B} = \frac{M}{A} \times \frac{N}{B}$. ∎

Definition 1.5.39

(i) An R-module A is said to be a **faithful module** if $rA \neq 0$ for any $0 \neq r \in R$.

(ii) A is said to be **irreducible** if A contains exactly two submodules (0) and itself.

Definition 1.5.40

(i) If K, A are submodules of M, and K is a submodule of M maximal with respect to $K \cap A = (0)$, then K is said to be a **complement** of A (or a **complement submodule** in M).

(ii) A nonzero submodule K of M is called **essential** (or **large**) in M (or M is an **essential extension** of K) if A is a submodule of M and $K \cap A = (0)$ implies $A = (0)$.

(iii) If A is an essential submodule of M, and $A \neq M$, then we say that M is a **proper essential extension** of A. If a submodule K is essential in M, then we denote this fact by $K \leq_e M$.

Goldie (1972) introduced the concept of a finite Goldie dimension (FGD) in modules.

We know that the dimension of a vector space is defined as the number of elements in the basis. One can define a basis of a vector space as a **maximal** set of linearly independent vectors or a **minimal** set of vectors, which span the space. The former case, when generalized to modules over rings, becomes the concept of the Goldie dimension.

As in the theory of vector spaces, for any submodule K, H of M such that $K \cap M = (0)$, the condition $\dim(K + H) = \dim(K) + \dim(H)$ holds. Also, if K and H are isomorphic, then $\dim(K) = \dim(H)$.

Definition 1.5.41

An R-module M is said to have an FGD if M does not contain a direct sum of an infinite number of nonzero submodules.

Definition 1.5.42

A nonzero submodule K of M is said to be a **uniform submodule** if every nonzero submodule of K is essential in K.

Lemma 1.5.43 (Theorem 0.1 (a) of Satyanarayana, 1989)

Let M be a nonzero module with an FGD. Then every nonzero submodule of M contains a uniform submodule.

Theorem 1.5.44 (Goldie, 1972)

Let M be a module with an FGD.

(i) (Existence) There exist uniform submodules $U_1, U_2, ..., U_n$ whose sum is direct and essential in M.
(ii) (Uniqueness) If there exist uniform submodules $V_1, V_2, ..., V_k$ whose sum is direct and essential in M, then $k = n$.

Definition 1.5.45

Let M be a module with FGD. Then by Theorem 1.5.44, there exist uniform submodules U_i, $1 \le i \le n$, whose sum is direct and essential in M. The number n is independent of the choice of the uniform submodules. This number n is called the **Goldie dimension** of M, and it is denoted by $\dim(M)$.

Remark 1.5.46

(i) A submodule B is a complement in M if and only if there exists a submodule A of M such that $B \cap A = (0)$ and $K^1 \cap A \ne (0)$ for any submodule K^1 of M with $B \subsetneq K^1$. In this case, $B + A \le_e M$.

(ii) If $A \cap B = (0)$ and C is a submodule of M, which is maximal with respect to the property $C \supseteq A$, $C \cap B = (0)$, then $C \oplus B$ is essential in M, where $\oplus$ denotes the direct sum. Moreover, C is a complement of B containing A.

Result 1.5.47

(i) If I, J, K are submodules of M such that I is essential in J and J is essential in K, then I is essential in K.

(ii) If $I \subseteq J \subseteq K$, then I is essential in K, if and only if I is essential in J and J is essential in K.

(iii) Let $G_1, G_2, \ldots, G_n, H_1, H_2, \ldots, H_n$ be submodules of M such that the sum $G_1 + G_2 + \ldots + G_n$ is direct and $H_i \subseteq G_i$ for $1 \le i \le n$. Then $H_1 + H_2 + \ldots + H_n$ is essential in $G_1 + G_2 + \ldots + G_n$, if and only if each H_i is essential in G_i for $1 \le i \le n$.

Note 1.5.48

Let A, B, and C be submodules of M such that

(i) $B \cap A = (0)$ and $K^1 \cap A \ne (0)$ for any submodule K^1 of M with $B \subsetneq K^1$.

(ii) C is a submodule of M, which is maximal with respect to the property $C \supseteq A$, $C \cap B = (0)$. Then we have the following:

(a) By Remark 1.5.46 (i), B is a complement in M.

(b) $A \oplus B$ is essential in M, and $A \oplus B \subseteq C \oplus B \subseteq M$.
   Using Result 1.5.47 (ii), we get that $A \oplus B$ is essential in $C \oplus B$.

(c) By Result 1.5.47 (iii), it follows that A is essential in C.

(d) C is a complement submodule, which is also an essential extension of A.

When we observe the following example, we learn that the condition $\dim(M/K) = \dim(M) - \dim(K)$ does not hold for a general submodule K of M. For this, consider $\mathbb{Z}$, the ring of integers. Since $\mathbb{Z}$ is a uniform $\mathbb{Z}$-module, it follows that $\dim(\mathbb{Z}) = 1$. Suppose $p_1, p_2, \ldots, p_k$ are distinct primes, and consider K, the submodule generated by the product of these primes. Now, $\mathbb{Z}/K$ is isomorphic to the external direct sum of the modules $\mathbb{Z}/(p_i)$, where (p_i) denotes the submodule of $\mathbb{Z}$ generated by pi (for $1 \leq i \leq k$), and so $\dim(\mathbb{Z}/K) = k$. For $k \geq 2$, $\dim \mathbb{Z} - \dim(K) = 1 - 1 = 0 \neq k = \dim(\mathbb{Z}/K)$. Hence, there arises a type of submodule K that satisfies the condition $\dim(M/K) = \dim(M) - \dim(K)$. In this connection, Goldie obtained the following result.

Theorem 1.5.49 (Goldie, 1972)

If M has an FGD and K is a complement submodule, then $\dim(M/K) = \dim(M) - \dim(K)$.

The converse of this theorem is as follows.

Theorem 1.5.50 (Theorem 2 of Reddy and Satyanarayana, 1987)

Let M be an R-module with FGD, and let K be a submodule of M such that $\dim(M) = \dim(K) + \dim(M/K)$. Then K is a complement submodule of M.

1.6 Graphs

Graph theory was postulated in 1736 with Euler's paper in which he solved the Kongsberg bridges problem. The last three decades have witnessed more interest in graph theory, particularly among applied mathematicians and engineers. Graph theory has a surprising number of applications in many developing areas. It is also intimately related to many branches of mathematics, including group theory, matrix theory, automata, and combinatorics. One of the features of graph theory is that it depends very little on the other branches of mathematics. Graph theory serves as a mathematical model for any system involving a binary relation. One of the attractive features of graph theory is its inherent pictorial character. The development

of high-speed computers is also one of the reasons for the recent growth of
interest in graph theory.

Definition 1.6.1

A **graph** $G = (V, E)$ consists of a nonempty set of objects $V = \{v_1, v_2, \ldots\}$
called **vertices** and another set $E = \{e_1, e_2, \ldots\}$ of elements called **edges** such
that each edge e_k is identified with an unordered pair $\{v_i, v_j\}$ of vertices. The
vertices v_i, v_j associated with the edge e_k are called the **end vertices** of e_k.
A vertex is also called a **node** or **point**; an edge is also called as a **branch**
or **line**. An edge associated with a vertex pair $\{v_i, v_i\}$ is called a **loop** or
self-loop. If there is more than one edge associated with a given pair of
vertices, then these edges are called **parallel edges** or **multiple edges**.

Example 1.6.2

Consider the graph in Figure 1.1.
    This is a graph with five vertices and seven edges. Here, $G = (V, E)$, where

$$V = \{v_1, v_2, v_3, v_4, v_5\} \text{ and } E = \{e_1, e_2, e_3, e_4, e_5, e_6, e_7\}.$$

The identification of edges with the unordered pairs of vertices is given by

$e_1 \leftrightarrow \{v_2, v_2\}, e_2 \leftrightarrow \{v_2, v_4\}, e_3 \leftrightarrow \{v_1, v_2\},$
$e_4 \leftrightarrow \{v_1, v_3\}, e_5 \leftrightarrow \{v_1, v_3\}, e_6 \leftrightarrow \{v_3, v_4\}.$

Here, "e_1" is a loop, and e_4 and e_5 are parallel edges.

Definition 1.6.3

A graph that has neither self-loops nor parallel edges is called a **simple
graph**. A graph containing either parallel edges or loops is also referred to
as a **general graph**.

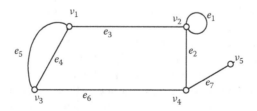

FIGURE 1.1
Graph.

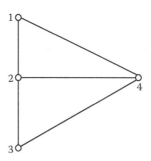

FIGURE 1.2
Graph.

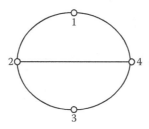

FIGURE 1.3
Graph.

Example 1.6.4

It can be observed that the two graphs given in Figures 1.2 and 1.3 are one and the same.

Graph theory has a wide range of applications in the engineering, medical, physical, social, and biological sciences. A graph can be used to represent almost any physical situation involving discrete objects and a relationship among them. In the following, we present a few such examples.

Problem 1.6.5: Konigsberg Bridges Problem

This is one of the best known examples of graph theory.

This problem was solved by Leonhard Euler (1707–1783) in 1736 using the concepts of graph theory. He is the originator of graph theory.

[**Problem**: There were two islands C and D connected to each other and to the banks A and B with seven bridges, as shown in Figure 1.4.

The problem was to start at any of the four land areas of the city A, B, C, and D, walk over each of the seven bridges once and only once, and return to the starting point.] Euler represented this situation by means of a graph as given in Figure 1.4. Euler proved that a solution for this problem does not exist.

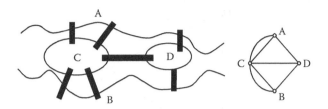

FIGURE 1.4
Konigsberg bridges.

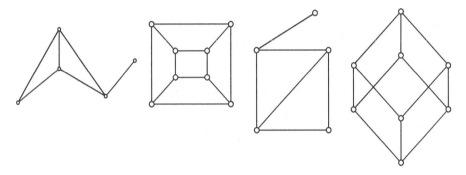

FIGURE 1.5
Graphs that cannot be drawn without lifting the pen or retracing.

Note 1.6.6

The Konigsberg bridges problem is the same as drawing figures without lifting the pen from the paper and without retracing a line. The same situation can be observed in Figure 1.5 for graphs/figures.

Problem 1.6.7: Utility Problem

There are three houses, H_1, H_2, H_3, each to be connected to each of the three utilities—water (W), gas (G), electricity (E)—by means of conduits, as shown in Figure 1.6.

Problem 1.6.8: Seating Arrangement Problem

Nine members of a new club meet each day for lunch at a round table. They decide to sit in such a way that "each member has different neighbors at each lunch." How many days can this arrangement last?

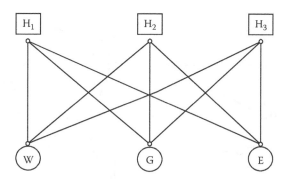

FIGURE 1.6
Graph of utility problem.

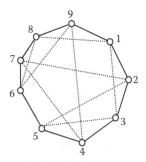

FIGURE 1.7
Graph of seating arrangement problem.

This situation can be represented by a graph with nine vertices such that each vertex represents a member, and an edge joining two vertices represents the relationship of sitting next to each other. Figure 1.7 shows two possible arrangements (1234567891, 1352749681) at the dinner table. From this figure, we can observe that there are two possible seating arrangements, which are 1234567891 and 1352749681.

It can be shown by the graph-theoretic considerations that there are only two more arrangements possible, which are 1573928461 and 1795836241. In general, for n people, the number of such possible arrangements is $\frac{n-1}{2}$ if n is odd and $\frac{n-2}{2}$ if n is even.

Definition 1.6.9

A graph G with a finite number of vertices and a finite number of edges is called a **finite graph**. A graph G that is not a finite graph is said to be an **infinite graph**.

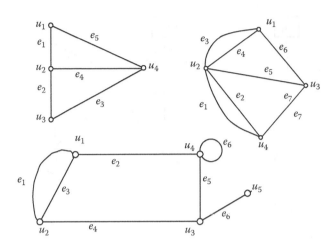

FIGURE 1.8
Finite graphs.

Example 1.6.10

(i) Consider the graphs given in Figure 1.8. It can be observed that the number of vertices and the number of edges are finite. Hence, the three graphs in Figure 1.8 are finite graphs.

(ii) Consider the graphs given in Figure 1.9. It can be understood that the number of vertices of the two graphs is not finite. So, we conclude that these two figures represent infinite graphs.

Henceforth, we place our attention on the study of finite graphs. So, we use the term "graph" for "finite graph."

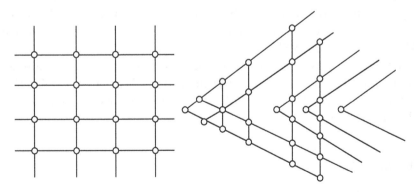

FIGURE 1.9
Infinite graphs.

Definition 1.6.11

If a vertex v is an end vertex of some edge e, then v and e are said to be **incident** with (or on, or to) each other. Two nonparallel edges are said to be **adjacent** if they are incident on a common vertex. Two vertices are said to be **adjacent** if they are the end vertices of the same edge. The number of edges incident on a vertex v is called the **degree** (or **valency**) of v. The degree of a vertex v is denoted by $d(v)$. It is to be noted that a self-loop contributes two to the degree of the vertex.

Example 1.6.12

(i) Consider the graph shown in Figure 1.10. Here, the edges e_2, e_6, e_7 are incident with the vertex u_4.
(ii) Consider the graph given in Figure 1.11. Here, the vertices u_4, u_5 are adjacent. The vertices u_1 and u_4 are not adjacent. The edges e_2 and e_3 are adjacent.

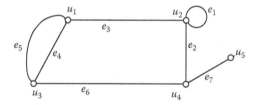

FIGURE 1.10
Graph (illustration for incidence).

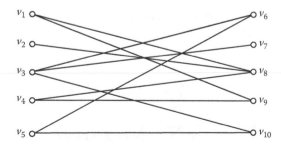

FIGURE 1.11
Graph (illustration degree of vertices).

(iii) Consider the graph given in Figure 1.11. Here $d(v_1) = 2$; $d(v_2) = 1$; $d(v_3) = 3$; $d(v_4) = 2$; $d(v_5) = 2$; $d(v_6) = 2$; $d(v_7) = 1$; $d(v_8) = 3$; $d(v_9) = 2$; $d(v_{10}) = 2$.

Theorem 1.6.13

The sum of the degrees of the vertices of a graph G is twice the number of edges. That is, $\sum_{v_i \in V} d(v_i) = 2e$. (Here e is the number of edges.)

Proof

(The proof is by induction on e.)

Case (i): Suppose $e = 1$. Suppose f is the edge in G with $f = uv$. Then $d(v) = 1$, $d(u) = 1$.

Therefore,

$$\sum_{x \in V} d(x) = \sum_{x \in V \setminus \{u, v\}} d(x) + d(u) + d(v).$$

$$= 0 + 1 + 1$$

$$= 2$$

$$= 2 \times 1$$

$$= 2 \times (\text{number of edges}).$$

Hence, the given statement is true for $n = 1$.

Now we can assume that the result is true for $e = k - 1$. Take a graph G with k edges. Now consider an edge f in G whose endpoints are u and v. Remove f from G. Then we get a new graph $G^* = G - \{f\}$.

Suppose $d^*(v)$ denotes the degree of vertices v in G^*.

For any $x \notin \{u, v\}$, we have $d(x) = d^*(x)$, and $d^*(v) = d(v) - 1$, $d^*(u) = d(u) - 1$. Now G^* has $k - 1$ edges, and so by the induction hypothesis, $\sum_{v_i \in V} d^*(v_i) = 2(k - 1)$.

Therefore,

$$2(k - 1) = \sum_{v_i \in V} d^*(v_i)$$

$$= \sum_{v_i \notin \{u, v\}} d^*(v_i) + d^*(u) + d^*(v)$$

$$= \sum_{v_i \notin \{u, v\}} d(v_i) + (d(u) - 1) + (d(v) - 1)$$

$$= \sum_{v_i \notin \{u, v\}} d(v_i) + d(u) + d(v) - 2$$

$$= \sum_{v_i \in V} d^*(v_i) - 2.$$

Hence, $2(k-1) + 2 = \sum_{v_i \in V} d^*(v_i)$, and so $2k = \sum_{v_i \in V} d(v_i)$.

Thus, by induction we get that "the sum of the degrees of the vertices of the graph G is twice the number of edges." ∎

Theorem 1.6.14

The number of vertices of odd degrees is always even.

Proof

By Theorem 1.6.13,

$$\sum_{i=1}^{n} d(v_i) = 2e. \tag{1.5}$$

If we consider the vertices of odd and even degrees separately, then

$$\sum_{i=1}^{n} d(v_i) = \sum_{v_j \text{ is even}} d(v_j) + \sum_{v_k \text{ is odd}} d(v_k). \tag{1.6}$$

Since the LHS of (ii) is even (from Equation 1.5) and the first expression on the RHS is even, it follows that the second expression on the RHS is always even.

Therefore, $\sum_{v_k \text{ is odd}} d(v_k)$ is an even number. $\qquad (1.7)$

In Equation 1.7, each $d(vk)$ is odd. The number of terms in the sum must be even to make the sum an even number. Hence, the number of vertices of odd degree is even. ∎

Definition 1.6.15

A **subgraph** of $G = (V, E)$ is a graph $G^1 = (V^1, E^1)$ such that $V^1 \subseteq V$, $E^1 \subseteq E$, and each edge of G^1 has the same end vertices in G^1 as in G.

Example 1.6.16

Consider the graph G given in Figure 1.12.

In the graph shown in Figure 1.12, we have $V(G) = \{v_1, v_2, v_3, v_4\}$ and $E(G) = \{e_1, e_2, e_3, e_4, e_5\}$. Now a subgraph G^1 is given by Figure 1.13. In this graph, $V(G^1) = \{v_2, v_3, v_4\}$, $E(G^1) = \{e_2, e_3, e_5\}$. Now $V(G^1) \subseteq V(G)$, $E(G^1) \subseteq E(G)$, and the endpoints of all the edges e_2, e_3, e_5 in G^1 are in $V(G^1)$. The edges in G^1 have the same endpoints as in G. So, G^1 is a subgraph of G.

Definition 1.6.17

  (i) If $d(v) = k$ for every vertex v of a given graph G, for some positive integer k, then the graph G is called a k-**regular graph** (or a **regular graph**, or a **regular graph of degree** k).
 (ii) A **complete graph** is a simple graph in which each pair of distinct vertices is joined by an edge. The complete graph on k vertices is denoted by C_k.

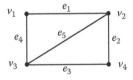

FIGURE 1.12
Graph.

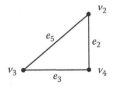

FIGURE 1.13
Subgraph.

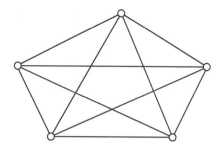

FIGURE 1.14
Complete graph.

Example 1.6.18

The graph given in Figure 1.14 is a complete graph on five vertices. It is also a regular graph of degree 4.

Note 1.6.19

A complete graph is a regular graph of degree $(p - 1)$, where p is the number of vertices.

Definition 1.6.20

Two graphs G and G^1 are said to be **isomorphic** if there is a one-to-one correspondence f between their vertices and a one-to-one correspondence g between their edges such that the incidence relationship must be preserved. (In other words, two graphs $G = (V, E)$ and $G^1 = (V^1, E^1)$ are said to be **isomorphic** if there exist bijections $f: V \to V^1$ and $g: E \to E^1$ such that $g(\overline{v_i v_j}) = \overline{f(v_i)f(v_j)}$ for any edge $v_i v_j$ in G.)

Example 1.6.21

Consider the graphs given in Figures 1.15 and 1.16.
The two graphs G_1 and G_2 are isomorphic (the mapping $f(u_i) = v_i$ is an isomorphism between these two graphs).

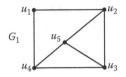

FIGURE 1.15
Isomorphic graph.

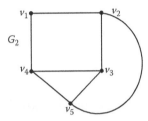

FIGURE 1.16
Isomorphic graph.

Definition 1.6.22

Let $G = (V, E)$ be a graph and $\emptyset \neq X \subseteq V$. Write $E_1 = \{xy \in E \mid x, y \in X\}$. Then $G_1 = (X, E_1)$ is a subgraph of G, and it is called the **subgraph generated by** X. It is also called the **maximal subgraph** with vertex set X.

Example 1.6.23

The graph shown in Figure 1.13 is a subgraph generated by the set $\{v_2, v_3, v_4\}$ of the graph in Figure 1.12.

Definition 1.6.24

In a graph G, a subset S of $V(G)$ is said to be a **dominating set** if every vertex not in S has a neighbor in S. The **domination number**, denoted by $\gamma(G)$, is $\min\{|S| \mid S \text{ is a dominating set in } G\}$.

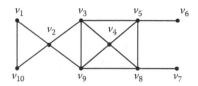

FIGURE 1.17
Graph (illustration domination number).

Example 1.6.25

Consider the graph G given in Figure 1.17.

In this graph, $\{v_1, v_3, v_6, v_7\}$ is a dominating set of size 4 and $\{v_2, v_5, v_8\}$ is a dominating set of size 3. There is no dominating set for G with one element or two elements. Hence, the domination number, $\gamma(G) = 3$.

Definition 1.6.26

A graph $G = (V, E)$ is said to be a **star graph** (Figure 1.18) if there exists a fixed vertex v such that $E = \{vu \mid u \in V \text{ and } u \neq v\}$. A star graph is said to be an **n-star graph** if the number of vertices of the graph is n.

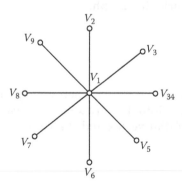

FIGURE 1.18
Star graph.

Example 1.6.27

Consider the graph $G = (V, E)$, where $V = \{v_1, v_2, v_3, v_4, v_5, v_6, v_7, v_8, v_9\}$ and $v_i \neq v_j$ for $1 \leq i, j \leq 9$, shown in Figure 1.18. In this graph, the vertex v_1 is the fixed vertex, and it is adjacent to all the other vertices of the graph. So, this graph is a star graph. Also, it is a nine-star graph.

Definition 1.6.28

A connected graph without circuits is called a *tree*. It is clear that every star graph is a tree.

Definition 1.6.29

A **directed graph** or **digraph** G consists of a nonempty set V (the elements of V are normally denoted by $v_1, v_2, \ldots$), a set E (the elements of E are normally denoted by $e_1, e_2, \ldots$), and a mapping φ that maps every element of E onto an ordered pair (v_i, v_j) of elements from V. The elements of V are called **vertices** or **nodes** or **points**. The elements of E are called **edges** or **arcs** or **lines**. If $e \in E$ and $v_i, v_j \in V$ such that $\varphi(e) = (v_i, v_j)$, then we write $e = \overrightarrow{v_i v_j}$. In this case, we say that e is an edge between v_i and v_j (or e is an edge from v_i to v_j). (An edge from v_i to v_j is denoted by a line segment with an arrow directed from v_i and v_j.) We also say that e originates at v_i and terminates at v_j. A directed graph is also called an **oriented graph**.

Note 1.6.30

Let G be a directed graph and $e = \overrightarrow{vu}$. Then we say that e is **incident out** of the vertex v and **incident into** the vertex u. In this case, we also say that the vertex v is called the **initial vertex** and the vertex u is called the **terminal vertex** of e.

Example 1.6.31

The graph given in Figure 1.19 is a digraph with 5 vertices and 10 edges. Here, v_5 is the initial vertex and v_4 is the terminal vertex for the edge e_7. The edge e_5 is a self-loop.

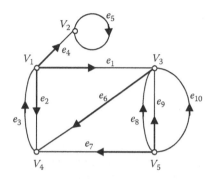

FIGURE 1.19
Digraph.

Definition 1.6.32

(i) The number of edges incident out of a vertex v is called the **out-degree** or **out-valency** of v. The out-degree of a vertex v is denoted by $d^+(v)$.

(ii) The number of edges incident into v is called the **in-degree** or **in-valency** of v. The in-degree of a vertex v is denoted by $d^-(v)$.

(iii) The degree of v is defined as the sum of the in-degree and out-degree of v, for any vertex v in a graph. (In symbols, we can write it as $d(v) = d^+(v) + d^-(v)$ for all vertices v.)

Example 1.6.33

Consider the graph given in Figure 1.19.

(i) Here, $d^+(v_1) = 3$, $d^+(v_2) = 1$, $d^+(v_3) = 1$, $d^+(v_4) = 1$, $d^+(v_5) = 4$.

(ii) $d^-(v_1) = 1$, $d^-(v_2) = 2$, $d^-(v_3) = 4$, $d^-(v_4) = 3$, $d^-(v_5) = 0$.

(iii) $d(v_1) = d^+(v_1) + d^-(v_1) = 3 + 1 = 4$.

1.7 Introduction to Fuzzy Sets

Zadeh (1965), a professor of electrical engineering and computer science, University of California, Berkeley, introduced in 1964 the notion of a "fuzzy subset" as a method for representing uncertainty. Zadeh has defined a fuzzy set as a generalization of a characteristic function, wherein the degree of membership of an element is more general than merely "yes" (denoted by 1) or "no" (denoted by 0).

Definition 1.7.1

Let U be a universal set. Consider $U \times [0, 1]$.

(i) Let $\mu: U \to [0, 1]$ be a function. Now $\{(x, \mu(x)) \mid x \in U\}$ (which is a subset of $U \times [0, 1]$) is called a **fuzzy set** (or **fuzzy subset**) of (or on) U. The fuzzy set defined here will be represented as μ.

(ii) A fuzzy subset μ is said to be a **nonempty fuzzy subset** if there exists $x \in U$ such that $\mu(x) \neq 0$.

(iii) A fuzzy subset μ is said to be a nonconstant fuzzy subset if there exists $x, y \in U$ such that $\mu(x) \neq \mu(y)$.

(iv) Let $t \in [0, 1]$. The set $\mu_t = \{x \in U \mid \mu(x) \geq t\}$ is called a **level set** (or **level subset**) of μ.

Example 1.7.2

Let $U = \{a, b, c, d, e\}$. Then $S = \{(a, 0.2), (b, 0.3), (c, 1), (d, 0.7), (e, 0)\}$ is a fuzzy set on U.

Here, $\mu(a) = 0.2$, $\mu(b) = 0.3$, and so on.

Note that $\mu_{0.45} = \{x \in U \mid \mu(x) \geq 0.45\} = \{c, d\}$ and $\mu_0 = U$.

Note 1.7.3

Consider Example 1.7.2.

(i) $(a, 0.2) \in S$. In this case, we say that "the degree of membership of a in S is 0.2."

(ii) $(c, 1) \in S$ indicates that the degree of membership of c in S is 1.

(iii) $(c, 0) \in S$. So, the degree of membership of e in S is 0.

Example 1.7.4

Define $\mu: \mathbb{Z}_6 \to [0, 1]$ by $\mu(x) = \begin{cases} 1 & \text{if } x = 0 \\ 3/4 & \text{if } x \in \{2, 4\} \\ 1/4 & \text{otherwise} \end{cases}$. Then μ is a fuzzy subset of $\mathbb{Z}_6$.

Definition 1.7.5

Let U be a universal set and $A \subseteq U$. Then the **characteristic function** $\chi_A \colon U \to \{0, 1\}$ is defined by

$$\chi_A(x) = \begin{cases} 1 & \text{if } x \in A \\ 0 & \text{otherwise} \end{cases}.$$

Note that the characteristic function on U is a fuzzy set on U. Here, the level sets are $(\chi_A)_1 = A$, $(\chi_A)_0 = U$, and $(\chi_A)_t = A$ for $0 < t$.

Example 1.7.6

Let $U = \{a, b, c, d, e, f\}$ and $A = \{a, b, c\}$.

Then $\chi_A \colon U \to \{0, 1\}$ is the characteristic function given by $\chi_A(a) = 1$, $\chi_A(b) = 1$, $\chi_A(c) = 1$, $\chi_A(d) = 0$, $\chi_A(e) = 0$, and $\chi_A(f) = 0$.

We can represent A by $\chi_A = \{(a, 1), (b, 1), (c, 1), (d, 0), (e, 0), (f, 0)\}$.

Definition 1.7.7

Let $\alpha, \beta \in [0, 1]$ such that $\beta > \alpha$, U is a universal set, and $A \subseteq U$. Then the function $\mu \colon U \to [0, 1]$ defined by

$$\mu(x) = \begin{cases} \beta & \text{if } x \in A \\ \alpha & \text{otherwise} \end{cases}$$

is called the **generalized characteristic function** of A.

Here the level sets are $\mu_t = U$ if $0 \le t \le \alpha$ and $\mu_t = A$ if $t > \alpha$.

Note 1.7.8

(i) Define $\mu \colon \mathbb{Z} \to [0, 1]$ by $\mu(x) = \begin{cases} 0.4 & \text{if } x \in 2\mathbb{Z} \\ 0.2 & \text{otherwise} \end{cases}$.

Then μ is a generalized characteristic function of $2\mathbb{Z}$.

(ii) Every characteristic function is a generalized characteristic function. Take $\alpha > 0$ and $\beta < 1$ in Definition 1.7.7. Then μ is a generalized characteristic function but not a characteristic function.

Notation 1.7.9

Let μ and σ be two fuzzy subsets of a set X. Then we write $\mu \subseteq \sigma$ if $\mu(x) \leq \sigma(x)$ for all $x \in X$.

Note 1.7.10

Let μ be a fuzzy subset of U and $s, t \in [0, 1]$ be such that $s < t$. Then $\mu_s \supseteq \mu_t$.

Definition 1.7.11

Let S and T be two fuzzy sets on U. The **union** of fuzzy sets S and T is the fuzzy set $S \cup T$ (on U), where the degree of membership of an element in $S \cup T$ is the maximum of the degrees of membership of this element in S and in T.

In other words, if μ_1 and μ_2 are two fuzzy sets, then $\mu_1 \cup \mu_2$ (the union of μ_1 and μ_2) is defined by $(\mu_1 \cup \mu_2)(x) = \max\{\mu_1(x), \mu_2(x)\}$.

Example 1.7.12

Suppose that $S = \{(x_1, f(x_1)), (x_2, f(x_2))\}$ and $T = \{(x_1, g(x_1)), (x_2, g(x_2)), (x_3, g(x_3))\}$ are two fuzzy sets. The degree of membership of x_3 in S is 0. Here, $S \cup T = \{(x_1, \max\{f(x_1), g(x_1)\}), (x_2, \max\{f(x_2), g(x_2)\}), (x_3, g(x_3))\}$.

The complement of a fuzzy subset μ of a set S, denoted by $\sim\mu$, is a fuzzy subset of S defined by

$$(\sim\mu)(x) = 1 - \mu(x) \text{ for all } x \in S.$$

Definition 1.7.13

Let S and T be two fuzzy sets on U. The **intersection** of two fuzzy sets S and T is the fuzzy set $S \cap T$ (on the set U), where the degree of membership of an element in $S \cap T$ is the minimum of the degrees of membership of this element in S and in T.

In other words, if μ_1 and μ_2 are two fuzzy sets, then $\mu_1 \cap \mu_2$ (the intersection of μ_1 and μ_2) is defined by $(\mu_1 \cap \mu_2)(x) = \min\{\mu_1(x), \mu_2(x)\}$.

Note 1.7.14

If $S = \{(x, f(x)), (y, f(y))\}$ and $T = \{(x, g(x)), (y, g(y)), (z, g(z))\}$, then $S \cap T$ is given by $S \cap T = \{(x, \min\{f(x), g(x)\}), (y, \min\{f(y), g(y)\}), (z, g(z))\}$.

1.7.15 Some Elementary Properties of ~, $\cup$, and $\cap$

Let μ, σ, and θ be any fuzzy subsets of a set S. Then the following hold:

(i) $\mu \cup \sigma = \sigma \cup \mu$ and
   $\mu \cap \sigma = \sigma \cap \mu$ (commutative laws)
(ii) $\mu \cup (\sigma \cup \theta) = (\mu \cup \sigma) \cup \theta$ and
   $\mu \cap (\sigma \cap \theta) = (\mu \cap \sigma) \cap \theta$ (associative laws)
(iii) $\mu \cup \mu = \mu$ and $\mu \cap \mu = \mu$ (idempotent laws)
(iv) $\mu \cup (\sigma \cap \theta) = (\mu \cup \sigma) \cap (\mu \cup \theta)$ and
   $\mu \cap (\sigma \cup \theta) = (\mu \cap \sigma) \cup (\mu \cap \theta)$ (distributive laws)
(v) $\mu \cup \chi_\Phi = \mu$ and $\mu \cap \chi_s = \mu$ (identity laws)
   where χ_A is the characteristic function of A for any set A.
(vi) $\mu \cup (\mu \cap \sigma) = \mu$ and
   $\mu \cap (\mu \cup \sigma) = \mu$ (absorption laws)
(vii) $\sim (\mu \cap \sigma) = (\sim\mu) \cup (\sim\sigma)$ and
   $\sim(\mu \cup \sigma) = (\sim\mu) \cap (\sim\sigma)$ (De Morgan's laws)
(viii) $\sim(\sim\mu) = \mu$ (involution law)

Note 1.7.16

In general, the union and intersection of any family $\{\mu_i \mid i \in \Omega\}$ of fuzzy subsets of a set S are defined by

$$\left(\bigcup_{i \in \Omega} \mu_i\right)(x) = \sup_{i \in \Omega} \mu_i(x) \text{ for all } x \in S$$

$$\left(\bigcap_{i \in \Omega} \mu_i\right)(x) = \inf_{i \in \Omega} \mu_i(x) \text{ for all } x \in S$$

Definition 1.7.17

Let X and Y be two nonempty sets, and let f be a function of X into Y. Let μ be a fuzzy subset of X. Then $f(\mu)$, the **image** of μ under f, is a fuzzy subset of Y,

defined by $(f(\mu))(y) = \begin{cases} \sup_{f(x)=y} \mu(x) & \text{if } f^{-1}(y) \neq \emptyset \\ 0 & f^{-1}(y) \neq \emptyset \end{cases}$.

Example 1.7.18

Let $X = \{a, b, c, d\}$ and $Y = \{u, v, w, z\}$.
   Define $f: X \to Y$ by $f(a) = v, f(b) = u, f(c) = w$, and $f(d) = w$.
   Consider $\mu = \{(a, 0.8), (b, 0.3), (c, 0.6), (d, 0.2)\}$. Then μ is a fuzzy subset of X.
   Now $f(\mu)(u) = \sup\{\mu(x) \mid f(x) = u\} = \sup\{\mu(b)\} = \sup\{0.3\} = 0.3$

$$f(\mu)(v) = \sup\{\mu(x) \mid f(x) = v\} = \sup\{\mu(a)\} = \sup\{0.8\} = 0.8$$

$$f(\mu)(w) = \sup\{\mu(x) \mid f(x) = w\} = \sup\{\mu(c), \mu(d)\} = \sup\{0.6, 0.2\} = 0.6$$

$$f(\mu)(z) = 0 \text{ (since } f^{-1}(z) = \mu)$$

   Therefore, $f(\mu) = \{(u, 0.3), (v, 0.8), (w, 0.6), (z, 0)\}$ is the image of μ under f, and $f(\mu)$ is a fuzzy subset of Y.

Definition 1.7.19

Let X and Y be two nonempty sets, and let f be a function of X into Y. Let σ be a fuzzy subset of Y. Then $f^{-1}(\sigma)$, the **preimage** of σ under f, is a fuzzy subset of X defined by $(f^{-1}(\sigma))(x) = \sigma(f(x))$ for all $x \in X$.

Example 1.7.20

Let $X = \{a, b, c, d\}$ and $Y = \{u, v, w, z\}$.
   Define $f: X \to Y$ by $f(a) = v, f(b) = u, f(c) = w$, and $f(d) = w$.
   Take $\sigma = \{(u, 0.3), (v, 0.7), (w, 0.2), (z, 0.1)\}$, a fuzzy subset of Y.
   Now $f^{-1}(\sigma)(a) = \sigma(f(a)) = \sigma(v) = 0.7$

$$f^{-1}(\sigma)(b) = \sigma(f(b)) = \sigma(u) = 0.3$$

$$f^{-1}(\sigma)(c) = \sigma(f(c)) = \sigma(w) = 0.2$$

$$f^{-1}(\sigma)(d) = \sigma(f(d)) = \sigma(w) = 0.2.$$

   Therefore, $f^{-1}(\sigma) = \{(a, 0.7), (b, 0.3), (c, 0.2), (d, 0.2)\}$ is a fuzzy subset of X.

Definition 1.7.21

Let f be any function from a set S to a set T, and let μ be any fuzzy subset of S. Then μ is called f-invariant if $f(x) = f(y)$ implies $\mu(x) = \mu(y)$, where x, $y \in S$.

Example 1.7.22

Let μ be a fuzzy subset of $\mathbb{Z}^+$ (set of positive integers) defined by

$$\mu(x) = \begin{cases} 1 & \text{if } x \in \{2, 4, 6, 8, \ldots\} \\ 0 & \text{otherwise} \end{cases}$$

Define $f\colon \mathbb{Z}^+ \to \{0, 1, 2, 3\}$ as follows:
$f(x) = x \bmod 4$ for all $x \in \mathbb{Z}^+$. Then μ is f-invariant.

Definition 1.7.23

Let "·" be a binary composition in a set S. The product $\mu\theta$ of any two fuzzy subsets μ, θ of S is defined by

$$(\mu\theta)(x) = \begin{cases} \sup \left\{\min\left(\mu(y), \theta(z)\right)\right\}, \; x = yz & \text{where } y, z \in S \\ 0 \; \text{if } x \text{ is not expressible as } x = yz \text{ for } y, z \in S \end{cases}$$

Definition 1.7.24

Let S be a nonempty set and μ be a fuzzy subset of S. For
$t \in [0, 1]$, $\mu_t = \{s \in S \mid \mu(x) \geq t\}$ is called a level subset of μ.

1.8 Some Fuzzy Algebraic Systems

Rosenfeld introduced the notion of fuzzy groups and showed that many group-theoretic results can be extended in an elementary manner to develop the theory of fuzzy groups.

Definition 1.8.1

A fuzzy subset μ of a group G is called a **fuzzy subgroup** of G if for all x, $y \in G$, the following conditions are satisfied:

(i) $\mu(xy) \geq \min\{\mu(x), \mu(y)\}$
(ii) $\mu(x^{-1}) \geq \mu(x)$, where xy stands for $x \cdot y$ (the usual product in G).

Theorem 1.8.2

Let μ be any fuzzy subgroup of a group G with identity e. Then the following statements are true:

   (i) $\mu(x^{-1}) = \mu(x) \le \mu(e)$ for all $x \in G$.
   (ii) $\mu(xy^{-1}) = \mu(e) \Rightarrow \mu(x) = \mu(y)$, where $x, y \in G$.
   (iii) The level subsets μ_t, $t \in [0, 1]$, $t \le \mu(e)$ are subgroups of G.
   (iv) $\mu(xy) = \mu(y)$ for all $y \in G$ if and only if $\mu(x) = \mu(e)$, where $x \in G$.

Definition 1.8.3

Let μ be a fuzzy subgroup of a group G and let $t \in [0, 1]$ be such that $t \le \mu(e)$, where e is the identity of G. The subgroup μ_t is called a **level subgroup** of μ.

Example 1.8.4

Let $G = \{1, -1, i, -i\}$ be the group, with respect to the usual multiplication. Define $\mu: G \to [0,1]$ by

$$\mu(x) = \begin{cases} 1 & \text{if } x = 1 \\ 0.5 & \text{if } x = -1 \\ 0 & \text{if } x \in \{i, -i\} \end{cases}$$

Then μ is a fuzzy subgroup of G.

Theorem 1.8.5

Two level subgroups μ_s and μ_t (with $s < t$) of a fuzzy subgroup μ of a group G are equal if and only if there is no x in G such that $s \le \mu(x) < t$.

Definition 1.8.6

A fuzzy subgroup μ of a group G is called **fuzzy normal subgroup** if $\mu(xy) = \mu(yx)$ for all $x, y \in G$.

Theorem 1.8.7

Let μ be any fuzzy subgroup of G. If $\mu(x) < \mu(y)$ for some $x, y \in G$, then $\mu(xy) = \mu(x) = \mu(yx)$.

Proof

Consider $\mu(xy) \geq \min\{\mu(x), \mu(y)\}$ (since μ is a fuzzy subgroup)

$= \mu(x)$ (by the given condition).

   Also $\mu(x) = \mu(xyy^{-1}) \geq \min\{\mu(xy), \mu(y)\} = \mu(xy)$.
   Therefore, $\mu(xy) = \mu(x)$.
Similarly, we can verify that $\mu(yx) = \mu(x)$.   ∎

Theorem 1.8.8

Let θ be any fuzzy subgroup of G such that $\text{Im}(\theta) = \{0, t\}$, where $t \in (0, 1)$. If $\theta = \mu \cup \sigma$ for some fuzzy subgroups μ and σ of G, then either $\mu \subseteq \sigma$ or $\sigma \subseteq \mu$. (Observe that $\mu \subseteq \sigma$ means $\mu(x) \leq \sigma(x)$ for all $x \in G$.)

Definition 1.8.9

A fuzzy subset μ of a ring R is called a **fuzzy subring** of R, if for all $x, y \in R$, the following conditions hold:

   (i) $\mu(x - y) \geq \min\{\mu(x), \mu(y)\}$ and
   (ii) $\mu(xy) \geq \min\{\mu(x), \mu(y)\}$

Definition 1.8.10

A fuzzy subset μ of a ring R is called a **fuzzy ideal** of R if for all $x, y \in R$, the following conditions hold:

   (i) $\mu(x - y) \geq \min\{\mu(x), \mu(y)\}$ and
   (ii) $\mu(xy) \geq \max\{\mu(x), \mu(y)\}$

Example 1.8.11

Let R be the ring of real numbers under the usual operations of addition and multiplication. Then the fuzzy subset μ of R, defined by

$$\mu(x) = \begin{cases} 1 & \text{if } x \text{ is rational} \\ 0.6 & \text{otherwise} \end{cases},$$

is a fuzzy subring, but not a fuzzy ideal of R.

Theorem 1.8.12

If μ is a fuzzy ideal of a ring R, then the following are true:

   (i) $\mu(x) = \mu(-x) \leq \mu(0)$ for all $x \in R$, where 0 is the identity in R.
  (ii) $\mu(x - y) = \mu(0) \Rightarrow \mu(x) = \mu(y)$ for all $x, y \in R$.
 (iii) The level subsets, μ_t, $t \in \text{Im}(\mu)$, are ideals of R.

Theorem 1.8.13

Two level subrings (level ideals) μ_s and μ_t (with $s < t$) of a fuzzy subring (fuzzy ideal) μ of a ring R are equal if and only if there is no x in R such that $s \leq \mu(x) < t$.

Theorem 1.8.14

If A is any subring (ideal) of a ring R, $A \neq R$, then the fuzzy subset μ of R, defined by

$$\mu(x) = \begin{cases} s & \text{if } x \in A \\ t & \text{if } x \in R - A \end{cases}$$

where $s, t \in [0, 1]$, $s > t$, is a fuzzy subring (fuzzy ideal) of R.

Theorem 1.8.15

A nonempty subset S of a ring R is a subring (ideal) of R if and only if the characteristic function χ_S is a fuzzy subring (fuzzy ideal) of R.

Definition 1.8.16

A fuzzy ideal μ of a ring R (not necessarily nonconstant) is called a **fuzzy prime ideal,** *if for any two fuzzy ideals σ and θ of R, the condition $\sigma\,\theta \subseteq \mu$ implies that $\sigma \subseteq \mu$ or $\theta \subseteq \mu$.*

Theorem 1.8.17

If μ and θ are any two fuzzy prime ideals of a ring R, then $\mu \cap \theta$ is a fuzzy prime ideal of R, if and only if $\mu \subseteq \theta$ or $\theta \subseteq \mu$.

Theorem 1.8.18

If $\{\mu_i \mid i \in Z^+\}$ is any collection of nonconstant fuzzy prime ideals of a ring R such that $\mu_1 \subseteq \mu_2 \subseteq \dots \subseteq \mu_n \subseteq \dots$, then

  (i) $\cup\,\mu_i$ is a fuzzy prime ideal of R.
 (ii) $\cap\,\mu_i$ is a fuzzy prime ideal of R.

Theorem 1.8.19

If f is a homomorphism from a ring R onto a ring R^1 and μ is any f-invariant fuzzy prime ideal of R, then $f(\mu)$ is a fuzzy prime ideal of R^1.

Theorem 1.8.20

If f is a homomorphism from a ring R onto a ring R^1 and μ^1 is any fuzzy prime ideal of R^1, then $f^{-1}(\mu^1)$ is a fuzzy prime ideal of R.

Theorem 1.8.21: Correspondence Theorem

If f is a homomorphism from a ring R onto a ring R, then the mapping $\mu \to f(\mu)$ defines a one–one correspondence between the set of all f-invariant fuzzy prime ideals of R and the set of all fuzzy prime ideals of R^1.

Pan (1987, 1988) introduced the concept of "fuzzy submodules," and the fuzzy finitely generated modules were introduced and studied by Pan (1987, 1988), Golan (1989), and Satyanarayanaand Mohiddin Shaw (2010).

Let M be an R-module and $\mu\colon M \to [0, 1]$ is a mapping. μ is said to be a fuzzy submodule if the following conditions hold:

(i) $\mu(m + m^1) \geq \min\{\mu(m), \mu(m^1)\}$ for all m and $m^1 \in M$ and
(ii) $\mu(am) \geq \mu(m)$ for all $m \in M, a \in R$

It is clear that if M is a unitary R-module and $\mu\colon M \to [0, 1]$ is a fuzzy set with $\mu(am) \geq \mu(m)$ for all $m \in M, a \in R$, then

(i)   For all $0 \neq a \subset R$, $\mu(am) = \mu(m)$ if a is left invertible.
(ii) $\mu(-m) = \mu(m)$.
(iii) If $\mu\colon M \to [0, 1]$ is a fuzzy submodule and $m, m^1 \in M$, then $\mu(m - m^1) \geq \min\{\mu(m), \mu(m^1)\}$.
(iv) If $\mu\colon M \to [0, 1]$ is a fuzzy submodule, then $\mu(0) \geq \mu(m)$ for all $m \in M$; and $\mu(0) = \sup_{m \in M} \mu(m)$.

Result 1.8.22

Let $M_1 \subseteq M$. Define $\mu(x) = 1$, if $x \in M_1$, $= 0$ otherwise.
    Then the following conditions are equivalent:

(i) μ is a fuzzy submodule and
(ii) M_1 is a submodule of M.

2

Fundamentals of Near Rings

In recent years, interest has arisen in algebraic systems with binary operations addition and multiplication satisfying all the ring axioms except possibly one of the distributive laws and the commutativity of addition. Such systems are called "near rings."

The first step toward near rings was the axiomatic research done by Dikson in 1905. He showed that there do exist "fields with only one distributive law" (called near fields). Some years later, these near fields showed up again and proved to be useful in coordinating certain important classes of geometric planes.

Many parts of the well-established theory of rings were transferred to near rings, and new near ring precise features were discovered, building up a theory of near rings step-by-step.

Every ring is a near ring. A natural example of a near ring (which is not a ring) is given by the set $M(G)$ of all mappings of an additive group G (not necessarily Abelian) into itself, with addition defined by $(f + g)(a) = f(a) + g(a)$ and multiplication by $(fg)(a) = f(g(a))$ (here, fg is the composition of mapping f and g) for all $a \in G, f, g \in M(G)$.

2.1 Definitions and Examples

In this section, we present necessary fundamental definitions and results that will be useful in the later sections and chapters.

Definition 2.1.1

A right near ring is a set N together with two binary operations "+" and "·" such that

  (i) $(N, +)$ is a group (not necessarily Abelian)
  (ii) $(N, \cdot)$ is a semigroup
  (iii) $(n_1 + n_2)n_3 = n_1 n_3 + n_2 n_3$ for all $n_1, n_2, n_3 \in N$.

Note 2.1.2

(i) In view of condition (iii) of Definition 2.1.1, N is called a right near ring.
(ii) If N satisfies the left distributive law (i.e., $a(b + c) = ab + ac$, for all a, b, $c \in N$) instead of condition (iii) of Definition 2.1.1, then N is called a left near ring.

Notation 2.1.3

N denotes a near ring. The class of all near rings will be denoted by η.

There exist right near rings which are not left near rings. We provide an example in the following.

Example 2.1.4

Let Γ be an additive group with 0 as its identity element; then the following set is a near ring under the usual addition and composition of mappings:

$$M(\Gamma) = \{f \colon \Gamma \to \Gamma\}.$$

The additive identity of $M(\Gamma)$, that is, the zero mapping, is also denoted by 0. The difference between additive identities of the group Γ and the near ring $M(\Gamma)$ can be understood depending on the situation.

Note 2.1.5

Consider Example 2.1.4. In this near ring, the left distributive law fails to hold. To verify this fact, take a, b, $c \in \Gamma$ and $\underline{a} \neq 0$.

Define $f_a \colon \Gamma \to \Gamma$ by $f_a(\gamma) = a$ for all $\gamma \in \Gamma$

$f_b \colon \Gamma \to \Gamma$ by $f_b(\gamma) = b$ for all $\gamma \in \Gamma$

$f_c \colon \Gamma \to \Gamma$ by $f_c(\gamma) = c$ for all $\gamma \in \Gamma$

Let $\gamma \in \Gamma$. Now $[f_a \circ (f_b + f_c)](\gamma) = f_a[(f_b + f_c)(\gamma)] = f_a[f_b(\gamma) + f_c(\gamma)] = f_a(b + c) = a$.

Also $[(f_a \circ f_b) + (f_a \circ f_c)](\gamma) = (f_a \circ f_b)(\gamma) + (f_a \circ f_c)(\gamma) = f_a[f_b(\gamma)] + f_a[f_c(\gamma)] = f_a(b) + f_a(c) = a + a \neq a$ (since $a \neq 0$). Therefore, $f_a \circ (f_b + f_c) \neq (f_a \circ f_b) + (f_a \circ f_c)$.

This shows that N fails to satisfy the left distributive law. This provides an example of a right near ring that is not a left near ring. Also, this provides an example of a near ring that is not a ring.

TABLE 2.1

Addition Table

+	0	a	b	c
0	0	a	b	c
a	a	0	c	b
b	b	c	0	a
c	c	b	a	0

TABLE 2.2

Multiplication Table

·	0	a	b	c
0	0	0	0	0
a	0	0	a	a
b	0	a	b	b
c	0	a	c	c

Example 2.1.6

Consider the set $N = \{0, a, b, c\}$ given by the explicit addition and multiplication in Tables 2.1 and 2.2. Here, the additive group is the Klein's four group.

It can be verified that $(N, +, \cdot)$ is a near ring but not a ring.

Observe that $b \cdot a + b \cdot c = a + b = c$, whereas $b \cdot (a + c) = bb = b$.

Therefore, $b \cdot a + b \cdot c = c \neq b = b \cdot (a + c)$.

This shows that N fails to satisfy the left distributive law. This provides an example of a right near ring that is not a left near ring. Also, this provides an example of a near ring that is not a ring.

Example 2.1.7

(i) Let $(G, +)$ be a group. Define the multiplication operation on G as $a \cdot b = 0$ for all $a, b \in G$. Then $(G, +, \cdot)$ is a near ring. If G is a non-Abelian group, then $(G, +, \cdot)$ is a near ring that is not a ring. The multiplication operation used here is called trivial multiplication.

(ii) Every ring is a near ring. In particular, the set of real numbers, the set of complex numbers, and the set of integers are rings as well as near rings with respect to the usual addition and multiplication.

(iii) Consider $\mathbb{Z}_8 = \{0, 1, 2, 3, 4, 5, 6, 7\}$ with respect to addition and multiplication modulo 8. This is a ring as well as a near ring.

(iv) We know that every ring is a near ring. Note 2.1.5 and Example 2.1.6 show that the converse need not be true.

Note 2.1.8

(i) A near ring N with identity is Abelian if $n(-1) = -n$ for all $n \in N$.
(ii) The statement given in (i) fails if the near ring N does not have the identity element.

Verification for (i): Suppose $(N, +, \cdot)$ is a near ring with identity.
Let $a, b \in N$. To show N is Abelian, we show that $a + b = b + a$.
We know that $(a + b) + (-b - a) = 0$. This means $(a + b) + [b(-1) + a(-1)] = 0$, which implies that $(a + b) + [(b + a)](-1) = 0$. Therefore, $a + b = b + a$. Hence, N is Abelian.

Example for (ii): Let $(G, +)$ be a non-Abelian group. Write $N = G$. Then $(N, +, \cdot)$ is a near ring with the trivial multiplication mentioned in Example 2.1.7. Note that there is no identity element in N, and also $(N, +)$ is non-Abelian.

Proposition 2.1.9

For all $n, n^1 \in N$,

(i) $0n = 0$, and $(-n)n^1 = -nn^1$.
(ii) $-(a + b) = -b - a$ for all $a, b \in N$.

Proof

(i) Let $n \in N$. Now $0n = (0 + 0)n = 0n + 0n$.
    This implies $0 + 0n = 0n + 0n \Rightarrow 0 = 0n$ for all $n \in N$.
    Let $n, n^1 \in N$. Now $(-n)n^1 + nn^1 = (-n + n)n^1 = 0n^1 = 0$.
    This means $(-n)n^1 = -nn^1$.
(ii) Take $a, b \in N$. Now $(a + b) + (-b + (-a)) = (a + (b - b)) - a) = (a + (-a)) = 0$.
    Therefore, $-(a + b) = -b - a$. ∎

Definition 2.1.10

$N_0 = \{n \in N \mid n0 = 0\}$ is called the **zero-symmetric part** of N.
$N_c = \{n \in N \mid n0 = n\} = \{n \in N \mid \text{for all } n^1 \in N, nn^1 = n\}$ is called the **constant part** of N. N_0 and N_c are known as a zero-symmetric near ring and a constant near ring if N_0 and N_c are themselves near rings.
Note that $\{n \in N \mid n0 = n\} = \{n \in N \mid \text{for all } n^1 \in N, nn^1 = n\}$.

Verification: Let $n \in \{n \in N \mid n0 = n\}$. We show that $n \in \{n \in N \mid \text{for all } n^1 \in N, nn^1 = n\}$.
Let $n^1 \in N$. Now $nn^1 = (n0)n^1 = n(0n^1) = n0 = n$. Therefore,

$$\{n \in N \mid n0 = n\} \subseteq \{n \in N \mid \text{for all } n^1 \in N, nn^1 = n\}.$$

Let $n \in \{n \in N \mid$ for all $n^1 \in N, nn^1 = n\}$. Since $0 \in N$, we have $n \in \{n \in N \mid n0 = n\}$.

Therefore $\{n \in N \mid$ for all $n^1 \in N, nn^1 = n\} \subseteq \{n \in N \mid n0 = n\}$. Hence

$$\{n \in N \mid n0 = n\} = \{n \in N \mid \text{ for all } n^1 \in N, nn^1 = n\}.$$

Note 2.1.11

$n0$ need not be equal to 0 and $n(-n^1)$ need not be equal to $-nn^1$.

Justification
(i) For this, consider $M_c(\Gamma)$, where Γ is a nonzero additive group.
   Let $0 \neq a \in \Gamma$. Define a mapping $f_a: \Gamma \to \Gamma$ by $f_a(x) = a$, for all $x \in \Gamma$. Now $f_a \in M_c(\Gamma)$ and $(f_a 0)(x) = f_a(0(x)) = f_a(0) = a \neq 0 = 0(x)$. This shows that $f_a 0 \neq 0$. So, if we write $n = f_a$, then $n0 \neq 0$.
(ii) Let $(\Gamma, +)$ be a group containing an element $0 \neq a \in G$ with $a + a \neq 0$ (Z_3 is one of such groups.) Take $n = f_a$. Now $(n(-n))(x) = (f_a(-f_a))\ (x) = (f_a(-f_a(x)) = f_a(-a) = a$. On the other hand, $-(nn)(x) = -(f_a f_a)(x) = -(f_a(f_a(x))) = -f_a(a) = -a$. Now $(n(-n))(x) = a \neq -a = -(nn)(x)$.
   Therefore, $n(-n^1) \neq -nn^1$, in general.

Example 2.1.12

Let $(\Gamma, +)$ be a group. Then

(i) $(M(\Gamma))_0 = M_0(\Gamma)$
(ii) $(M(\Gamma))_c = M_c(\Gamma)$

Verification
(i) Let $f \in (M(\Gamma))_0$. This implies $f: \Gamma \to \Gamma$ such that $f0 = 0$.
   We show that $f \in M_0(\Gamma)$. Now $f(0) = f(0(0)) = (f0)(0) = 0(0) = 0$. So, $f \in M_0(\Gamma)$.
   Conversely, suppose that $f \in M_0(\Gamma) \Rightarrow f(0) = 0$.
   Now we show that $f \in (M(\Gamma))_0$.
   Let $\gamma \in \Gamma$. Now $(f0)(\gamma) = f(0(\gamma)) = f(0) = 0 = 0(\gamma)$.
   This shows that $(f0)(\gamma) = 0(\gamma)$ for all $\gamma \in \Gamma$, which yields $f0 = 0$.
   Therefore, $f \in (M(\Gamma))_0$. Hence $(M(\Gamma))_0 = M_0(\Gamma)$.
(ii) Let $f \in (M(\Gamma))_c$. Then $f0 = f$. Now we show that $f \in M_c(\Gamma)$.
   Let $\gamma \in \Gamma$. Now $f(\gamma) = f0(\gamma) = f(0(\gamma)) = f(0)$ and for $\gamma \neq \delta \in \Gamma$ we have $f(\delta) = f0(\delta) = f(0(\delta)) = f(0)$. This means that f is a constant mapping, and so $f \in M_c(\Gamma)$.
   Therefore, the inclusion $(M(\Gamma))_c \subseteq M_c(\Gamma)$ is clear.
   Let $f \in M_c(\Gamma)$. Then f is a constant mapping for any $\gamma \in \Gamma$.

Now $(f0)(\gamma) = f(0(\gamma)) = f(0) = f(\gamma)$. This shows that $(f0)(\gamma) = f(\gamma)$ for all $\gamma \in \Gamma$.
This implies that $f0 = f$. This means $f \in (M(\Gamma))_c$. Therefore, $M_c(\Gamma) \subseteq (M(\Gamma))_c$.
Hence $(M(\Gamma))_c = M_c(\Gamma)$.

Definition 2.1.13

A near ring N is called

(i) **a zero-symmetric near ring** if $N = N_0$ and
(ii) **a constant near ring** if $N = N_c$

Definition 2.1.14

Let N be a near ring, and $e \in N$. The element e is said to be

(i) a **left identity** if $e \cdot n = n$, for all $n \in N$
(ii) a **right identity** if $n \cdot e = n$, for all $n \in N$
(iii) an **identity** if it is both right identity and left identity

Definition 2.1.15

An element $n \in N$ is said to be

(i) **right invertible** if there exists an element $m \in N$ such that $nm = e$
(ii) **left invertible** if there exists an element $m \in N$ such that $mn = e$
(iii) **invertible** if it is both right and left invertible

Definition 2.1.16

An element $n \in N$ is said to be **left cancelable** (**right cancelable**, respectively) if $a, b \in N$, $na = nb \Rightarrow a = b$ ($an = bn \Rightarrow a = b$, respectively).

Definition 2.1.17

A nonzero element n is said to be a **right zero divisor** (**left zero divisor**, respectively) if there exists a nonzero element $a \in N$ such that $an = 0$ ($na = 0$, respectively).

Theorem 2.1.18 (Proposition 1.111 of Pilz, 1983)

Let N be a near ring. An element $n \in N$ is right cancelable if and only if n is not a right zero divisor.

Proof

Suppose that $n \in N$ is right cancelable. Let us also suppose that n is a right zero divisor. Then there exists $n_1 \neq 0$ such that $n_1 n = 0$. This implies $n_1 n = 0 = 0n$. Now, by right cancellation, we get $n_1 = 0$, which is a contradiction. Therefore, n is not a right zero divisor.

Conversely, suppose that n is not a right zero divisor. Let $n, n_1, n_2 \in N$ such that $n_1 n = n_2 n$. This implies $n_1 n - n_2 n = 0$. This means that $(n_1 - n_2)n = 0$. Since n is not a right zero divisor, we get that $n_1 - n_2 = 0$. This implies that $n_1 = n_2$. Therefore, n is right cancelable. ∎

Theorem 2.1.19

If $n \in N_0$ is left cancelable, then n is not a left zero divisor.

Proof

Suppose $n \in N_0$ is left cancelable. Let us also suppose that n is a left zero divisor. Then there exists $0 \neq n_1 \in N$ such that $nn_1 = 0$. Since N is zero-symmetric, $nn_1 = 0 = n0$. Again, since n is left cancelable, we have $n_1 = 0$, which is a contradiction. ∎

Definition 2.1.20

An element $n \in N$ is said to be an **idempotent** if $n^2 = n$. An element $n \in N$ is said to be a **nilpotent** element if there exists a least positive integer k such that $n^k = 0$.

Example 2.1.21

(i) In the near ring $(\mathbb{Z}_6, +, \cdot)$, the element 3 is idempotent (because $3^2 = 3$).
(ii) In the near ring $(\mathbb{Z}_8, +, \cdot)$, the element 2 is nilpotent (because $2^3 = 8 = 0$).

Definition 2.1.22

An element $d \in N$ is said to be **distributive** if for all $m, n \in N, d(m + n) = dm + dn$. We denote $N_d = \{d \in N \mid d \text{ is distributive}\}$.

Example 2.1.23

 (i) If N is a ring, then $N = N_d$.

 (ii) If N has an identity 1, then $1 \in N_0$.

Proposition 2.1.24 (Proposition 1.13 of Pilz, 1983) (Pierce Decomposition)

If $e \in N$ is idempotent, then for any $n \in N$ there corresponds exactly one $n_0 \in \{x \in N \mid xe = 0\}$ and there corresponds exactly one $n_1 \in Ne = \{xe \mid x \in N\}$ such that $n = n_0 + n_1$. Taking $e = 0$, for any $n \in N$ there corresponds exactly one $n_0 \in N_0$ and exactly one $n_c \in N_c$ such that $n = n_0 + n_c$.

 Hence $(N, +) = (N_0, +) + (N_c, +)$ and $N_0 \cap N_c = \{0\}$.

Proof

Part 1: Suppose $e \in N$ is an idempotent element and let $n \in N$.

 Now $n = (n - ne) + ne$. Consider $(n - ne)e = ne - nee = ne - ne = 0$.

 So $(n - ne) \in \{x \in N \mid xe = 0\}$ and also $ne \in Ne$.

 Suppose that

$$n = n_0 + n_1 = n_0^1 + n_1^1 \tag{2.1}$$

where $n_0, n_0{}^1 \in \{x \in N \mid xe = 0\}$, $n_1, n_1{}^1 \in Ne$.

 Now

$$ne = (n_0 + n_1)e = (n_0^1 + n_1^1)e$$
$$\Rightarrow n_0 e + n_1 e = n_0^1 e + n_1^1 e$$
$$\Rightarrow n_1 e = n_1^1 e \tag{2.2}$$

 For $n_1, n_1^1 \in Ne$, there exists $y_1, y_1^1 \in N$ such that $n_1 = y_1 e$ and $n_1^1 = y_1^1 e$.

 Now $n_1 e = (y_1 e)e = y_1 ee = y_1 e = n_1$, and also $n_1^1 e = (y_1^1 e)e = y_1^1 e = n_1^1$. From Equation 2.2, we have $n_1 = n_1^1$. Also, from Equation 2.1, $n_0 = n_0^1$.

 Therefore, for all $n \in N$, there exists exactly one $n_0 \in \{x \in N \mid xe = 0\}$ and there exists exactly one $n_1 \in Ne$ such that $n = n_0 + n_1$.

Part 2: Suppose $e = 0$ and $n \in N$.

 From Part 1, we can have exactly one $n_0 \in \{x \in N \mid x0 = 0\}$ and exactly one $n_1 \in N_0$ such that $n = n_0 + n_1$. Now $n_0 \in \{x \in N \mid x0 = 0\} = N_0$. We show that $N_0 = N_c$.

 Let $n \in N_c$. Then $n0 = n$. This means that $n = n0 \in N_0$.

 Therefore, $N_c \subseteq N_0$. Let $y \in N_0$. Then there exists some $n^1 \in N$ such that $y = n^1 0$.

Now we show that $y \in N_c$.

Consider $y0 = (n^10)0 = n^1(00) = n^10 = y$.

Therefore, $y \in N_c$ and hence $N_0 = N_c$. Hence, for all $n \in N$, there exists exactly one $n_0 \in N_0$ and exactly one $n_c \in N_c$ such that $n = n_0 + n_c$.

Now we show that $(N, +) = (N_0, +) + (N_c, +)$ and $N_0 \cap N_c = \{0\}$.

From the preceding result, it follows that for each $n \in N$, there exists $n_0 \in N_0$ and $n_c \in N_c$ such that $n = n_0 + n_c$.

This implies

$$(N, +) \subseteq (N_0, +) + (N_c, +) \tag{2.3}$$

Let $n_0 \in N_0$ and $n_c \in N_c$. Now

$$(N_0, +) + (N_c, +) \subseteq (N, +) \tag{2.4}$$

From Equations 2.3 and 2.4, we have $(N, +) = (N_0, +) + (N_c, +)$.

Next we show that $N_0 \cap N_c = \{0\}$.

Clearly, $\{0\} \subseteq N_0$ and $\{0\} \subseteq N_c \Rightarrow \{0\} \subseteq N_0 \cap N_c$. Let $n \in N_0 \cap N_c$. This implies $n \in N_0$ and $n \in N_c$. This implies $n0 = 0$ and $n0 = n$, and so $n = 0$.

Therefore, $N_0 \cap N_c = \{0\}$. ∎

Definition 2.1.25

Let N be a near ring.

  (i) N is said to be an **Abelian near ring** if $a + b = b + a$ for all $a, b \in N$.

  (ii) N is said to be a **commutative near ring** if $a \cdot b = b \cdot a$ for all $a, b \in N$.

  (iii) If $N = N_d$, then we say that N is a **distributive near ring**.

  (iv) If all nonzero elements of N are left (right) cancelable, then we say that N fulfills the **left (right) cancellation** law.

  (v) N is said to be an **integral near ring** if N has no nonzero divisors of zero.

  (vi) If $(N^* = N\backslash\{0\}, .)$ is a group, then N is said to be a **near field**.

  (vii) A near ring with the property that the set N_d generates $(N, +)$ is called a **distributively generated near ring** (denoted by **dgnr**).

Example 2.1.26

$M(\Gamma)$ is Abelian if and only if Γ is Abelian.

Verification: Suppose $M(\Gamma)$ is Abelian. Let $a, b \in \Gamma$. Define the constant mappings $f_a\colon \Gamma \to \Gamma$ and $f_b\colon \Gamma \to \Gamma$ by $f_a(x) = a$, $f_b(x) = b$ for all $x \in \Gamma$. Clearly, $f_a, f_b \in M(\Gamma)$. Since $M(\Gamma)$ is Abelian, we have that $f_a + f_b = f_b + f_a$.

Now $(f_a + f_b)(x) = (f_b + f_a)(x)$ implies that $f_a(x) + f_b(x) = f_b(x) + f_a(x)$.

Therefore, $a + b = b + a$. This shows that Γ is Abelian. The converse is clear.

2.2 Substructures of Near Rings and Quotient Near Rings

Every algebraic system will have its own subsystem(s). Subgroups and normal subgroups are two different types of subsystems in the theory of groups. Subfields are the subsystems of fields. Subspaces are the subsystems of vector spaces. Subrings and ideals are the two different types of subsystems of rings. Submodules are the subsystems of modules. In a similar way, we study two different types of substructures of near rings, namely, subnear rings and ideals. In this section, we present the results related to these two subsystems.

Definition 2.2.1

(i) An additive subgroup M of a near ring N with $MM \subseteq M$ is called a **subnear ring** of N. It is denoted by $M \leq N$.
(ii) A subnear ring M of N is called **left invariant (right invariant,** respectively) if $MN \subseteq M$, $(NM \subseteq M$, respectively). If M is both left invariant and right invariant, then we say that it is **invariant.**

Definition 2.2.2

A normal subgroup I of $(N, +)$ is called

(i) a right ideal if $IN \subseteq I$
(ii) a left ideal if $n(m + i) - nm \in I$ for all $n, m \in N$ and for all $i \in I$
(iii) an ideal if it is both right and left ideal.

Example 2.2.3

(i) N_0 and N_c are subnear rings of N.
   Verification: First we show that N_0 is a subnear ring of N.
   We show that N_0 is a subgroup of N. Let $x, y \in N_0$. Then $x0 = 0$ and $y0 = 0$.
   Now $(x - y)0 = x0 - y0 = 0$. Therefore, $x - y \in N_0$.
   Therefore $(N_0, +)$ is a subgroup of $(N, +)$. Take $n_1, n_2 \in N_0$.
   Now $(n_1 n_2)0 = n_1(n_2 0) = n_1 0 = 0$.
   Therefore, $n_1 n_2 \in N_0$ and so $N_0 N_0 \subseteq N_0$. Hence, N_0 is a subnear ring of N.
   Next we show that N_c is a subnear ring of N.
   Let $x, y \in N_c$. This implies $(x - y)0 = x0 - y0 = x - y$. This means $x - y \in N_c$.
   So $(N_c, +)$ is a subgroup of $(N, +)$.
   Let $n_c, n_c^1 \in N_c$. This implies $(n_c n_c^1)0 = n_c(n_c^1 0) = n_c n_c^1$, and so $n_c n_c^1 \in N_c$.
   Hence, $N_c N_c \subseteq N_c$. Therefore, N_c is a subnear ring of N.

TABLE 2.3

Addition Table

+	0	1	2	3	4	5
0	0	1	2	3	4	5
1	1	0	3	2	5	4
2	2	4	0	5	1	3
3	3	5	1	4	0	2
4	4	2	5	0	3	1
5	5	3	4	1	2	0

TABLE 2.4

Multiplication Table

·	0	1	2	3	4	5
0	0	0	0	0	0	0
1	1	1	1	1	1	1
2	1	1	1	2	1	2
3	0	0	0	3	0	3
4	0	0	0	4	0	4
5	1	1	1	5	1	5

(ii) Let $N = \{0, 1, 2, 3, 4, 5\}$.

Definitions of the addition and multiplication operations on N are shown in Tables 2.3 and 2.4.
Then $(N, +, \cdot)$ is a near ring and $I = \{0, 3, 4\}$ is an ideal of N.

(ii) It is clear that every ideal of a near ring is a subnear ring, but the converse is not true in general.
(iii) Consider the two near rings $(\mathbb{Z}, +, \cdot)$ and $(\mathbb{Q}, +, \cdot)$, where $\mathbb{Z}$ and $\mathbb{Q}$ are the set of integers and the set of rational numbers, respectively. $\mathbb{Z}$ is a subnear ring of $\mathbb{Q}$. But $\mathbb{Z}$ is not an ideal of the near ring $\mathbb{Q}$.

Definition 2.2.4

A near ring N is called **simple** if N has no nontrivial ideals.

Example 2.2.5

(i) Let $N = \{0, 1, 2\}$. Define addition (+) and multiplication ($\cdot$) as in Tables 2.5 and 2.6.
Then $(N, +, \cdot)$ is a near ring. It is a simple near ring.
(ii) For any prime number p, the near ring $(\mathbb{Z}_p, +, \cdot)$ is a simple near ring.

TABLE 2.5

Addition Table

+	0	1	2
0	0	1	2
1	1	2	0
2	2	0	1

TABLE 2.6

Multiplication Table

·	0	1	2
0	0	0	0
1	0	1	1
2	0	0	2

Remark 2.2.6

If I, J are ideals of N, then

   (i) $I + J = \{a + b \mid a \in I, b \in J\}$ is an ideal of N
  (ii) $I \cap J$ is an ideal of N
 (iii) $I \cup J$ is an ideal of N, provided $I \subseteq J$ or $J \subseteq I$

Lemma 2.2.7

If I and J are two ideals, $a \in I, b \in J$, then

  (i) $a + b = c + a$ for some $c \in J$
 (ii) $a + b = b + d$ for some $d \in J$

Proof

  (i) $a + b = (a + b - a) + a = c + a$, where $c = a + b - a \in J$ and $a \in I$
 (ii) $a + b = b + (-b + a + b) = b + d$, where $d = -b + a + b \in I$ ■

Definition 2.2.8

Let $\{I_k\}_{k \in K}$ (here K is an index set) be a collection of ideals of a near ring N. The set A of all finite sums of elements from $\cup\, I_k$ is called the sum of the ideals $\{I_k\}_{k \in K}$, and the sum is denoted by $\sum_{k \in K} I_k$ or $I_1 + I_2 + \ldots$.

Definition 2.2.9

Let $\{I_k\}_{k \in K}$ be a collection of ideals of N. Their sum $\sum_{k \in K} I_k$ is called an (internal) direct sum of each element. $\sum_{k \in K} I_k$ has a unique representation as a finite sum of elements of different I_k's.

In this case, we write the sum $\sum_{k \in K} I_k$ (or $I_1 + I_2 + \ldots$).

Proposition 2.2.10 (Proposition 2.5 of Pilz, 1983)

For each family $\{I_k\}_{k \in K}$ of ideals of N, the following conditions are equivalent:

(i) The sum of the ideals I_k's is direct.

(ii) The sum of the normal subgroups $(I_k, +)$ is direct.

(iii) For all $k \in K$: $I_k \cap \left(\sum_{\substack{l \in k \\ l \neq k}} I_l \right) = \{0\}$.

Proof

Suppose the sum of ideals I_k's is direct.

Then the sum of the normal subgroup $(I_k, +)$ is direct.

Therefore, the condition (i) $\Rightarrow$ (ii) is proved.

Assume (ii).

Let $x \in I_k \cap \left(\sum_{l \in K} I_l \right)$; now x can be written as $x = x_{l_1} + x_{l_2} + \ldots + x_{l_n}$, where $l_1, l_2, \ldots, l_n$ are different from k and $x_{l_i} \in I_i$.

So, we can write x as $x = x_{l_1} + x_{l_2} + \ldots + x_{l_n} + 0_k$. Also, $x = 0_{l_1} + 0_{l_2} + \ldots + 0_{l_n} + x$.

Since the sum is direct, it follows that these two representations must be the same.

So, $x = 0 \Rightarrow I_k \cap \sum I_l = \{0\}$.

Therefore, the condition (ii) $\Rightarrow$ (iii) is proved.

Assume the condition (iii).

Now we have to show that the sum of I_k's is direct. Suppose some x in the sum has the following representation:

$$x = x_1 + x_2 + \ldots + x_n = y_1 + y_2 + \ldots + y_n$$

(Without loss of generality, we may assume that indices are the same for both representations by adding zero if necessary, at appropriate places.)

Now $x_1 - y_1 = (y_2 + y_3 + \ldots + y_n) - (x_2 + x_3 + \ldots + x_n)$

$\Rightarrow x_1 - y_1 = y_2 - x_2 + y_3 - x_3 + \ldots + y_n - x_n \in I_1 \cap \sum_{\substack{l \in K \\ l \neq 1}} I_l = \{0\}$.

This implies $x_1 = y_1$. Similarly, we can prove that $x_i = y_i$ for all i. Therefore, x has a unique representation. Hence, the sum is direct. ∎

Proposition 2.2.11 (Proposition 2.6 of Pilz, 1983)

Let $\sum_{k \in K} I_k$ be direct, $a, a' \in I_i, b, b' \in I_j, i \neq j$. Then we have the following:

(i) $a + b = b + a$
(ii) $a'(a + b) = a'a$
(iii) $ab = a0$
(iv) If $N = N_0$, then $ab = 0$

Proof

(i) Since I_i, I_j are ideals, we have $a + b - a - b \in I_i$, and $a + b - a - b \in I_j$. Since the sum is direct, we have $J_i \cap I_j = \{0\}$, and so $a + b - a - b \in I_i \cap I_j = \{0\}$, which implies that $a + b = b + a$.

(ii) $a'(a + b) - a'a \in I_j$ (since I_j is an ideal). Also, since $a'(a + b) \in I_i$ and $a'a \in I_i$, and I_i is an ideal, we have $a'(a + b) - a'a \in I_i$. Therefore, $a'(a + b) - a'a \in I_i \cap I_j = \{0\}$.
Hence, $a'(a + b) = a'a$.

(iii) $ab - a0 = a(b + 0) - a0 \in I_i \cap I_j = \{0\}$. This implies $ab = a0$.

(iv) Suppose $N = N_0$. Then for all $a \in N$, $a0 = 0$. Now $ab = a0 = 0$. This implies $ab = 0$. ∎

Remark 2.2.12

Let N be a near ring and I an ideal of N.

(i) Define a relation $\sim$ on N as $a \sim b$ if and only if $a - b \in I$ for all $a, b \in N$. Now we verify that this relation is an equivalence relation.
Reflexive: Since $a - a = 0 \in I$, we have that $a \sim a$.
Symmetric: Suppose $a \sim b$. Then $a - b \in I \Rightarrow b - a \in I \Rightarrow b \sim a$.
Transitive: Suppose $a \sim b, b \sim c$. Then $a - b \in I, b - c \in I \Rightarrow a - c = (a - b) + (b - c) \in I$, which implies that $a \sim c$.
Therefore, the relation $\sim$ is an equivalence relation on N.

(ii) Write $a + I = \{a + x \mid x \in I\}$ for $a \in N$.
Now $a + I$ is the equivalence class containing a.

(iii) Write $N/I = \{a + I/a \in N\}$, the set of all equivalence classes.

(iv) Define $+$ and $\cdot$ on N/I as $(a + I) + (b + I) = (a + b) + I$ and $(a + I) \cdot (b + I) = (a \cdot b) + I$. Then it follows that $(N/I, +, \cdot)$ is a near ring. In this near ring, $0 + I$ is the additive identity and $(-a) + I$ is the additive inverse of $a + I$. This near ring N/I is called the quotient near ring of N modulo I. Now we have the following conclusions.

Theorem 2.2.13

If N is a near ring and I is an ideal of N, then N/I is also a near ring.

Result 2.2.14

The ideals of N/I are of the form J/I, where J is an ideal of N and $I \subseteq J \subseteq N$.

Sketch of Proof: Let K be an ideal of N/I. Write $J = \{j \mid j + I \in K\}$. It is easy to verify that J is an ideal of N such that $I \subseteq J$. Also, $J/I = \{j + I \mid j \in J\} = K$.

2.3 Homomorphism and Isomorphism

In this section, we define homomorphism between near rings, and we present fundamental homomorphism theorems.

Definition 2.3.1

Let N and N^1 be near rings. A mapping $h: N \to N^1$ is called

   (i) a **homomorphism** (or **near ring homomorphism**)
       if $h(m + n) = h(m) + h(n)$ and $h(mn) = h(m)h(n)$ for all $m, n \in N$.
       Moreover, if $N = N^1$, then a homomorphism h is called an **endomorphism**. The set of all endomorphisms of N is denoted by $\text{End}(N)$.
  (ii) a **monomorphism** if h is a homomorphism and one–one.
 (iii) an **epimorphism** if h is a homomorphism and onto.
  (iv) an **isomorphism** if h is a homomorphism, one–one, and onto.
   (v) an **automorphism** if h is an isomorphism and $N = N^1$. The set of all automorphisms of N is denoted by $\text{Aut}(N)$.

Theorem 2.3.2 (Homomorphism Theorem for Near Rings, Pilz, 1983)

   (i) If I is an ideal of N, then the canonical mapping $\pi: N \to N/I$ (defined by $\pi(n) = n + I$) is a near ring epimorphism. Also, N/I is a homomorphic image of N.

(ii) Conversely, if $h: N \rightarrow N^1$ is an epimorphism, then ker(h) is an ideal of N, and $N/\mathrm{ker}(h) \cong N^1$.

Proof

(i) Suppose I is an ideal of N. Define the canonical mapping $\pi: N \rightarrow N/I$ by $\pi(n) = n + I$.

Now we show that π is an epimorphism.

Let $n_1, n_2 \in N$.

Now $\pi(n_1 + n_2) = (n_1 + n_2) + I$

$$= (n_1 + I) + (n_2 + I)$$
$$= \pi(n_1) + \pi(n_2).$$

And $\pi(n_1 n_2) = (n_1 n_2) + I$

$$= (n_1 + I)(n_2 + I)$$
$$= \pi(n_1)\pi(n_2).$$

Therefore, π is a near ring homomorphism.

Let $x \in N/I$. Then there exists $n_1 \in N$ such that $x = n_1 + I$.

Now $\pi(n_1) = n_1 + I = x$. So π is onto. Thus, we verified that $\pi: N \rightarrow N/I$ is a near ring epimorphism. Hence, N/I is a homomorphic image of N.

(ii) Suppose $h: N \rightarrow N^1$ is an epimorphism.

Part 1: Now we show that ker(h) is an ideal of N and $N/\mathrm{ker}(h) \cong N^1$.

First we show that ker(h) is a normal subgroup of N.

Clearly, $0 \in \mathrm{ker}(h)$, and so $\mathrm{ker}(h) \neq \emptyset$.

Let $n \in N$ and $x \in \mathrm{ker}(h)$.

Now $h(n + x - n) = h(n) + h(x) - h(n)$ (since h is a homomorphism)

$$= h(n) + 0 - h(n) \qquad (\text{since } h(x) = 0)$$
$$= h(n) - h(n) = 0.$$

This shows that $n + x - n \in \mathrm{ker}(h)$.

Part 2: Now we show that ker(h) is a right ideal of N.

Let $n \in N$ and $x \in \mathrm{ker}(h)$.

Now $h(xn) = h(x)h(n) = 0h(n) = 0$. Therefore, $xn \in \mathrm{ker}(h)$. Hence (ker(h)) $N \subseteq \mathrm{ker}(h)$.

Part 3: In this part, we show that ker(h) is a left ideal of N.

Let $n, n^1 \in N$ and $x \in \mathrm{ker}(h)$.

Now $h(n(n^1 + x) - nn^1) = h(n)h(n^1 + x) - h(n)h(n^1)$

$$= h(n)(h(n^1) + h(x)) - h(n)h(n^1)$$
$$= h(n)(h(n^1) + 0) - h(n)h(n^1)$$
$$= h(n)h(n^1) - h(n)h(n^1) = 0.$$

Therefore, $n(n^1 + x) - nn^1 \in \mathrm{ker}(h)$. Hence, ker($h$) is an ideal of N.

Part 4: Define a mapping $\varphi: N/\mathrm{ker}(h) \rightarrow N^1$ as follows:

$\varphi(n + \mathrm{ker}(h)) = h(n)$ for all $n + \mathrm{ker}(h) \in N/\mathrm{ker}(h)$. Now we show that φ is a near ring isomorphism.

Let $n_1 + \ker(h)$, $n_2 + \ker(h) \in N/\ker(h)$.
Now, $\varphi(n_1 + \ker(h)) = \varphi(n_2 + \ker(h))$
　　if and only if $h(n_1) = h(n_2)$
　　if and only if $h(n_1) - h(n_2) = 0$
　　if and only if $h(n_1 - n_2) = 0$
　　if and only if $(n_1 - n_2) \in \ker(h)$
　　if and only if $n_1 + \ker(h) = n_2 + \ker(h)$.
Therefore, φ is one–one and well-defined.
Let $n^1 \in N^1$. Since h is onto, there exists $n \in N$ such that $h(n) = n^1$.
Now $\varphi(n + \ker(h)) = h(n) = n^1$, and so φ is onto.
Part 5: Let $n_1 + \ker(h)$, $n_2 + \ker(h) \in N/\ker(h)$.
Now $\varphi((n_1 + \ker(h)) + (n_2 + \ker(h)))$

$$= \varphi((n_1 + n_2) + \ker(h))$$
$$= h(n_1 + n_2)$$
$$= h(n_1) + h(n_2) \quad \text{(since } h \text{ is a homomorphism)}$$
$$= \varphi(n_1 + \ker(h)) + \varphi(n_2 + \ker(h)).$$

Also, $\varphi((n_1 + \ker(h))(n_2 + \ker(h))) = \varphi((n_1 n_2) + \ker(h))$
$$= h(n_1 n_2)$$
$$= h(n_1)h(n_2)$$
$$= \varphi(n_1 + \ker(h))\varphi(n_2 + \ker(h)).$$

So, φ is a near ring homomorphism.
Hence, $\varphi: N/\ker(h) \to N^1$ is a near ring isomorphism.

Definition 2.3.3

A proper ideal I of N is called **maximal** if $I \subseteq J \subseteq N$, and J is an ideal of N implies that either $I = J$ or $J = N$.

Proposition 2.3.4

An ideal I of N is maximal if and only if N/I is simple.

Proof

Suppose I is a proper ideal of N, and suppose I is a maximal ideal in N. Let K be an ideal of N/I. To show N/I is simple, we have to show that either $K = (0)$ or $K = N/I$. Then $K = J/I$ for some ideal J of N such that $I \subseteq J \subseteq N$. Since $I \subseteq J \subseteq N$ and I is a maximal ideal of N, we have that $I = J$ or $J = N$. This implies that $J/I = I$ or $J/I = N/I$. Therefore, $K = J/I = 0 + I$, the zero element in N/I or $K = N/I$. Hence, N/I is a simple near ring.

Converse

Suppose that N/I is simple. We prove that I is a maximal ideal of N.

Let J be any ideal of N such that $I \subseteq J$. This implies that J/I is an ideal of N/I. Since N/I is simple, we have either $J/I = I$ or $J/I = N/I$. Next we show that $J/I = I \Rightarrow J = I$. Suppose $J/I = I$.

It is obvious that $I \subseteq J$. Take $x \in J$. This implies $x + I \in J/I = I \Rightarrow x \in I$. Now we verified that $J/I = I \Rightarrow J = I$.

Next we show that $J/I = N/I \Rightarrow J = N$. Suppose that $J/I = N/I$. Since J is an ideal of N, we have $J \subseteq N$.

Take $n \in N$. Now $n + I \in N/I = J/I \Rightarrow n + I = i_1 + I$ for some $i_1 \in J$.

This implies that $n - i_1 \in I \subseteq J$, and hence $n - i_1 + i_1 \in J$. So, $n \in J$.

Now we verified that either $J = I$ or $J = N$. Hence, I is a maximal ideal of N.   ∎

Theorem 2.3.5 (First Isomorphism Theorem) (Pilz, 1983)

If I and J are ideals of N, then $I \cap J$ is an ideal of J and $I + J / I \cong J / I \cap J$.

Sketch: We know that $I \cap J$ is an ideal of J. The mapping $\varphi: I + J \to J / I \cap J$ defined by $\varphi (a + b) = b + (I \cap J)$ is an epimorphism, and $\ker(\varphi) = I$. Hence, $I + J / I \cong J / I \cap J$.

Definition 2.3.6

An ideal I of N is called a **direct summand** (of N) if there exists an ideal J of N such that $N = I \oplus J$. In this case, the ideal J is called a **direct complement** of I in N.

Example 2.3.7

Consider the near ring $\mathbb{Z}_6 = \{0, 1, 2, 3, 4, 5\}$. Then $I = \{0, 3\}$ and $J = \{0, 2, 4\}$ are ideals of $\mathbb{Z}_6$. It is clear that $I \cap J = \{0\}$, and $I + J = \{x + y \mid x \in I, y \in J\} = \{0, 2, 4, 3, 5, 1\} = \mathbb{Z}_6$. Hence, the ideal I is a direct summand, and the ideal J is the complement of I in the near ring $\mathbb{Z}_6$. Also, the sum $I + J$ is a direct sum.

Proposition 2.3.8

An ideal of N is a direct summand if and only if for all $\alpha \in \mathrm{Aut}(I)$, α can be extended to an epimorphism of N onto I.

Proof

Suppose I is a direct summand and J is its direct complement. Then $N = I + J$.

Let $\alpha \in \text{Aut}(I)$. Then $\alpha: I \to I$ is an isomorphism.

Define $\bar{\alpha}: N \to I$ by $\bar{\alpha}(n) = \bar{\alpha}(i + j) = \alpha(i)$.

Now $\bar{\alpha}$ is an epimorphism of N onto I. Let $n_1, n_2 \in N$.

$\bar{\alpha}(n_1 + n_2) = \bar{\alpha}(i_1 + j_1 + i_2 + j_2) = \bar{\alpha}(i_1 + j_1 + i_2 + j_2) = \alpha(i_1 + i_2) = \alpha(i_1) + \alpha(i_2) = \bar{\alpha}(i_1 + j_1) + \bar{\alpha}(i_2 + j_2) = \bar{\alpha}(n_1) + \bar{\alpha}(n_2)$ $\bar{\alpha}(n_1 n_2) = \bar{\alpha}((i_1 + j_1)(i_2 + j_2)) = \bar{\alpha}(i_1 i_2 + j_1 j_2) = \alpha(i_1 i_2) = \alpha(i_1)\,\alpha(i_2) = \bar{\alpha}\,(i_1 + j_1)\,\alpha(i_2 + j_2) = \bar{\alpha}(n_1)\,\bar{\alpha}(n_2)$.

Let $i \in I$; then there exists $i' \in I$ such that $\alpha(i') = i$.

Consider $i' = i' + 0 \in N$ and $\bar{\alpha}(i') = \alpha(i' + 0) = \alpha(i') = i$.

Therefore, $\bar{\alpha}$ is an epimorphism.

Conversely, assume that α is the identity map of I onto I. By hypothesis, α can be extended to an epimorphism of N onto I, say $\bar{\alpha}$.

Let J be the kernel of $\bar{\alpha}$. Now J is an ideal of N.

Take $x \in I \cap J$. This implies $x \in I$ and $\bar{\alpha}(x) = 0$. Also,

$x \in I \Rightarrow \alpha(x) = \bar{\alpha}(x) \Rightarrow \alpha(x) = 0 \Rightarrow x = \alpha(x) = 0$ (since $\alpha: I \to I$ is the identity map).

Therefore, $I \cap J = \{0\}$.

Let $n \in N$. To show that $n - \bar{\alpha}(n) \in J$, consider $\bar{\alpha}(n - \bar{\alpha}(n))$. Now $\bar{\alpha}(n - \bar{\alpha}(n)) = \bar{\alpha}(n) - \bar{\alpha}(\alpha(n)) = \bar{\alpha}(n) - \bar{\alpha}(n)$ (since $\bar{\alpha}(n) \in I$ and $\alpha = \bar{\alpha}$ on I) $= 0$. This implies $n - \bar{\alpha}(n) \in J$, and so $n = n - J(n) \in J + I$. Hence, $N = I \oplus J$. ∎

Theorem 2.3.9

Let I be an ideal of N. If I is a direct summand, then each ideal of I is an ideal of N.

Proof

Suppose that the ideal I of N is a direct summand. Let I_1 be its direct complement.

Let J be any ideal of I; then J is a normal subgroup of I.

To show J is normal in N, let $x \in J$ and $n \in N$. Since $N = I + I_1$, we have $n = i + i_1$ for some $i \in I$, $i_1 \in I_1$. Now

$n + x - n = (i + i_1) + x - (i + i_1)$

$\qquad = i + i_1 + x - i_1 - i$ (where $x \in J \subseteq I$ and $i_i \in I_1$; also, since $I \cap I_1 = \{0\}$,

$\qquad\qquad$ we have $x + i_1 = i_1 + x$)

$\qquad = i + x + i_1 - i_1 - i$

$\qquad = i + x - i$ (since J is normal in I).

Let $n, n' \in N$ and $x \in J$. Now $n = i + i_1$ and $n' = i' + i_1'$.

Consider $n(n' + x) - nn'$.

Now

$$n(n'+x) - nn' = (i + i_1)(i' + i_1' + x) - (i + i_1)(i' + i_1')$$

$$= i(i' + i_1' + x) + i_1(i' + i_1' + x) - (i(i' + i_1') + i_1(i' + i_1'))$$

$$= i(i' + x + i_1') + i_1(i' + x + i_1') - (i(i' + i_1') + i_1(i' + i_1'))$$

$$= i(i' + x) + i_1 i_1' - (ii' - ii')$$

(since $I \oplus J$, we have $a'(a+b) = a'a$, for a, $a' \in J$, $b \in J$)

$$= i(i' + x) - ii' \in J.$$

Let $x \in J$ and $n \in N$. Then $n = i + i_1$.

Since J is an ideal of I, we have that $xn = x(i + i_1) = xi \in J$.

Therefore, J is a right ideal of N. Hence, J is an ideal of N.　∎

2.4 Introduction to Matrix Near Rings

Matrix near rings are introduced in Meldrum and Van der Walt (1986). The matrix near ring of a given near ring N is denoted by $M_n(N)$. Some results about the correspondence between the two-sided ideals in the given near ring N and the two-sided ideals in the matrix near ring $M_n(N)$ are studied by Meldrum and Van der Walt (1986). In this section, we provide some elementary results on matrix near rings from the works of Meldrum and Van der Walt (1986).

Let n be an arbitrary natural number and $(N, +, \cdot)$ be a near ring with identity 1. We denote

N^n as the direct sum of n-copies of $(N, +)$.

Define

$$e_i = \left(0, \ldots, \underset{ith}{1}, \ldots, 0\right).$$

The injection mapping $i_j \colon N \to N^n$ is defined as

$$i_j(a) = \left(0, \ldots, \underset{ith}{a}, \ldots, 0\right).$$

The projection mapping $\pi_j \colon N^n \to N$ is defined by $\pi_j(a_1, \ldots, a_n) = a_j$. Define $f^r \colon N \to N$ as $f^r(s) = rs$ for all $s \in N$. Now define $f_{ij}^r \colon N^n \to N^n$ by $f_{ij}^r = i_i f^r \pi_j$ for all $1 \le i, j \le n$. Now

$$f_{ij}^r(a_1, \ldots, a_n) = i_i f^r \pi_j(a_1, \ldots, a_n) = i_i f^r(a_j) = i_i(ra_j) = \left(0, \ldots, \underset{ith}{ra_j}, \ldots, 0\right).$$

Definition 2.4.1

The near ring of $n \times n$ matrices over N is the subnear ring of $M(N^n)$ generated by the set $\{f_{ij}^r : r \in N, 1 \leq i, j \leq n\}$, and it is denoted by $M_n(N)$.

Some authors such as Booth and Groenewald (1991b), Groenewald (1989), Meldrum and Van der Walt (1986), and Satyanarayana, Lokeswara Rao, and Prasad (1996), Meldrum and Meyer (1996) have studied prime ideals and related concepts in matrix near rings.

Proposition 2.4.2 (Meldrum and Van der Walt, 1986)

The matrix near ring $M_n(N)$ is a right near ring with identity.

Proof

Since $M_n(N)$ is the subnear ring of the right near ring $M(N^n)$, we have that $M_n(N)$ is a right near ring. Since $f_{ii}^1 \in M_n(N)$, it follows that $f_{11}^1 + f_{22}^1 + \ldots + f_{nn}^1 \in M_n(N)$.

For any $(a_1, \ldots, a_n) \in N^n$, we have

$$(f_{11}^1 + f_{22}^1 + \ldots + f_{nn}^1)(a_1, \ldots, a_n) = f_{11}^1(a_1, \ldots, a_n) + \ldots + f_{nn}^1(a_1, \ldots, a_n)$$
$$= (a_1, \ldots, a_n) = I(a_1, \ldots, a_n),$$

where $I : N^n \to N^n$ is the identity map.

This implies $I = f_{11}^1 + f_{22}^1 + \ldots + f_{nn}^1$, and it acts as an identity in $M_n(N)$.

Hence, $M_n(N)$ is a right near ring with identity. ∎

Proposition 2.4.3 (Meldrum and Van der Walt, 1986)

If N is a ring with identity, then $M_n(N)$ is the ring of $n \times n$ matrices over N.

Proof

Suppose N is a ring. Let N^1 be the ring of $n \times n$ matrices over N. Then we may consider every element of N^1 as a mapping from $N^n \to N^n$. Since $M(N^n)$ is a near ring and N^1 is a ring with respect to the same operations, we have that N^1 is a subnear ring of $M(N^n)$.

Therefore, $N^1 \subseteq M(N^n)$.

Now each f_{ij}^r is a matrix of order $n \times n$, and we have $f_{ij}^r \in N^1$ for all $r \in N$, $1 \leq i, j \leq n$.

Therefore, $\{f_{ij}^r : r \in N, 1 \leq i, j \leq n\} \subseteq N^1$.

Therefore, $M_n(N) = \{f_{ij}^r : r \in N, 1 \leq i, j \leq n\} \subseteq N^1$.

Therefore, $M_n(N) \subseteq N^1$. Let $A = [a_{ij}]_{n \times n} \in N^1$. Then

$$A = [a_{ij}]_{n \times n} = \sum_{i,j} f_{ij}^{a_{ij}} \in \{f_{ij}^r : r \in N\} \subseteq M_n(N).$$

Therefore, $N^1 \subseteq M_n(N)$. Hence, $N^1 = M_n(N)$. ∎

Definition 2.4.4

For $1 \leq i, j \leq n$, f_{ij}^1 are defined as the **matrix units**, and f_{ij}^1 is denoted by E_{ij}.

Definition 2.4.5

The **identity matrix** I is defined as $E_{11} + E_{22} + \dots + E_{nn}$.

Definition 2.4.6

The ith row of matrix A is the function $\pi_i A : N^n \to N$. It is denoted by $A(i)$.

Definition 2.4.7

The product of a scalar $r \in N$ (scalar on the left of a matrix A) and the given matrix A are defined by $\sum_{i=1}^n i_i f^r A(i)$.

Remark 2.4.8

(i) If N is zero-symmetric, then $rE_{ij} = f_{ij}^r$ for all $r \in N$.
     [**Verification:** Take $r \in N$. Then $rE_{ij} (a_1, \dots, a_n) = r f_{ij}^1 (a_1, \dots, a_n) = r(0, \dots, a_j, 0, \dots, 0)$ (here a_j is in ith place) $= (r0, \dots, ra_j, \dots r0) = (0, \dots, ra_j, \dots 0)$ (since N is zero-symmetric) $= f_{ij}^r (a_1, \dots, a_n)$.]
(ii) If N is not zero-symmetric, then rE_{ij} is not equal to f_{ij}^r in general is not true.
     [**Verification:** Since N is not zero-symmetric, there exists $r \in N$ such that $r0 \neq 0$.
     Now $rE_{ij}(a_1, \dots, a_n) = r f_{ij}^1(a_1, \dots, a_n) = r(0, \dots, a_j, 0, \dots, 0)$ (here a_j is in ith place) $= (r0, \dots, ra_j, \dots r0) \neq (0, \dots, ra_j, \dots 0)$ (since $r0 \neq 0$) $= f_{ij}^r(a_1, \dots, a_n)$.]

Definition 2.4.9

Scalar multiplication on the right of a matrix A by an element $r \in N$ is defined by $Ar = A(f_{11}^r + f_{22}^r + \ldots + f_{nn}^r)$.

Definition 2.4.10

A matrix of the form $E_{11}r_1 + E_{22}r_2 + \ldots + E_{nn}r_n$ is called a **diagonal matrix**.

Definition 2.4.11

A matrix of the form $E_{11}r + \ldots + E_{nn}r$ is called a **scalar matrix**.

We now provide some useful matrix calculations with proofs.

Lemma 2.4.12 (Meldrum and Van der Walt, 1986)

$f_{ij}^r + f_{ij}^s = f_{ij}^{r+s}$, where r and s are positive integers and $1 \le j \le n$.

Proof

Take $(a_1, \ldots, a_n) \in N^n$. Now
$$(f_{ij}^r + f_{ij}^s)(a_1, \ldots, a_n) = f_{ij}^r(a_1, \ldots, a_n) + f_{ij}^s(a_1, \ldots, a_n)$$
$$= f_{ij}^r(a_1, \ldots, a_n) + f_{ij}^s(a_1, \ldots, a_n)$$
$$= i_i f^r \pi_j(a_1, \ldots, a_n) + i_i f^s \pi_j(a_1, \ldots, a_n)$$
$$= i_i f^r(a_j) + i_i f^s(a_j)$$
$$= \left(0, \ldots, \underset{\text{ith}}{ra_j}, \ldots, 0 \right) + \left(0, \ldots, \underset{\text{ith}}{sa_j}, \ldots, 0 \right)$$
$$= \left(0, \ldots, \underset{\text{ith}}{ra_j + sa_j}, \ldots, 0 \right) = \left(0, \ldots, \underset{\text{ith}}{(r+s)a_j}, \ldots, 0 \right)$$
$$= i_i f^{r+s}(a_j) = i_i f^{s+s} \pi_j(a_1, \ldots, a_n) = f_{ij}^{r+s}(a_1, \ldots, a_n). \qquad \blacksquare$$

Lemma 2.4.13 (Meldrum and Van der Walt, 1986)

$f_{ij}^r + f_{kl}^s = f_{kl}^s + f_{ij}^r$ if $i \ne k$, where r and s are positive integers, $1 \le i, j, k, l \le n$.

Proof

$$(f_{ij}^r + f_{kl}^s)(a_1,\ldots,a_n) = f_{ij}^r(a_1,\ldots,a_n) + f_{kl}^s(a_1,\ldots,a_n)$$

$$= \left(0,\ldots,\underset{ith}{ra_j},\ldots,0\right) + \left(0,\ldots,\underset{kth}{sa_l},\ldots,0\right)$$

$$= \left(0,\ldots,\underset{ith}{ra_j},\ldots,\underset{kth}{sa_l},\ldots,0\right) \text{ if } i<k.$$

Similarly, we can verify that

$$(f_{kl}^s + f_{ij}^r)(a_1,\ldots,a_n) = \left(0,\ldots,\underset{ith}{ra_j},\ldots,\underset{kth}{sa_l},\ldots,0\right)$$

if $i < k$.   ■

Lemma 2.4.14 (Meldrum and Van der Walt, 1986)

$$f_{ij}^r\, f_{kl}^s = \begin{cases} f_{il}^{rs} & \text{if } j = k \\ f_{il}^{r0} & \text{if } j \neq k \end{cases}, \text{ where } r \text{ and } s \text{ are positive integers, } 1 \leq i, j, k, l \leq n.$$

Proof

Suppose $i = j$. Now

$$(f_{ij}^r\, f_{kl}^s)(a_1,\ldots,a_n) = f_{ij}^r(f_{kl}^s(a_1,\ldots,a_n))$$

$$= f_{ij}^r\left(0,\ldots,\underset{kth}{sa_l},\ldots,0\right)$$

$$= f_{ij}^r\left(0,\ldots,\underset{jth}{sa_l},\ldots,0\right) (\text{since } j = k)$$

$$= i_i f^r \pi_j (0,\ldots,sa_l,\ldots 0)$$

$$= i_i f^r(sa_l) = i_i r(sa_l)$$

$$= i_i (rs)a_l$$

$$= \left(0,\ldots,\underset{ith}{(rs)a_l},\ldots,0\right)$$

$$= f_{il}^{rs}(a_1,\ldots,a_n).$$

Suppose $j \neq k$. $(f_{ij}^r \, f_{kl}^s)(a_1,\ldots,a_n) = f_{ij}^r \, (f_{kl}^s \, (a_1,\ldots,a_n))$

$$= f_{ij}^r \left(0,\ldots,\underset{kth}{sa_l},\ldots,0\right)$$

$$= i_i f^r \pi_j (0,\ldots,sa_l,\ldots 0)$$

$$= i_i f^r (0) \text{ (since } j \neq k)$$

$$= i_i(r0) = \left(0,\ldots,\underset{ith}{r0},\ldots,0\right).$$

$$f_{il}^{r0} \, (a_1,\ldots,a_n) = i_i f^{r0} \pi_l (a_1,\ldots,a_n) = i_i f^{r0} \, (a_l) = i_i f^{r0}(a_l)$$

$$= i_i(r0(a_l)) = i_i(r0) = \left(0,\ldots,\underset{ith}{r0},\ldots,0\right).$$

The proof is complete. ∎

Lemma 2.4.15 (Meldrum and Van der Walt, 1986)

$$f_{ij}^r \left(f_{1k_1}^{r_1} + f_{2k_2}^{r_2} + \ldots + f_{nk_n}^{r_n}\right) = f_{ij}^r \left(f_{jk_j}^{r_j}\right) = f_{ik_j}^{rr_j}.$$

Proof

$$f_{ij}^r \left(f_{1k_1}^{r_1} + f_{2k_2}^{r_2} + \ldots + f_{nk_n}^{r_n}\right)(a_1,\ldots,a_n) = f_{ij}^r \left(f_{1k_1}^n(a_1,\ldots,a_n) + \ldots + f_{nk_n}^n \, (a_1,\ldots,a_n)\right)$$

$$= f_{ij}^r \left((r_1 a_{k_1}, 0,\ldots,0) + \ldots + (0,\ldots,r_n a_{k_n})\right)$$

$$= f_{ij}^r \left((r_1 a_{k_1},\ldots,r_n a_{k_n})\right)$$

$$= \left(0,\ldots,\underset{ith}{rr_j a_{k_j}},\ldots,0\right)$$

$$= f_{ik_j}^{rr_j} \, (a_1,\ldots,a_n).$$

This is true for all $(a_1,\ldots,a_n) \in N^n$. ∎

Lemma 2.4.16 (Meldrum and Van der Walt, 1986)

An element $r \in N$ is zero-symmetric if and only if f_{ij}^r is zero-symmetric in $M_n(N)$.

Proof

Suppose r is zero-symmetric in N. This implies $r0 = 0$. Consider $f_{ij}^r O$ where O is the zero element in $M_n(N)$.

Now $f_{ij}^r O(a_1, \ldots, a_n) = f_{ij}^r (0, \ldots, 0) = \left(0, \ldots, \underset{ith}{r0}, \ldots, 0\right) = (0, \ldots, 0) = O(0, \ldots, 0)$.

Therefore, $f_{ij}^r O = 0$. Hence, f_{ij}^r is zero-symmetric in $M_n(N)$.

Converse

Suppose f_{ij}^r is zero-symmetric in $M_n(N)$.
  This implies $f_{ij}^r O = O$,

$\Rightarrow f_{ij}^r O(a_1, \ldots, a_n) = O(a_1, \ldots, a_n)$

$\Rightarrow \left(0, \ldots, \underset{ith}{r0}, \ldots, 0\right) = (0, \ldots, 0)$, and so $r0 = 0$.

Therefore, r is zero-symmetric in N.   ∎

Theorem 2.4.17 (Meldrum and Van der Walt, 1986)

An element r is distributive in N if and only if f_{ij}^r is distributive in $M_n(N)$.

Proof

Suppose r is distributive in N. Then $r(s + t) = rs + rt$ for all $s, t \in N$.
  To show f_{ij}^r is distributive in $M_n(N)$,
  Let $A, B \in M_n(N)$. Now

$$f_{ij}^r (A + B)X = f_{ij}^r \left(\sum_{i=1}^{n} i_i A(i) + \sum_{i=1}^{n} i_i B(i) \right) X$$

$$= f_{ij}^r \left(\sum_{i=1}^{n} i_i A(i)X + \sum_{i=1}^{n} i_i B(i)X \right)$$

$$= (0, \ldots, rA(j)X + rB(j)X, \ldots, 0).$$

Now $(f_{ij}^r A + f_{ij}^r B)X = f_{ij}^r AX + f_{ij}^r BX$

$$= f_{ij}^r \sum_{i=1}^{n} i_i A(i)X + f_{ij}^r \sum_{i=1}^{n} i_i B(i)X$$

$$= (0, \ldots, rA(j)X + rB(j)X, \ldots 0).$$

Therefore, $f_{ij}^r (A + B)X = (f_{ij}^r A + f_{ij}^r B)X$.

Converse

Suppose that f_{ij}^r is distributive in $M_n(N)$.

Consider $f^r_{ij}(f^s_{ji} + f^t_{ji})(1,\ldots,1) = f^r_{ij}(f^s_{ji}(1,\ldots,1) + f^t_{ji}(1,\ldots,1))$

$$= f^r_{ij}\left(\left(0,\ldots,\underset{jth}{s},\ldots,0\right) + \left(0,\ldots,\underset{jth}{t},\ldots,0\right)\right)$$

$$= f^r_{ij}\left(0,\ldots,\underset{jth}{s+t},\ldots,0\right)$$

$$= \left(0,\ldots,\underset{jth}{r(s+t)},\ldots,0\right).$$

On the other hand,

$$(f^r_{ij}f^s_{ji} + f^r_{ij}f^t_{ji})(1,\ldots,1) = f^r_{ij}f^s_{ji}(1,\ldots,1) + f^r_{ij}f^t_{ji}(1,\ldots,1)$$

$$= f^{rs}_{ii}(1,\ldots,1) + f^{rt}_{ii}(1,\ldots,1)$$

$$= \left(0,\ldots,\underset{ith}{rs},\ldots,0\right) + \left(0,\ldots,\underset{ith}{rt},\ldots,0\right)$$

$$= \left(0,\ldots,\underset{ith}{rs+rt},\ldots,0\right).$$

Therefore, $r(s+t) = rs + rt$. Hence, r is distributive in N. ∎

Theorem 2.4.18 (Meldrum and Van der Walt, 1986)

An element $r \in N$ is constant if and only if f^r_{ij} is constant in $M_n(N)$.

Proof

Suppose r is a constant element in N. This implies $r0 = r$. We show that f^r_{ij} is constant in $M_n(N)$.

Consider $(f^r_{ij}O)X = f^r_{ij}(OX)$

$$= f^r_{ij}\left(\sum_{i=1}^{n} O(i)\right) = f^r_{ij}(0,\ldots,0)$$

$$= \left(0,\ldots,\underset{ith}{r0},\ldots,0\right) = \left(0,\ldots,\underset{ith}{r},\ldots,0\right).$$

$$f^r_{ij}(X) = f^r_{ij}(a_1,\ldots,a_n)$$

$$= i_i f^r \pi_j (a_1,\ldots,a_n) = \left(0,\ldots,\underset{ith}{ra_j},\ldots,0\right)$$

$$= \left(0,\ldots,\underset{ith}{r},\ldots,0\right).$$

Therefore $(f^r_{ij}O)(X) = (f^r_{ij})X$ for all $X \in N^n$. Hence, $f^r_{ij}O = f^r_{ij}$. Hence, f^r_{ij} is constant in $M_n(N)$.

Converse

Suppose that f_{ij}^r is constant in $M_n(N)$. This means that $f_{ij}^r O = f_{ij}^r$. This implies that $f_{ij}^r O(1, \ldots, 1) = f_{ij}^r (1, \ldots, 1)$.

Now $f_{ij}^r (0, \ldots, 0) = (0, \ldots, r, \ldots, 0)$, and so $(0, \ldots, r0, \ldots, 0) = (0, \ldots, r, \ldots, 0)$.
Therefore, r is a constant element in N. ∎

Lemma 2.4.19 (Meldrum and Van der Walt, 1986)

If $r = s + t$ is the standard decomposition for r into a zero-symmetric part (s) and a constant part (t), then $f_{ij}^r = f_{ij}^s + f_{ij}^t$ is the corresponding decomposition for f_{ij}^r.

Proof

Suppose $r = s + t$ is the standard decomposition for r, where $s \in N_0$ and $t \in N_c$.

Therefore, $f_{ij}^r = f_{ij}^{s+t} = f_{ij}^s + f_{ij}^t$. Since $s \in N_0$, $t \in N_c$, we have $f_{ij}^s \in (M_n(N))_0$ and $f_{ij}^t \in (M_n(N))_c$. Therefore, $f_{ij}^r = f_{ij}^s + f_{ij}^t$ is the standard decomposition for f_{ij}^r. ∎

Corollary 2.4.20

N is zero-symmetric if and only if $M_n(N)$ is zero-symmetric.

Proof

Suppose N is zero-symmetric. This implies r is zero-symmetric for every $r \in N$. This implies $\left\{ f_{ij}^r : r \in N \right\}$ is zero-symmetric (by Lemma 2.4.16).

Therefore $\left\{ f_{ij}^r : r \in N \right\} \subseteq (M(N^n))_0$, and $(M(N^n))_0$ is a subnear ring of $M(N^n)$.

Therefore, $M_n(N)$ is the subnear ring generated by $\left\{ f_{ij}^r : r \in N \right\} \subseteq (M(N^n))_0$, and hence $M_n(N)$ is zero-symmetric.

Conversely, suppose $M_n(N)$ is zero-symmetric. This implies f_{ij}^r is zero-symmetric for all $r \in N$, $1 \le i, j \le n$. This means r is zero-symmetric for all $r \in N$. Hence, N is zero-symmetric. ∎

Theorem 2.4.21 (Meldrum and Van der Walt, 1986)

For any matrix A, any $1 \le k \le n$, and any $x, y, \ldots, z \in N$, there are $a, b, \ldots, c \in N$ such that $A(f_{1k}^x + f_{2k}^y + \ldots + f_{nk}^z) = f_{1k}^a + f_{2k}^b + \ldots + f_{nk}^c$.

Proof

We use induction on $w(A)$, the weight of the matrix A. Suppose $w(A) = 1$, then $A = f_{ij}^r$ for some $r \in N$, $1 \leq i, j \leq n$.

Consider $f_{ij}^r (f_{1k}^{x_1} + f_{2k}^{x_2} + \ldots + f_{nk}^{x_n}) = f_{ik}^{rx_j} = f_{1k}^0 + \ldots + f_{ik}^{rx_j} + \ldots + f_{nk}^0$.

Suppose the lemma holds for all matrices with weight less than m ($m \geq 2$).

If $w(A) = m$, then $A = B + C$ or $A = BC$, where $w(B) < m$, $w(C) < m$.

By the induction hypothesis, there are $p, q, \ldots, r$ and $s, t, \ldots, u$ such that $B(f_{1k}^x + f_{2k}^y + \ldots + f_{nk}^z) = f_{1k}^p + f_{2k}^q + \ldots + f_{nk}^r$ and $C(f_{1k}^x + f_{2k}^y + \ldots + f_{nk}^z) = f_{1k}^s + f_{2k}^t + \ldots + f_{nk}^u$.

Consider $A (f_{1k}^x + f_{2k}^y + \ldots + f_{nk}^z)$.

Now
$$
\begin{aligned}
A(f_{1k}^x + f_{2k}^y + \ldots + f_{nk}^z) &= (B+C)(f_{1k}^x + f_{2k}^y + \ldots + f_{nk}^z) \\
&= B(f_{1k}^x + f_{2k}^y + \ldots + f_{nk}^z) + C(f_{1k}^x + f_{2k}^y + \ldots + f_{nk}^z) \\
&= f_{1k}^p + f_{2k}^q + \ldots + f_{nk}^r + f_{1k}^s + f_{2k}^t + \ldots + f_{nk}^u \\
&= f_{1k}^{p+s} + f_{2k}^{q+t} + \ldots + f_{nk}^{r+u}.
\end{aligned}
$$

Similarly, if $A = BC$, then
$$
\begin{aligned}
A(f_{1k}^x + f_{2k}^y + \ldots + f_{nk}^z) &= BC(f_{1k}^x + f_{2k}^y + \ldots + f_{nk}^z) \\
&= B(f_{1k}^s + f_{2k}^t + \ldots + f_{nk}^u) \\
&= f_{1k}^{s^1} + \ldots + f_{nk}^{n^1}.
\end{aligned}
$$
∎

Corollary 2.4.22

For any matrix A and any $x, y, \ldots, z \in N$, there is $a \in N$ such that $E_{11}A(f_{11}^x + f_{21}^y + \ldots + f_{n1}^z) = f_{11}^a$.

Proof

$$
\begin{aligned}
E_{11}A(f_{11}^x + f_{21}^y + \ldots + f_{n1}^z) &= E_{11}(f_{11}^a + f_{21}^b + \ldots + f_{n1}^c) \\
&= f_{11}^1(f_{11}^a + f_{21}^b + \ldots + f_{n1}^c) = f_{11}^a
\end{aligned}
$$
∎

Lemma 2.4.23

For any $\alpha \in N^n$, there exists $B \in M_n(N)$ such that $\alpha = Be_1$.

Proof

Let $\alpha = (a, b, \ldots, c) \in N^n$. Take $B = f_{11}^a + f_{21}^b + \ldots + f_{n1}^c$.

Therefore, $Be_1 = (f_{11}^a + f_{21}^b + \ldots + f_{n1}^c)(1, 0, \ldots, 0) = (a, b, \ldots, c) = \alpha$.
∎

3

More Concepts on Near Rings

3.1 N-Groups

The well-known concept of a "vector space over an arbitrary field" becomes a "module over a ring" when it is generalized. When we generalize further, this concept becomes a "module over a near ring." Modules over near rings N are also known as N-groups. In this section, we present some fundamental results on N-groups, some of them drawn from Pilz (1983). Throughout this chapter, G denotes an N-group.

Fields, vector spaces, rings, modules, and near rings are all N-groups. Hence, the study of N-groups becomes quite important.

Definition 3.1.1

Let $(N, +, \cdot)$ be a near ring, and $(G, +)$ be a group with additive identity 0. G is said to be an N-group if there exists a mapping $N \times G \to G$ (the image of $(n, g) \in N \times G$ is denoted by ng), satisfying the following conditions:

(i) $(n + m)g = ng + mg$ and
(ii) $(nm)g = n(mg)$

for all $g \in G$ and $n, m \in N$.
We denote this N-group by $_N G$.

Example 3.1.2

(i) Let N be a near ring. Then the mapping $N \times N \to N$ (multiplication in N) with respect to the operation $(N, +)$ becomes an N-group. This N-group is denoted by $_N N$.
(ii) Each (left) module M over a ring R is an N-group where $N = R$.

(iii) Let G be a group. Then G is an $M(G)$-group ($_{M(G)}G$). Here, the mapping $M(G) \times G \to G$ $(f, a) \mapsto f(a)$ for all $f \in M(G)$ and $a \in G$. If G is a non-Abelian group, then this $M(G)$-group ($_{M(G)}G$) cannot be a module over a ring.

Proposition 3.1.3

Let G be an N-group.

(i) $0x = 0$ for all $x \in G$.
(ii) $(-n)x = -nx$ for all $x \in G$ and $n \in N$.
(iii) $n0 = 0$ for all $n \in N_0$.
(iv) $nx = n0$ for all $x \in G$ and $n \in N_c$.

Proof

(i) Let $x \in G$.
    Consider $0x = (0 + 0)x$. Now $0x = 0x + 0x \Rightarrow 0 + 0x = 0x + 0x \Rightarrow 0x = 0$.
(ii) From (i), $0 = 0x = (-n + n)x$. This implies $(-n)x + nx = 0$, and hence $(-n)x = -nx$.
(iii) Take $n \in N_0$. Then $n0 = n(0 \cdot 0) = (n0)0 = 0 \cdot 0$ (since $n \in N_0$) $= 0$.
(iv) Let $x \in G$ and $n \in N_c$. Then $n0 = n$.
     Now $nx = (n0)x = n(0x) = n0$.
     Therefore, $nx = n0$ for all $n \in N_c$ and $x \in G$.   ∎

Definition 3.1.4

Let G be an N-group. The N-group G is said to be a **unitary N-group** if

(i) N is a near ring with unity 1
(ii) $1 \cdot x = x$ for all $x \in G$.

Example 3.1.5

Every unitary module M over a ring R with unity is a unitary N-group.

Definition 3.1.6

Let G be an N-group. A subgroup $(H, +)$ of $(G, +)$ is said to be an **N-subgroup** of G if $NH \subseteq H$ (this is denoted by $H \leq {_N}G$).

Remark 3.1.7

If G is an Abelian group, N is a ring, and $_NG$ is a module over the ring N, then the concepts of N-subgroups and submodules are one and the same.

Example 3.1.8

Let $(H, +)$ be a subgroup of a group $(G, +)$. Then $M_H(G) = \{f \in M(G) \mid f(H) \subseteq H\}$ is a subnear ring of $M(G)$.

Verification

Part 1: First, we show that $M_H(G)$ is a subgroup of $M(G)$.
  Let $f, g \in M_H(G)$ and $x \in H$.
  Clearly, $f(x) \in f(H) \subseteq H$ and $g(x) \in g(H) \subseteq H$.
  Now $(f - g)(x) = f(x) - g(x) \in H$.
  Therefore, $f - g \in M_H(G)$. This shows that $M_H(G)$ is a subgroup of $(M(G), +)$.
Part 2: We show that $M_H(G) \, M_H(G) \subseteq M_H(G)$.
  Let $f_1, f_2 \in M_H(G)$ and $x \in H$.
  Now $(f_1 f_2)(x) = f_1(f_2(x)) \in f_1(f_2(H)) \subseteq f_1(H) \subseteq H$.
  This is true for all $x \in H$.
  Therefore $(f_1 f_2)(H) \subseteq H$. Hence, $f_1 f_2 \in M_H(G)$ for all $f_1, f_2 \in M_H(G)$.
  Hence, $M_H(G)$ is a subnear ring of $M(G)$.

Example 3.1.9

Let H be a normal subgroup of a group $(G, +)$. Then $M_{G/H}(G) = \{f \in M(G) \mid f(x + H) \subseteq f(x) + H$, for all $x \in G\}$ is a subnear ring of $M(G)$.

Verification

Part 1: Given that H is a normal subgroup of $(G, +)$.
  Now we show that $M_{G/H}(G)$ is a subgroup of $M(G)$.
  Let $f, g \in M_{G/H}(G)$, and let $x \in G$. Then
  $f(x + H) \subseteq f(x) + H$ and $g(x + H) \subseteq g(x) + H$, and so
  $(f - g)(x + H) \subseteq f(x + H) - g(x + H)$

$$\subseteq (f(x) + H) - (g(x) + H)$$

$$= (f(x) - g(x)) + H \text{ (since } H \text{ is a normal subgroup)}$$

$$= (f - g)(x) + H.$$

  So $(f - g)(x + H) \subseteq (f - g)(x) + H$ for all $x \in G$.
  This shows that $f - g \in M_{G/H}(G)$ for all $f, g \in M_{G/H}(G)$.
  Therefore, $M_{G/H}(G)$ is a subgroup of $M(G)$.

Part 2: We show that $M_{G/H}(G) \, M_{G/H}(G) \subseteq M_{G/H}(G)$.
  For this, take $f, g \in M_{G/H}(G)$.
  Now $(fg)(x + H) = f(g(x + H)) \subseteq f(g(x) + H) \subseteq f(g(x)) + H$.
  This shows that $fg \in M_{G/H}(G)$.
  So, $M_{G/H}(G)$ is a subnear ring of $M(G)$.

3.2 Homomorphisms in N-Groups

In this section, we prove some elementary results on homomorphisms on N-groups.

Definition 3.2.1

Let N be a near ring and G, G^1 be N-groups. Then a mapping $h: G \rightarrow G^1$ is called an **N-homomorphism** (or **N-group homomorphism**) if it satisfies

  (i) $h(x + y) = h(x) + h(y)$ and (ii) $h(ng) = nh(g)$ for all $n \in N$, and $x, y \in G$.

Note 3.2.2

  (i)   A one–one (onto, respectively) homomorphism is called a **monomorphism** (epimorphism, respectively).
  (ii) A one–one and onto homomorphism is called an **isomorphism**.
  (ii) A homomorphism from $G \rightarrow G$ is called an **endomorphism** on G.
  (iii) An endomorphism that is a bijection is called an **automorphism**.

Notation 3.2.3

  (i)   A near ring monomorphism from N into N^1 may be represented as $N \hookrightarrow N^1$. In this case, we also say that N can be embedded in N^1.
  (ii) Hom $(N, N^1) = \{f \mid f: N \rightarrow N^1$ is a homomorphism$\}$.
  (iii) $\text{Hom}_N(G, G^1) = \{f \mid f: G \rightarrow G^1$ is an N-homomorphism$\}$.
  (iv) If there exists an N-isomorphism between G and G^1, then we say that G and G^1 are **N-isomorphic**. We denote this fact by $G \cong {}_N G^1$.

Example 3.2.4

Let G be an N-group. Consider N as an N-group. Fix $x \in G$. Define $h_x: N \rightarrow G$ by $h_x(n) = nx$ for all $n \in N$. Then $h_x \in \text{Hom}_N(N, G)$.

Verification

Given that $x \in G$ and $h_x: N \to G$ by $h_x(n) = nx$ for all $n \in N$.

We show that h_x is N-homomorphism. Let $n_1, n_2 \in N$.

Now $h_x(n_1 + n_2) = (n_1 + n_2)x = n_1 x + n_2 x = h_x(n_1) + h_x(n_2)$.

Let $a \in G$. Then $h_x(na) = (na)x = n(ax) = n(h_x(a))$. Therefore, $h_x \in \text{Hom}_N(N, G)$.

Definition 3.2.5

A normal subgroup H of an N-group $(G, +)$ is called an **ideal** of G (denoted by $H \trianglelefteq {}_N G$) if $n(x + a) - nx \in H$ for all $n \in N$, $x \in G$, and $a \in H$.

Remark 3.2.6

Let N be a near ring and $I \subseteq N$. From Definition 3.2.5, it follows that I is a left ideal of N if and only if I is an ideal of the N-group ${}_N N$.

The concepts of "left ideal of the near ring N" and "ideal of the N-group N" are one and the same.

Thus, the set of all left ideals of the near ring N is equal to the set of all ideals of the N-group ${}_N N$.

Remark 3.2.7

Let H be a subgroup of $(G, +)$.

Then the following two conditions are equivalent:

(i) H is an ideal of the N-group G; and
(ii) $x \equiv y \pmod{H}$, $a \equiv b \pmod{H} \Rightarrow x + a \equiv y + b \pmod{H}$ and $nx \equiv ny \pmod{H}$.

Verification

(i) $\Rightarrow$ (ii): Suppose that $x_1 \equiv x_1^1 \pmod{H}$ and $x_2 \equiv x_2^1 \pmod{H}$.

This implies that $x_1 - x_1^1 \in H$, $x_2 - x_2^1 \in H$.

We show that $x_1 + x_2 \equiv x_1^1 + x_2^1 \pmod{H}$ and $nx_1 \equiv nx_1^1 \pmod{H}$ for all $n \in N$.

Now $(x_1 + x_2) - (x_1^1 + x_2^1) = x_1 + (x_2 - x_2^1) - x_1^1$

$$= x_1 + (x_2 - x_2^1) + x_1 - x_1 - x_1^1$$

$$= (x_1 + (x_2 - x_2^1) + x_1) - (x_1^1 - x_1) \in H$$

(since H is normal and $x_2 - x_2^1 \in H$).

This implies $(x_1 + x_2) \equiv (x_1^1 + x_2^1) \pmod{H}$.

Now $nx_1 - nx_1^1 = n(x_1 - x_1^1 + x_1^1) - nx_1^1 \in H$ (since $x_1 - x_1^1 \in H$ and H is an ideal of G). This means that $nx_1 \equiv nx_1^1 \pmod{H}$.

(ii) $\Rightarrow$ (i): In this part we show that H is an ideal of G.
First, we show that H is a normal subgroup of G.
Let $x \in G$ and $\delta \in H$.
We know that $x \equiv x \pmod{H}$ and $\delta \equiv 0 \pmod{H}$. By the assumed condition, $x + \delta \equiv x + 0 \pmod{H}$. This implies $x + \delta \equiv x \pmod{H}$. That is, $x + \delta - x \in H$. Hence, H is a normal subgroup of G.
Let $n \in N$, $x \in G$, and $\delta \in H$.
We know that $n \equiv n \pmod{H}$ and $x + \delta \equiv x \pmod{H}$.
By the assumed condition, $n(x + \delta) \equiv nx \pmod{H}$.
This implies that $n(x + \delta) - nx \in H$.
Hence, H is an ideal of G.

Remark 3.2.8

Let G be an N-group and H a normal subgroup of $(G, +)$.
  Then the following two conditions are equivalent:

  (i) $n(x + a) - nx \in H$, for all $n \in N$, $x \in G$, and $a \in H$; and
  (ii) $n(b + x) - nx \in H$, for all $n \in N$, $x \in G$, and $b \in H$.

Verification

(i) $\Rightarrow$ (ii): Suppose (i).
  Now $n(b + x) - nx = n(x - x + b + x) - nx$

$$= n(x + a) - nx \text{ (where } a = -x + b + x \in H) \in H \qquad \text{(by (i))}.$$

The other part (ii) $\Rightarrow$ (i) is similar.
The verification for some results in N-groups is similar to that of near rings. In such cases, we have not provided the proofs.

Definition 3.2.9

  (i) Let $I \unlhd N$. Then we know that $N/I = \{n + I \mid n \in N\}$ is called a **factor near ring** or **quotient near ring** (refer to Remark 2.2.12).
  (ii) Let $H \unlhd {}_N G$, then $G/H = \{g + H \mid g \in G\}$ is an N-group and it is called a **factor N-group** or **quotient N-group**.

Remark 3.2.10

  (i) If L is a left ideal of N (denoted as, $L \unlhd_l N$) then we may consider L as an ideal of the N-group N.
       In this case, we can consider the factor N-group N/L, the N-group N modulo L.
  (ii) Clearly $\{0\}$, N are ideals of N; and $\{0\}$, G are ideals of ${}_N G$.
       These ideals are known as **trivial ideals**.

Theorem 3.2.11 (Homomorphism Theorem for N-Groups)

(i) If H is an ideal of an N-group G, then the canonical mapping π: $G \rightarrow G/H$ is an N-epimorphism. So G/H is a homomorphic image of G.

(ii) Conversely, if $h: G \rightarrow G^1$ is an N-epimorphism, then ker(h) is an ideal of G and $G/\ker(h) \cong G^1$.

Theorem 3.2.12 (Second Isomorphism Theorem)

Let $h: N \rightarrow N^1$ be an epimorphism. Then

(i) h induces a one–one correspondence between the subnear rings (ideals, respectively) of N containing ker(h) and the subnear rings of (ideals, respectively) N^1.

(ii) $N/I \cong h(N)/h(I)$ for all ideals I of N with $\ker(h) \subseteq I$.

(iii) If $\pi: N \rightarrow N/I$ is the canonical epimorphism, then we get that $(N/I)/(J/I) \cong N/J$ for all ideals J of N with $J \supseteq I$.

Proof

Part 1: Let $h: N \rightarrow N^1$ be an epimorphism.

Let $\mathscr{S}$ be the set of all ideals of N containing ker h and $\mathscr{S}^1$ be the set of all ideals of N^1.

Define $f: \mathscr{S} \rightarrow \mathscr{S}^1$ by $f(I) = \{h(i) \mid i \in I\} = h(I)$ for all $I \in \mathscr{S}$.

Now we show that $f(I)$ is an ideal of N^1. Let $x \in f(I)$. Then there exists $i \in I$ such that $h(i) = x$.

Take $n^1 \in N^1$. Since $h: N \rightarrow N^1$ is an epimorphism, there corresponds $n \in N$ such that $h(n) = n^1$.

Consider $n^1 + x - n^1 = h(n) + h(i) - h(n)$

$$= h(n + i - n) \in f(I).$$

Hence, $f(I)$ is normal in N^1.

Let $x \in f(I)$ and $n^1 \in N^1$.

Now $xn^1 = h(i)h(n) = h(in) \in f(I)$. Therefore, $f(I)N^1 \subseteq f(I)$.

Let $n_1^1, n_2^1 \in N^1$ and $x \in f(I)$. Then there exist $n_1, n_2 \in N$ such that $h(n_1) = n_1^1$, $h(n_2) = n_2^1$ and there exists $i \in I$ such that $h(i) = x$. Now

$n_1^1(n_2^1 + x) - n_1^1 n_2^1 = h(n_1)(h(n_2) + h(i)) - h(n_1)h(n_2)$

$$= h(n_1)(h(n_2 + i)) - h(n_1 n_2) \quad \text{(since h is a homomorphism)}$$

$$= h(n_1(n_2 + i) - n_1 n_2) \in f(I).$$

Hence, $f(I)$ is an ideal of N^1.

Part 2: Now define $g: \mathscr{S}^1 \rightarrow \mathscr{S}$ by $g(I^1) = \{i \in N \mid h(i) \in I^1\}$ for all $I^1 \in \mathscr{S}^1 = h^{-1}(I^1)$.

Now we show that $g(I^1)$ is an ideal of N containing ker h.

Let $y \in g(I^1)$. Then $y \in N$ and $h(y) \in I^1$.

Let $n \in N$. Since $h: N \to N^1$ is a function, it follows that $h(n) \in N^1$ for all $n \in N$.

Now $h(n + y - n) = h(n) + h(y) - h(n) \in I^1$. Therefore, for $y \in N, n \in N, n + y - n \in N$, and $h(n + y - n) \in I^1$. This implies that $n + y - n \in g(I^1)$.

Hence, $g(I^1)$ is normal in N.

Let $y \in g(I^1)$ and $n \in N$.

Now $h(yn) = h(y)h(n) \in I^1$ (since $I^1 \trianglelefteq N^1 \Rightarrow I^1 N^1 \subseteq I^1$). Therefore, $yn \in g(I^1)$.

Hence, $g(I^1)N \subseteq g(I^1)$.

Let $n_1, n_2 \in N$ and $y \in g(I^1)$.

$h(n_1(n_2 + y) - n_1 n_2) = h(n_1)h(n_2 + y) - h(n_1)h(n_2)$

$$= h(n_1)(h(n_2) + h(y)) - h(n_1)h(n_2) \in I^1 \text{(since } I^1 \trianglelefteq_1 N^1).$$

Therefore, $n_1(n_2 + y) - n_1 n_2 \in g(I^1)$ for all $n_1, n_2 \in N$ and $y \in g(I^1)$.

Hence, $g(I^1)$ is an ideal of N.

Now we show that $\ker(h) \subseteq g(I^1)$. Let $x \in \ker(h)$. Then $h(x) = 0^1 \in I^1$, which implies that $x \in g(I^1)$. Therefore, $\ker(h) \subseteq g(I^1)$. Hence, $g(I^1) \trianglelefteq N$ and $\ker(h) \subseteq g(I^1)$.

Part 3: Here $f: \mathscr{S} \to \mathscr{S}^1$ and $g: \mathscr{S}^1 \to \mathscr{S}$. Therefore, $g \circ f: \mathscr{S} \to \mathscr{S}$ and $f \circ g: \mathscr{S}^1 \to \mathscr{S}^1$.

Let $I \in \mathscr{S}$. Now $(g \circ f)(I) = g(f(I)) = g(h(I)) = h^{-1}(h(I)) = (h^{-1} \circ h)(I) = I$.

Therefore, $g \circ f = \mathscr{I}_{\mathscr{S}}$, the identity function on $\mathscr{S}$.

Clearly, $f \circ g = \mathscr{I}_{\mathscr{S}^1}$, the identity function on $\mathscr{S}^1$.

Therefore, $g = f^{-1}$, and f and g are bijective functions.

Hence, there exists a one–one correspondence between the ideals of N containing $\ker(h)$ and the ideals of N^1.

Now we show that for all ideals I of N containing $\ker(h)$, $N/I \cong h(N)/h(I)$.

Define $\varphi: N \to h(N)/h(I)$ as $\varphi(n) = h(n) + h(I)$, for all $n \in N$.

Let $n_1, n_2 \in N$.

Now $\varphi(n_1 + n_2) = h(n_1 + n_2) + h(I)$

$$= (h(n_1) + h(n_2)) + h(I)$$

$$= (h(n_1) + h(I)) + (h(n_2) + h(I))$$

$$= \varphi(n_1) + \varphi(n_2).$$

And also $\varphi(n_1 n_2) = h(n_1 n_2) + h(I)$

$$= (h(n_1)h(n_2)) + h(I)$$

$$= (h(n_1) + h(I))(h(n_2) + h(I))$$

$$= \varphi(n_1)\varphi(n_2).$$

Let $y \in h(N)/h(I)$. Then there exists $n \in N$ such that $y = h(n) + h(I)$.

Now $\varphi(n) = h(n) + h(I) = y$, and so φ is onto. Hence, φ is an epimorphism.

By the first homomorphism theorem, we have $N/\ker(\varphi) \cong h(N)/h(I)$.

Now it is enough to prove that $\ker(\varphi) = I$.

Consider $n \in \ker(\varphi) \Leftrightarrow \varphi(n) = 0$

$$\Leftrightarrow h(n) + h(I) = 0$$
$$\Leftrightarrow h(n) \in h(I)$$
$$\Leftrightarrow h(n) = h(i) \text{ for some } i \in I$$
$$\Leftrightarrow h(n - i) = 0 \text{ for some } i \in I$$
$$\Leftrightarrow n - i \in \ker(h) \text{ for some } i \in I$$
$$\Leftrightarrow n - i \in \ker(h) \subseteq I, \ i \in I$$
$$\Leftrightarrow n \in I.$$

Therefore, $\ker(\varphi) = I$.

Suppose $\pi: N \to N/I$ is a canonical epimorphism.

Define $\psi: N \to (N/I)/(J/I)$ by $\psi(n) = (n + I) + (J/I)$ for all $n \in N$.

Now we show that ψ is an epimorphism.

Let $n_1, n_2 \in N$.

Now $\psi(n_1 + n_2) = ((n_1 + n_2) + I) + (J/I)$

$$= ((n_1 + I) + (n_2 + I)) + (J/I)$$
$$= ((n_1 + I) + (J/I)) + ((n_2 + I) + (J/I))$$
$$= \psi(n_1) + \psi(n_2).$$

And also $\psi(n_1 n_2) = (n_1 n_2 + I) + (J/I)$

$$= ((n_1 + I)(n_2 + I)) + (J/I)$$
$$= ((n_1 + I) + (J/I))((n_2 + I) + (J/I))$$
$$= \psi(n_1)\psi(n_2).$$

Hence, ψ is a homomorphism.

Let $x \in (N/I)/(J/I)$. Then there exists $n \in N$ such that $x = (n + I) + (J/I)$.

Now $\psi(n) = (n + I) + (J/I) = x$, so ψ is onto.

Hence, ψ is an epimorphism.

By the first homomorphism theorem, we have $N/\ker(\psi) \cong (N/I)/(J/I)$.

Now $\ker(\psi) = \{n \in N \mid \psi(n) = J/I\}$

$$= \{n \in N \mid (n + I) + (J/I) = J/I\}$$
$$= \{n \in N \mid (n + I) \in J/I\}$$
$$= J.$$

Hence, $N/J \cong (N/I)/(J/I)$.

Theorem 3.2.13

Let $h: G \to G^1$ be an N-epimorphism. Then

(i) h induces a one–one correspondence between the N-subgroups (ideals, respectively) of G containing ker(h) and the N-subgroup (ideals, respectively) of G^1 by $A \mapsto h(A)$, for $A \subseteq G$.

(ii) $G/H \cong h(G)/h(H)$ for all N-ideals H of G with ker(h) $\subseteq H$.

(iii) If $\pi: G \rightarrow G/H$ is the canonical epimorphism, then we get that $(G/H)/(H_1/H) \cong G/H_1$ for all ideals H_1 of G with $H \subseteq H_1$.

Observation 3.2.14

In the case of rings, the concepts of invariant subnear rings and ideals are one and the same. Every ideal of a near ring is an invariant subnear ring. But every ideal of a near ring need not be an invariant subnear ring.

Proposition 3.2.15

(i) N_0 is a left ideal of N, but N_0 is not generally an ideal of N.

(ii) N_c is an invariant subnear ring of N, but in general neither a right nor a left ideal.

Proof

(i) We show that N_0 is a left ideal of N. Clearly, N_0 is a normal subgroup of N.

For any $n_0 \in N_0$ and $n, m \in N$, we have

$$(n(m + n_0) - nm)0 = n(m + n_0)0 - (nm)0$$
$$= n(m0 + n_00) - (nm)0$$
$$= (nm)0 - (nm)0 = 0.$$

Therefore, $n(m + n_0) - nm \in N_0$. Hence, N_0 is a left ideal of N.

Let $N = M(\mathbb{R})$, $N_0 = M_0(\mathbb{R})$. Let $I_R \circ f = f \notin M_0(\mathbb{R})$ (here I_R is the identity mapping on $\mathbb{R}$). Therefore, N_0 is not an ideal of N.

(ii) We show that N_c is an invariant subnear ring of N. Clearly, N_c is a subnear ring of N.

Let $n_c n \in N_c N$. Now $(n_c n)0 = n_c(n0) = n_c = n_c n$. Therefore, $n_c n \in N_c$. This shows that $N_c N \subseteq N_c$.

On the other hand, take $nn_c \in NN_c$. Now $(nn_c)0 = n(n_c 0) = nn_c$. Therefore, $nn_c \in N_c$.

Hence, N_c is an invariant subnear ring of N.

Now we show that N_c is not a normal subgroup of N. Let G be a non-Abelian group.

So there exists $a, b \in G$ such that $a + b \neq b + a$.

Define $f_a: G \to G$ such that $f_a(x) = a$, for all $x \in G$.

So, $f_a \in M_c(G) = N_c$. Let $M(G) = N$. Consider $I_G \in M(G)$ (here I_G is the identity mapping on G).

Now $(I_G + f_a - I_G)(0) = a$, but $(I_G + f_a - I_G)(b) = b + a - b \ne a$.

This shows that $I_G + f_a - I_G \notin M_c(G)$. Hence, N_c is not a normal subgroup of N.   ∎

Note 3.2.16

$M_c(G)$ is a normal subgroup of $M(G)$, if and only if G is Abelian.

Verification

Suppose G is Abelian. Let $f_c \in M_c(G)$ and $f \in M(G)$, and $g \in G$.

Now $(f + f_c - f)(g) = f(g) + f_c(g) - f(g) = f_c(g)$. Therefore $(f + f_c - f) = f_c \in M_c(G)$.

Hence, $M_c(G)$ is a normal subgroup of $M(G)$.

Converse

Suppose that $M_c(G)$ is a normal subgroup of $M(G)$.

Let us suppose that G is not Abelian.

So, there exists $a, b \in G$ such that $a + b \ne b + a$.

Define $f: G \to G$ by $f(x) = a$, for all $x \in G$.

Consider the identity mapping $I: G \to G$.

Now $(I + f - I)(a) = a + a - a = a$.

$$(I + f - I)(b) = b + a - b \ne a.$$

Therefore, $I + f - I$ is not a constant mapping, and so $(I + f - I) \notin M_c(G)$, $f \in M_c(G)$, $I \in M(G)$.

Hence, $M_c(G)$ is not a normal subgroup in $M(G)$, a contradiction.

Proposition 3.2.17 (Proposition 1.34 of Pilz, 1983)

(i) If L is a left ideal of N, then $N_0 L \subseteq L$.

(ii) $N = N_0$ if and only if each left ideal of N is an N-subgroup of N.

(iii) $N = N_0$ implies that $(H \trianglelefteq {}_N G \Rightarrow H \le {}_N G)$ for all $G \in {}_N \mathscr{S}$.

Proof

(i) Suppose L is a left ideal of N. Take $n_0 \in N_0$ and $l \in L$.

Now $n_0 l = n_0(0 + l) - 0 = n_0(0 + l) - n_0 0 \in L$, and so $N_0 L \subseteq L$.

(ii) Suppose $N = N_0$. Let L be the left ideal of N.

From (i), we have $N_0 L \subseteq L$. Since $N = N_0$, we get that $NL \subseteq L$.

Hence, L is an N-subgroup of N.

Converse

Clearly $\{0\}$ is a left ideal of N. By our supposition $\{0\}$ is an N-subgroup of N, that is, $N\{0\} \subseteq \{0\}$. Now $n0 = 0$ for all $n \in N$.

Therefore, $n \in N_0$ for all $n \in N$. Hence, $N \subseteq N_0$.

This shows that $N = N_0$.

(iii) Suppose $N = N_0$ and $H \trianglelefteq {}_N G$. Let $n_0 \in N_0 = N$ and $\delta \in H$.

Now $n_0\delta = n_0(0 + \delta) - 0 = n_0(0 + \delta) - n_0 0 \in H$ (since $H \trianglelefteq {}_N G$ and $N = N_0$).

Therefore, $NH \subseteq H$, and hence $H \leq_N G$. ∎

Proposition 3.2.18

   (i) Ng is an N-subgroup of G for all $g \in G$.
   (ii) $N_0 = N_c 0 \subseteq H$ for all N-subgroups H of G.

Proof

   (i) Take $g \in G$. Clearly, Ng is a subgroup of G. Let $x \in Ng$.
      Then there exists some $n_1 \in N$ such that $x = n_1 g$, and let $n \in N$.
      Clearly, $n(n_1 g) = (nn_1)g \in Ng$.
      Therefore, $N(Ng) \subseteq Ng$.
      Hence, $Ng \leq_N G$.
   (ii) Let $H \leq_N G$. This implies H is a subgroup of G and $NH \subseteq H$.
      Since $0 \in H$, we have $N0 \subseteq H$.
      Now $N0 = (N_0 + N_c)0 = N_0 0 + N_c 0 = N_c 0$ (since $N_0 0 = \{0\}$).
      Therefore, $N0 = N_c 0 \subseteq H$. ∎

Observation 3.2.19

$N0$ is the smallest of all N-subgroups of G.
   We use the notation Ω for $N0 = N_c 0$.

Results 3.2.20

   (i) If $N = N_0$, then $\Omega = \{0\}$.
   (ii) For all $g \in G$, we have $\Omega = N_c g$. Take $g \in G$.
      We know $n_c g = n0$ for all $n_c \in N_c$ and for all $n \in N$.
      Now $\Omega = N0 = N_c g$. Therefore, $\Omega = N_c g$.
   (iii) To prove $\Omega \cong {}_N N_c$, define a mapping $\varphi \colon N_c \to \Omega = N_c 0$ by $\varphi(n_c) = n_c 0$ for
      all $n_c \in N_c$.
      Let $n_c, n_c^1 \in N_c$, and suppose $\varphi(n_c) = \varphi(n_c^1)$ if and only if $n_c 0 = n_c^1 0$, if
      and only if $n_c = n_c^1$.
      Therefore, φ is one–one and well-defined.

Now $\varphi(n_c + n_c^1) = (n_c + n_c^1)0 = n_c0 + n_c^10 = \varphi(n_c) + \varphi(n_c^1)$.
Further, for any $n \in N$, $\varphi(nn_c) = (nn_c)0 = n(n_c0) = n\varphi(n_c)$.
For any $n_c0 \in N_c0$, $\varphi(n_c) = n_c0$. Hence, φ is an N-isomorphism from N_c to Ω.

Proof

(i) $\Omega = N0 = N_00 = \{n_00 \mid n_0 \in N_0\} = \{0\}$. ∎

Observation 3.2.21

N is simple if and only if N has no nontrivial ideals. The N-group G is called **N-simple** if and only if G has no N-subgroups except Ω and G. The N-group G is **simple** if and only if G has no nontrivial ideals.

Proposition 3.2.22

If N is simple, then all homomorphic images are isomorphic either to $\{0\}$ or to N.

Proof

Suppose N is simple. Let N^1 be the homomorphic image of N under the epimorphism h (say).

By the first homomorphism theorem, $N/\ker(h) \cong N^1$, and $\ker(h)$ is an ideal of N.

Since N is simple, $\ker(h) = \{0\}$ or $\ker(h) = N$. This implies $N/\{0\} \cong N^1$ or $N/N \cong N^1$.

But $N/\{0\} \cong N$ and $N/N \cong \{0\}$. Therefore, either $N \cong N^1$ or $N^1 \cong \{0\}$. ∎

Proposition 3.2.23

If the N-group G is simple, then all N-homomorphic images are N-isomorphic either to $\{0\}$ or to G.

Proof

Suppose the N-group G is simple. Let G^1 be the N-homomorphic image of G under h (say). That is, $h: G \to G^1$ is an N-epimorphism.

By the first homomorphism theorem for N-groups, we have that $G/\ker(h) \cong G^1$ and $\ker(h)$ is an ideal of G.

Since G is simple, we have $\ker(h) = \{0\}$ or $\ker(h) = G$. This implies $G/\{0\} \cong G^1$ or $G/G \cong G^1$.

But $G/\{0\} \cong G$ and $G/G \cong \{0\}$, and so we conclude that $G \cong G^1$ or $G^1 \cong \{0\}$. ∎

Example 3.2.24

(i) $M_0(G)$ is a simple near ring.
(ii) If $N = N_0$, then every ideal of the N-group G is an N-subgroup (refer to Proposition 3.2.17).
(iii) Suppose $N = N_0$. If the N-group G is N-simple, then G is simple.

Verification

(i) Let $(0) \neq K$ be an ideal of $M_0(G)$. We show that $K = M_0(G)$.
Clearly, $K \subseteq M_0(G)$. Let $f \in M_0(G) \Rightarrow f(0) = 0$.
Now $f = f(I_G + \bar{0}) - f\bar{0}$ (since $M_0(G) = (M(G))_0) \in K$ (since K is an ideal of $M_0(G)$ and $I_G \in K$).
Therefore, $M_0(G) \subseteq K$. Hence, $K = M_0(G)$.
Thus, $M_0(G)$ is a simple near ring.

(iii) Suppose $N = N_0$ and G is N-simple.
G is N-simple $\Rightarrow G$ has no N-subgroups except Ω and G.
But $N = N_0$; by Proposition 3.2.17, $\Omega = \{0\}$. By (ii), each ideal of G is an N-subgroup of G.
Therefore, the only ideals of G are $\{0\}$ and G. Hence, G is simple.

Proposition 3.2.25

A proper ideal H of an N-group G is maximal in G if and only if G/H is simple.

Proof

Suppose H is a proper ideal of N-group G and suppose H is a maximal ideal of $_N G$.

Now we prove that G/H is simple. Let H_1/H be an ideal of G/H, where H_1 is an ideal of N-group G and $H \subseteq H_1$. Since H is a maximal ideal of N-group G, we have either $H = H_1$ or $H_1 = G$.

This implies that $H_1/H = H$ or $H_1/H = G/H$.

Therefore, H_1/H is the maximal ideal of G/H. This shows that G/H is simple.

Converse

Suppose G/H is simple. Now we show that H is a maximal ideal of G.

Suppose H_1 is any ideal of G such that $H \subseteq H_1$. Then H_1/H is an ideal of G/H. Since G/Δ is simple, we have $H_1/H = H$ or $H_1/H = G/H$.

Therefore, $H_1 = H$ or $H_1 = G$.
Hence, H_1 is the maximal ideal of G. ∎

Theorem 3.2.26 (Proposition 2.29 of Pilz, 1983)

Let N be a near-ring and G an N-group. Let $\{H_\alpha\}_{\alpha \in I}$ be a family of ideals of G such that the sum $\sum H_\alpha$ is direct. Then for any $n \in N$ and $\sum a_\alpha \in \sum_{\alpha \in I} H_\alpha$, we have $n(\sum a_\alpha) = \sum na_\alpha$.

Proposition 3.2.27

Let G be an N-group. If K is an ideal of G and if L is a N-subgroup of G, then every ideal Z of $L + K$ which contains K can be written as $Z = Z_1 + K$, where Z_1 is an ideal of L.

Proof

Let $\varphi: L \to (L + K)/K$ be the canonical epimorphism. Write $Z_1 = \varphi^{-1}(Z/K)$.
   Then $\varphi(Z_1) = (Z_1 + K)/K$ and hence $Z_1 + K = Z$. ∎

Proposition 3.2.28 (2.22 of Pilz, 1983)

If A and B are ideals of an N-group G, and A, B are N-subgroups of G, then for all $n \in N$, $\alpha \in A$, and $\beta \in B$, $n(\alpha + \beta) \equiv n\alpha + n\beta \pmod{A \cap B}$.

Proof

Since A is an ideal of G, we have $n(\alpha+\beta)-n\beta \in A$. Also since A is an N-subgroup of G, we have $n\alpha \in A$. Therefore $n(\alpha + \beta) - n\beta - n\alpha \in A + A = A$.
   This shows that $n(\alpha + \beta) \equiv n\beta - n\alpha \pmod{A}$.
   In a similar argument, we one can get that $n(\alpha + \beta) \equiv n\beta + n\alpha \pmod{B}$.
   Therefore $n(\alpha + \beta) = n\alpha + n\beta \pmod{A \cap B}$. ∎

Lemma 3.2.29

Let U and K be nonzero ideals of an N-group G such that $U \cap K = (0)$. Then the following are equivalent.

(i) The intersection of any two nonzero ideals of G/K which are contained in $(U + K)/K$ is nonzero.

(ii) The intersection of any two nonzero ideals of G which are contained in U is nonzero.

Proof

(i) $\Rightarrow$ (ii): Let U_1, U_2 be two nonzero ideals of G contained in U such that $U_1 \cap U_2 = (0)$. Now $(U_1 + K)/K$, $(U_2 + K)/K$ are two nonzero ideals of G/K contained in $(U + K)/K$. If $u_1 \in U_1$, $u_2 \in U_2$ are such that $u_1 + K = u_2 + K$, then $u_1 - u_2 \in K \cap (U_1 + U_2) \subseteq K \cap U = (0)$ and so $u_1 = u_2 \in U_1 \cap U_2 = (0)$. Hence $((U_1 + K)/K) \cap ((U_2 + K)/K) = (0)$. This is a contradiction to (i).

(ii) $\Rightarrow$ (i): Let K_1/K, K_2/K be two nonzero ideals of G/K contained in $(U + K)/K$ such that $(K_1/K) \cap (K_2/K) = (0)$. We first show that $K_i \cap U \neq (0)$ for $i = 1, 2$. Suppose $K_i \cap U = (0)$. If $x \in K_i$, then

$$x + K \in (K_i/K) \subseteq ((U + K)/K)$$

and hence $x = u_1 + k$ for some $u_1 \in U$, $k \in K$. Now $x - k = u_1 \in U \cap K_1 = (0)$ and $x = k \in K$. So $K_i \subseteq K$ and $(K_i/K) = (0)$, a contradiction. Therefore $K_i \cap U \neq (0)$ for $i = 1, 2$ and by (i), $(K_1 \cap U) \cap (K_2 \cap U) \neq (0)$. But $K \cap U = (K_1 \cap K_2) \cap U = (K_1 \cap U) \cap (K_2 \cap U) \neq (0)$, by (ii). This contradicts our hypothesis that $K \cap U = (0)$. Hence (i) holds. ∎

3.3 Chain Conditions on Ideals

As in the theory of rings, if the set of ideals of a near ring N fulfills the DCC (descending chain condition) on ideals of N, then we say that N has DCCI. Similar conventions will apply: DCCR for right ideals, DCCL for left ideals, and DCCN for N-subgroups. We use the notions ACCI, ACCR, ACCL, and ACCN (respectively) for ascending chain condition on ideals, right ideals, left ideals, and N-subgroups (respectively) of N.

Remark 3.3.1

(i) If N is zero-symmetric, then DCCN implies DCCI.
(ii) In N, we have DCCR or DCCL $\Rightarrow$ DCCI.
(iii) If $N = N_0$, then DCCN $\Rightarrow$ DCCL (the same condition holds for ACC).

Proof

(i) Let $I_1 \supseteq I_2 \supseteq \ldots$ be a descending chain of ideals of N.
Then I_k is a left ideal of N for all k.
Since $N = N_0$, by Proposition 3.2.17, it follows that each left ideal of N is an N-subgroup of N.

Therefore, $I_1 \supseteq I_2 \supseteq \ldots$ is a descending chain of N-subgroups of N. Since N has DCCN, we get $I_k = I_{k+1} = \ldots$, for some positive integer k. This shows that N has DCCI.

(ii) Since every ideal is a left (right) ideal, we have DCCL (or DCCR) $\Rightarrow$ DCCI.

(ii) Since every left ideal is an N-subgroup of $N = N_0$, we have DCCN $\Rightarrow$ DCCL. ∎

Examples 3.3.2

(i) Consider the near ring $(\mathbb{Z}, +, \cdot)$, where $\mathbb{Z}$ is the set of integers and $+$ and $\cdot$ are the usual addition and multiplication operations, on the set of integers. It is easy to verify that this near ring of integers satisfies ACCI. The condition DCCI does not hold in $\mathbb{Z}$ (because $(2) \supset (4) \supset (8) \supset \ldots$ is an infinite descending chain of ideals in $\mathbb{Z}$).

(ii) Consider $Z_{12} = \{0, 1, 2, 3, 4, 5, 6, 7, 8, 9, 10, 11\}$, the near ring of integers modulo 12 with the usual addition and multiplication of integers modulo 12. Z_{12} contains only a finite number of ideals. Hence, Z_{12} satisfies DCCI. The following are some descending chains of ideals in the near ring Z_{12}:

$$I = \{0, 2, 4, 6, 8, 10\} \supset \{0, 6\} \supset \{0\}$$
$$I = \{0, 2, 4, 6, 8, 10\} \supset \{0, 4, 8\} \supset \{0\}$$
$$I = \{0, 3, 6, 9\} \supset \{0, 6\} \supset \{0\}$$

Note 3.3.3

Let N be a near ring and I an ideal of N. If J_1 and J_2 are two ideals of N containing I such that $J_2/I \subseteq J_1/I$, then $J_2 \subseteq J_1$.

Verification

Suppose $J_2/I \subseteq J_1/I$. To verify that $J_2 \subseteq J_1$, let $x \in J_2$. Then $x + I \in J_2/I$.
This implies $x + I \in J_1/I$ (since $J_2/I \subseteq J_1/I$) $\Rightarrow x + I = x_1 + I$ for some $x_1 \in I$.
This implies $x - x_1 \in I \Rightarrow x - x_1 \in J_1$ (since J_1 is an ideal of N containing I) $\Rightarrow x \in J_1$, and so $J_2 \subseteq J_1$.

Theorem 3.3.4 (Theorem 2.35 of Pilz, 1983)

(i) If I is an ideal of N and N has the DCCI (DCCN, DCCL), then the same applies to N/I.

(ii) If I is an ideal of N and I is a direct summand, then N has the DCCI (DCCN, DCCL), if and only if I and N/I have the DCCI (DCCN, DCCL).

(iii) If an ideal H of an N-group G (where N is zero-symmetric) is a direct summand, then G has the DCCI (DCCN), if and only if H and G/H have this property.

Proof

(i) Suppose N has DCCI and I is an ideal of N.

Let $J_1/I \supseteq J_2/I \supseteq \ldots$ be the descending chain of ideals of N/I. By Note 3.3.3, we have that $J_1 \supseteq J_2 \supseteq J_3 \ldots$ is a descending chain of ideals of N.

Since N has DCCI, there exists an integer k such that $J_k = J_{k+1} = \ldots \Rightarrow J_k/I = J_{k+1}/I = \ldots$. Therefore, N/I has DCCI.

Verification for DCCN: Let $H_1/I \supseteq H_2/I \supseteq \ldots$ be a descending chain of N-subgroups of N/I.

Then $H_1 \supseteq H_2 \supseteq \ldots$ is a descending chain of N-subgroups of N.

Since N has DCCN, for some integer k, $H_k = H_{k+1} = \ldots$.

This implies that $H_k/I = H_{k+1}/I = \ldots$.

Therefore, N/I has DCCN. A similar verification holds for DCCL.

(ii) Let I be an ideal of N and I be a direct summand; and let N have DCCI.

By (i), N/I has DCCI. We have to show that I has DCCI.

Now, since each ideal of I is an ideal of N and since N has DCCI, it follows that I has DCCI.

(iii) Use Proposition 3.2.17 (ii) and similar to (ii).

Converse

Suppose I and N/I have DCCI.

To show that N has DCCI,

let $J_1 \supseteq J_2 \supseteq \ldots$ be a descending chain of ideals of N

$\Rightarrow J_1 + I \supseteq J_2 + I \supseteq \ldots$ is a descending chain of ideals of N containing I

$\Rightarrow (J_1 + I)/I \supseteq (J_2 + I)/I \supseteq \ldots$ is a descending chain of ideals of N/I.

Also, $J_1 \cap I \supseteq J_2 \cap I \ldots$ is a descending chain of ideals of I.

Since both I and N/I have DCCI, there exists a positive integer k such that $(J_k + I)/I = (J_{k+1} + I)/I = \ldots$ and

$J_k \cap I = J_{k+1} \cap I = \ldots$.

Now it is enough to show that $J_k \subseteq J_{k+1}$. Take $x \in J_k$. Now $x \in J_k + I = J_{k+1} + I$.

So, there exists $y \in J_{k+1}$, $i \in I$ such that $x = y + i$. Also, $y \in J_{k+1} \subseteq J_k$.

Therefore, $x - y \in I \cap J_k = J_{k+1} \cap I \subseteq J_{k+1} \Rightarrow x \in J_{k+1}$.

Therefore, $J_k \subseteq J_{k+1}$. Hence, $J_k = J_{k+1}$. In the same way, we can verify that $J_k = J_{k+1} = \ldots$.

Thus, N has DCCI.   ∎

Definition 3.3.5

(i) A finite sequence $N = N_0 \supset N_1 \supset \ldots \supset N_n = \{0\}$ of subnear rings N_i of N is called a **normal sequence** of N if N_i is an ideal of N_{i-1} for all $i \in \{1, 2, \ldots, n\}$. Here n is called the length of the sequence.

(ii) A normal sequence is said to be an **invariant sequence** if all N_is are ideals of N.

(iii) The near rings $N_{i-1}/N_i (i = \{1, 2, \ldots, n\})$ are called **factors** of the normal sequence.

Clearly, any invariant sequence is a normal sequence.

Definition 3.3.6

A sequence $N = M_0 \supset M_1 \supset \ldots \supset M_m = \{0\}$ is called a **refinement** of the sequence $N = N_0 \supset N_1 \supset \ldots \supset N_n = \{0\}$, if for all $i \in \{1, 2, \ldots, n\}$ there exists $j \in \{0, 1, 2, \ldots, m\}$ such that $N_i = M_j$.

The sequences are called **isomorphic** if $n = m$ and the factors are isomorphic in some order.

Definition 3.3.7

A normal (invariant, respectively) sequence is called a **composition sequence** (**principal sequence**, respectively) if it has no proper refinement.

Proposition 3.3.8 (Proposition 2.38 of Pilz, 1983)

A normal sequence is a composition sequence if and only if all factors are simple.

Proof

Let $N = N_0 \supset N_1 \supset \ldots \supset N_n = \{0\}$ be a normal sequence.

Suppose this sequence is a composition sequence. Then it has no proper refinement.

To show that all the factors are simple, let us suppose that the factor N_i/N_{i+1} is not simple for some i.

This means that there exists an ideal J/N_{i+1} in N_i/N_{i+1} such that $0 \neq J/N_{i+1} \neq N_i/N_{i+1}$. It is clear that $N_{i+1} \subseteq J \subseteq N_i$ and $N_{i+1} \neq J \neq N_i$.

Therefore, $N = N_0 \supset N_1 \supset \ldots \supset N_i \supset J \supset N_{i+1} \supset \ldots \supset N_n = \{0\}$ is a proper refinement of the given sequence, which is a contradiction.

Hence, all factors of the given sequence are simple. ∎

Converse

Suppose all factors of the given sequence are simple.

Let us suppose that the given sequence is not a composition sequence.

Then it has a proper refinement.

Therefore, there exists an ideal M of N_i such that $N_{i+1} \subseteq M \subseteq N_i$ and $N_{i+1} \neq M \neq N_i$ for some i.

So, M/N_{i+1} is a nontrivial ideal of N_i/N_{i+1}.

Therefore, N_i/N_{i+1} is not simple, a contradiction to the fact that all factors are simple.

Hence, the given sequence is a composition sequence. ∎

Corollary 3.3.9

A sequence that is isomorphic to a composition sequence (principal sequence, respectively) is also a composition sequence (principal sequence, respectively).

Proof

Let S_1 be a normal sequence and S_2 be a composition sequence such that S_1 and S_2 are isomorphic in some order. Since S_2 is a composition sequence, all factors of S_2 are simple. Hence, all factors of S_1 are simple, that is, S_1 is a composition sequence. The other part (proof for principal sequence) is similar. ∎

Zassenhaus Lemma 3.3.10

If A, B, C, D are ideals of N such that $A \subseteq B$ and $C \subseteq D$, then

$$\frac{A+(B \cap D)}{A+(B \cap C)} \cong \frac{C+(B \cap D)}{C+(A \cap D)}.$$

Proof

We show that both the LHS (left-hand side) and the RHS (right-hand side) are isomorphic to $(B \cap D)/((A \cap D)+(B \cap C))$.

Define $\varphi\colon (A + B) \cap D \to (B \cap D)/K$ as $\varphi(a + x) = x + K$, where $a \in A$, $x \in B$, and $K = (A \cap D) + (B \cap C)$.

To show φ is well-defined, suppose $a_1 + x_1 = a_2 + x_2$, $a_1, a_2 \in A$, and $x_1, x_2 \in B$.

Then $a_1 - a_2 = x_2 - x_1 \in K$.

This implies $x_1 + K = x_2 + K$, and so $\varphi(a_1 + x_1) = \varphi(a_2 + x_2)$.

Therefore, φ is well-defined.

$$\varphi((a_1 + x_1) + (a_2 + x_2)) = \varphi(a_1 + x_1 + a_2 - x_1 + x_1 + x_2)$$
$$= (x_1 + x_2) + K \text{ (since } a_1 + x_1 + a_2 - x_1 \in A)$$
$$= (x_1 + K) + (x_2 + K)$$
$$= \varphi(a_1 + x_1) + \varphi(a_2 + x_2)$$
$$\varphi((a_1 + x_1)(a_2 + x_2)) = \varphi(a_1(a_2 + x_2) + x_1(a_2 + x_2))$$
$$= \varphi(a_1(a_2 + x_2) + x_1(a_2 + x_2) - x_1 x_2 + x_1 x_2)$$
$$= x_1 x_2 + K \text{ (since } a_1(a_2 + x_2) + x_1(a_2 + x_2) - x_1 x_2 \in A)$$
$$= (x_1 + K)(x_2 + K)$$
$$= \varphi(a_1 + x_1)\varphi(a_2 + x_2).$$

Therefore, φ is a homomorphism.

Clearly, φ is onto, and so φ is an epimorphism.

Now we show that $\ker(\varphi) = A + (B \cap C)$.

For this, suppose $\varphi(a + x) = 0 + K$,

if and only if $x + K = 0 + K$,

if and only if $x \in K$,

if and only if $x = d + b$ for some $d \in A \cap D$ and $b \in B \cap C$.

This implies $a + x = a + (d + b) = (a + d) + b \in A + (B \cap C)$.

Therefore,

$$\frac{A + (B \cap D)}{A + (B \cap C)} \cong \frac{B \cap D}{(A \cap D) + (B \cap C)} \tag{3.1}$$

Similarly, we can prove that

$$\frac{C + (B \cap D)}{C + (A \cap D)} \cong \frac{B \cap D}{(A \cap D) + (B \cap C)} \tag{3.2}$$

From Equations 3.1 and 3.2, we get that

$$\frac{A + (B \cap D)}{A + (B \cap C)} \cong \frac{C + (B \cap D)}{C + (A \cap D)}. \qquad \blacksquare$$

Lemma 3.3.11

Given two normal sequences S_1 and S_2. They have refinements that are isomorphic.

Proof

Let $S_1: N = N_0 \supset N_1 \supset \ldots \supset N_n = \{0\}$ and $S_2: M = M_0 \supset M_1 \supset \ldots \supset M_m = \{0\}$ be two normal sequences.

For each $i = 0, 1, 2, \ldots, n-1$, we insert ideals (n in number) between N_i and N_{i+1} as follows:

$$N_i \supset [N_{i+1} + (N_i \cap M_0)] \supset [N_{i+1} + (N_i \cap M_1)] \supset \ldots \supset [N_{i+1} + (N_i \cap M_m)] \supset N_{i+1}.$$

Put $N_{ij} = N_{i+1} + (N_i \cap M_j)$, for $i = 0, 1, 2, \ldots, n-1$ and $j = 0, 1, 2, \ldots, m$, where $N_0 = M_0$.

(i) $N_{i_0} = N_{i+1} + (N_i \cap M_0) = N_{i+1} + (N_i \cap N_0) = N_{i+1} + N_i = N_i$

(ii) $N_{i_{j-1}} \supset N_{i_j}$

     Now $M_{j-1} \supset M_j \Rightarrow N_i \cap M_{j-1} \supset N_i \cap M_j$

$$\Rightarrow N_{i+1} + (N_i \cap M_j - 1) \supset N_{i+1} + (N_i \cap M_j)$$

$$\Rightarrow N_{i_{j-1}} \supset N_{i_j}$$

(iii) $N_{i_m} = N_{i+1} + (N_i \cap M_m) = N_{i+1} + \{0\} = N_{i+1}$

     Therefore, $N_i = N_{i_0} \supset N_{i_1} \supset N_{i_2} \supset \ldots \supset N_{i_m} = N_{i+1}$.

     Hence, $N_i = N_{i+1} + (N_i \cap M_0) \supset N_{i_2} \supset \ldots \supset N_{i+1} + (N_i \cap M_m) = N_{i+1}$.

     Therefore, the resulting refinement of S_1 is of length nm.

     Similarly, for each $j = 0, 1, 2, \ldots, m-1$, we insert ideals between M_j and M_{j+1}.

$$M_j = [M_{j+1} + (M_j \cap N_0)] \supset [M_{j+1} + (M_j \cap N_1)] \supset \ldots \supset [M_{j+1} + (M_j \cap N_n)] = M_{j+1}$$

so that the resulting refinement of S_2 is of length mn.

For each $i = 0, 1, 2, \ldots, n-1$ and $j = 0, 1, \ldots, m-1$, we have

$$\frac{N_{i+1} + (N_i \cap M_j)}{N_{i+1} + (N_i \cap M_{j+1})} \approx \frac{M_j + (N_i \cap M_j)}{M_{j+1} + (N_{i+1} \cap M_j)}.$$

Hence, these resulting refinements of S_1 and S_2 are isomorphic.   ∎

Jordon Holder Theorem 3.3.12 (Corollary of 2.40 Pilz, 1983)

If N has a composition (principal, respectively) sequence, then each normal (invariant, respectively) sequence can be refined to a composition (principal) sequence, and all these sequences are isomorphic.

Proof

Let S_1 be a normal sequence and S_2 be the given composition sequence.

By Lemma 3.3.11, S_1 and S_2 have refinements that are isomorphic. But S_2 has no proper refinement. This implies S_1 has a refinement that is isomorphic to S_2. Since S_2 is a composition sequence, the refinement of S_1 that is isomorphic to S_2 is also a composition sequence. The proof of the other part is similar.   ∎

3.4 Direct Systems in *N*-Groups

As in the theory of modules over rings, the notions of "essential ideals" and "uniform ideals" are important in the theory of *N*-groups. In this section, we consider zero-symmetric near rings only.

Definition 3.4.1

Let H_1 and H_2 be two ideals of an *N*-group *G*. We say that H_1 is **essential** in H_2 if it satisfies the following conditions:

(i) $H_1 \subseteq H_2$ and (ii) *L* is an ideal of *G* contained in H_2 and $H_1 \cap L = (0)$ implies $L = (0)$. We denote this fact by $H_1 \leq_e H_2$.

From this definition, it is clear that if *I* is an essential ideal of *G* and *K* is an ideal of *G* such that $I \subseteq K$, then *K* is also essential.

Remark 3.4.2

(i) The intersection of a finite number of essential ideals is essential.
(ii) If *I*, *J*, *K* are ideals of *G* such that $I \leq_e J$ and $J \leq_e K$, then $I \leq_e K$.
(iii) If $I \leq_e J$, then $I \cap K \leq_e J \cap K$ for all ideals *K* of *G*.
(iv) If $I \subseteq J \subseteq K$, then $I \leq_e K$ if and only if $I \leq_e J$ and $J \leq_e K$.

Proof

The proofs of (i) and (ii) are straightforward.
(iii) Let *L* be an ideal of *G* such that $L \subseteq J \cap K$ and $(I \cap K) \cap L = (0)$.
This implies that $I \cap (K \cap L) = (0)$.
Since $I \leq_e J$ and $K \cap L \subseteq L \subseteq J$, we have that $L = K \cap L = (0)$.
(iv) Suppose $I \subseteq J \subseteq K$ and $I \leq_e K$.
Let *L* be an ideal of *G* such that $L \subseteq J$ and $I \cap L = (0)$.
Since $I \leq_e K$ and $L \subseteq J \subseteq K$, we have that $L = (0)$. This shows that $I \leq_e J$.
Next we show that $J \leq_e K$.
Let *A* be an ideal of *G* such that $A \subseteq K$ and $J \cap A = (0)$.
Now $I \cap A \subseteq J \cap A = (0) \Rightarrow I \cap A = (0)$.
Since $I \leq_e K$, we have that $A = (0)$.

Converse

Suppose that $I \leq_e J$ and $J \leq_e K$. We show that $I \leq_e K$.
Let *L* be an ideal of *G* such that $L \subseteq K$ and $I \cap L = (0)$.
This implies that $I \cap (L \cap J) = (0)$. Since $I \leq_e J$, we have that $L \cap J = (0)$.
Again, since $J \leq_e K$, we get that $L = (0)$. This shows that $I \leq_e K$.

Remark 3.4.3

Let H_1, H_2, and H_3 be ideals of G. Then

(i) H_1 and H_2 are essential in H_3, which implies $H_1 \cap H_2$ is essential in H_3.
(ii) Suppose $H_1 \subseteq H_2 \subseteq H_3$. Then H_1 is essential in H_3 if and only if H_1 is essential in H_2 and H_2 is essential in H_3.

Proof

Similar to that of Remark 3.4.2. ∎

Lemma 3.4.4

Let G_1, G_2, H_1, H_2 be ideals of an N-group G such that $G_1 \cap G_2 = (0)$ and $H_i \subseteq G_i$ for $i = 1, 2$. If $H_1 + H_2$ is essential in $G_1 + G_2$, then H_i is essential in G_i for $i = 1, 2$.

Proof

Suppose $H_1 + H_2$ is essential in $G_1 + G_2$ and H_1 is not essential in G_1.
   Then there exists a nonzero ideal A of G such that $A \subseteq G_1$ and $A \cap H_1 = (0)$.
   To show $A \cap (H_1 + H_2) = (0)$, let $x \in A \cap (H_1 + H_2)$.
   Then $x \in A$ and $x = h_1 + h_2$ for some $h_1 \in H_1$, $h_2 \in H_2$.
   Now $-h_1 + x = h_2 \in (H_1 + A) \cap H_2 \subseteq G_1 \cap G_2 = (0)$.
   This means that $h_2 = 0$ and $x = h_1 \in H_1 \cap A = (0)$. Therefore, $x = 0$.
   Hence, $A \cap (H_1 + H_2) = (0)$.
   Since $H_1 + H_2$ is essential in $G_1 + G_2$ and $A \subseteq G_1 + G_2$, we have $A = (0)$, a contradiction.
   Thus, H_1 is essential in G_1. Similarly, we can show that H_2 is essential in G_2. ∎
   By the principle of mathematical induction, we have the following.

Corollary 3.4.5

Let G_1, G_2, ..., G_n, H_1, H_2, ..., H_n be ideals of an N-group G such that $H_i \subseteq G_i$ for $1 \leq i \leq n$, and the sum $G_1 + G_2 + ... + G_n$ is direct. If the sum $H_1 + H_2 + ... + H_n$ is essential in $G_1 + G_2 + ... + G_n$, then H_i is essential in G_i for $1 \leq i \leq n$.
   Next we establish the converse of Lemma 3.4.4 and Corollary 3.4.5. Before proving this, we first prove a couple of lemmas. We introduce the following notation.

Notation 3.4.6

For any nonempty subset A of an N-group G, we write

$$A^* = \{g + x - g \mid x \in A, \ g \in G\}$$
$$A^+ = \{n(g + x) - ng \mid x \in A, \ g \in G, \ n \in N\}$$
$$A^0 = \{x - y \mid x, \ y \in A\}$$

Note that $A \subseteq A^*$. Let X be a nonempty subset of G.
Define $X_0 = X$, and for any $i \geq 1$, $X_i = X_{i-1}^* \cup X_{i-1}^+ \cup X_{i-1}^0$.
Now for any $i \geq 1$, $X_{i-1} \subseteq X_{i-1}^* \subseteq X_i$. Therefore, $X_0 \subseteq X_1 \subseteq X_2 \subseteq \ldots$.
A straightforward verification gives us the following.

Lemma 3.4.7

If X is any nonempty subset of an N-group G, then the ideal generated by X is $<X> = \bigcup_{i=0}^{\infty} X_i$.

Lemma 3.4.8 (Reddy and Satyanarayana, 1988)

Let H_1 and H_2 be ideals of G such that $H_1 \cap H_2 = (0)$. Let $h_1 \in H_1$ and $h_2 \in H_2$. Then for any $x_1 \in <h_1>$, there exists an element $x_2 \in <h_2>$ such that $x_1 + x_2 \in <h_1 + h_2>$.

Proof

Write $X = \{h_1\}$, $Y = \{h_2\}$, and $Z = \{h_1 + h_2\}$.

Following Notation 3.4.6, by Lemma 3.4.7, we have $<h_1> = \bigcup_{k=0}^{\infty} X_k$, $<h_2> = \bigcup_{k=0}^{\infty} Y_k$, and $<h_1 + h_2> = \bigcup_{k=0}^{\infty} Z_k$.

For each integer $k \geq 0$, let $P(k)$ be the statement: for each $x \in X_k$, there exists an element $y \in Y_K$ such that $x + y \in Z_k$.

$P(0)$ is trivially true. Suppose $P(k-1)$ is true for some positive integer $k \geq 1$.

Let $x \in X_k$. Then either $x = x_1 - x_2$ for some $x_1, x_2 \in X_{k-1}$ or $x = g + x_1 - g$ for some $x_1 \in X_{k-1}$ and $g \in G$, or $x = n(g + x_1) - ng$ for some $g \in G, n \in N, x_1 \in X_{k-1}$.

Since $x_1, x_2 \in X_{k-1}$, by our supposition, there exist $y_1, y_2 \in Y_{k-1}$ such that $x_1 + y_1, x_2 + y_2 \in Z_{k-1}$.

If $x = x_1 - x_2$, then take $y = y_1 - y_2$.

If $x = g + x_1 - g$, then take $y = g + y_1 - g$.

If $x = n(g + x_1) - ng$, then take $y = n(g + x_1 + y_1) - n(g + x_1)$.

In all three cases, we have $y \in Y_k$. In the first case, since $H_1 \cap H_2 = (0)$, we have $x + y = (x_1 - x_2) + (y_1 - y_2) = (x_1 + y_1) - (x_2 + y_2)$.

In the second case, $x + y = (g + x_1 - g) + (g + y_1 - g) = g + (x_1 + y_1) - g$, and in the third case, we get $x + y = y + x = n(g + x_1 + y_1) - ng$. Therefore, $x + y \in Z_k$. Thus, $P(k)$ is true. ∎

We now prove the converse of Lemma 3.4.4.

Lemma 3.4.9

Let G_1, G_2, H_1, H_2 be ideals of G such that $G_1 \cap G_2 = (0)$ and $H_i \subseteq G_i$ for $i = 1, 2$. Then H_i is essential in G_i for $i = 1, 2$, which implies $H_1 + H_2$ is essential in $G_1 + G_2$.

Proof

Suppose H_1 and H_2 are essential in G_1 and G_2, respectively. Write $A_1 = H_1 + G_2$ and $A_2 = G_1 + H_2$. We now show that A_1 is essential in $G_1 + G_2$.
Let $0 \neq a \in G_1 + G_2$. Then $a = a_1 + a_2$ for some $a_1 \in G_1, a_2 \in G_2$.
   If $a_1 = 0$, then $a \in A_1$, and hence $<a> \cap A_1 \neq (0)$.
   If $a_1 \neq 0$, then since H_1 is essential in G_1 and $(0) \neq <a_1> \subseteq G_1$, there exists a nonzero element $x_1 \in <a_1> \cap H_1$.
   By Lemma 3.4.8, there exists an element x_2 in $<a_2>$ such that $x_1 + x_2 \in <a_1 + a_2>$.
   Since $x_1 \neq 0$, $x_1 + x_2$ is nonzero, and it belongs to $<a_1 + a_2> \cap A_1 = <a> \cap A_1$.
   Thus, A_1 is essential in $G_1 + G_2$.
   Similarly, we can show that A_2 is essential in $G_1 + G_2$.
   Now by Remark 3.4.3, $H_1 + H_2 = A_1 \cap A_2$ is essential in $G_1 + G_2$. ∎
   By finite mathematical induction, we have the following.

Corollary 3.4.10

Let $G_1, G_2, \ldots, G_n, H_1, H_2, \ldots, H_n$ be ideals of G such that the sum $G_1 + G_2 + \ldots + G_n$ is direct, and $H_i \subseteq G_i$ for $1 \leq i \leq n$. If H_i is essential in G_i for $1 \leq i \leq n$, then $H = H_1 + H_2 + \ldots + H_n$ is essential in $G_1 + G_2 + \ldots + G_n$.
   Combining Corollaries 3.4.5 and 3.4.10, we have the following.

Theorem 3.4.11

Let $G_1, G_2, \ldots, G_n, H_1, H_2, \ldots, H_n$ be ideals of G such that the sum $G_1 + G_2 + \ldots + G_n$ is direct and $H_i \subseteq G_i$ for $1 \leq i \leq n$. Then $H_1 + H_2 + \ldots + H_n$ is essential in $G_1 + G_2 + \ldots + G_n$, if and only if each H_i is essential in G_i for $1 \leq i \leq n$.
   A trivial verification shows us the following:

Proposition 3.4.12

Let H be a nonzero ideal of an N-group G. Then the following are equivalent.

(i) Every nonzero ideal of G that is contained in H is essential in H.
(ii) X_1, X_2 are two nonzero ideals of G such that $X_1 \subseteq H$, $X_2 \subseteq H$ implies $X_1 \cap X_2 \neq (0)$.

Definition 3.4.13

Let H be a nonzero ideal of an N-group G. Then H is said to be **uniform** if it satisfies any one of the two conditions of Proposition 3.4.12.

Definition 3.4.14

An ideal H of G is said to have **finite Goldie dimension (FGD)** if H contains no infinite direct sum of nonzero ideals of G.

Theorem 3.4.15

Let G be an N-group and H be an ideal of G. Then the following are equivalent.

(i) H has FGD.
(ii) For any increasing sequence $H_0 \subseteq H_1 \subseteq H_2 \subseteq \ldots$ of ideals of G with each $H_i \subseteq H$, there exists an integer k such that H_i is essential in H_{i+1} for $i \geq k$.

Proof

(i) $\Rightarrow$ (ii): Suppose (ii) fails. Then there exists a subsequence $\{H_{k_i}\}_{i=1}^{\infty}$ of $\{H_i\}_{i=1}^{\infty}$ such that for each I, H_{k_i} is not G-essential in $H_{k_{i+1}}$. Hence, for each i, there exists a nonzero ideal A_i of G such that $A_i \subseteq H_{k_{i+1}}$ and $H_{k_i} \cap A_i = (0)$.

One can easily verify that the sum $\sum_{1} A_i$ is a direct sum of nonzero ideals, which is contained in H, which is a contradiction to (i).

(ii) $\Rightarrow$ (i) is clear. ∎

Definitions 3.4.16

(i) A nonempty family $\{G_i\}_{i \in 1}$ of proper ideals of G is said to be a **direct system** if for any finite number of elements $i_1, i_2, \ldots, i_k$ of I, there is an element i_0 in I such that $G_{i_0} \supseteq G_{i_1} + \ldots + G_{i_k}$.

(ii) An ideal I of G is said to be **finitely generated** if there exists a finite number of elements $a_1, a_2, \ldots, a_n$ in I such that the ideal generated by $\{a_1, a_2, \ldots, a_n\}$ is equal to I.

Theorem 3.4.17 (Theorem 1.2 by Satyanarayana and Prasad, 2000)

Let G be an N-group. The following are equivalent:

(i) G has ACCI.
(ii) For any ideal J of G, the condition is that every direct system of ideals of G that are contained in J is bounded above by an ideal J^* of G, where $J^* \not\subseteq J$.
(iii) Every ideal J of G is finitely generated.

Proof

(i) $\Rightarrow$ (ii): Suppose G has ACCI. Let us suppose that there exists a direct system $\{J_i\}_{i \in I}$ of ideals of G that are properly contained in J, which is not bounded above by any ideal J^* of G with $J^* \not\subseteq J$. Without loss of generality, we may assume that each $J_i \neq (0)$.

Then $\Sigma_{i \in I} J_i$ is an ideal of G that is contained in J. Let $i_1 \in I$. Now $J_{i_1} \neq (0)$.

Since $\Sigma_{i \in I} J_i \not\subseteq J_{i_1}$, there exists $i_2 \in I$ such that $J_{i_2} \not\subseteq J_{i_1} \Rightarrow J_{i_1} \not\subseteq J_{i_1} + J_{i_2}$.

By the definition of a direct system, we get that $\Sigma_{i \in I^*} J_i$ is a proper subset of J for any finite subset I^* of I. Therefore, $J_{i_1} + J_{i_2}$ is properly contained in J.

Now, $\Sigma_{i \in I} J_i \not\subseteq J_{i_1} + J_{i_2} \Rightarrow$ there exists $i_3 \in I$ such that $J_{i_1} + J_{i_2} \not\subseteq J_{i_1} + J_{i_2} + J_{i_3}$.

This process can be continued, and eventually we get $J_{i_1} \not\subseteq J_{i_1} + J_{i_2} \not\subseteq J_{i_1} + J_{i_2} + J_{i_3} \not\subseteq \ldots$, which is a strict ascending chain of ideals of G, a contradiction to (i).

(ii) $\Rightarrow$ (iii): Let us suppose J is an ideal of G that is not finitely generated. Write

$S^* = \{I/I$ is an ideal of G, I is finitely generated, and $I \subseteq J\}$.

By (ii), S^* is bounded above by an ideal J^* of G such that $J^* \not\subseteq J$

$\Rightarrow$ there exists $x \in J\backslash J^*$

$\Rightarrow <x> \subseteq J$ and $<x> \not\subseteq J^* \Rightarrow <x> \in S^*$ and $<x> \not\subseteq J^*$, which is a contradiction.

(iii) $\Rightarrow$ (i): Suppose that every ideal of G is finitely generated.

Let us suppose G has no ACCI.

Then there exists a strict ascending chain $I_1 \not\subseteq I_2 \not\subseteq \ldots$ of ideals of G.

Consider $\cup_{i=1}^{\infty} I_i$. Now $\cup_{i=1}^{\infty} I_i$ is an ideal of G that is also finitely generated. Therefore, there exist $a_1, a_2, \ldots, a_k \in G$ such that $\cup_{i=1}^{\infty} I_i = <a_1, \ldots, a_k>$.

Now $a_1 \in I_{i_1}, \ldots, a_k \in I_{i_k}$ for some $i_1, i_2, \ldots, i_k$. Write $i^* = \max \{i_1, i_2, \ldots, i_k\}$.

Clearly, $I_{i_j} \subseteq I_{i^*}$ for $1 \leq j \leq k \Rightarrow a_1, a_2, \ldots, a_k \in I_{i^*}$

$\Rightarrow \cup_{i=1}^{\infty} I_i \subseteq I_{i^*} \Rightarrow I_{i^*+1} \subseteq I_{i^*}$, which is a contradiction.

Hence, G has ACCI. This completes the proof. ∎

3.5 Some Results on $M_n(N)$-Group N^n

Let N be a near ring and let N^n denote the Cartesian product of n copies of N. Then $N^n = \{(x_1, x_2, \ldots, x_n) \mid x_i \in N, 1 \leq i \leq n\}$ may be regarded as an $M_n(N)$-group N^n.

To verify this, take $f \in M_n(N)$ and $x \in N^n$. We know that $f \in M_n(N) \subseteq M(N^n)$ (the near ring of the set of all mappings from N^n to N^n). Now $f: N^n \to N^n$. So, $fx \in N^n$. Now we verified that $fx \in N^n$ for any two elements $f \in M_n(N)$ and $x \in N^n$. With respect to this operation, N^n becomes an $M_n(N)$-group N^n.

Result 3.5.1 (Proposition 4.1 of Meldrum and Van der Walt, 1986)

If L is a left ideal of N, then L^n is an ideal of the $M_n(N)$-group N^n.

Proof

First, we have to show that L^n is a normal subgroup of N^n.

Let $\alpha = (a_1, \ldots, a_n), \beta = (b_1, \ldots, b_n) \in L^n$. Now

$$\alpha - \beta = (a_1 - b_1, \ldots, a_n - b_n) \in L^n \text{ (since } L \text{ is a left ideal of } N).$$

Let $r = (r_1, \ldots, r_n) \in N^n$.

Now $r + \alpha - r = (r_1 + a_1 - r_1, \ldots, r_1 + a_1 - r_1) \in L^n$.

Therefore, $r + \alpha - r \in L^n$ for all $\alpha \in L^n$ and $r \in N^n$. Thus, L^n is a normal subgroup of N^n.

Suppose $\alpha = (a_1, \ldots, a_n) \in L^n$ and $r = (r_1, \ldots, r_n) \in N^n$.

Now we have to show that for any matrix $A \in M_n(N)$, there is $\alpha^1 \in L^n$ such that $A(r + \alpha) = Ar + \alpha^1$.

We will show this by induction on $w(A)$, the weight of A.

Let $w(A) = 1$ and $A = f_{ij}^s$.

Now $A(r + \alpha) = f_{kl}^s(r + \alpha)$

$$= (0, \ldots, s(r_j + a_j), \ldots, 0)$$

$$= (0, \ldots, sr_j - sr_j + s(r_j + a_j) - sr_j + sr_j, \ldots, 0)$$

$$= (0, \ldots, sr_j + a_j^1, \ldots, 0), \text{ where } a_j^1 \in L$$

$$= (0, \ldots, sr_j, \ldots, 0) + (0, \ldots, a_j^1, \ldots, 0)$$

$$= \left(f_{ij}^s\right)(r_1, \ldots, r_n) + \alpha^1, \text{ where } \alpha^1 \in L^n$$

$$= Ar + \alpha^1.$$

Suppose the preceding statement is true for all the matrices with weight less than m $(m \geq 2)$.

Let $w(m) = m$ and $A = B + C$ or $A = BC$, where $w(B), w(C) < m$.

Now $A(r+\alpha) = (B+C)(r+\alpha)$

$\qquad\qquad = B(r+\alpha)+C(r+\alpha)$

$\qquad\qquad = B(r+\alpha^1)+C(r+\alpha^{11})$, where $\alpha^1, \alpha^{11} \in L^n$

$\qquad\qquad = Br+Cr-Cr+\alpha^1+Cr+\alpha^{11} \in L^n$

$\qquad\qquad = (B+C)\,r+\alpha^{111}$, where $\alpha^{111} = -Cr+\alpha^1+Cr+\alpha^{11} \in L^n$

$\qquad\qquad = Ar+\alpha^{111}$.

Let $A = BC$.

Then $A(r+\alpha) = (BC)(r+\alpha)$

$\qquad\qquad = B[C\,(r+\alpha)]$

$\qquad\qquad = B[Cr+\alpha^{11}]$ (where $\alpha^{11} \in I^n_\cdot$)

$\qquad\qquad = B(Cr)+\alpha^{111}$ (where $\alpha^{111} \in L^n$)

$\qquad\qquad = (BC)r+\alpha^{111}$ (where $\alpha^{111} \in L^n$)

$\qquad\qquad = Ar+\alpha^{111}$, where $\alpha^{111} \in L^n$.

Let $A \in M_n(N)$, $r \in N^n$, and $\alpha \in L^n$. Now
$A(r + \alpha) - Ar = Ar + \alpha^1 - Ar \in L^n$ (since $\alpha^1 \in L^n$ and $Ar \in N^n$).
Therefore, L^n is an ideal of the $M_n(N)$-group N^n. ∎

Notation 3.5.2

For any ideal I of N^n, we write $I_* = \{x \in N \mid x = \pi_j A$ for some $A \in I, 1 \le j \le n\}$, where π_j is the jth projection map from N^n to N.

Lemma 3.5.3 (Lemma 1.2 of Satyanarayana and Prasad, 2005)

Let I be an ideal of N^n. Then

$$I_{**} = \{x \in N \mid (x, 0, ..., 0) \in I\}.$$

Proof

Let $x \in$ RHS. This implies $(x, 0, ..., 0) \in I$. Then $\pi_1(x, 0, ..., 0) = x \in I_{**}$.

Take $x \in I_{**} \Rightarrow x = \pi_j A$ for some $A = (x_1, ..., x_n) \in I \Rightarrow x = x_j$. Since $f^1_{1j} \in M_n(N)$ and $A \in I$, by Proposition 3.2.17 $(x_j, 0, ..., 0) = f^1_{1j} A \in I$, which implies $x_j \in$ RHS. The proof is complete. ∎

Lemma 3.5.4 (Lemma 1.3 of Satyanarayana and Prasad, 2005)

I_{**} is a left ideal of N.

Proof

Let $n \in N$ and $x \in I_{**}$. Consider $(n + x - n, 0, ..., 0)$.

Since I is an ideal of N^n, we have

$(n + x - n, 0, ..., 0) = (n, 0, ..., 0) + (x, 0, ..., 0) - (n, 0, ..., 0) \in I$.

Therefore, $n + x - n \in I_{**}$. This shows that I_{**} is a normal subgroup of N.

Now for any $n_1, n_2 \in N$ and $x \in I_{**}$, we have

$(n_1(n_2 + x) - n_1 n_2, 0, ..., 0) = f_{11}^m((n_2, 0, ..., 0) + (x, 0, ..., 0)) - f_{11}^m(n_2, 0, ..., 0) \in I$,

which implies that $n_1(n_2 + x) - n_1 n_2 \in I_{**}$. Hence, I_{**} is a left ideal of N. ∎

Proposition 3.5.5 (Theorem 1.4 of Satyanarayana and Prasad, 2005)

Suppose L is a subset of N. Then L is an ideal of N-group N, if and only if L^n is an ideal of N^n.

Proof

Suppose L^n is an ideal of N^n.

Now $x \in L$, if and only if $(x, 0, ..., 0) \in L^n$, if and only if $x \in (L^n)_{**}$ (by Lemma 3.5.3).

Therefore, $L = (L^n)_{**}$. By Lemma 3.5.4, $L = (L^n)_{**}$ is an ideal of N. ∎

The converse follows from Result 3.5.1.

Lemma 3.5.6 (Lemma 1.5 of Satyanarayana and Prasad, 2005)

If I is an ideal of N^n, then $(I_{**})^n = I$.

Proof

Let $(x_1, x_2, ..., x_n) \in (I_{**})^n \Rightarrow x_i \in I_{**}$ for $1 \le i \le n$.

Then $(x_i, 0, ..., 0) \in I$ (by Lemma 3.5.3).

Since I is an ideal of N^n, we have $f_{i1}^1(x_i, 0, ..., 0) \in I$.

This implies $(0, ..., x_i, ..., 0) \in I$ for $1 \le i \le n$.

Therefore $(x_1, x_2, ..., x_n) = \sum_i (0, ..., \overset{ith}{x_i}, ..., 0) \in I$.

Hence $(I_{**})^n \subseteq I$.

The other part is clear from Notation 3.5.2. ∎

Remark 3.5.7

Suppose I, J are ideals of N. Then

(i) $(I \cap J)^n = I^n \cap J^n$ and
(ii) $I \cap J = (0) \Leftrightarrow (I \cap J)^n = (0) \Leftrightarrow I^n \cap J^n = (0)$.

Proof

(i) Clearly, $I \cap J \subseteq I$, $I \cap J \subseteq J$, and so $(I \cap J)^n \subseteq I^n$, $(I \cap J)^n \subseteq J^n$.
     Hence $(I \cap J)^n \subseteq I^n \cap J^n$. On the other hand, take $(x_1, x_2, \ldots, x_n) \in I^n \cap J^n$.
     Then $(x_1, \ldots, x_n) \in I^n$ and $(x_1, \ldots, x_n) \in J^n$.
     This implies $x_i \in I$, $x_i \in J$ for $1 \leq i \leq n \Rightarrow x_i \in I \cap J$ for $1 \leq i \leq n$
     $\Rightarrow (x_1, x_2, \ldots, x_n) \in (I \cap J)^n$. Therefore, $I^n \cap J^n \subseteq (I \cap J)^n$.
     Hence $(I \cap J)^n = I^n \cap J^n$.
(ii) $I \cap J = (0) \Leftrightarrow (I \cap J)^n = (0)$, and the rest follows from (i).   ∎

Note 3.5.8

(i)   If u is an element of N, then

$$<(0, \ldots, \underset{ith}{u}, \ldots, 0)> = <(u, 0, \ldots, 0)>.$$

(ii) If U is an ideal of N and I is an ideal of N^n such that $I \subseteq U^n$, then
      $I_{**} \subseteq U$.
(iii) If $A \leq_e N$, then $A^n \leq_e N^n$.

Verification

(i) Suppose u is an element of N. Then $(u, 0, \ldots, 0) \in N^n$.
     Now $(u, 0, \ldots, 0) = f_{1i}^1 (0, \ldots, \underset{ith}{u}, 0, \ldots, 0) \in <(0, \ldots, \underset{ith}{u}, \ldots, 0)>$.
     Therefore, $<(u, 0, \ldots, 0)> \subseteq <(0, \ldots, \underset{ith}{u}, 0, \ldots, 0)>$.
     Consider $(0, \ldots, \underset{ith}{u}, \ldots, 0)$.
     Now $(0, \ldots, \underset{ith}{u}, \ldots, 0) = f_{i1}^1 (u, 0, \ldots, 0) \in <(u, 0, \ldots, 0)>$.
     Therefore, $<(0, \ldots, \underset{ith}{u}, \ldots, 0)> \subseteq <(u, 0, \ldots, 0)>$.
     Thus, $<(0, \ldots, \underset{ith}{u}, \ldots, 0)> = <(u, 0, \ldots, 0)>$.
(ii) Suppose U is an ideal of N and I is an ideal of N^n such that $I \subseteq U^n$.
     Now $x \in I_{**}$.
     By Lemma 3.5.3 $(x, 0, 0, \ldots, 0) \in I$.
     Since $I \subseteq U^n$, we have $(x, 0, 0, \ldots, 0) \in U^n$.
     This implies that $x \in U$. Therefore, $I^{**} \subseteq U$.
(ii) Suppose $A^n \cap I = (0)$ and I is an ideal of N^n. By Lemma 3.5.6, $I = (I_{**})^n$.
     Therefore, $A^n \cap (I_{**})^n = (0)$. Now by Remark 3.5.7, we get $(A \cap I_{**})^n = (0)$.
     Again, by Remark 3.5.6, $A \cap I_{**} = (0)$. Since $A \leq_e N$, we get that $I_{**} = (0)$,
     and hence $I = (0)$.

Lemma 3.5.9 (Lemma 1.8 of Satyanarayana and Prasad, 2005)

If x, u are elements of N and $x \in <u>$, then $(x, 0, \ldots, 0) \in <(u, 0, \ldots, 0)>$.

Proof

Following Notation 3.4.6, we have $< u > = \bigcup_{i=1}^{\infty} A_{i+1}$, where $A_{i+1} = A_i^* \cup A_i^\circ \cup A_i^+$ with $A_o = \{u\}$. Here

$$A_i^* = \{m + y - m \mid m \in N, \ y \in A_i\};$$
$$A_i^\circ = \{a - b \mid a, \ b \in A_i\}; \text{ and}$$
$$A_i^+ = \{n_1(n_2 + a) - n_1 n_2 \mid n_1, \ n_2 \in N \text{ and } a \in A_i\}.$$

Since $x \in <u>$, we have $x \in A_k$ for some k. Now we show (by induction on k) that $x \in A_k \Rightarrow (x, 0, ..., 0) \in <(u, 0, ..., 0)>$.

Suppose $k = 0$. Then $x \in A_k = A_o = \{u\} \Rightarrow x = u$.

Therefore $(x, 0, ..., 0) = (u, 0, ..., 0) \in <(u, 0, ..., 0)>$.

This shows that the statement is true for $k = 0$.

Suppose the induction hypothesis for $k = i$, that is,

$$x \in A_i \Rightarrow (x, 0, ..., 0) \in <(u, 0, ..., 0)>. \text{ Let } x \in A_{i+1} = A_i^* \cup A_i^\circ \cup A_i^+.$$

Case (i): Suppose $x \in A_i^*$.

$x \in A_i^* \Rightarrow x = m + y - m$ for some $m \in N, y \in A_i$. Now

$y \in A_i \Rightarrow (y, 0, ..., 0) \in <(u, 0, ..., 0)>$, and so

$(x, 0, ..., 0) = (m, 0, ..., 0) + (y, 0, ..., 0) - (m, 0, ..., 0) \in <(u, 0, ..., 0)>$.

Case (ii): Suppose $x \in A_i^\circ$.

$x \in A_i^\circ \Rightarrow x = a - b$ for some $a, b \in A_i$. Now $a, b \in A_i$

$\Rightarrow (a, 0, ..., 0), (b, 0, ..., 0) \in <(u, 0, ..., 0)>$, and so

$(x, 0, ..., 0) = (a - b, 0, ..., 0) = (a, 0, ..., 0) - (b, 0, ..., 0) \in <(u, 0, ..., 0)>$ (by the induction hypothesis).

Case (iii): Suppose $x \in A_i^+$

$x \in A_i^+ \Rightarrow x = n_1(n_2 + a) - n_1 n_2$ for some $n_1, n_2 \in N$ and $a \in A_i$.

Now $(x, 0, ..., 0) = f_{11}^m ((n_2, 0, ..., 0) + (a, 0, ..., 0)) - f_{11}^m (n_2, 0, ..., 0) \in <(u, 0, ..., 0)>$.

Therefore $(x, 0, ..., 0) \in < (u, 0, ..., 0) >$.

Thus, by mathematical induction we have that $(x, 0, ..., 0) \in <(u, 0, ..., 0)>$ for any $x \in <u>$. ∎

Theorem 3.5.10 (Theorem 1.9 of Satyanarayana and Prasad, 2005)

If u is an element of N, then $< u >^n = <(u, 0, ..., 0)>$.

Proof

Since $(u, 0, ..., 0) \in <u>^n$, we have $<(u, 0, ..., 0)> \subseteq <u>^n$.

Take $(x_1, x_2, ..., x_n) \in <u>^n$

$\Rightarrow x_i \in <u>$ for $1 \leq i \leq n$.

By Lemma 3.5.9 $(x_i, 0, ..., 0) \in <(u, 0, ..., 0)>$ for all i.

Now $(x_1, x_2, ..., x_n) = (x_1, 0, ..., 0) + f_{21}^1 (x_2, 0, ..., 0) + ... + f_{n1}^1 (x_n, 0, ..., 0) \in <(u, 0, ..., 0)>$ (by Proposition 3.2.17).

Therefore, $<u>^n \subseteq <(u, 0, ..., 0)>$. This completes the proof. ∎

Proposition 3.5.11 (Lemma 2.1 of Satyanarayana and Prasad, 2005)

For any $x_1, x_2, ..., x_n \in N$, we have $(<x_1> + <x_2> + ... + <x_n>)^n = <x_1>^n + <x_2>^n + ... + <x_n>^n$.

Proof

Since $<x_i> \subseteq <x_1> + ... + <x_n>$, we have

$<x_i>^n \subseteq (<x_1> + ... + <x_n>)^n$ for $1 \leq i \leq n$.

Therefore, $<x_1>^n + ... + <x_n>^n \subseteq (<x_1> + ... + <x_n>)^n$.

Let $(a_1, a_2, ..., a_n) \in (<x_1> + ... + <x_n>)^n$

$\Rightarrow a_i \in <x_1> + ... + <x_n>$ for $1 \leq i \leq n$

$\Rightarrow a_i = a_{i_1} + ... + a_{i_n}$, where $a_{ij} \in <x_j>$ for $1 \leq i \leq n, 1 \leq j \leq n$.

Now $(a_1, a_2, ..., a_n) = (a_{1_1} + ... + a_{1_n}, a_{2_1} + ... + a_{2_n}, ..., a_{n_1} + ... + a_{n_n}) = (a_{1_1}, a_{2_1}, ..., a_{n_1}) + (a_{1_2}, a_{2_2}, ..., a_{n_2}) + ... + (a_{1_n}, a_{2_n}, ..., a_{n_n}) \in <x_1>^n + <x_2>^n + ... + <x_n>^n$.

Therefore $(<x_1> + <x_2> + ... + <x_n>)^n \subseteq <x_1>^n + <x_2>^n + ... + <x_n>^n$.

The proof is complete. ∎

4

Prime and Semiprime Ideals

4.1 Fundamental Definitions and Results

We begin this section with the following notation.

Notation 4.1.1

If S and T are subsets of N, then we write $ST = \{st \mid s \in S \text{ and } t \in T\}$. For any natural number n, we write $S^n = S \times S \times \ldots \times S$ (n times).

Proposition 4.1.2 (Proposition 2.57 of Pilz, 1983)

Let N be a near ring. Then

   (i) For subsets R, S, T of N, $(RS)T = R(ST)$.
  (ii) If $h: N \to \bar{N}$ is a homomorphism, then for all subsets S, T of N, $h(ST) = h(S) \, h(T)$, and for all subsets $\bar{S}, \bar{T}$ of $\bar{N}$, $h^{-1}(\overline{ST}) \supseteq h^{-1}(S)h^{-1}(T)$.
 (iii) For all ideals I of N, and for all subsets S, T of N, $(S + I)(T + I) = ST + I$.

Proof

   (i) Take $x \in (RS)T$. Then $x = (rs)t$ for some $r \in R$, $s \in S$, and

$$t \in T = r(st) \in R(ST).$$

       Therefore, $(RS)T \subseteq R(ST)$.
       In a similar way, we can verify that $R(ST) \subseteq (RS)T$.
       Therefore, $(RS)T = R(ST)$.
  (ii) Since h is a homomorphism, for all $s \in S$ and $t \in T$, it follows that $h(st) = h(s)h(t)$.
       Let $\bar{s} \in \bar{S}$ and $\bar{t} \in \bar{T}$. Since h is a homomorphism, it follows that $h^{-1}(\bar{s})h^{-1}(t) = h^{-1}(\bar{s} \, \bar{t}) \in h^{-1}(\overline{ST})$.

(iii) Let I be an ideal of N. Define a mapping $\pi\colon N \to N/I$ by $\pi(n) = n + I$; then π is a canonical epimorphism. Now $\pi(st) = \pi(s) \cdot \pi(t)$ for all $s, t \in N$. Therefore, $st + I = (s + I)(t + I)$ for all $s, t \in N$. Hence, $ST + I = (S + I)$ $(T + I)$. ∎

Definition 4.1.3

(i) An ideal P of N is a called a **prime ideal**, if for all ideals I, J of N, $IJ \subseteq P \Rightarrow I \subseteq P$ or $J \subseteq P$.

(ii) An ideal S of N is said to be a **semiprime,** if for any ideal I of N, $I^2 \subseteq S \Rightarrow I \subseteq S$.

(iii) An ideal P of N is said to be a **completely prime** ideal, if for $a, b \in N$, $ab \in P \Rightarrow$ either $a \in P$ or $b \in P$.

(iv) An ideal S of N is said to be a **completely semiprime** ideal, if for any element $a \in N$, $a^2 \in S \Rightarrow a \in S$.

Definition 4.1.4

A near ring N is said to be

(i) a prime near ring if (0) is a prime ideal;

(ii) a completely prime near ring (or integral near ring) if (0) is completely a prime ideal;

(iii) a semiprime near ring if (0) is a semiprime ideal;

(iv) a completely semiprime near ring if (0) is completely a semiprime ideal.

Notation 4.1.5

If S is a subset of N, then $<S>$ (or (S)) denotes the ideal generated by S. If $S = \{a\}$, then we write $<a>$ (or (a)) for $<S>$.

Proposition 4.1.6 (Proposition 2.61 of Pilz, 1983)

For an ideal P of N, and ideals I, J of N, the following conditions are equivalent:

(i) P is a prime ideal.

(ii) $IJ \subseteq P \Rightarrow I \subseteq P$ or $J \subseteq P$.

(iii) $a \notin P$ and $b \notin P \Rightarrow (a)(b) \not\subseteq P$, for all $a, b \in N$.

(iv) If I contains P properly, and J contains P properly, then $IJ \not\subseteq P$.

(v) $I \not\subseteq P$ and $J \not\subseteq P \Rightarrow IJ \not\subseteq P$.

Proof

We prove the following implications:

(i) $\Leftrightarrow$ (ii) $\Leftrightarrow$ (iii), (i) $\Rightarrow$ (iii),
(iii) $\Rightarrow$ (iv), (iv) $\Rightarrow$ (v).

To prove (i) $\Rightarrow$ (ii), assume (i).
Suppose that $IJ \subseteq P$. Since P is prime, we have $I \subseteq P$ or $J \subseteq P$. Therefore,
(i) $\Rightarrow$ (ii).
(ii) $\Rightarrow$ (i) is clear.

To prove (ii) $\Rightarrow$ (iii), assume (ii).
Suppose $a \notin P$ and $b \notin P$. Then $(a) \not\subseteq P$ and $(b) \not\subseteq P$.
If $(a)(b) \subseteq P$, then by (ii), it follows that either $(a) \subseteq P$ or $(b) \subseteq P$, which is
a contradiction.
Hence, $(a)(b) \not\subseteq P$. The proof for (ii) $\Rightarrow$ (iii) is complete.

To prove (iii) $\Rightarrow$ (ii), assume (iii).
Suppose I, J are ideals of N such that $(IJ) \subseteq P$. If possible, assume that I
$\not\subseteq P$ and $J \not\subseteq P$.
Then by (iii), $(IJ) \not\subseteq P$, which is a contradiction. Therefore, $I \subseteq P$ or $J \subseteq P$.
Therefore, the proof for (iii) $\Rightarrow$ (ii) is complete.

To prove (i) $\Rightarrow$ (iii), assume (i).
Suppose $a \notin P$ and $b \notin P$, where $a, b \in N$.
If possible, suppose that $(a)(b) \subseteq P$. Then since P is a prime ideal, we
have $(a) \subseteq P$ or $(b) \subseteq P$. This implies that $a \in P$ or $b \in P$, which is a
contradiction.
Therefore, $(a)(b) \not\subseteq P$. The proof for (i) $\Rightarrow$ (iii) is complete.

To prove (iii) $\Rightarrow$ (iv), assume (iii).
Suppose I and J are ideals of N such that $I \supset P$ and $J \supset P$.
Choose $a \in I \backslash P$ and $b \in J \backslash P$. This means that $a \notin P$ and $b \notin P$.
By condition (iii), we get $(a)(b) \not\subseteq P$. Hence, $IJ \not\subseteq P$.
This proves (iii) $\Rightarrow$ (iv).

To prove (iv) $\Rightarrow$ (v), assume (iv).
Suppose $I \not\subseteq P$ and $J \not\subseteq P$. Take $a \in I \backslash P$ and $b \in J \backslash P$.
Then $(a) + P$ contains P properly, and $(b) + P$ contains P properly.
By condition (iv), it follows that $((a) + P) ((b) + P) \not\subseteq P$.
Therefore, there exist $a^1 \in (a), b^1 \in (b), p, p^1 \in P$ such that $(a^1 + p) (b^1 + p^1) \in P$.
This implies that $a^1(b^1 + p^1) + p(b^1 + p^1) \notin P$.
Therefore, $a^1(b^1 + p^1) - a^1b^1 + p(b^1 + p^1) \notin P$.
Since P is an ideal, we have $a^1(b^1 + p^1) - a^1b^1 \in P$, $p(b^1 + p^1) \in P$, and $a^1b^1 \notin P$.
Hence, $IJ \not\subseteq P$. The proof for (iv) $\Rightarrow$ (v) is complete. ∎

Theorem 4.1.7 (Proposition 2.62 of Pilz, 1983)

Let $(P_\alpha)_{\alpha \in A}$ be a family of prime ideals totally ordered by inclusion. Then $\bigcap_{\alpha \in A} P_\alpha = P$ is also a prime ideal.

Proof

We define an ordering on the set A as follows.
  For any $\alpha, \beta \in A, \alpha \leq \beta \Rightarrow P_\alpha \subseteq P_\beta$.
  Clearly, $\bigcap_{\alpha \in A} P_\alpha$ is an ideal.
  Let I, J be ideals of N such that $IJ \subseteq P$.
  Now $IJ \subseteq \bigcap_{\alpha \in A} P_\alpha$. That is, $IJ \subseteq P_\alpha$ for all $\alpha \in A$.
  Since each P_α is prime, it follows that $I \subseteq P_\alpha$ or $J \subseteq P_\alpha$ for each α.

$$\text{Suppose } I \not\subseteq P_{\alpha_1} \text{ for some } \alpha_1 \in A. \tag{4.1}$$

  Since $P_{\alpha 1}$ is prime, it follows that $J \subseteq P_{\alpha 1}$. Now for any $\beta \geq \alpha_1$, we have $J \subseteq P_\beta$.
  Next we show that for all $r < \alpha_1, J \subseteq P_r$. Let us suppose that there exists $r \in A$ such that $r < \alpha_1$ and $J \subseteq P_r$. Since P_r is prime, it follows that $I \subseteq P_r$, and so $I \subseteq P_{\alpha_1}$, which is a contradiction to the assumption (4.1).
  Therefore, $J \subseteq P_r$ for all $r < \alpha_1$. This shows that $J \subseteq P_\alpha$ for all $\alpha \in A$.
  That is, $J \subseteq \bigcap_{\alpha \in A} P_\alpha = P$. Hence, P is a prime ideal. ∎

Proposition 4.1.8 (Proposition 2.63 of Pilz, 1983)

If I is an ideal of N that is a direct summand, and P is a prime ideal of N, then $P \cap I$ is a prime ideal in I.

Proof

Suppose J_1 and J_2 are ideals of I (by considering I as a near ring) such that $J_1 J_2 \subseteq P \cap I$. This implies that $J_1 J_2 \subseteq P$ and $J_1 J_2 \subseteq I$.
  Since I is a direct summand of N, by Theorem 2.3.9, it follows that J_1, J_2 are ideals of N.
  Since P is a prime ideal of N and $J_1 J_2 \subseteq P$, it follows that $J_1 \subseteq P$ or $J_2 \subseteq P$.
  Therefore, $J_1 \subseteq P \cap I$ or $J_2 \subseteq P \cap I$. Hence, $P \cap I$ is a prime ideal in I. ∎

Proposition 4.1.9

Suppose I and P are ideals of N such that $I \subseteq P$, and if $\pi: N \to N/I = \bar{N}$ is the canonical epimorphism, then P is a prime ideal of N if and only if $\pi(P)$ is a prime ideal of $\bar{N}$.

Proof

Suppose P is prime. Let I and J be ideals of N/I such that $IJ \subseteq \pi(P)$.
Substitute $I_1 = \pi^{-1}(I)$ and $J_1 = \pi^{-1}(J)$.
Now $I_1 J_1 = \pi^{-1}(I)\pi^{-1}(J) \subseteq \pi^{-1}(IJ) \subseteq \pi^{-1}(\pi(P)) = P$.
This shows that $I_1 J_1 \subseteq P$. Since P is prime, it follows that $I_1 \subseteq P$ or $J_1 \subseteq P$.
Therefore, $J = \pi(\pi^{-1}(J)) = \pi(J_1) \subseteq \pi(P)$ or $I = \pi(\pi^{-1}(I)) = \pi(I_1) \subseteq \pi(P)$.
Thus, we have proved that $IJ \subseteq \pi(P)$ implies either $I \subseteq \pi(P)$ or $J \subseteq \pi(P)$.
Hence, $\pi(P)$ is a prime ideal of $\bar{N}$. ∎

Definition 4.1.10

(i) A near ring without zero divisors is called an **integral near ring**.
(ii) A near ring N is called a **zero near ring** if $NN = \{0\}$.

Example 4.1.11

Every integral near ring is a prime near ring.

Verification

Suppose N is an integral near ring.
To show N is a prime near ring, it is enough to show (0) is a prime ideal.
Let I and J be ideals of N such that $IJ \subseteq (0)$.
If either $I = (0)$ or $J = (0)$, then there is nothing to prove.
If possible, suppose that $I \neq \{0\}$ and $J \neq \{0\}$. Then we can choose $0 \neq a \in I$ and $0 \neq b \in J$ such that $ab = 0$, which is a contradiction to the fact that N is an integral.
Therefore, either $I = (0)$ or $J = (0)$.
Thus, we have proved that (0) is a prime ideal of N. Hence, N is a prime near ring.

Corollary 4.1.12 (Proposition 2.67 of Pilz, 1983)

Let I be an ideal of N. Then I is a prime ideal if and only if N/I is a prime near ring.

Proof

We know that $\pi: N \to N/I$ is a canonical epimorphism.
By Proposition 4.1.9, we come to a conclusion that I is prime if and only if $\pi(I)$ is prime, that is, I is a prime ideal if and only if N/I is a prime near ring. ∎

Example 4.1.13

If N is a constant near ring, then each normal subgroup of $(N, +)$ is a prime ideal.

Proof

Suppose N is a constant near ring. Then $nk = n$ for all $n, k \in N$.
Let M be a normal subgroup of $(N, +)$.
Let $n \in N$. Since N is constant, we have $mn = m \in M$ for all $m \in M$.
Therefore, $MN \subseteq M$, and so M is a right ideal of N.
Let $n, n^1 \in N$ and $m \in M$.
Now $n(n' + m) - nn' = n - n = 0 \in M$ (since N is constant).
This shows that M is a left ideal of N. Therefore, M is an ideal of N.
Next we show that M is a prime ideal of N.
Let I and J be ideals of N such that $IJ \subseteq M$. Now $I = IJ \subseteq M$.
Hence, M is a prime ideal of N.   ∎

Corollary 4.1.14

Each constant near ring is a prime near ring.

Proof

We know that (0) is a normal subgroup of N. Since N is a constant near ring, by Example 4.1.13, it follows that (0) is a prime ideal. Thus, N is a prime near ring.   ∎

Proposition 4.1.15 (Proposition 2.70 of Pilz, 1983)

If the near ring N is simple, then either N is prime or N is a zero near ring.

Proof

Suppose N is not a zero near ring. Then $NN \neq (0)$. We have to prove that (0) is a prime ideal of N. Suppose I and J are ideals of N such that $IJ \subseteq (0)$. Since I and J are ideals of N, and N is simple, it follows that $I, J \in \{(0), N\}$. If $I = N$ and $J = N$, and then $NN = IJ \subseteq (0)$, which is a contradiction. Therefore, $I = (0)$ or $J = (0)$. Thus, (0) is a prime ideal of N. Hence, N is a prime near ring.   ∎

Proposition 4.1.16 (Proposition 2.71 of Pilz, 1983)

If I is a maximal ideal of N, then either I is a prime ideal or $N^2 \subseteq I$.

Proof

Suppose I is a maximal ideal of N; then N/I is simple. Therefore, the only ideals of N/I are $0 + I = I$ and N/I. By Proposition 4.1.15, N/I is a prime near ring or a zero near ring. If N/I is a prime near ring, then I is a prime ideal. If N/I is a zero near ring, then $(N/I)(N/I) = I$, and this implies that $N^2 \subseteq I$. The proof is complete. ∎

Definition 4.1.17

Let H and K be subsets of an N-group G. Then we define $(H : K)$ as follows: $(H : K) = \{n \in N \mid nK \subseteq H\}$. The set $(0 : H)$ is called the **annihilator** of H.

Note 4.1.18

From Definition 4.1.17, if $H = \{h\}$, then $(H : K) = (h : K)$. Similarly, for $(H : k)$ if $K = \{k\}$.

Theorem 4.1.19 (Proposition 1.42 of Pilz, 1983)

Let H and K be subsets of G. If H is a subgroup (normal subgroup, N-subgroup, ideal, respectively) of an N-group G, then $(H : K)$ is a subgroup (normal subgroup, N-subgroup, ideal, respectively) in the N-group N.

Proof

Part 1: Suppose that H is a subgroup of G. To show that $(H : K)$ is a subgroup of N, take $n_1, n_2 \in (H : K)$. Then $n_1 k \subseteq H$ and $n_2 k \subseteq H$, for all $k \in K$.

Now $(n_1 - n_2)k = n_1 k - n_2 k \in H$. Therefore, $(H : K)$ is a subgroup of N.

Part 2: Suppose H is normal. To show that $(H : K)$ is normal, take $a \in (H : K)$ and $n \in N$. Now for all $k \in K$ $(n + a - n)k = nk + ak - nk \in H$ (since $nk \in G$, $ak \in H$, and H is normal).

Therefore, $n + a - n \in (H : K)$. Hence, $(H : K)$ is a subgroup of N.

Part 3: Suppose H is an N-subgroup of G. To show $(H : K)$ is an N-subgroup of N, we have to show that $N(H : K) \subseteq (H : K)$. Let $n \in N$ and $a \in (H : K)$.

Since $a \in (H : K)$, it follows that $ak \in H$ for all $k \in K$. Since H is an N-subgroup, it follows that $nak \in H$. This implies that $na \in (H : K)$. This shows that $(H : K)$ is an N-subgroup of N.

Part 4: Suppose that H is an ideal of an N-group G. To show that $(H : K)$ is an ideal of N, take $n_1, n_2 \in N$, $a \in (H : K)$. Since $a \in (H : K)$, it follows that $ak \in H$ for all k.

Now for any $k \in K$, we have

$$(n_1(n_2 + a) - n_1 n_2)k = n_1(n_2 + a)k - n_1 n_2 k$$
$$= n_1(n_2 k + ak) - n_1 n_2 k \in H$$

(since H is an ideal of G, and $ak \in H$).

This shows that $(H : K)$ is an ideal of N.   ∎

Corollary 4.1.20 (Corollary 1.43 of Pilz, 1983)

Let G be an N-group. Then
$(0 : a)$ is a left ideal of N for all $a \in G$.

Proof

Take $a \in G$. By Theorem 4.1.19, it is clear that $(0 : a)$ is a normal subgroup of N. Now we verify the condition for a left ideal.

Let $x \in (0 : a)$, $n_1, n_2 \in N$.

Now $(n_1(n_2 + x) - n_1 n_2)a = n_1(n_2 + x)a - n_1 n_2 a$
$$= n_1(n_2 a + xa) - n_1 n_2 a$$
$$= n_1(n_2 a + 0) - n_1 n_2 a = 0.$$

Therefore, $(n_1(n_2 + x) - n_1 n_2) \in (0 : a)$. Hence, $(0 : a)$ is a left ideal of N.   ∎

Theorem 4.1.21 (Proposition 1.45 of Pilz, 1983)

(i) Let H be a subset of an N-group G. Then $(0 : H) = \bigcap_{h \in H}(0 : h)$, and $(0 : H)$ is a left ideal.

(ii) $(0 : H)$ is an ideal of N for all N-subgroups H of G.

Proof

(i) Take $h \in H$. Let $x \in (0 : h)$. Then $xh = 0$. This is true for all $h \in H$.
    Therefore, $xH = 0$. Hence, $x \in (0 : H)$. On the other hand, take $x \in (0 : H)$.
    Then $xH = (0)$. This means $xh = 0$ for all $h \in H$.
    This implies that $x \in (0 : h)$ for all $h \in H$, that is, $x \in \bigcap_{h \in H}(0 : h)$.
    Therefore, $(0 : H) = \bigcap_{h \in H}(0 : h)$.
    By Corollary 4.1.20, each $(0 : h)$ is a left ideal and $(0 : H) = \bigcap_{h \in H}(0 : h)$ is a left ideal.

(ii) By (i), $(0 : H)$ is a left ideal. To show that $(0 : H)$ is a right ideal, take
   $n \in N$ and $x \in (0 : H)$.
   This implies that $n \in N$ and $xH \subseteq (0)$
   $\Rightarrow xnH \subseteq xH$ (since H is an N-subgroup) $\subseteq (0)$
   $\Rightarrow xn \in (0 : H)$. This shows that $(0 : H)$ is a right ideal. ∎

Now we present the definitions and important results of "nilpotent elements" and "nilpotent ideals" in near rings.

Definition 4.1.22

(i) An element $n \in N$ is called **nilpotent** if there exists a positive integer k such that $n^k = 0$.
(ii) A subset S of N is called **nilpotent** if there exists a positive integer k such that $S^k = 0$.
(iii) A subset S of N is called **nil** if every element of S is a nilpotent element.

Result 4.1.23 (Remark 2.97 of Pilz, 1983)

(i) Let S be a subset of N. If S is nilpotent, then S is nil.
(ii) If $S \subseteq T \subseteq N$ and T is nil (nilpotent, respectively), then S is nil (nilpotent, respectively).

Proof

(i) Suppose S is nilpotent. Then there exists a positive integer k such that $S^k = (0)$. This implies that $s^k = 0$ for all $s \in S$. Therefore, s is nilpotent for all $s \in S$, and hence S is nil.
(ii) Suppose $S \subseteq T \subseteq N$ and T is nil. Then t is nilpotent for all $t \in T$. This implies that t is nilpotent for all $t \in S$ (since $S \subseteq T$). Therefore, S is nil.

The other part of the proof is similar. ∎

Theorem 4.1.24 (Theorem 2.100 of Pilz, 1983)

Let I be an ideal of N. Then N is nil (nilpotent, respectively) if and only if I and N/I are nil (nilpotent, respectively).

Proof

Suppose that N is nilpotent. Since $I \subseteq N$ and N is nilpotent, I is also nilpotent. Also, since N is nilpotent, there exists a positive integer k such that $N^k = (0)$.

Now $(N/K)^k = N^k/I = \{0 + I\}$ (since $N^k = \{0\}$). Therefore, N/I is nilpotent.

Converse

Suppose that I and N/I are nilpotent. Since N/I is nilpotent, there exists a positive integer t such that $(N/I)^t = \{0 + I\}$, that is, $N^t \subseteq I$. Since I is nilpotent, there exists a positive integer s such that $I^s = \{0\}$. Therefore, $(N^t)^s \subseteq I^s = \{0\}$. Therefore, $N^{ts} = \{0\}$. Therefore, N is nilpotent.   ∎

Now we introduce the notion of a "prime left ideal" in near rings.

Definition 4.1.25 (Satyanarayana, Lokeswararao, and Prasad, 1996)

A left ideal P of N is said to be a prime left ideal if it satisfies the following condition:

  I, J are left ideals of N such that $IJ \subseteq P$ implies that $I \subseteq P$ or $J \subseteq P$.

Proposition 4.1.26 (Proposition 1.2 of Satyanarayana, Lokeswararao, and Prasad, 1996)

The following conditions are equivalent:

  (i) P is a prime left ideal.
  (ii) $<a>_1<b>_1 \subseteq P \Rightarrow a \in P$ or $b \in P$.
  (iii) $a<b>_1 \subseteq P \Rightarrow a \in P$ or $b \in P$.
  (iv) $<a>_1(P + <b>_1) \subseteq P \Rightarrow a \in P$ or $b \in P$.
  (v) $A(P + B) \subseteq P \Rightarrow A \subseteq P$ or $B \subseteq P$, where $a, b \in N$, and A and B are left ideals of N.

Here $<a>_1$ and $<b>_1$ denote the left ideals generated by elements $a, b \in N$, respectively.

Proof

(i) $\Rightarrow$ (ii): Follows from the definition.
  Proof for (ii) $\Rightarrow$ (iii):
  Suppose $a<b>_1 \subseteq P$. By Theorem 4.1.19, it follows that $<a>_1<b>_1 \subseteq P$.
  Now by (ii), it follows that either $a \in P$ or $b \in P$. This proves (iii).
  Proof for (iii) $\Rightarrow$ (iv): Suppose $<a>_1(P + <b>_1) \subseteq P$.
  Now $<a>_1<b>_1 \subseteq <a>_1(P + <b>_1) \subseteq P$.
  By condition (iii), it follows that $a \in P$ or $b \in P$. This proves (iv).
  Proof for (iv) $\Rightarrow$ (v): Suppose $A(P + B) \subseteq P$.
  Let us assume that $A \not\subseteq P$ and $B \not\subseteq P$.
  Then there exist $a \in A\backslash P$ and $b \in B\backslash P$.
  Now $<a>_1(P + <b>_1) \subseteq A(P + B) \subseteq P$.

By condition (iv), it follows that either $a \in P$ or $b \in P$, which is a contradiction.
Therefore, either $A \subseteq P$ or $B \subseteq P$. This proves (v).
Proof for (v) $\Rightarrow$ (i): Suppose A and B are two left ideals of N such that $AB \subseteq P$.
Let $a \in A$, $p \in P$, and $b \in B$.
Since P is a left ideal of N, we have $a(p + b) - ab \in P$, and also $ab \in P$.
Therefore, $a(p + b) - ab + ab \in P$. This implies that $A(P + B) \subseteq P$.
Now by condition (v), it follows that either $A \subseteq P$ or $B \subseteq P$. ■

Remark 4.1.27 (Remark 1.3 of Satyanarayana, Lokeswararao, and Prasad, 1996)

Suppose P is a left ideal, which is not a right ideal. Then for any $b \in N$ such that $Pb \subseteq P$, the following are equivalent:

(i) $<a>_1<b>_1 \subseteq P \Rightarrow a \in P$.
(ii) $<a>_1(P + <b>_1) \subseteq P \Rightarrow a \in P$.
(iii) A and B are left ideals, $A(P + B) \subseteq P$, $a \in P$.

Note 4.1.28

If P is a left ideal, then the following conditions are equivalent:

(i) $PL \subseteq P$, L is a left ideal $\Rightarrow L \subseteq P$.
(ii) $P<b>_1 \subseteq P$, $b \in N \Rightarrow b \in P$.
(iii) P is the largest left ideal contained in the set X, where $X = \{x \in N \mid P<x>_1 \subseteq P\}$.

Note 4.1.29

If P is a prime left ideal that is not a two-sided ideal, then P is the largest left ideal contained in $X = \{x \in N \mid P<x>_1 \subseteq P\}$.

Proof

Suppose the left ideal P is not a two-sided ideal.
    Then there exists an element $b \in N$ such that $Pb \nsubseteq P$.
    Suppose that L is a left ideal of N such that $PL \subseteq P$.
    Then $(Pb)L \subseteq P(bL) \subseteq PL \subseteq P$. By Theorem 4.1.19, we obtain $<Pb>_1L \subseteq P$
    $\Rightarrow <Pb>_1 \subseteq P$ or $L \subseteq P$
    $\Rightarrow Pb \subseteq P$ or $L \subseteq P$. Since $Pb \nsubseteq P$, it follows that $L \subseteq P$.
    Therefore, for any left ideal L of P, we have $PL \subseteq P \Rightarrow L \subseteq P$.
    Now by Note 4.1.28, it follows that P is the largest left ideal that is contained in X. ■

Definition 4.1.30 (Van der Walt, 1964)

$M \subseteq N$ is called an **m-system** if for all $a, b \in M$ there exists $a_1 \in <a>$, $b_1 \in <b>$ such that $a_1 b_1 \in M$.

Example 4.1.31

   (i) The empty set $\emptyset$ and near ring N are m-systems.
  (ii) For all $n \in N$, the set $\{n, n^2, \ldots\}$ is an m-system.

Theorem 4.1.32 (Corollary 2.80 of Pilz, 1983)

If P is an ideal of N, then P is a prime ideal if and only if $N \backslash P$ is an m-system.

Proof

Suppose P is a prime ideal of N. To show that $N \backslash P$ is an m-system, let $a, b \in N \backslash P \Rightarrow a \notin P$ and $b \notin P$. This means $<a><b> \not\subseteq P$. This implies that $a_1 \in <a>$ and $b_1 \in <b> \Rightarrow a_1 b_1 \in N \backslash P$. This implies that $N \backslash P$ is an m-system.

Converse

Suppose $N \backslash P$ is an m-system. Suppose $a \notin P$ and $b \notin P$. By Proposition 4.1.6, it is enough to show that $<a><b> \not\subseteq P$. Since $N \backslash P$ is an m-system and $a, b \in N \backslash P$, there exists $a_1 \in <a>$ and $b_1 \in <b>$ such that $a_1 b_1 \in N \backslash P$. This implies that $a_1 b_1 \notin P$. Therefore, $<a><b> \not\subseteq P$. Hence, P is a prime ideal of N.

Example 4.1.33

Consider the near ring $\mathbb{Z}$ of integers.

   (i) (0) is a prime ideal in $\mathbb{Z}$, and so $\mathbb{Z} \backslash (0)$ is an m-system.
  (ii) For any prime number p, it follows that $p\mathbb{Z}$ is a prime ideal, and so $\mathbb{Z} \backslash p\mathbb{Z}$ is an m-system.

Proposition 4.1.34 (Proposition 2.81 of Pilz, 1983)

Let $M \subseteq N$ be a nonvoid m-system in N and I an ideal of N with $I \cap M = \emptyset$. Then I is contained in a prime ideal $P \neq N$ with $P \cap M = \emptyset$.

Proof

Take $\Im = \{J \mid J$ as an ideal of N and $J \supseteq I, J \cap M = \emptyset\}$. Since I is an ideal of N, and $I \supseteq I$ and $I \cap M = \emptyset$, it follows that $I \in \Im$. Therefore, $\Im$ is nonempty.

Let $\{J\}_{i \in I}$ be a chain of ideals of $\Im$. This chain has an upper bound, say $\cup J_i \supseteq I$, and so $\cup J_i \in \Im$.

Therefore, by Zorn's Lemma, $\Im$ contains a maximal element, say P. Since P is maximal, $P \neq N$. Suppose $J_1 \supseteq P$ and $J_2 \supseteq P$. Then there exists $j_1 \in J_1 \cap M$ and $j_2 \in J_2 \cap M$ such that $<j_1><j_2> \subseteq J_1 J_2$. Since M is an m-system, there exists $a \in <j_1>$ and $b \in <j_2>$ such that $ab \in M$. This implies that $<j_1 j_2> \cap M \neq \emptyset$. Since $j_1 \in J_1 \backslash P$ and $j_2 \in J_2 \backslash P$, it follows that $<j_1><j_2> \not\subseteq P$. This shows that $J_1 J_2 \not\subseteq P$. Hence, P is a prime ideal of N. ∎

Now we introduce the concepts of left m-system-1 and left m-system-2.

Definition 4.1.35

(i) Let $M \subseteq N$. M is said to be a **left m-system-1** if either $M = \emptyset$ or $(a, b \in M \Rightarrow a^1 b^1 \in M$ for some $a^1 \in <a>_1$ and $b^1 \in <b>_1)$.

(ii) M is said to be a **left m-system-2** if either $M = \emptyset$ or $(a, b \in M \Rightarrow ab^1 \in M$ for some $b^1 \in <b>_1)$.

Note 4.1.36

It is clear that if M is a left m-system-2, then it is a left m-system-1.

Proposition 4.1.37 (Proposition 1.8 of Satyanarayana, Lokeswararao, and Prasad, 1996)

If P is a left ideal of N, then the following conditions are equivalent:

(i) P is a prime left ideal.
(ii) $N \backslash P$ is a left m-system-2.
(iii) $N \backslash P$ is a left m-system-1.

Proof

(i) $\Rightarrow$ (ii):

Suppose P is prime. Consider $N \backslash P$. If $P = N$, then $N \backslash P = \emptyset$ is a left m-system-2. Suppose $N \backslash P \neq \emptyset$. To show that $N \backslash P$ is a left m-system-2, take a, $b \in N \backslash P \Rightarrow a, b \notin P$. Since P is a prime left ideal, by Proposition 4.1.26, it follows that $a<b>_1 \not\subseteq P$.

Therefore, there exists $b^1 \in <b>_1$ such that $ab^1 \notin P$, which implies that $ab^1 \in N\backslash P$.

(ii) $\Rightarrow$ (iii): follows from Note 4.1.36.

(iii) $\Rightarrow$ (i): Suppose $N\backslash P$ is a left m-system-1. If $N\backslash P = \varnothing$, then $P = N$, and hence P is a prime left ideal. Suppose $<a>_1<b>_1 \subseteq P$. Now we have to show that $a \in P$ or $b \in P$.

Let us suppose that $a \notin P$ and $b \notin P$. Then $a \in N\backslash P$ and $b \in N\backslash P \Rightarrow a^1b^1 \in N\backslash P$ for some $a^1 \in <a>_1$ and $b^1 \in <b>_1 \Rightarrow a^1b^1 \notin P \Rightarrow <a>_1<b>_1$ is not a subset of P, which is a contradiction to the supposition. Hence, P is a prime left ideal. ∎

Definition 4.1.38

If S is an ideal of N, it is **semiprime** if and only if for all ideals I of N, $I^2 \subseteq S$ implies that $I \subseteq S$.

Theorem 4.1.39 (Proposition 2.82 of Pilz, 1983)

For an ideal S of N, the following conditions are equivalent:

(i) S is semiprime.
(ii) For all ideals I of N, $<I^2> \subseteq S \Rightarrow I \subseteq S$.
(iii) For all $n \in N$, $<n>^2 \subseteq S \Rightarrow n \in S$.
(iv) For all ideals I of N, $I \supset S \Rightarrow I^2 \not\subseteq S$.
(v) For all ideals I of N, $I \not\subseteq S \Rightarrow I^2 \not\subseteq S$.

Proof

Similar to the proof for Proposition 4.1.6. ∎

Proposition 4.1.40

Let I be an ideal of N that is a direct summand and S an ideal of N that is semiprime; then $S \cap I$ is semiprime in I.

Proof

Let J be an ideal of I. Suppose $J^2 \subseteq S \cap I$. Then $J^2 \subseteq S$ and $J^2 \subseteq I$. Since I is a direct summand of N, it follows that J is an ideal of N. Since S is an ideal of N,

S is semiprime, and $J^2 \subseteq S$, it follows that $J \subseteq S$. Therefore, $J \subseteq S \cap I$. Hence, $S \cap I$ is semiprime in I. ■

Proposition 4.1.41

Let I be an ideal of N and $I \subseteq S$, where S is an ideal of N. Then S is semiprime if and only if $\pi(S) \subseteq N/I$ is semiprime.

Proof

Similar to the proof of Proposition 4.1.9. ■

Definition 4.1.42

A subset S of N is called an **sp-system**, if for all $s \in S$, there exist $s_1, s_2 \in {<}s{>}$ such that $s_1 s_2 \in S$.

Theorem 4.1.43

(i) Each m-system is an sp-system.
(ii) For every ideal S of N, the ideal S is semiprime if and only if $N \backslash S$ is an sp-system.

Proof

(i) Let M be an m-system.
This implies that for all $a, b \in M$, there exist $a_1 \in {<}a{>}$ and $b_1 \in {<}b{>}$ such that $a_1 b_1 \in M$. This means that for all $a \in M$ there exists $a_1, a_2 \in {<}a{>}$ such that $a_1 a_2 \in M$. This shows that M is an sp-system.
(ii) Suppose that S is a semiprime ideal of N. To show that $N \backslash S$ is an sp-system, take $a \in N \backslash S$. Then $a \notin S$. Since S is semiprime, by Theorem 4.1.39, it follows that ${<}a{>}^2 \not\subseteq S$. This implies that $a_1 a_2 \notin S$ for some a_1, $a_2 \in {<}a{>}$. This implies that $a_1 a_2 \in N \backslash S$. Therefore, $N \backslash S$ is an sp-system.

Converse

Suppose that $N \backslash S$ is an sp-system. Let $a \in N \backslash S$. Since $N \backslash S$ is an sp-system, there exist $a_1, a_2 \in (a)$ such that $a_1 a_2 \in N \backslash S$. This means that $a_1 a_2 \notin S$. This implies that ${<}a{>}^2 \not\subseteq S$. Therefore, by Theorem 4.1.39, it follows that S is semiprime. ■

Proposition 4.1.44 (Proposition 2.90 of Pilz, 1983)

Let S be a nonvoid sp-system in N. Let I be an ideal of N with $I \cap S = \varnothing$. Then I is contained in a semiprime ideal $\neq N$.

Proof

Let $\mathfrak{I} = \{J \mid J$ be an ideal of N and $J \supseteq I, J \cap S = \varnothing\}$. Since I is an ideal of N, and $I \supseteq I$ and $I \cap S = \varnothing$, it follows that $I \in \mathfrak{I}$. Therefore, $\mathfrak{I}$ is nonempty.

Let $\{J\}_{i \in I}$ be a chain of ideals of $\mathfrak{I}$. This chain has an upper bound, say $\cup J_i \supset I$, and so $\cup J_i \in \mathfrak{I}$.

Therefore, by Zorn's lemma, $\mathfrak{I}$ contains a maximal element, say P. Since P is maximal, $P \neq N$. Now we show that P is a semiprime ideal of N. Suppose $J \supset P$. Then $<j> \subseteq J$ for some $j \in J \cap S$. Since S is an sp-system, there exist $j_1, j_2 \in J$ such that for every $j \in S, <j>^2 \cap S \neq \varnothing$. This implies that $J^2 \not\subset P$. Hence, P is a semiprime ideal of N.   ∎

Proposition 4.1.45

Any intersection of prime ideals is a semiprime ideal.

Proof

Let $\{P_i\}$ be a collection of prime ideals. Write $S = \bigcap_{i \in I} P_i$. To show that S is semiprime, take an ideal I of N such that $I^2 \subseteq S$. Now $I^2 \subseteq S \Rightarrow I^2 \subseteq \bigcap_{i \in I} P_i \Rightarrow I^2 \subseteq P_i$ for all $i \in I \Rightarrow I \subseteq P_i$ for all $i \in I$ (since P_i is a prime ideal) $\Rightarrow I \subseteq \bigcap_{i \in I} P_i = S$. This shows that I is a semiprime ideal.   ∎

Proposition 4.1.46 (Proposition 2.92 of Pilz, 1983)

Let S be an sp-system and $s \in S$; then there is some m-system M with $s \in M \subseteq S$.

Proof

Let $s \in S$. Since S is an sp-system, there exist $s_1, s_2 \in S$ such that $s_1 s_2 \in S$. This implies that there exist $s_1^1, s_2^1 \in (s_1 s_2)$ such that $s_1^1, s_2^1 \in S$. Continuing in this manner, we get a sequence $s, s_1 s_2, s_1^1 s_2^1, s_1^k s_2^k \ldots$ for all positive integers k such that $s_1^{(k)} s_2^{(k)} \in S$ and $<s> \supseteq (s_1 s_2) \supseteq s_1^k s_2^k \supseteq \ldots$.

Take $M = \{s, s_1 s_2, s_1^k s_2^k, \ldots\}$. We show that this is a required m-system. Without loss of generality, suppose that $l \leq k$ and $s_1^{(k)} s_2^{(k)}, s_1^{(l)} s_2^{(l)} \in M$. Then

$s_1^{(k)}s_2^{(k)} \supseteq s_1^{(l)}s_2^{(l)}$. Take $s_1^{(l+1)}s_2^{(l+1)} \in <s_1^{(l)}s_2^{(l)}> \subseteq <s_1^{(k)}s_2^{(k)}>$. Therefore, $s_1^{(l+1)}s_2^{(l+1)} \in M$. The proof is complete. ∎

4.2 Prime Radicals in Near Rings

Definition 4.2.1 (Sambasivarao and Satyanarayana, 1984)

(i) A subset M of N is called an m^*-**system** if for all $a, b \in M$, we have $a<b> \cap M \neq \varnothing$ (i.e., there exists an element $b_1 \in <b>$ such that $ab_1 \in M$). (This definition generalizes the definition of an m-system, which was introduced by Van der Walt in 1964.)

(ii) A subset S of N is called an sp^*-**system** if for all $a \in S$, we have $a<a> \cap M \neq \varnothing$ (i.e., there exists an element a_1 in $<a>$ such that $aa_1 \in S$). (This definition generalizes the definition of an sp-system, which was introduced by Mason (1980).)

(iii) A sequence $a_0, a_1, a_2, \ldots, a_n, \ldots$ of elements in N is called an m^*-**sequence** if $a_n \in a_{n-1}<a_{n-1}>$ for all $n \geq 1$.

Remark 4.2.2

(i) An m^*-system is an m-system, and an sp^*-system is an sp-system.
(ii) Every m^*-system is an sp^*-system, but the converse is not true.

Example 4.2.3

Let $N = \mathbb{Z}$, the ring of integers. Then $N\backslash<10>$ is an sp^*-system, but not an m^*-system.

Remark 4.2.4

Let P be an ideal of N. Then

(i) P is prime if and only if for all $a, b \in N$ ($a<b> \subseteq P \Rightarrow$ either $a \in P$ or $b \in P$).
(ii) P is semiprime if and only if $a \in N$ ($a<a> \subseteq P \Rightarrow a \in P$).
(iii) P is prime if and only if $N\backslash P$ is an m^*-system.
(iv) P is semiprime if and only if $N\backslash P$ is an sp^*-system.

Remark 4.2.5

If M is a nonempty m^*-system and if I is an ideal of N such that $M \cap I = \emptyset$, then there exists a prime ideal P containing I such that $P \cap M = \emptyset$.
    The verification is the same as in the proof of Proposition 4.1.34.

Remark 4.2.6

If $\{a_n\}_{n \geq 0}$ is an m^*-sequence in N, then the set $\{a_n \mid n \geq 0\}$ is an m^*-system.

Lemma 4.2.7 (Sambasivarao and Satyanarayana, 1984)

Let S be an sp^*-system and $a \in S$.

  (i) There exists an m^*-sequence $\{a_n\}_{n \geq 0}$ in S starting with a (i.e., $a = a_0$).
  (ii) Then there exists an m^*-system M such that $a \in M \subseteq S$.

Proof

  (i) Write $a_0 = a$. Since S is an sp^*-system and $a_0 \in S$, it follows that $a_0 <a_0> \cap$
       $S \neq \emptyset$. So, there exists $b_0 \in <a_0>$ such that $a_0 b_0 \in S$. Write $a_1 = a_0 b_0$.
       Since $a_1 \in S$, it follows that $a_1 <a_1> \cap S \neq \emptyset$. Therefore, there exists $b_1 \in$
       $<a_1>$ such that $a_1 b_1 \in S$. Write $a_2 = a_1 b_1$.
       If we continue this process, we get a sequence of elements $a_0, a_1, a_2, \ldots$
       such that $a_{n+1} = a_n b_n$ with $b_n \in <a_n>$ for all $n \geq 0$. Therefore, $a_0, a_1, a_2, \ldots$
       is an m^*-sequence in S starting with $a = a_0$.
  (ii) Now $a_{n+1} = a_n b_n \in a_n <a_n> \subseteq <a_n>$, which implies that $<a_{n+1}> \subseteq <a_n>$.
       Write $M = \{a_0, a_1, \ldots\}$. We wish to show that M is an m^*-system. Let
       $x, y \in M$. Then $x = a_s$ and $y = a_t$ for some positive integers s and t.
       Without loss of generality, we suppose that $s \geq t$.
       Now $<a_1> \supseteq <a_2> \supseteq \ldots \supseteq <a_t> \supseteq \ldots \supseteq <a_s> \ldots$. Also, $a_{s+1} = a_s b_s$, $b_s \in$
       $<a_s> \subseteq <a_t>$. Hence, M is an m^*-system.   ∎

Definition 4.2.8

Let I be an ideal of N. Then a prime (completely prime, respectively) ideal of N containing I is called a **minimal prime (minimal completely prime**, respectively) **ideal** of I if P is minimal in the set of all prime (completely prime, respectively) ideals containing I.

Theorem 4.2.9 (Pilz, 1983 or Sambasivarao, 1982)

Let I be an ideal of N and M a nonempty m-system in N such that $I \cap M = \varnothing$. Then

   (i) There exists an m-system M^* maximal relative to the properties $M \subseteq M^*$ and $I \cap M^*$, which is empty.
   (ii) If M^* is an m-system maximal relative to the properties $M \subseteq M^*$ and $I \cap M^* = \varnothing$, then $N \backslash M^*$ is a minimal prime ideal of I.

Theorem 4.2.10 (Van der Walt, 1964; Sambasivarao, 1982; and Groenewald, 1984)

Let I be a semiprime ideal of a near ring N. Then I is the intersection of all minimal prime ideals of I.

Proof

Let I be any semiprime ideal of N. Then by Remark 4.2.4 (iv), $N \backslash I$ is an sp-system. So, by Lemma 4.2.7 (ii), it follows that for each $s \in N \backslash I$, there exists an m-system M in N such that $s \in M \subseteq N \backslash I$. So, $M \cap I$ is empty. Now by Theorem 4.2.9, there exists a minimal prime ideal P of I such that $P \cap M$ is empty. But $s \in M$. Hence, $s \notin P$. Therefore, $I = \bigcap_{P \supseteq I} P$, where P ranges over all minimal prime ideals of I. Thus, every semiprime ideal I is the intersection of all minimal prime ideals of I. ∎

Corollary 4.2.11

If I is a semiprime ideal in a near ring N, then I is the intersection of all prime ideals containing I.

Proof

By Theorem 4.2.10, it follows that $I = \cap P$, where P ranges over all minimal prime ideals and I contains $\bigcap_{P \supseteq I} P$, where P ranges over all prime ideals containing I. Therefore, I is the intersection of all prime ideals containing I. ∎

Result 4.2.12

If S is a semiprime ideal and I a nilpotent ideal, then I is contained in S.

Proof

Let S be a semiprime ideal and I be a nilpotent ideal of N. Since I is nilpotent, there exists a positive integer k such that $I^k = 0$. Without loss of generality, we assume that k is the least positive integer with $I^k = 0$. We have to show that $k = 1$.

Suppose $k > 1$. Then $I^{(k-1)} \cdot I = 0 \Rightarrow I^{k-1} \subseteq (0 : I) \Rightarrow <I^{k-1}> \subseteq (0 : I)$ (since $(0 : I)$ is an ideal by Theorem 4.1.21 (ii)) $\Rightarrow <I^{k-1}>I = 0$. Since S is semiprime, by Proposition 4.1.45, it follows that $S = \cap_{i \in I} P_i$, where $\{P_i\}_{i \in I}$ is the collection of all prime ideals containing S.

Now $<I^{k-1}>I = 0 \subseteq S = \cap_{i \in I} P_i \Rightarrow <I^{k-1}>I \subseteq P_i$ for all $I \in I \Rightarrow I^{k-1} \subseteq P_i$ or $I \subseteq P_i$ for all $i \in I \Rightarrow I^{k-1} \subseteq P_i$ for all $i \in I \Rightarrow I^{k-1} \subseteq \cap_{i \in I} P_i = S$, which is a contradiction to the minimality of k. Hence, $I \subseteq S$.   ∎

Definition 4.2.13

   (i) The intersection of all prime ideals of N is called the **prime radical of N**, and it is denoted by P-rad(N).

  (ii) For any proper ideal I of N, the intersection of all prime ideals of N containing I is called the **prime radical of I** and is denoted by P-rad(I).

 (iii) The intersection of all completely prime ideals of N is called the **completely prime radical of N**, and it is denoted by C-rad(N).

 (iv) For any proper ideal I of N, the **completely prime radical** of I is defined as the intersection of all completely prime ideals of N containing I, and it is denoted by C-rad(I).

Lemma 4.2.14

Let $M \subseteq N$ be a nonvoid m^*-system in N and I an ideal of N with $I \cap M = \emptyset$. Then I is contained in a prime ideal $P \neq N$ with $P \cap M = \emptyset$.

Proof

A straightforward verification as in the proof of Proposition 4.1.34.   ∎

Theorem 4.2.15

If I is a proper ideal of N, then
$$P\text{-rad}(I) = \{a \in N \mid \text{every } m^*\text{-sequence } \{a_n\}_{n \geq 0} \text{ starting with } a \text{ meets } I\}.$$

Proof

Let X be the right-hand side set. Let $a \in N$ and $a \notin P$-rad(I). Then there exists a prime ideal P containing I such that $a_0 = a \notin P$. Since P is prime, $N \backslash P$ is an

m^*-system and $N\backslash P$ is an sp^*-system. Now there exists an m^*-sequence in $N\backslash P$ starting with a_0 that does not meet P, and so, evidently it does not meet I. Thus, $a \notin X$. Hence, $X \subseteq P\text{-rad}(I)$.

Converse

Suppose $a \in N$ with $a \notin X$; then there is an m^*-sequence $\{a_n\}_{n \geq 0}$ such that $a_0 = a$ and $a_n \notin I$ for all n. Write $M = \{a_n \mid n \geq 0\}$. Since $M \cap I$ is empty and M is an m^*-system, by Lemma 4.2.14, there is a prime ideal P such that $P \supseteq I$, and $P \cap M$ is empty. Clearly, $a \notin P$ and $a \notin P\text{-rad}(I)$. Thus, $X \supseteq P\text{-rad}(I)$. The proof is complete. ∎

Definition 4.2.16

An element a in a near ring N is said to be **strongly nilpotent** if every m^*-sequence $\{a_n\}_{n \geq 0}$ with $a_0 = a$ vanishes (i.e., there is a positive integer n such that $a_k = 0$ for all $k \geq n$).

Clearly, every strongly nilpotent element in N is nilpotent. But the following example shows that the converse is not true.

Example 4.2.17

Consider the ring of all 3×3 matrices over the ring of integers. Write $a_0 =$
$\begin{pmatrix} 0 & 1 & 1 \\ 0 & 0 & 1 \\ 0 & 0 & 0 \end{pmatrix}$ and $x = \begin{pmatrix} 1 & 1 & 1 \\ 1 & 1 & 1 \\ 1 & 1 & 1 \end{pmatrix}$. Define $a_1 = a_0 x a_0, \ldots, a_n = a_{n-1} x a_{n-1}, \ldots$. Now $a_n \neq 0$, and hence a_0 is not strongly nilpotent. But $a_0^3 = 0$. This shows that a_0 is nilpotent.

Corollary 4.2.18

(i) The prime radical of N is the set of all strongly nilpotent elements in N.

(ii) (**Theorem 2.105 of Pilz, 1983**): $P\text{-rad}(N)$ is contained in the set of all nilpotent elements of N, and hence $P\text{-rad}(N)$ is nil.

Proof

(i) This follows from Theorem 4.2.15.

(ii) This follows since every strongly nilpotent element is nilpotent. ∎

Theorem 4.2.19

Let I be an ideal of N.
   Then P-rad$(I) = \{a \in N \mid$ every m^*-system that contains a must contain an element of $I\}$.

Proof

Let X be the right-hand side set. Let $a \in P$-rad(I). If $a \notin X$, then there exists an m^*-system M with $a \in M$ and $M \cap I = \emptyset$. This implies that there exists a prime ideal P such that $P \supseteq I$ and $M \cap P = \emptyset$. So, $a \notin P$. Therefore, $a \notin P$-rad(I), which is a contradiction.

Converse

Suppose $a \in X$. If $a \notin P$-rad(I), then there exists a prime ideal P containing I such that $a \notin P$. This implies that $a \in N\backslash P$, and $N\backslash P$ is an m^*-system. Since $a \in X$, we have $(N\backslash P) \cap I \neq \emptyset$, which is a contradiction. The proof is complete. ∎

Corollary 4.2.20

If N is a near ring, then
   P-rad$(0) = \{a \in N \mid$ every m^*-system that contains a contains $0\}$.

Theorem 4.2.21 (Theorem 2.94 of Pilz, 1983)

If $n \in P$-rad(I), then there exists a natural number k such that $n^k \in I$.

Proof

Suppose $n \in P$-rad(I); then $n \in P$ for all prime ideals P such that $P \supseteq I$. We know that $M = \{n, n^2, n^3, \ldots\}$ is an m-system. Suppose that $n^k \notin I$ for all positive integers k. This implies that $M \cap I = \emptyset$, that is, there is a prime ideal $P \subseteq I$ such that $P \cap M = \emptyset$, which is a contradiction to the fact that $n \in P$-rad(I). The proof is complete. ∎
   Murata, Kurata, and Marubayashi (1969) introduced and investigated a generalization of prime rings, called f-prime rings. Groenewald and Potgieter (1984) extended the results to f-prime and f-semiprime near rings. The f-prime radical of a fixed near ring N was defined by Reddy and Satyanarayana (1986) as the intersection of all f-prime ideals of N, and it was described as the subset of N consisting of 0 and all f-strongly nilpotent elements. Note that f-primeness is defined by a mapping f_N assigning an ideal $f_N(a)$ of N to each element $a \in N$, which is subject to certain constraints. This mapping f_N is not uniquely determined; so, each such mapping f_N defines an f-primeness as well as an f-strong nilpotence.

In the following, we shall study f-P-rad(I), the properties of the assignment f-P-rad, belonging to a fixed mapping f_N and designating the f-prime radical f-P-rad(N) to each near ring N.

We shall refer to this assignment f-P-rad briefly as to the f-radical.

Let us recall that we are working on the variety of all zero-symmetric right near rings. Sometimes, we shall restrict our considerations to a universal class U of near rings, which is a subclass of near rings closed under taking ideals and homomorphic images.

Let P-rad be a mapping that assigns to each near ring N an ideal P-rad(N) of N. Such a mapping P-rad may satisfy some of the following conditions.

(i) $\varphi(P\text{-rad}(N)) \subseteq P\text{-rad}(\varphi(N))$ for each homomorphism $\varphi\colon N \to \varphi(N)$
(ii) P-rad($N/(P$-rad(N))) $= 0$ for all N
(iii) P-rad is an idempotent: P-rad $(P\text{-rad}(N)) = P\text{-rad}(N)$ for all N
(iv) P-rad is complete: P-rad(I) $= I$ is an ideal of N implies that P-rad(I) $\subseteq$ P-rad (N).

Definition 4.2.22

Consider the preceding four points.

(a) The mapping P-rad is said to be a **Hoehnke radical** if it satisfies (i) and (ii). P-rad is called a **Plotkin radical** if it satisfies (i), (iii), and (iv).
(b) An idempotent and complete Hoehnke radical is called a **Kurosh–Amitsur radical**. To any radical P-rad, we associate two classes of near rings, the **radical class**.

$$\mathbb{R}_{p\text{-rad}} = \{N \mid P\text{-rad}(N) = N\}$$

and the **semisimple class**

$$S_{p\text{-rad}} = \{N \mid P\text{-rad}(N) = 0\}.$$

Notation 4.2.23

The notions of f-primeness, f-nilpotence, and f-radical of a near ring N are based on an ideal mapping f_N designating to each element $a \in N$ an ideal $f_N(a)$ of N such that

(i) $a \in f_N(a)$
(ii) I is an ideal of N and $x \in f_N(a) + I$ implies that $f_N(x) \subseteq f_N(a) + I$

An ideal mapping f_N is, in fact, uniquely determined by the ideal $f_N(0)$ as seen from the following lemma.

Lemma 4.2.24 (Lemma 1 of Satyanarayana and Wiegandt, 2005)

An ideal mapping f_N satisfies conditions (i) and (ii) of Notation 4.2.23 if and only if $f_N(a) = f_N(0) + (a)_N$ for all $a \in N$.

Proof

Note that for $I = 0$, condition (ii) yields $f_N(x) \subseteq f_N(a)$ for all $x \in f_N(a)$.

Since $a \in f_N(0) + (a)_N$ by (ii) and (i), we have $f_N(a) \subseteq f_N(0) + (a)_N \subseteq f_N(0) + f_N(a) \subseteq f_N(a)$.

Converse

Suppose that $f_N(a) = f_N(0) + (a)_N$. Condition (i) is trivially fulfilled.

If $x \in f_N(a) + I$, then $x \in f_N(0) + (a)_N + I$.

Hence, $f_N(x) = f_N(0) + (x)_N \subseteq f_N(0) + (a)_N + I = f_N(a) + I$, and so (ii) is also satisfied. ∎

Note 4.2.25

We may define an ideal mapping f_N by assigning to each near ring N an ideal $f_N(0)$ of N. In view of Lemma 4.2.24, no relation exists between $f_N(0)$ and $f_M(0)$, where N and M are arbitrary distinct near rings. Dealing, however, simultaneously with several near rings, it is reasonable and not restrictive to demand the following additional requirement:

(iii) If $N \cong M$, then $f_N(0) \cong f_N(a) = f_M(0)$ by the same isomorphism.

An element $a \in N$ is said to be strongly nilpotent in N if for any given sequence $a_0, a_1, \dots$ of elements of N with $a_0 = a$, $a_i = a_{i-1} a_{i-1}^*$ where $a_{i-1}^* \in (a_{i-1})_N$, there corresponds an integer m such that $an = 0$ for all $n \geq m$. The importance of the notion of strongly nilpotent is shown in Corollary 4.2.18. The prime radical P-rad(N) of a near ring N is the set of all strongly nilpotent elements of N.

Definition 4.2.26

A radical P-rad is said to be **hereditary** if P-rad(N) $\cap I \subseteq P$-rad(I) for every ideal I of N and all near rings N.

Kaarli and Kriis (1987) proved that the prime radical of near rings is hereditary. One can easily see that Corollary 4.2.18 provides an alternative proof for the hereditariness of P-rad(N).

Definition 4.2.27

An element $a \in N$ is said to be *f*-**strongly nilpotent** in N, if every element of the ideal $f_N(a)$ is strongly nilpotent in N, that is, $f_N(a) \subseteq P$-rad(N).

Clearly, every f-strongly nilpotent element is strongly nilpotent. The converse is not true: let $N \neq P\text{-rad}(N) \neq 0$, and define $f_N(0) = N$. Then the elements of $P\text{-rad}(N)$ are strongly nilpotent in N but not f-strongly nilpotent in N. Moreover, by definition, the element $0 \in N$ need not be f-strongly nilpotent in N.

The f-radical f-P-rad—called the f-prime radical—was introduced and characterized by Reddy and Satyanarayana (1986).

f-P-rad$(N) = \{$all f-strongly nilpotent elements of $N\} \cup \{0\}$.

Definition 4.2.28

(i) A subset S of N is an f-system if it contains an m-system S^*, called the kernel of S, such that $S = \emptyset$ or $f(s) \cap S^* \neq \emptyset$ for every $s \in S$.

(ii) An ideal P of N is said to be f-prime if $C(P)$ is an f-system.

(iii) The f-prime radical f-P-rad(N) of N is defined as the intersection of all f-prime ideals of N; equivalently, f-P-rad(N) is the set of all those elements $x \in N$ for which every f-system that contains x must contain 0.

(iv) We say that a sequence $x_1, x_2, \ldots$ of elements of N vanishes (or terminates) after a certain stage if there exists a positive integer k such that $x_s = 0$ for all $s \geq k$.

(v) We know that an element x in N is said to be strongly nilpotent if every sequence $x_1, x_2, \ldots$ of elements of N with $x_1 = x$ and $x_i = x_{i-1} x_{i-1}^*$, for some $x_{i-1}^* \in < x_{i-1} >$ must terminate after a certain stage.

(vi) An element a in N is said to be f-**strongly nilpotent** if every element of $f(a)$ is strongly nilpotent.

Theorem 4.2.29

Let N be a near ring in which $0 \neq P\text{-rad}(N) \neq N$ and $x \in C(P\text{-rad}(N))$.

Define $f(a) = < \{a, x\} >$ for every $a \in N$. Then N contains no f-strongly nilpotent elements. In particular, the element 0 is not strongly nilpotent.

Proof

Let us suppose there exists an f-strongly nilpotent element y in N; then every element of $f(y)$ is strongly nilpotent, and so by Corollary 4.2.18, it follows that $f(y) \subseteq P\text{-rad}(N)$, which implies $x \in P\text{-rad}(N)$, which is a contradiction. Thus, N contains no f-strongly nilpotent elements.

In general, a strongly nilpotent element may not be f-strongly nilpotent. ∎

Example 4.2.30

Let $X = \{0, b\}$ and define $+$ and $\cdot$ on X as $0 + 0 = 0 + b = b + b = b + 0 = 0$.

$0 \cdot b = b \cdot b = b \cdot 0 = 0 \cdot 0 = 0$. Now $(X, +, \cdot)$ is a near ring. Write $N = Y \oplus X$, the direct sum of near rings Y and X, where $Y = Z_6$, the ring of integers modulo 6. The fact that $< \{0\} \oplus X >^2 = \{0\}$ implies that (0) is not a semiprime ideal of N, and since P-rad(N) is semiprime, we have $(0) \neq P$-rad(N). Secondly, it can be easily verified that the ideal $\{0,2,4\} + X$ of N is a prime ideal of N, and so P-rad$(N) \subseteq \{0,2,4\} \oplus X \subseteq N$. Thus, $(0) \neq P$-rad$(N) \neq N$. Let $x \in C(P$-rad$(N))$ and define $f(a) = < \{a, x\} >$ for all $a \in N$. Let $0 \neq y \in P$-rad(N). Then by Theorem 4.2.29, we have that y is not f-strongly nilpotent; however, by Corollary 4.2.18, we have y is strongly nilpotent.

Theorem 4.2.31

Let N be a near ring. Then
$\quad$ f-P-rad$(N) = \{x \in N \mid x$ is f-strongly nilpotent$\}$.

Proof

Let $0 \neq x \in f$-P-rad(N). Suppose x is not strongly nilpotent, that is, there is a sequence $a_1, a_2, \ldots$ of elements of N such that $a_1 = y \in f(x)$ and $a_i = a_{i-1} a_{i-1}^*$ for some $a_{i-1}^* \in < a_{i-1} >$ and $a_{i-1} \neq 0$ for all $i \geq 2$. By Remark 4.2.6, it follows that $S = \{a_i \mid i \geq 1\}$ is an m-system. Now $x \notin f$-P-rad(N), which is a contradiction. Thus, x is f-strongly nilpotent.

Converse

Suppose that x is an f-strongly nilpotent element. If $x \notin f$-P-rad(N), then there exists an f-prime ideal P such that $x \notin P$. Now $K = C(P)$ is an f-system. Let K^* be a kernel of the f-system K. Since $P \cap K = \varnothing$ and K^* is an m-system by Proposition 4.1.34, there exists a prime ideal P^* such that $P \subseteq P^*$ and $P^* \cap K^* = \varnothing$. Since $x \in K$, there exists an element y in $f(x) \cap K^*$. So, $y \in K^* \subseteq C(P^*) \subseteq C(P$-rad$(N))$, which implies that $y \notin P$-rad(N). Now $y \in f(x)$, and by Corollary 4.2.18, it follows that y is not strongly nilpotent. This is a contradiction to the fact that x is f-strongly nilpotent. The proof is complete. ∎

Theorem 4.2.32

(i) Let P be an ideal of N such that $f(0) \subseteq P$. Then P is prime if and only if P is f-prime.

(ii) If $f(0) \subseteq f$-P-rad(N), then
  (a) An ideal P is prime if and only if P is f-prime.
  (b) f-P-rad$(N) = P$-rad(N).

(iii) (a) If $f(0) = (0)$, then P-rad$(N) = f$-P-rad(N).
  (b) Suppose $f(0) \neq (0)$. Then P-rad$(N) = f$-P-rad$(N) \neq (0)$ if and only if $f(0) \subseteq f$-P-rad(N).

Proof

(i) It is trivial that every prime ideal is f-prime.

To prove the converse, suppose P is f-prime. It suffices to show $<a>$ $<b> \subseteq P$, $a, b \in N$ implies that either $a \in P$ or $b \in P$. Suppose $<a>$ $<b> \subseteq P$.

Now by definition, we have that $a \in f(0) + <a>$ implies $f(a) \subseteq f(0) + <a>$ and $b \in f(0) + <b>$ and $f(b) \subseteq f(0) + <b>$.

Now $f(a)f(b) \subseteq (<a> + f(0)) (<B> + f(0)) \subseteq <a> <b> + f(0) \subseteq P$.

Since P is f-prime by Lemma 1.2 (a) of Groenewald and Potgieter (1984), either $a \in P$ or $b \in P$.

(ii) (a) If P is f-prime, then P contains f-P-rad(N), and so $P \supseteq f(0)$.

Now the result follows from (i).

(b) Follows from (ii(a)).

(iii) (a) If $f(0) = (0)$, then $f(a) = (a)$ for all $a \in N$, and so the two concepts f-prime and prime are the same. Thus, f-P-rad(N) = P-rad(N).

(b) Let $f(0) \neq (0)$. Suppose P-rad(N) = f-P-rad(N) $\neq 0$.

Let $0 \neq x \in f$-P-rad(N). By Theorem 4.2.31, it follows that x is f-strongly nilpotent. Now by Corollary 4.2.18, it follows that $f(x) \subseteq P$-rad(N). Since $0 \in f(x) + (0)$, $f(0) \subseteq f(x) + (0) = f(x)$, and hence $f(0) \subseteq f(x) \subseteq r(N) \subseteq f$-$P$-rad($N$). The converse follows from (ii)(b). ∎

Proposition 4.2.33

The f-radical f-P-rad (N) of a near ring N is the intersection

f-P-rad(N) = $\cap$ {I is an ideal of N | I is an f-prime ideal of N}

= $\cap$ {I is an ideal of N | N/I is an f-prime}.

(I is an f-prime ideal if and only if N/I is an f-prime near ring.)

Lemma 4.2.34 (Lemma 2 of Satyanarayana and Wiegandt, 2005)

Let f be an ideal mapping and N a near ring. Then f-P-rad(N) = P-rad(N) if and only if $f_N(0) \subseteq P$-rad(N). If f-P-rad(N) $\neq 0$, then f-P-rad(N) = P-rad(N). If P-rad(N) = N, then f-P-rad(N) = N.

Proof

Suppose that f-P-rad(N) = P-rad(N) $\neq 0$. Then each strongly nilpotent element is f-strongly nilpotent. Hence, for any nonzero element $a \in P$-rad(N), we have $f_N(0) + (a)_N = f_N(a) \subseteq P$-rad($N$), which implies that $f_N(0) \subseteq P$-rad(N).

Assume that f-P-rad(N) $\subset P$-rad(N), and let $a \in P$-rad(N)\$f$-$P$-rad($N$). Then the element is not f-strongly nilpotent, and so $f_N(0) + (a)_N = f_N(a) \not\subseteq P$-rad($N$). Since $(a)_N \subseteq P$-rad(N), by $a \in P$-rad(N), we conclude that $f_N(0) \not\subseteq P$-rad(N).

If f-P-rad$(N) \neq 0$, then there exists an element $a \neq 0$ that is f-strongly nilpotent, that is, $f_N(0) + (a)_N = f_N(a) \subseteq P$-rad$(N)$. Hence, $f_N(0) \subseteq P$-rad(N), and so f-P-rad$(N) = P$-rad(N).   ∎

Let P-rad $(N) = N$. Since $f_N(0) \subseteq N = P$-rad(N) is always true, the last assertion also holds.

Theorem 4.2.35 (Theorem 4 of Satyanarayana and Wiegandt, 2005)

In a universal class U of 0-symmetric near rings, any f-radical f-P-rad is an idempotent Hoehnke radical, which is complete if and only if P-rad is complete in U and then f-P-rad $= P$-rad.

For a near ring N, f-P-rad$(N) = P$-rad(N) if and only if $f_N(0) \subseteq P$-rad(N). If $f_N(0) \not\subseteq P$-rad(N), then f-P-rad$(N) = 0$.

The radical classes $\mathbb{R}_f$ and $\mathbb{R}_\beta$ coincide for every f-radical r_f. Every near ring N with f-P-rad$(N) = 0$ is a subdirect sum of f-prime near rings.

Proof

By Proposition 4.2.33, the f-radical f-P-rad(N) of a near ring N is the intersection f-P-rad$(N) = \cap \{I$ is an ideal of $N \mid N/I$ is f-prime$\}$.

The semisimple class corresponding to the f-radical f-P-rad is $s_f = \{N \mid f$-P-rad$(N) = 0\}$, which is obviously closed under taking subdirect sums. Hence, f-P-rad is a Hoehnke radical.

For the idempotence of f-P-rad, we have to prove that f-P-rad$(f$-P-rad$(N)) = f$-P-rad(N). As in Lemma 4.2.34, we have two cases.

Case (i): Suppose f-P-rad$(N) = 0$.

Then f-P-rad $(f$-P-rad$(N)) = f$-P-rad$(0) = 0 = f$-P-rad(N).

Case (ii): Suppose f-P-rad$(N) = P$-rad(N).

Then f-P-rad$(f$-P-rad$(N)) = f$-P-rad$(P$-rad$(N))$.

Suppose that f-P-rad$(P$-rad$(N)) = 0$.

Now f-P-rad$(P$-rad$(N)) \neq P$-rad$(P$-rad$(N))$; so, by Lemma 4.2.34, it follows that $f_{P\text{-rad}(N)}(0) \not\subseteq P$-rad$(P$-rad$(N)) = P$-rad$(N)$, which is a contradiction. Hence, f-P-rad$(P$-rad$(N)) \neq 0$, and so again by Lemma 4.2.34, it follows that f-P-rad$(P$-rad$(N)) = P$-rad$(P$-rad$(N)) = P$-rad$(N) = f$-P-rad(N). Thus, f-P-rad$(f$-P-rad$(N)) = f$-P-rad(N) for all near rings N.

Assume that f-P-rad is a complete Hoehnke radical in the universal class considered.

Then f-P-rad is a Kurosh–Amitsur radical, and so f-P-rad is determined by its radical class

$\mathbb{R}_f = \{N \mid f$-$P$-rad$(N) = N\}$.

Since every prime near ring is f-prime, it follows that f-P-rad $(N) \subseteq P$-rad(N). This shows us that the radical class $\mathbb{R}_f$ is contained in the radical class $\mathbb{R}_{p\text{-rad}}$ of the prime radical P-rad. If $0 \neq N \in \mathbb{R}_{p\text{-rad}}$, then $f_N(0) \subseteq N = P$-rad(N), and so by Lemma 4.2.34, it follows that f-P-rad$(N) = P$-rad$(N) = N$, proving the containment $\mathbb{R}_{p\text{-rad}} \subseteq \mathbb{R}_f$.

Hence, we get Σ_I (I is an ideal of N such that P-rad $(N) = I$) $= \Sigma_I$ (I ideal of N such that P-rad $(N) = I$) $= f$-P-rad$(N) = P$-rad(N).

Thus, f-P-rad as well as P-rad is complete in the universal class considered. The rest follows from Lemma 4.2.34. ∎

Concerning f-P-rad –semisimple near rings, we have obviously the following.

Corollary 4.2.36

f-P-rad$(N) = 0$ for a near ring N if and only if either P-rad$(N) = 0$ or N possesses nonzero strongly nilpotent elements, one of which is f-strongly nilpotent in N.

Corollary 4.2.37

In the variety of all zero-symmetric near rings, an f-radical f-P-rad is never complete, and hence never a Kurosh–Amitsur radical.

Proof

We note that Kaarli and Kriis (1987) proved that the prime radical P-rad is not complete; so, the statement follows directly from the theorem. ∎

We shall give a sufficient, but not necessary condition for the hereditariness of an f-radical f-P-rad.

Proposition 4.2.38 (Proposition 3 of Satyanarayana and Wiegandt, 2005)

If an ideal mapping f satisfies $f_I(0) \subseteq f_N(0)$ for every ideal I of N and near ring N, then the corresponding f-radical f-P-rad is hereditary.

Proof

Let $a \in f$-P-rad$(N) \cap I$ be a nonzero element. Then the element a is f-strongly nilpotent in N; so $f_N(a) \subseteq P$-rad(N) and

$$f_I(a) \subseteq f_I(a) + (a)_I$$
$$\subseteq (f_N(0) \cap I) + ((a)_N \cap I)$$
$$\subseteq (f_N(0) + (a)_N) \cap I = f_N(a) \cap I$$
$$\subseteq P\text{-rad}(N) \cap I.$$

Since by the result of Kaarli and Kriis (1987), the *P*-rad is hereditary, we have $f_I(a) \subseteq P\text{-rad}(N) \cap I \subseteq P\text{-rad}(I)$, and so $a \in I$ is *f*-strongly nilpotent in *I*. Thus, $a \in f\text{-}P\text{-rad}(I)$, proving that *f*-*P*-rad(*N*) $\cap I \subseteq f$-*P*-rad(*I*). ∎

We conclude this section with two examples.

Example 4.2.39

The prime radical *P*-rad is a hereditary *f*-radical in the variety of all zero-symmetric near rings, but $f_I(0) \not\subseteq f_N(0) \cap I$ for some ideal *I* of *N*. Let us define $f_N(0) = P\text{-rad}(N)$. Then *f*-*P*-rad(*N*) = *P*-rad(*N*) for all *N*, and *f*-strong nilpotence coincides with strong nilpotence. As proved by Kaarli and Kriis (1987), the prime radical *P*-rad is not complete; so there exists a near ring *N* possessing an ideal *I* such that *P*-rad(*I*) = $I \not\subseteq P\text{-rad}(N)$. Hence, taking into account the hereditariness of *P*-rad, we have *P*-rad(*I*) $\not\subseteq P\text{-rad}(N) \cap I \subseteq P\text{-rad}(I)$.

Therefore, there exists an element $a \in P\text{-rad}(I) \backslash (P\text{-rad}(N) \cap I)$.

Now $a \in f_I(0) \backslash (f_N(0) \cap I)$, and so $f_I(0) \not\subseteq (f_N(0) \cap I)$.

Example 4.2.40

In the subvariety of all associative rings, the assignment $\tau : N \to \tau(N)$, where $\tau(N)$ is the ideal of *N* consisting of all additively torsion elements and 0, defines a hereditary Kurosh–Amitsur radical, called the **torsion ideal**. We may define $f_N(0) = \tau(N)$ for all rings. Then by Lemma 4.2.34, $\tau(N) = f_N(0) \subseteq P\text{-rad}(N)$ if and only if *f*-*P*-rad(*N*) = *P*-rad(*N*). In this case, a nonzero element $a \in N$ is *f*-strongly nilpotent in *N* precisely when the element *a* as well as each additively torsion element is strongly nilpotent in *N*. If $\tau(N) \not\subseteq P\text{-rad}(N)$, that is, if there exists an additively torsion element in *N* that is not strongly nilpotent, then *N* does not contain any *f*-strongly nilpotent elements at all.

4.3 Insertion of Factors' Property Ideals in Near Rings and *N*-Groups

In this section, we study the concepts of insertion of factors property (IFP) in near rings and some related results. We also introduce the IFP *N*-group and the IFP ideal of *N*-group. We present some results on these concepts.

Definition 4.3.1

(i) A near ring N is said to have IFP if $a, b \in N$, $ab = 0 \Rightarrow anb = 0$ for all $n \in N$ (in other words, $ab = 0 \Rightarrow aNb = (0)$).

(ii) N is said to have **strong IFP** if every homomorphic image of N has IFP.

(iii) If an ideal I of N has IFP, then I is called an IFP ideal.

Result 4.3.2 (Proposition 9.2 of Pilz, 1983)

N has strong IFP if and only if for all ideals I of N, for all $a, b, n \in N$, $ab \in I \Rightarrow anb \in I$.

Proof

Suppose I has strong IFP and let I be an ideal of N.

Let $\pi: N \to N/I$ be the canonical epimorphism. Suppose that $a, b, n \in N$ and $ab \in I$.

To show that $anb \in I$, consider $\pi(a)\pi(b) = \pi(ab) = ab + I = 0 + I$.

Since N/I has IFP and π is onto, we have $\pi(a)\pi(n)\pi(b) = 0 \Rightarrow \pi(anb) = 0 \Rightarrow anb \in I$.

Therefore, $ab \in I \Rightarrow anb \in I$ for all $a, b, n \in N$.

Converse

Suppose that for all ideals I of N, and for all $a, b, n \in I \Rightarrow anb \in I$.

We show that N has strong IFP. Let $f: N \to N^1$ be a homomorphism.

Now we show that $f(N)$ has IFP.

By the first isomorphism theorem, $N/I \equiv f(N)$, where $I = \ker f$.

Therefore, we have to show that N/I has IFP.

Take $a + I, b + I, n + I \in N/I$ and $(a + I)(b + I) = 0$

$$\Rightarrow ab + I = 0 \Rightarrow ab \in I$$
$$\Rightarrow anb \in I \text{ (by converse hypothesis)}$$
$$\Rightarrow (a + I)(n + I)(b + I) = 0.$$

Hence, N has strong IFP. ∎

Result 4.3.3 (Theorem 9.3 of Pilz, 1983)

The following assertions are equivalent:

(i) N has IFP.

(ii) for all $n \in N$, $(0 : n)$ is an ideal of N.

(iii) for all $S \subseteq N$, $(0 : S)$ is an ideal of N.

Proof

(i) $\Rightarrow$ (ii): Take $x, y \in (0 : n)$. Then $xn = 0$ and $yn = 0$.
By the right distributive law, $(x + y)n = xn + yn = 0 + 0 = 0$.
Therefore, $x + y \in (0 : n)$. Clearly $(0 : n)$ is a normal subgroup of N and a left ideal of N.
To show $(0: n)$ is a right ideal of N, take $n^1 \in (0: n)$ and $n^{11} \in N$. Now $n^1 n = 0$. Since N has IFP, we have $n^1 mn = 0$ for all $m \in N$. In particular, $n^1 n^{11} n = 0$, where $n^{11} \in N$. This implies that $n^1 n^{11} \in (0 : n)$. Therefore, $(0 : n)N \subseteq N$. Hence, $(0 : n)$ is an ideal of N.

(ii) $\Rightarrow$ (iii): Since $(0 : S) = \bigcap_{s \in S}(0 : s)$ and each $(0: s)$ is an ideal, we have that $(0 : S)$ is an ideal of N.

(iii) $\Rightarrow$ (i): Assume that for all $S \subseteq N$, $(0 : S)$ is an ideal of N. Let $a, b \in N$.
Suppose $ab = 0$. This implies that $a \in (0 : b) = (0: \{b\})$. Since $\{b\} \subseteq N$ and $(0 : \{b\})$ is an ideal (by assumption), we have $an \in (0 : \{b\})$ for all $n \in N$. This implies $anb = 0$ for all $n \in N$. Hence, N has IFP. ∎

Definition 4.3.4

An N-group G is said to be an **IFP N-group** if it satisfies the following condition:

$$n \in N, g \in G, ng = 0 \Rightarrow nmg = 0 \text{ for all } m \in N.$$

By this definition, it is clear that every IFP near ring N is an IFP N-group.

Proposition 4.3.5 (Result 1.1 of Satyanarayana, Prasad, and Pradeepkumar, 2004)

(i) If G is an IFP N-group with a torsion-free element $a \in G$, then N is an IFP near ring.
   (Here, an element $0 \neq a \in G$ is said to be **torsion-free** if $n \in N, na = 0 \Rightarrow n = 0$.)

(ii) If G is a faithful IFP N-group, then N is an IFP near ring.
   (Here, an N-group G is called a **faithful** N-group if $(0 : G) = 0$.)

(iii) If N is not an IFP near ring, then it has no faithful IFP N-groups.

(iv) If G is an IFP N-group and I is an ideal of N such that $I \subseteq (0 : G)$, then G is an IFP N^*-group where $N^* = N/I$ is the quotient near ring.

(v) If G is an IFP N-group, then the quotient near ring $N/(0 : G)$ is an IFP near ring.

Proof

(i) Take $x, y \in N$ such that $xy = 0$. Now $xy = 0 \Rightarrow xya = 0$. Since G has IFP, we have $xnya = 0$ for all $n \in N$. Therefore, $xny \in (0 : a) = 0$ (since a is torsion-free). This is true for all $n \in N$. Hence, N is an IFP near ring.

(ii) Suppose $x, y \in N$ such that $xy = 0$. Then $xyG = 0$. Since G has IFP, we have $xnyG = 0$ for all $n \in N$. This implies that $xny \in (0: G) = (0)$, since G is faithful. This is true for all $n \in N$. Hence, N is an IFP near ring.

(iii) Follows from (ii).

(iv) Let $n + I, m + I \in N/I, a \in G$ with $(n + I)a = 0$. This implies $na = 0$. Since G has IFP, we have $nma = 0$. This implies $(n + I)(m + I)a = 0$. This is true for all $m + I \in N/I$. Hence, G is an IFP (N/I)-group.

(v) Taking $I = (0: G)$ in (iv), we have that G is an IFP N/I-group. Since $I = (0: G)$, it follows that G is a faithful N/I-group; and so by (ii), we have N/I is IFP. ∎

Theorem 4.3.6 (Theorem 1.2 of Satyanarayana, Prasad, and Pradeepkumar, 2004)

Let G be an IFP N-group such that $NG \neq 0$ and every monogenic N-subgroup of G has an ascending chain condition on ideals (ACCI). (Here, an N-group G is said to be monogenic if there exists $a \in G$ such that $Na = G$.) Then there exists an element $g \in G$ such that

(i) $(0: g)$ is a prime left ideal of N.

(ii) $(0: g)$ is a prime ideal of N.

(iii) There exists a monogenic N-subgroup Ng of G such that Ng is isomorphic to N/P, where P is a prime (left) ideal of N. Moreover, Ng is N-isomorphic to a prime near ring N^*.

Proof

Write $A = \{(0: x) \mid 0 \neq x \in G, Nx \neq 0\}$.

First we show that A contains a maximal element. Let us suppose that A contains no maximal elements. Then there exists an infinite strict increasing sequence $(0: x_1) \subset (0: x_2) \subset \ldots$ of elements of A.

Now $(0: x_2)/(0: x_1) \subset (0: x_3)/(0: x_1) \subset \ldots$ is also a strict increasing infinite sequence of ideals of the N-group $N/(0: x_1)$. This implies that $N/(0: x_1)$ has no ACCI.

Consider the monogenic N-subgroup Nx_1 of G. By the given condition, Nx_1 has an ACCI. We know that $(0: x_1)$ is a left ideal of N. So, $N/(0: x_1)$ is N-isomorphic to Nx_1 under the map $n + (0: x_1) \to nx_1$.

Since Nx_1 has an ACCI, we have $N/(0: x_1)$ has an ACCI, which is a contradiction. Therefore, A contains a maximal element, say $(0: g)$. Write $P = (0: g)$.

(i) Now we show that P is a prime left ideal. If P is not a prime left ideal, then there exist $a \notin P$ and $b \notin P$ such that $ab \in P$.

Since $P = (0: g)$, it follows that $ag \neq 0$, $bg \neq 0$, and $abg = 0$. Since G is an IFP N-group, it follows that $(0: g) \subseteq (0: bg)$, and so $(0: bg) = (0: g)$ by the maximality of $(0: g)$.

Now $a \in (0 : bg) = (0 : g) \Rightarrow ag = 0$, which is a contradiction. Hence, $P = (0 : g)$ is a prime left ideal.

(ii) Since G has an IFP, by Result 4.3.3, it follows that $P = (0 : g)$ is a two-sided ideal. Now, as in part (i), we can show that P is a prime ideal.

(iii) Now Ng is a monogenic N-subgroup of G such that Ng is isomorphic to N/P. Since P is prime, it follows that $N^* = N/P$ as a prime near ring. ∎

4.4 Finite Spanning Dimension in N-Groups

It is known that the concept of a "finite spanning dimension" (FSD), which was introduced by Fleury (1974) in modules over rings, is a generalization of the dimension of finite-dimensional vector spaces. In this chapter, we generalize this concept to N-groups.

We present the notions of FSD, hollow, and supplement in N-groups, and obtain some important results concerning these notions.

For N-groups satisfying the condition, "every hollow N-subgroup is an ideal," we prove a structure theorem: If G is an N-group having an FSD, then there exists an integer t and hollow N-subgroups $H_1, H_2, ..., H_t$ such that $G = H_1 + H_2 + ... + H_t$ and $G \neq \sum_{\substack{i=1}}^{t} H_i$ for every $1 \leq j \leq t$. Moreover, if there exist hollow N-subgroups $H_1^1, H_2^1, \overset{i \neq j}{...}, H_q^1$ such that $G = H_1^1 + H_2^1 + ... + H_q^1$ and none of the terms can be deleted from the sum, then $q = t$. The integer t determined here is called the **spanning dimension** of G and is denoted by Sd(G).

Definition 4.4.1

(i) Let G be an N-group. A subset S of G is said to be **small** in G if $S + K = G$ and K is an ideal of G imply $K = G$.

(ii) G is said to be **hollow** if every proper ideal of G is small in G.

(iii) G is said to have an **FSD**, if for any strictly decreasing sequence of N-subgroups
$X_0 \supset X_1 \supset X_2 \supset X_3 \supset ...$
of G such that X_i is an ideal of X_{i-1}, there exists an integer k such that X_j is small in G for all $j \geq k$.

Remark 4.4.2

Clearly, every N-group with a descending chain condition (DCC) on N-subgroups has an FSD.

Definition 4.4.3

Let H be an N-subgroup of G. Then an N-subgroup K of G is said to be a **supplement** for H if $H + K = G$ and $H + K^1 \neq G$ for every proper ideal K^1 of K.

Lemma 4.4.4 (Lemma 0.1 of Reddy and Satyanarayana, 1988)

Let G be an N-group. If H is an N-subgroup and K is an ideal of G such that $H + K = G$, then for any ideal H_1 of H, $H_1 + K$ is an ideal of G.

Proof

Since H_1 is an N-subgroup of G and K is an ideal of G, it follows that $H_1 + K$ is an N-subgroup of G. Let $a \in H_1 + K$, $g \in G$, and $n \in N$. Then there exist $h_1 \in H_1$, $k, k_1 \in K$, $h \in H$ such that $a = h_1 + k_1$ and $g = h + k$.

Since K is normal in G, there exists an element $k_2 \in K$ such that $k + h_1 = h_1 + k_2$. Now we have

$$
\begin{aligned}
g + a - g &= (h + k) + (h_1 + k_1) - (h - k) \\
&= h + (k + h_1) + k_1 - k - h \\
&= h + (h_1 + k_2) + k_1 - k - h \\
&= h + h_1 + (k_2 + k_1 - k) - h \\
&= (h + h_1 - h) + h + (k_2 + k_1 - k) - h.
\end{aligned}
$$

Since $h + h_1 - h \in H_1$ and $h + (k_2 + k_1 - k) - h \in K$, we have $g + a - g \in H_1 + K$. So, $H_1 + K$ is normal in G. To show that $H_1 + K$ is an ideal of G, consider

$$
\begin{aligned}
n(g + a) - ng &= n((h + k) + (h_1 + k_1)) - n(n + k) \\
&= n(h + (k + h_1) + k_1) - n(h + k) \\
&= n(h + (h_1 + k_2) + k_1) - n(h + k) \\
&= n((h + h_1) + (k_2 + k_1)) - n(h + h_1) + n(h + h_1) - nh + nh - n(h + k).
\end{aligned}
$$

Since $n((h + h_1) + (k_2 + k_1)) - n(h + h_1) \in K$, $n(h + h_1) - nh \in H_1$,

$nh - n(h + k) \in K$ and $K + H_1 + K = H_1 + K$, and we have

$n(g + a) - ng \in H_1 + K$. Thus, $H_1 + K$ is an ideal of G. ∎

Definition 4.4.5

A strictly decreasing sequence of N-subgroups $X_1 \supset X_2 \supset X_3 \supset \dots$ of G is said to be a *-**sequence** (an ideal sequence) if X_i is an ideal of X_{i-1} for $i \geq 2$.

Remark 4.4.6

G has a DCC on N-subgroups
    $\Rightarrow G$ contains no *-sequence of N-subgroups
    $\Rightarrow G$ has a DCC on ideals.

Proposition 4.4.7 (Proposition 0.2(c) of Reddy and Satyanarayana, 1988))

Let G be an N-group having an FSD and K an ideal of G that is nonsmall in G. Then G/K contains no *-sequences, and hence G/K has an FSD.

Proof

Let $(X_1/K) \supset (X_2/K) \supset (X_3/K) \supset \dots$ be a *-sequence in G/K. Then by Theorem 3.3.4, $X_1 \supset X_2 \supset \dots$ is a *-sequence. Since K is nonsmall and $X_i \supseteq K$, for each i, X_i is also nonsmall in G. Thus, we arrive at a contradiction to the fact that G has an FSD. Hence, G/K contains no strictly decreasing *-sequences. It is now evident that G/K has an FSD.   ■

Proposition 4.4.8 (Proposition 0.3 of Reddy and Satyanarayana, 1988)

Let G be an N-group and K an ideal of G. Suppose that L is an N-subgroup of G, not contained in K, and $f: G \to G/K$ is the canonical epimorphism. Then

  (i) If L is hollow, then so is $f(L)$.
  (ii) If L is a supplement for K and if G/K is hollow, then L is hollow.

Proof

  (i) Suppose L is hollow and $f(L) = (L + K)/K$ is not hollow. Then there exist two proper ideals X/K, Y/K of $f(L)$ such that $(X/K) + (Y/K) = f(L)$. By Proposition 3.2.27, there exist two proper ideals X_1, Y_1 of L such that $X = X_1 + K$ and $Y = Y_1 + K$.
    So, $X + Y = X_1 + Y_1 + K$. Since $(X/K) + (Y/K) = f(L)$,

$X + Y = L + K$. By modular law,

$$L = L \cap (L + K) = L \cap (X + Y)$$
$$= L \cap (X_1 + Y_1 + K)$$
$$= X_1 + Y_1 + (L \cap K).$$

Since X_1 and Y_1 are proper ideals of L and L is hollow, we have $L = L \cap K$, and so $L \subseteq K$, which is a contradiction. Therefore, $f(L)$ is hollow.

(ii) Suppose L is a supplement for K and $f(L)$ is hollow. Let X and Y be ideals of L such that $X + Y = L$. Then $f(X) + f(Y) = f(L)$. Since $f(L)$ is hollow, we have either $f(X) = f(L)$ or $f(Y) = f(L)$. Suppose $f(X) = f(L)$, then $X + K = L + K$. By modular law, $L = X + (L \cap K)$. Now $X + K = X + (L \cap K) + K = L + K = G$, and X is an ideal of L. Since L is a supplement for K, we have $X = L$. Similarly, if $f(Y) = f(L)$, then we can show that $Y = L$. Hence, the result. ∎

Corollary 4.4.9

Let H be a nonsmall N-subgroup of G. If every proper ideal of H is small in G, then H is hollow.

Proof

Since H is nonsmall in G, there exists a proper ideal K of G such that $H + K = G$. Since every proper ideal of H is small in G, it is evident that H is a supplement for K. By Proposition 4.4.8, it is enough if we show that G/K is hollow.

To show this, let X and Y be ideals of G containing K such that $X + Y = G$. By modular law, we have $X = (X \cap H) + K$ and $Y = (Y \cap H) + K$. Therefore, $(X \cap H) + (Y \cap H) + K = X + Y = G$, and so $X \cap H = H$ or $Y \cap H = H$. Hence, either $X = G$ or $Y = G$. This completes the proof. ∎

Lemma 4.4.10 (Lemma 1.1 of Reddy and Satyanarayana, 1988)

If G has an FSD, then every nonsmall N-subgroup H of G contains a hollow N-subgroup, which is nonsmall in G.

Proof

If H is hollow, then there is nothing to prove. Suppose that H is not hollow. By Corollary 4.4.9, there exists a proper ideal H_1 of H, which is nonsmall in G. If H_1 is hollow, our result is true. If not, again using Corollary 4.4.9, there exists a proper ideal H_2 of H_1 such that H_2 is nonsmall in G.

Continuing in this manner, we obtain a strictly decreasing sequence of N-subgroups $H_0 = H \supset H_1 \supset H_2 \supset \ldots$ of G such that H_{i+1} is an ideal of H_i for $i \geq 1$. Since each H_i is nonsmall and G has an FSD, our process must terminate at a finite stage, say at the kth step. Then H_k is hollow and nonsmall in G, which is contained in H.   ∎

Lemma 4.4.11 (Lemma 1.2 of Reddy and Satyanarayana, 1988)

Suppose G has an FSD. Let H be an ideal of G and K an N-subgroup such that $H + K = G$. Then there exists a strict finite chain $K = K_0 \supset K_1 \supset K_2 \supset \ldots \supset K_s$ of N-subgroups such that K_i is an ideal of K_{i-1} for $1 \leq i \leq s$ and K_s is a supplement for H. Further, if $H \neq G$, then each K_i is nonsmall in G.

Proof

If $H = G$, then $K_1 = (0)$ is a supplement for H and the result is true in this case. Suppose $H \neq G$. If K is not a supplement for H, then there exists a proper ideal K_1 of K such that $H + K_1 = G$. Again, if K_1 is not a supplement for H, then there exists a proper ideal K_2 of K_1 such that $H + K_2 = G$. If we continue this process, we get a strict chain $K = K_0 \supset K_1 \supset K_2 \supset \ldots$ of N-subgroups with K_i as an ideal of K_{i-1} for $i \geq 1$. Since each K_i is nonsmall and G has an FSD, this chain must terminate at a finite step, and so there exists an integer S such that K_s is a supplement for H. The proof is complete.   ∎

Note 4.4.12

Since the sum of two N-subgroups need not be an N-subgroup from now onward, we consider only those N-groups G in which every hollow N-subgroup is an ideal of G. Clear examples of near rings belonging to this class of N-groups are given in the following example.

Example 4.4.13

If N is a Boolean near ring (i.e., every element of N is an idempotent) or N is a strongly regular near ring (i.e., for every $a \in N$, there is an $x \in N$ such that $a = xa^2$), then every N-subgroup of N is an ideal of N (by Remark 2.2 of Satyanarayana, 1984). Therefore, every hollow N-subgroup of N is an ideal of N.

Theorem 4.4.14 (Theorem 1.3 of Reddy and Satyanarayana, 1988)

Let G be an N-group having an FSD. Then

(i) There exist hollow N-subgroups $H_1, H_2, ..., H_t$ such that $G = H_1 + H_2 + ... + H_t$ and $G \neq \sum_{\substack{i=1 \\ i \neq j}}^{t} H_i$ for every $1 \leq j \leq t$.

(ii) If there exist hollow N-subgroups $H_1^1, H_2^1, ..., H_q^1$ such that $H_1^1 + H_2^1 + ... + H_q^1 = G$ and none of the terms can be deleted from the summation, then $q = t$.

Proof

(i) If G is hollow, take $H_1 = G$. Suppose G is not hollow. Then G contains a nonsmall proper ideal H. By Lemma 4.4.10, H contains a hollow N-subgroup H_1 that is nonsmall in G. Since H_1 is nonsmall, there exists a proper ideal K of G such that $H_1 + K = G$, and by Lemma 4.4.11, there exists a strict chain of N-subgroups

$$K = K_0 \supset K_1 \supset K_2 \supset ... \supset K_{s_1}$$

of G such that K_i is an ideal of K_{i-1} for $1 \leq i \leq s_1$ and K_{s_1} is a supplement for H_1. If K_{s_1} is hollow, take $H_2 = K_{s_1}$. Then $G = H_1 + H_2$, and our result is true. If K_{s_1} is not hollow, since K_{s_1} is nonsmall in G, by Lemma 4.4.10, K_{s_1} contains a hollow N-subgroup H_2 that is nonsmall in G. Since H_2 is nonsmall, there exists a proper ideal K^1 of G such that $H_2 + K^1 = G$. By modular law, $K_{s_1} = K_{s_1} \cap G = K_{s_1} \cap (H_2 + K^1) = H_2 + (K_{s_1} \cap K^1)$. Now $H_1 + H_2 + (K_{s_1} \cap K^1) = G$, and it is clear that none of the terms can be deleted from the sum. Write $K_{s_{1+1}} = K_{s_1} \cap K^1$. Since K^1 is an ideal of G, $K_{s_{1+1}}$ is an ideal of K_{s_1}. Again using Lemma 4.4.11, there exists a strict chain of N-subgroups

$$K_{s_{1+1}} \supset K_{s_{1+2}} \supset K_{s_{1+3}} \supset ... \supset K_{s_1+s_2}$$

such that $K_{s_{1+i+1}}$ is an ideal of $K_{s_{1+i}}$ for $1 \leq i \leq s_2-1$ and $K_{s_1+s_2}$ is a supplement for $H_1 + H_2$. If $K_{s_1+s_2}$ is hollow, take $H_3 = K_{s_1+s_2}$. Now $G = H_1 + H_2 + H_3$, and none of the terms can be deleted from the summation. If we continue this process, at the tth step, we have hollow N-subgroups $H_1, H_2, ..., H_t$ and a strict chain of N-subgroups $K = K_0 \supset K_1 \supset K_2 \supset ... \supset K_{s_1} \supset K_{s_{1+1}} \supset ... \supset K_{s_1+s_2} \supset ... \supset K_{s_1+s_2+...+s_t}$ such that each term is an ideal of its proceeding term, and $G = H_1 + H_2 + ... + H_t + K_{s_1+s_2+...+s_t}$, and none can be deleted from the sum. Since each K_i is nonsmall and G has an FSD, this process must stop at the $(t-1)$th step only if $K_{s_1+s_2+...+s_{(t-1)}}$ is hollow. Now take $H_t = K_{s_1+s_2+...+s_{(t-1)}}$. Thus, we get hollow N-subgroups $H_1, H_2, ..., H_t$ such that $H_1 + H_2 + ... + H_t = G$, and no term can be deleted from the sum. Even though the proof for (ii) is parallel to ring theory, for the sake of completeness, we present the proof here.

(ii) Suppose $G = H_1^1 + H_2^1 + ... + H_q^1$, where H_i^1 is hollow and none of the terms from the summation may be deleted. Without loss of generality, we may assume that $q > t$. Consider $H_2 + H_3 + ... + H_t$. This is a proper ideal of G by construction.

We show that for some i $(1 \le i \le q)$, $G = H_1^1 + H_2 + ... + H_t$ and none of the terms from the summation can be deleted. Suppose for $1 \le i \le q - 1$ that $G \ne H_1^1 + H_2 + ... + H_t > t$.

Then for $1 \le i \le q - 1$, by modular law,

$$H_1^1 + H_2 + ... + H_t = (H_1^1 + H_2 + ... + H_t) \cap (H_1 + H_2 + ... + H_t)$$
$$= ((H_1^1 + H_2 + ... + H_t) \cap H_1) + (H_2 + ... + H_t)$$
$$= U_i^1 + H_2 + ... + H_t$$

where $U_i^1 = (U_i^1 + H_2 + ... + H_t) \cap H_1$. Since $G \ne H_1^1 + H_2 + ... + H_t$, U_i^1 is a proper ideal of H_1, and hence U_i^1 is small in G for $1 \le i \le q - 1$.

Now $H_1^1 + H_1^1 + ... + H_q^1 + H_2 + ... + H_t = G$

$\Rightarrow U_i^1 + H_2 + ... + H_t + H_2^1 + ... + H_q^1 = G$

$\Rightarrow H_2 + ... + H_t + H_2^1 + ... + H_q^1 = G$

$\Rightarrow U_2^1 + H_2 + ... + H_t + H_3^1 + ... + H_q^1 = G$

$\Rightarrow H_2 + ... + H_t + H_3^1 + ... + H_q^1 = G.$

If we continue this, finally we get $H_2 + ... + H_t + H_q^1 = G$. Thus, there exists a j such that $H_2 + ... + H_t + H_j^1 = G$. From the construction H_j^1 cannot be deleted from the sum. Suppose $H_j^1 + H_3 + ... + H_t = G$. Using the modular law as earlier, there exists a proper ideal $U = H_j^1 \cap (H_2 + ... + H_t)$ or H_j^1 such that $H_2 + ... + H_t = U + H_3 + ... + H_t$. So, $U + H_1 + H_3 + ... + H_t = H_1 + H_2 + ... + H_t = G$, and hence $H_1 + H_3 + ... + H_t = G$, which is a contradiction.

So, H_2 cannot be deleted from the sum $H_j^1 + H_2 + ... + H_t = G$. Similarly, we can show that no H_i can be deleted.

Actually, we have replaced H_1 by some H_j^1 and studied the problem of deletion. Next we replaced H_2 by another H_j^1 and showed that the sum is equal to G and that no term can be deleted.

When we continue in this manner, if $q > t$, we find that after replacing all of the H_j's that we have actually deleted some H_j's. This is a contradiction. Therefore, $p = q$.   ∎

Definition 4.4.15

The integer determined in Theorem 4.4.14 is denoted by Sd(G), and is called the **spanning dimension** of G.

Proposition 4.4.16

Let G be an N-group with an FSD. If H is an ideal of G and K is an N-subgroup of G, which is a supplement of H, then K has an FSD.

Proof

If X is an N-subgroup of K that is nonsmall in K, then there exists a proper ideal J of K such that $X + J = K$.

Since $K + H = G$ and H is an ideal of G, by Lemma 4.4.4, $J + H$ is an ideal of G. Since K is a supplement of H, $J + H$ is a proper ideal of G. But $X + (J + H) = G$, which implies that X is not small in G. Thus, it follows that since G has an FSD, K also has an FSD. ∎

Lemma 4.4.17 (Lemma 2.1 of Satyanarayana, 1991)

Suppose G has an FSD. Then

(i) Every N-subgroup having an FSD is an ideal.
(ii) If K and L are two ideals of G such that K is a supplement of L and L is a supplement of K, then $Sd(G) = Sd(L) + Sd(K)$.
(iii) Let K be a supplement of an ideal H of G; then $Sd(K) = Sd(G)$ if and only if $K = G$.
(iv) If K is a supplement of an ideal H of G, then K is an ideal of G.

Proof

(i) If J is an N-subgroup of G having an FSD, then by Theorem 4.4.14, there exist hollow N-subgroups $H_1, H_2, \ldots, H_t$ such that $H_1 + H_2 + \ldots + H_t = J$, and by our assumption each H_i is an ideal of G, and hence J is an ideal of G.

(ii) Suppose K and L are two ideals of G such that K is a supplement of L and L is a supplement of K. By Proposition 4.4.16, K and L have an FSD, and by Theorem 4.4.14, there exist hollow N-subgroups $H_1, H_2, \ldots, H_t$, $H_1^1, H_2^1, \ldots, H_t, H_1^1, H_2^1, \ldots, H_s^1$ such that $K = H_1 + H_2 + \ldots + H_t$ and $L = H_1^1 + H_2^1 + \ldots + H_s^1$, and none of the terms can be deleted either from the summation $K = H_1 + H_2 + \ldots + H_t$ or from $L = H_1^1 + H_2^1 + \ldots + H_s^1$. Now $G = K + L = H_1 + H_2 + \ldots + H_t + H_1^1 + H_2^1 + \ldots + H_s^1$. Now since K and L are supplements to each other, it is clear that no term can be deleted from the summation

$$G = H_1 + H_2 + \ldots + H_t + H_1^1 + H_2^1 + \ldots + H_s^1.$$

Thus, $Sd(G) = t + s = Sd(K) + Sd(L)$.

(iii) Let K be a supplement of an ideal H of G. Suppose $Sd(K) = Sd(G)$. Since G has an FSD, by Lemma 4.4.11, there exists an N-subgroup J contained in H such that J is a supplement for K. Now by Proposition 4.4.16, J has an FSD, and by (i) J is an ideal of G. Now it is clear that K is also a supplement for J. Now by (ii), we have $Sd(K) + Sd(J) = Sd(G)$, which implies that $Sd(J) = 0$. Now it is evident that $J = (0)$. The rest is clear.

(iv) If K is a supplement, then by Proposition 4.4.16, K has an FSD and by (i), K is an ideal. ■

Theorem 4.4.18 (Theorem 2.3 (iv) of Satyanarayana, 1991)

If G has an FSD and if $(0) \neq K$ is an ideal of G that is a supplement of an ideal H of G, then K and G/K have an FSD and $Sd(G/K) = Sd(G) - Sd(K)$.

Proof

By Proposition 4.4.16, K has an FSD. Since K is a supplement, it follows that K is nonsmall, and by Proposition 4.4.7, G/K has an FSD. By Lemma 4.4.11, there exists a supplement L for K that is contained in H. Now K is also a supplement for L. By Lemma 4.4.17 (ii), $Sd(G) = Sd(K) + Sd(L)$. By Theorem 4.4.14, there exist hollow N-subgroups $L_1, L_2, ..., L_s$ such that $L = L_1 + L_2 + ... + L_s$, none of the terms can be deleted from the summation, and $s = Sd(L)$. Let f: $G \rightarrow G/K$ be the canonical epimorphism.

If $L_i \subseteq K$ for some i, then we have a contradiction. So $L_i \nsubseteq K$ for each i, and by Proposition 4.1.8, $f(L_1), f(L_2), ..., f(L_s)$ are hollow N-subgroups of G/K. Clearly,

$$f(L) = f(L_1) + f(L_2) + ... + f(L_s) = G/K.$$

Suppose $f(L_2) + ... + f(L_s) = G/K$.

Then $L_2 + L_3 + ... + L_s + K = G$. Since L is a supplement of K and $L_2 + L_3 + ... + L_s$ is an ideal of L, we have $L = L_2 + ... + L_s$, which is a contradiction. Similarly, we may show that no term can be omitted from the sum $f(L_1) + f(L_2) + ... + f(L_s) = G/K$.

Thus, $Sd(G/K) = s = Sd(L)$ and $Sd(L) = Sd(G) - Sd(K)$.

Therefore, $Sd(G/K) = Sd(G) - Sd(K)$.

From the proof just given, we obtain the following. ■

Remark 4.4.19

If G has an FSD, H is a nonsmall ideal of G, and L is a supplement of H, then $Sd(G/H) = Sd(L)$.

Lemma 4.4.20 (Theorem 2.3 (ii) of Satyanarayana, 1991)

Let G be an N-group having an FSD and H be a nonzero ideal satisfying

(i) H and G/H have an FSD; and
(ii) $\text{Sd}(G) = \text{Sd}(G/H) + \text{Sd}(H)$

Then H is nonsmall in G.

Proof

Suppose H is small in G. Let $f: G \to G/H$ be the canonical epimorphism. If $\text{Sd}(G) = 1$, then $\text{Sd}(G/H) = 0$, and hence $G = H$. So, in this case H is trivially nonsmall in G. Assume that $\text{Sd}(G) > 1$. By Theorem 4.4.14, there exist hollow N-subgroups $Z_1, Z_2, \dots, Z_k$ such that $G = Z_1 + Z_2 + \dots + Z_k$ and $X = \text{Sd}(G)$. Since Z_i is nonsmall, $Z_i \not\subseteq H$, and by Proposition 4.4.8, $f(Z_1), f(Z_2), \dots, f(Z_k)$ are hollow N-subgroups of G/H. Clearly, $f(Z_1) + f(Z_2) + \dots + f(Z_k) = G/H$. Since H is small, none can be deleted from the summation, and if $f(Z_1) + f(Z_2) + \dots + f(Z_{i-1}) + f(Z_{i+1}) + \dots + f(Z_k) = G/H$ for some $1 \le i \le k$, then

$$Z_1 + \dots + Z_{i-1} + Z_{i+1} + \dots + Z_k + H = G.$$

Since H is small, $Z_1 + \dots + Z_{i-1} + Z_{i+1} + \dots + Z_k = G$. This is a contradiction since $\text{Sd}(G) = k$. Therefore, $\text{Sd}(G/H) = k$, and by hypothesis $\text{Sd}(G/H) = 0$ and so $H = (0)$, which is a contradiction. Thus, H is nonsmall in G. ∎

Lemma 4.4.21

If G has an FSD and H is a nonsmall ideal of G, then the following are equivalent:

(i) H is a supplement of an ideal of G.
(ii) Every N-subgroup of H that is small in G is small in H.
(iii) Every ideal of G contained in H that is small in G is small in H.

Proof

(i) $\Rightarrow$ (ii): Let X be an N-subgroup of H that is small in G. Suppose $X + H^1 = H$ and H^1 is an ideal of H. If H is a supplement for an ideal K of G, then
$X + H^1 + K = H + K = G$.
By Lemma 4.4.4, $H^1 + K$ is an ideal of G, and since X is small, we have $H^1 + K = G$. Since H is a supplement for K, by definition $H^1 = H$. This shows that X is small in H.

(ii) $\Rightarrow$ (iii) is trivial.

(iii) $\Rightarrow$ (i): Since H is nonsmall, there exists a proper ideal A of G such that $H + A = G$. By Lemma 4.4.11, there exists an ideal K^* of G such that $K^* \subseteq A$ and K^* is a supplement for H. Now by Lemma 4.4.17 (i), K^* is an ideal.

Again, using Lemma 4.4.17, there exists an ideal H^* of G such that $H^* \subseteq H$ and H^* is a supplement of K^*. Now it is evident that K^* is also a supplement for H^*. We first show that $H \cap K^*$ is small in G. To show this, let Z be an ideal of G with $(H \cap K^*) + Z = G$. Now

$$K^* = G \cap K^* = ((H \cap K^*) + Z) \cap K^*$$
$$= (H \cap K^*) + (Z \cap K^*) \text{ and}$$

$(H \cap K^*) + (Z \cap K^*) + H = K^* + H = G$. Since $(H \cap K^*) \subseteq H$, we have $(Z \cap K^*) + H = G$. The fact that K^* is a supplement for H shows that $Z \cap K^* = K^*$, and so $Z \supseteq K^*$. Therefore, $Z = G$ and hence $H \cap K^*$ is small in G. By assumption, $H \cap K^*$ is small in H. Now by modular law,

$$H = H \cap G = H \cap (H^* + K^*) = H^* + (K^* \cap H).$$

Since $K^* \cap H$ is small in H, we have $H = H^*$, which shows that H is a supplement of K^*. By Lemma 4.4.17 (i), K^* is an ideal. Therefore, H is a supplement of an ideal of G. ∎

Theorem 4.4.22 (Theorem 2.3 (iv) of Satyanarayana, 1991)

Let G be an N-group having an FSD and H an ideal of G satisfying

(i) H and G/H have an FSD; and
(ii) $\mathrm{Sd}(G/H) = \mathrm{Sd}(G) - \mathrm{Sd}(H)$

Then H is supplement of an ideal of G.

Proof

If $H = (0)$, there is nothing to prove. Suppose $H \neq (0)$. By Lemma 4.4.20, H is nonsmall. By Lemma 4.4.11, there exists an N-subgroup K^* that is a supplement for H. Since H is an ideal, by Proposition 4.4.16, K^* has an FSD, and by Lemma 4.4.17 (i), K^* is an ideal. Since H is nonsmall, we note that K^* is a proper ideal of G. By Remark 4.4.19, $\mathrm{Sd}(G/H) = \mathrm{Sd}(K^*)$. Now suppose H is not a supplement. Then by Lemma 4.4.21, there exists an ideal S of G contained in H such that S is small in G and S is not small in H. Since H has an FSD, by Lemma 4.4.11, there exists a proper N-subgroup K of H such that K is a supplement for S in H. Again by Lemma 4.4.16, K has an FSD, and by Lemma 4.4.17 (i), K is an ideal of G. Now by Lemma 4.4.11, there exists an N-subgroup $(0) \neq S^*$ of H such that S^* is a supplement for K in H that is contained in S. Now, $H = S^* + K$, and by Lemma 4.4.17 (ii), $\mathrm{Sd}(H) = \mathrm{Sd}(S^*) + \mathrm{Sd}(K)$ and $\mathrm{Sd}(S^*) \geq 1$. Now $S^* + K + K^* = H + K^* = G$, and since S^* is small in G, $K + K^* = G$. Let K^1 be a supplement of K^* with $K^1 \subseteq K$. By Lemma 4.4.17 (ii), $\mathrm{Sd}(G) = \mathrm{Sd}(K^1) + \mathrm{Sd}(K^*)$. Since $K^1 \subseteq K$ and both K and K^1 have an FSD, it follows that $\mathrm{Sd}(K^1) \leq \mathrm{Sd}(K)$. Thus, we have

$$
\begin{aligned}
\mathrm{Sd}(G) &= \mathrm{Sd}(K^1) + \mathrm{Sd}(K^*) \\
&\leq \mathrm{Sd}(K) + \mathrm{Sd}(K^*) \\
&< \mathrm{Sd}(S^*) + \mathrm{Sd}(K) + \mathrm{Sd}(K^*) \\
&= \mathrm{Sd}(H) + \mathrm{Sd}(K^*) \\
&= \mathrm{Sd}(H) + \mathrm{Sd}(G/H) \\
&= \mathrm{Sd}(G).
\end{aligned}
$$

So we get $\mathrm{Sd}(G) < \mathrm{Sd}(G)$, and this is a contradiction. The proof is complete. ∎
Combining Theorems 4.4.18 and 4.4.22, we have the following.

Theorem 4.4.23

Let G be an N-group having an FSD. Then for an ideal H of G, the following conditions are equivalent:

(i) H is a supplement for an ideal of G.
(ii) H and G/H have FSD and $\mathrm{Sd}(G) = \mathrm{Sd}(G/H) + \mathrm{Sd}(H)$.

5

Dimension and Decomposition Theory

It is well known that the dimension of a vector space is defined as the number of elements in its basis. The basis of a vector space is a maximal set of linearly independent vectors or a minimal set of vectors that spans the space. The former, when generalized to modules over rings, becomes the concept of Goldie dimension. Goldie (1972) introduced the concept finite Goldie dimension (FGD) in modules over rings. A module is said to have FGD if it contains no infinite direct sum of nonzero submodules. The concept of FGD in N-groups was introduced by Reddy and Satyanarayana (1988). It was later studied by Satyanarayana and Prasad (1998, 2000, 2005b, 2005c); Satyanarayana, Prasad and Pradeep Kumar (2004).

5.1 Finite Goldie Dimension

We begin this section with the following lemma.

Lemma 5.1.1 (Lemma 1.2.5 of Satyanarayana, 1984)

Let G be an N-group having FGD. Then every nonzero ideal H of G contains a uniform ideal.

Proof

Suppose H contains no uniform ideal of G. Then H is not uniform, and so there exist two nonzero ideals H_1 and H_1^1 of G such that $H_1 \cap H_1^1 = (0)$ and $H_1 + H_1^1 \subseteq H$. By our supposition, H^1 is not uniform and this implies that there exist two nonzero ideals H_2, H_2^1 of G such that $H_2 \cap H_2^1 = (0)$ and $H_2 + H_2^1 \subseteq H_1^1$. If we continue this process, we get two infinite sequences $\{H_i\}_1^\infty$, $\{H_i^1\}_1^\infty$ of nonzero ideals of G such that $H_i \cap H_i^1 = (0)$ and $H_i + H_i^1 \subseteq H_{i-1}^1$ for $i \geq 2$. Now the sum $\Sigma_{i=1}^\infty H_i$ is an infinite direct sum of nonzero ideals of G, which is a contradiction. The proof is complete. ∎

Theorem 5.1.2 (Theorem 2.4 of Reddy and Satyanarayana, 1988a)

Let G be an N-group.

(i) If G has FGD, then there exists a finite number of uniform ideals of G, whose sum is direct and essential in G.
(ii) If G has uniform ideals $U_1, U_2, ..., U_n$ such that the sum $U_1 + U_2 + ... + U_n$ is direct and essential in G, then G has FGD. Moreover, the positive integer n is independent of the choice of the U_i's.

Proof

(i) By Lemma 5.1.1, G contains a uniform ideal U_1. If U_1 is essential in G, we are through. Suppose U_1 is not essential; then there exists a nonzero ideal H_1 of G with $U_1 \cap H_1 = (0)$. Again, using Lemma 5.1.1, H_1 contains a uniform ideal U_2. If $U_1 + U_2$ is essential in G, then the theorem is true. Otherwise, there exists a nonzero ideal H_2 of G such that $(U_1 + U_2) \cap H_2 = (0)$. Now there exists a uniform ideal $U_3 \subseteq H_2$. If we continue this process, we get a chain
$U_1 \subseteq (U_1 + U_2) \subseteq (U_1 + U_2 + U_3) \subseteq ...$, which must terminate.
Hence, for some n, $U_1 + U_2 + ... + U_n$ must be essential in G.
(ii) Suppose G has uniform ideals $U_1, U_2, ..., U_n$ such that the sum $S = U_1 + U_2 + ... + U_n$ is direct and essential in G. Let $V_1, V_2, ..., V_m$ be non-zero ideals of G, whose sum $T = V_1 + V_2 + ... + V_m$ is direct. We shall show that $m \leq n$, from which (ii) follows.
First, we show that if J is an ideal of G such that $J \cap U_i \neq (0)$ for all i, then J is essential in G. If $J \cap U_i \neq (0)$, we have that $J \cap U_i$ is essential in U_i, and by Corollary 3.4.5, $(J \cap U_1) + ... + (J \cap U_n)$ is essential in S, and by Remark 3.4.2, $(J \cap U_1) + ... + (J \cap U_n)$ is essential in G. Again by Remark 3.4.2 and since $(J \cap U_1) + ... + (J \cap U_n) \subseteq J$, it follows that J is essential in G.
Now consider $T_1 = V_2 + V_3 + ... + V_m$. Since T_1 is not essential in G, there exists an i ($1 \leq i \leq n$) such that $T_1 \cap U_i = (0)$, and we may suppose that $T_1 \cap U_1 = (0)$.
Thus, the sum $U_1 + V_2 + ... + V_m$ is direct.
Again, since $T_2 = U_1 + V_3 + ... + V_m$ is not essential, there exists an i ($2 \leq i \leq n$) such that $T_2 \cap U_i = (0)$, say $i = 2$. Therefore, $U_1 + U_2 + V_3 + ... + V_m$ is direct. If we continue this process, we arrive at the result that $m \leq n$. ∎

Definition 5.1.3

Let G be an N-group with FGD. Then the integer n determined in Theorem 5.1.2 is called the **dimension** of G. We shall denote it by dim G.

Corollary 5.1.4

Suppose G has FGD and $n = \dim G$. Then

(i) The number of summands in any decomposition of an ideal I of G as a direct sum of nonzero ideals of G is at most n.
(ii) An ideal H of G is essential in G if and only if H contains a direct sum of n uniform ideals.

Proof

(i) Suppose $I = K_1 + K_2 + \ldots + K_s$ for some nonzero ideals $K_1, K_2, \ldots, K_s$ of G and the sum $K_1 + K_2 + \ldots + K_s$ is direct. Now, by the proof of the second part of Theorem 5.1.2, we have $s \leq n$.
(ii) Suppose H is an essential ideal of G. Since G has FGD, by Theorem 5.1.2, there exist uniform ideals $U_1, U_2, \ldots, U_n$ in G such that the sum $U_1 + U_2 + \ldots + U_n$ is direct and essential in G. Since H is essential, $H \cap U_i$ is nonzero and so $H \cap U_i$ is a uniform ideal of G, for $1 \leq i \leq n$. Thus, H contains the sum $(H \cap U_1) + \ldots + (H \cap U_n)$ of n uniform ideals.

Converse

Suppose that H contains a direct sum $V_1 + V_2 + \ldots + V_m$ of uniform ideals $V_1, V_2, \ldots, V_n$. If H is not essential, by Remark 3.4.2, $V_1 + V_2 + \ldots + V_n$ is not essential, and so there exists a nonzero ideal L of G such that $(V_1 + V_2 + \ldots + V_n) \cap L = (0)$. Now $L + V_1 + V_2 + \ldots + V_n$ is a direct sum of $n + 1$ nonzero ideals, which is a contradiction to part (i) of this corollary. The proof is complete. ∎

Corollary 5.1.4 (ii) may be restated as "Let G be an N-Group with FGD. Then an ideal H of G is essential in G if and only if $\dim (H) = \dim (G)$."

Notation 5.1.5

Let G be an N-group having FGD and H be an ideal of G. Since G contains no infinite direct sum of nonzero ideals, H also does not contain an infinite direct sum of nonzero ideals of G, and if we proceed as in Theorem 5.1.2, there exist uniform ideals $U_1, U_2, \ldots, U_k$ in G such that the sum $U_1 + U_2 + \ldots + U_k$ is direct and G-essential in H. Further, the integer k is independent of the choice of U_is. This number k is called the dimension of H relative to G and is denoted by $\dim_G(H)$ (or simply by $\dim H$).

Note 5.1.6

A nonzero ideal U of G is uniform if and only if $\dim U = 1$.

Proof

Suppose U is uniform. Then U is uniform and U is essential in U. Hence, dim $U = 1$.

Converse

Suppose dim $U = 1$. Then there exists a uniform ideal V of G such that V is essential in U. Now we have to show that U is uniform. Let us suppose U is not uniform. Then there exist two nonzero ideals I and J of G contained in U such that $I \cap J = (0)$. Since V is essential in U, it follows that $V \cap I \neq (0)$ and $V \cap J \neq (0)$. Now $V \cap I$ and $V \cap J$ are two nonzero ideals in V whose intersection is (0), which is a contradiction (refer to Proposition 3.4.12).   ∎

Proposition 5.1.7

Let G be an N-group with FGD. Then an ideal H of G is essential in G if and only if $\dim(H) = \dim(G)$. (This is a restatement of Corollary 5.1.4 (ii).)

Theorem 5.1.8

If G has FGD and G_1, G_2 are two ideals of G such that $G_1 \cap G_2 = (0)$, then

$$\dim(G_1 + G_2) = \dim(G_1) + \dim(G_2).$$

Proof

Write $m_1 = \dim(G_1)$ and $m_2 = \dim(G_2)$. Now there exist uniform ideals $U_1, U_2, \ldots, U_{m_1}, V_1, V_2, \ldots, V_{m_2}$ of G such that the sums $U_1 + \ldots + U_{m_1}$ and $V_1 + \ldots + V_{m_2}$ are direct and G-essential in G_1, G_2, respectively. Since $G_1 \cap G_2 = (0)$, the sum $U_1 + \ldots + U_{m_1} + V_1 + \ldots + V_{m_2}$ is direct and by Corollary 3.4.10,

$$U_1 + U_2 + \ldots + U_{m_1} + V_1 + V_2 + \ldots + V_{m_2}$$

is essential in $G_1 + G_2$. Therefore,

$$\dim(G_1 + G_2) = m_1 + m_2 = \dim G_1 + \dim G_2.   ∎$$

Corollary 5.1.9

If G has FGD and H is an ideal of G with dim $H <$ dim G, then there exist uniform ideals $U_1, U_2, \ldots, U_k$ such that the sum $H + U_1 + U_2 + \ldots + U_k$ is direct and essential in G. Moreover, $k = \dim G - \dim H$.

Proof

Since dim $H <$ dim G, by Proposition 5.1.7, H is not essential. By Zorn's lemma, there is an ideal H^1 that is maximal subject to $H \cap H^1 = (0)$. Then, $H + H^1$ is essential in G. Let $k = $ dim H^1. Now there exist uniform ideals $U_1, U_2, ..., U_k$ of G such that the sum $U_1 + U_2 + ... + U_k$ is direct and G-essential in H^1. By Theorem 3.4.10,

$H + U_1 + U_2 + ... + U_k$ is direct and G-essential in $H + H^1$. By Remark 3.4.2, $H + U_1 + U_2 + ... + U_k$ is essential in G. By Proposition 5.1.7 and Theorem 5.1.8,

$$\dim G = \dim(H + U_1 + ... + U_k)$$
$$= \dim(H) + \dim(U_1) + ... + \dim(U_k)$$
$$= \dim(H) + k.$$

Hence, $k = $ dim $G -$ dim H.

In vector space theory, for a subspace W of a vector space V, we have $\dim(V/W) = $ dim $V -$ dim W. This condition does not hold for an ideal W of an N-group V over a near ring N, where "dim" denotes the "Goldie dimension." ∎

Definition 5.1.10

Let K, I be ideals of G. K is said to be a **complement** of I if it satisfies the following two conditions:

(i) $K \cap I = (0)$; and
(ii) K^1 is an ideal of G such that $K^1 \supseteq K$, $K^1 \neq K$ imply that $K^1 \cap I \neq (0)$.

Result 5.1.11 (Lemma 1.4 of Satyanarayana, 1991)

Let us fix an ideal I of G. By Zorn's lemma, the set of all ideals H of G satisfying $H \cap I = (0)$ contains a maximal element, say K. Again, by Zorn's lemma, the set of all ideals X of G satisfying $X \supseteq I$ and $X \cap K = (0)$ contains a maximal element, say K^*. Then we have

(i) K is a complement of I
(ii) K^* is a complement of K
(iii) $K + I$ and $K + K^*$ are essential ideals
(iv) I is essential in K^*.

Proof

(i) Consider the set of ideals H of G such that $H \cap I = (0)$. By Zorn's lemma, this set has a maximal element, say K. Then, it is evident that K is a complement of I.

(ii) Let K^* be an ideal that is maximal relative to $K^* \supseteq I$ and $K^* \cap K = (0)$. Clearly, K^* is a complement of K.

(iii) It is easy to verify that $K + K^*$ and $K + I$ are essential.

(iv) Since $I \subseteq K^*$, it follows that $K + I \subseteq K + K^*$. Also, $K + I$ is essential in $K + K^*$. By Corollary 3.4.5, I is essential in K^*. ∎

Definition 5.1.12

Let H, H^1 be two ideals of an N-group G such that $H \subseteq H^1$. Then H^1 is called a G-essential extension of H in G if H is G-essential in H^1. If, in addition, $H^1 \neq H$, we say that H^1 is a proper G-essential extension.

Proposition 5.1.13 (Proposition 2.1.4 of Satyanarayana, 1984)

If G has FGD and if an ideal H of G has no proper G-essential extensions, then G/H has FGD.

Proof

Let $H_1, H_2, \ldots, H_k$ be ideals of G such that $(H_1/H) + (H_2/H) + \ldots + (H_k/H)$ is a direct sum of nonzero ideals in G/H and $H \subseteq H_i$ for $1 \le i \le k$. By hypothesis, there exist nonzero ideals J_i in G such that $J_i \subseteq H_i$, $J_i \cap H = (0)$. We now show that the sum $J_1 + J_2 + \ldots + J_k$ is direct. Suppose $a_1 + a_2 + \ldots + a_k = 0$, where $a_i \in J_i$ for $1 \le i \le k$. Then $(a_1 + H) + (a_2 + H) + \ldots + (a_k + H)$ is the zero element in G/H, and since $a_i + H \in H_i/H$ and the sum $(H_1/H) + (H_2/H) + \ldots + (H_k/H)$ is direct, we have $a_i + H = 0 + H$ for each i. Hence, $a_i \in H$ and so $a_i \in H \cap J_i = (0)$. Thus, the sum $J_1 + J_2 + \ldots + J_k$ is direct. By Corollary 5.1.4 (i), $k \le n$, where $n = \dim(G)$. Therefore, any direct sum of nonzero ideals of G/H has at most n terms, and hence G/H has FGD. ∎

Proposition 5.1.14 (Remark 1.5 of Satyanarayana, 1991)

If G has FGD and if K is a complement ideal in G, then

(i) K has no proper G-essential extensions and
(ii) G/K has FGD.

Proof

(i) If K is a complement of I, then for any ideal K^1 of G containing K properly, $K^1 \cap I \neq (0)$ and $K \cap (K^1 \cap I) = (0)$. Thus, K has no proper G-essential extensions in G.

Now, (ii) follows from Proposition 5.1.13. ∎

Theorem 5.1.15 (Theorem 2.1.7 of Satyanarayana, 1984)

Let K be an ideal of G and $\pi: G \to G/K$ be a canonical epimorphism. Then the following conditions are equivalent.

(i) K is a complement.
(ii) For any ideal K^1 of G containing K, K^1 is a complement in G, if and only if $\pi(K^1)$ is complement in G/K.

Proof

(i) $\Rightarrow$ (ii): Suppose K^1 is an ideal of G containing K. If K^1 is a complement of an ideal S in G, then it is easy to verify that $\pi(K^1)$ is a complement of $\pi(S)$. Conversely, suppose $\pi(K^1)$ is a complement of an ideal I/K. We now show that K^1 is a complement ideal. Suppose K is a complement of Z, where Z is an ideal of G. Let X be a complement of K containing Z. Since $K^1 \cap I = K$, we have $K^1 \cap (I \cap X) = (0)$. Let Y be a complement of $I \cap Z$ containing K^1. Now $Y \cap I \supseteq K$ and $(Y \cap I) \cap Z = (0)$, and hence $Y \cap I = K$. Therefore, $\pi(Y) \cap \pi(I) = (0)$. Since $\pi(Y) \supseteq \pi(K^1)$ and $\pi(Y) \cap \pi(I) = (0)$, we have $\pi(Y) = \pi(K^1)$ and $Y = K^1$. Thus, K^1 is a complement ideal in G.
(ii) $\Rightarrow$ (i): Since $\pi(K)$ is a complement in G/K, it is evident that K is a complement in G. ∎

Example 5.1.16

If V is a vector space, then every subspace of V is a complement.

Theorem 5.1.17 (Theorem 2.1.10 of Satyanarayana, 1984)

Let K be an ideal of G and $\pi: G \to G/K$ be the canonical epimorphism. Then the following conditions are equivalent:

(i) K is a complement.
(ii) For any essential ideal S of G, $\pi(S)$ is essential in G/K.

Proof

(i) $\Rightarrow$ (ii): Let S be an essential ideal of G. To show $\pi(S)$ is essential in G/K, let Z/K be an ideal of G/K such that $\pi(S) \cap (Z/K) = (0)$. Suppose $Z \neq K$. Then by Proposition 5.1.14 (i), there exists a nonzero ideal A of G such that $A \cap K = (0)$ and $A \subseteq Z$. Now $x \in S \cap A \Rightarrow x + K \in \pi(S) \cap \pi(Z) = (0)$ and $x \in A \Rightarrow x \in K \cap A = (0)$.

Since S is an essential ideal and $S \cap A = (0)$, we have $A = (0)$, which is a contradiction. Thus, $Z = K$, and hence $\pi(S)$ is essential in G/K.

(ii) $\Rightarrow$ (i): Suppose K is not a complement. Then by Proposition 5.1.14, K has a proper G-essential extension K^*. Let X be a complement of K^* in G. Then $K^* + X$ is essential in G. Since K is G-essential in K^*, by Theorem 3.4.10, $K + X$ is G-essential in $K^* + X$, and so is essential in G. By hypothesis, $\pi(K + X)$ is essential in G/K. Since K^* contains K properly, $\pi(K^*)$ is a nonzero ideal of G/K. Now

$$(K + X) \cap K^* = K + (X \cap K^*) = K,$$

which shows that $\pi(K + X) \cap \pi(K^*) = (0)$. This is a contradiction to the fact that $\pi(K + X)$ is essential in G/K. The proof is complete. ∎

Definition 5.1.18

(i) A complement ideal in G is said to be a **maximal complement ideal** if it is maximal in the set of all complement ideals properly contained in G.

(ii) An ideal I of G is said to be meet-irreducible if it cannot be written as $J \cap K$, where J and K are ideals of G, $I \neq J$, $I \neq K$.

Lemma 5.1.19

Suppose K is a maximal complement ideal of G and I is an ideal of G with $K + I \nsubseteq K$. Then $K + I$ is essential.

Proof

Suppose $K + I$ is not essential. Then there exists a nonzero ideal A of G such that $(K + I) \cap A = (0)$. Let K^* be a complement of A containing $K + I$. Then $K^* \subseteq K$, and by the maximality of K, we have $K = K^*$. So, $K + I \subseteq K$. This is a contradiction. ∎

Proposition 5.1.20

Every maximal complement K in G is meet-irreducible.

Proof

Suppose K is a complement of the ideal J. Suppose K is not meet-irreducible. Then there exist two ideals I_1, I_2 of G properly containing K such that $K = I_1 \cap I_2$. Write $S_i = I_i \cap J$ for $i = 1, 2$. It is evident that $S_i \neq (0)$, $S_1 \cap S_2 = (0)$, and $(K + S_1) \cap S_2 = (0)$. Since $K + S_1 \nsubseteq K$, by Lemma 5.1.19, $K + S_1$ is essential in G, and hence $S_2 = (0)$. This is a contradiction. ∎

Proposition 5.1.21 (Proposition 2.1.14 of Satyanarayana, 1984)

Let I be a nonessential ideal of G. If I is meet-irreducible, then I is a complement. Moreover, if G has maximum condition on complement ideals, then every meet-irreducible ideal is a maximal complement.

Proof

Since I is not essential, there exists a nonzero ideal A of G such that $I \cap A = (0)$. Let K be a complement of A containing I. Now $K \cap (I + A) = I + (K \cap A) = I$. Since I is meet-irreducible and $I + A \neq I$, we have $K = I$. Thus, I is a complement ideal. Now suppose that G has maximum condition on complement ideals. Since I is not essential, there exists a proper ideal K containing I that is also a complement ideal. Since G has maximum condition on complement ideals, there exists a maximal element in the set $B = \{S \mid S$ is a proper ideal containing I, and S is a complement ideal$\}$, say K^*. Now K^* is a maximal complement ideal. Suppose K^* is a complement of an ideal B of G. Since K^* is proper, $B \neq (0)$. Now $K^* \cap (I + B) = I + (K^* \cap B) = I$. Since I is meet-irreducible and $I \neq I + B$, we have $K^* = I$. Hence, I is a maximal complement ideal. ∎

Theorem 5.1.22 (Theorem 2.1.15 of Satyanarayana, 1984)

(i) Every maximal complement K of G is meet-irreducible and nonessential.
(ii) Conversely, if G has maximum condition on complement ideals, then every meet-irreducible and nonessential ideal of G is a maximal complement in G.

Proof

(i) If K is a maximal complement in G, then by Proposition 5.1.14, K is not essential in G, and by Proposition 5.1.20, K is meet-irreducible.
(ii) Follows from Proposition 5.1.21. ∎

Now we present the notion "*E*-irreducible ideal" and prove that a nonessential ideal K of G is *E*-irreducible if and only if $\dim(G/K) = \dim G - \dim K$, where G is an N-group with FGD.

Remark 5.1.23

Let K_1 and K_2 be two ideals of G. Then for any ideal K of G, the following conditions are equivalent:

(i) $K = K_1 \cap K_2$ and either K is G-essential in K_1 or K is G-essential in K_2 imply $K = K_1$ or $K = K_2$.

(ii) $K = K_1 \cap K_2$, K is essential in K_1, $K \neq K_1$ imply $K = K_2$.

(iii) $K = K_1 \cap K_2$, K is essential in K_1 imply $K = K_1$ or $K = K_2$.

Definition 5.1.24

An ideal K of G satisfying any one of the conditions of Remark 5.1.23 is called an *E*-irreducible ideal.

Proposition 5.1.14 shows that a complement ideal has no proper G-essential extensions. Hence, every complement ideal is *E*-irreducible. But, the following example shows that the converse need not be true in general.

Example 5.1.25

Consider $N = \mathbb{Z}$, the ring of integers, and $G = \mathbb{Z}_{12}$, the ring of integers modulo 12. Now G is an N-group, and the principal ideal K generated by 2 in G is *E*-irreducible, but not a complement.

Clearly, every meet-irreducible ideal is an *E*-irreducible ideal. But, the following example establishes that every *E*-irreducible ideal need not be meet-irreducible.

Example 5.1.26

Consider $N = \mathbb{Z}$, the ring of integers, and $M = \mathbb{Z}_2 \times \mathbb{Z}_3 \times \mathbb{Z}_5$, where $\mathbb{Z}_i$ is the ring of integers modulo i for $i = 2, 3, 5$. Write $K = (0) \times (0) \times \mathbb{Z}_5$.

Since $K = (\mathbb{Z}_2 \times (0) \times \mathbb{Z}_5) \cap ((0) \times \mathbb{Z}_3 \times \mathbb{Z}_5)$, K is not meet-irreducible. But K is *E*-irreducible.

Now we give an example of an ideal that is not *E*-irreducible.

Example 5.1.27

Consider $N = \mathbb{Z}$, the ring of integers, and $G = \mathbb{Z}_8 \times \mathbb{Z}_3$, $\mathbb{Z}_k$ is the ring of integers modulo k for $k = 3, 8$. Now $K = <4> \times (0)$ is not an E-irreducible ideal of G.

(Since $K = (\mathbb{Z}_8 \times (0)) \cap (<4> \times \mathbb{Z}_3)$ and K is essential in $\mathbb{Z}_8 \times (0)$.)

Next we prove that if G has FGD and I is a nonessential ideal of G, then I is E-irreducible if and only if dim $G = \dim I + \dim(G/I)$.

We first begin with the following.

Lemma 5.1.28 (Lemma 2.2.6 of Satyanarayana, 1984)

If G has FGD and if K is an E-irreducible ideal of G, then G/K has FGD and $\dim(G/K) \le \dim G$.

Proof

Suppose $K_1, K_2, \ldots, K_n$ are ideals of G containing K such that $(K_1/K) + (K_2/K) + \ldots + (K_n/K)$ is a direct sum of nonzero ideals in G/K. Then $K_1 \cap \Sigma_{i \ne j} K_j = K$, and since K is E-irreducible, K is not G-essential in K_i for $1 \le i \le n$. Therefore, it is possible to pick up a nonzero ideal U_i of G contained in K_i such that $K \cap U_i = (0)$. We now show that the sum $K + U_1 + U_2 + \ldots + U_n$ is direct. Let $k + u_1 + u_2 + \ldots + u_n = 0$, where $k \in K$, $u_i \in U_i$ for $1 \le i \le n$. Then $u_1 + u_2 + \ldots + u_n \in K$, and hence $(u_1 + K) + (u_2 + K) + \ldots + (u_n + K)$ is the zero element in G/K. Since the sum $(K_1/K) + (K_2/K) + \ldots + (K_n/K)$ is direct, $u_i + K = 0 + K$ and $u_i \in K \cap U_i = (0)$. Thus, the sum $K + U_1 + U_2 + \ldots + U_n$ is direct. Now the result is trivial. ∎

Lemma 5.1.29 (Lemma 2.2.7 of Satyanarayana, 1984)

Let G be an N-group with FGD. If A is an ideal of G such that $\dim(G/A) = 1$ and A is nonessential, then $\dim A = \dim G - 1$.

Proof

Since A is not essential, there is a nonzero ideal I of G such that $I \cap A = (0)$. Let K be a complement of A containing I. Suppose $\dim K \ge 2$. Then K contains a direct sum of two uniform ideals U_1, U_2 of G. Clearly, $U_i \cap A = (0)$ for $i = 1, 2$. By Lemma 3.2.29, $((U_1 + A)/A)$, $((U_2 + A)/A)$ are two uniform ideals of G/A.

It is easy to verify that the sum $((U_1 + A)/A) + ((U_2 + A)/A)$ is direct, and hence $\dim(G/A) \geq 2$, which is a contradiction. Hence, $\dim K \not\geq 2$. Since $K \neq (0)$, we must have that $\dim K = 1$. Since K is a complement of A, the sum $K + A$ is essential and direct. So $\dim G = \dim(K + A) = \dim K + \dim A = 1 + \dim A$. ∎

Theorem 5.1.30 (Theorem 2.2.8 of Satyanarayana, 1984)

Let G be an N-group having FGD. If K is an E-irreducible ideal that is nonessential in G, then G/K has FGD and $\dim(G/K) = \dim G - \dim K$.

Proof

By Lemma 5.1.28, G/K has FGD. If $\dim(G/K) = 1$, then by Lemma 5.1.29, $\dim(G/K) = \dim G - \dim K$. Suppose $\dim(G/K) = m$, where $m \geq 2$. Then by Theorem 5.1.2, there exist ideals $K_1, K_2, \ldots, K_m$ of G containing K properly such that for $1 \leq i \leq m$, K_i/K is uniform and the sum $(K_1/K) + (K_2/K) + \ldots + (K_m/K)$ is direct and essential in G/K. Therefore, $K = K_i \cap K_j$ for $1 \leq i \neq j \leq m$. Since K is E-irreducible and $K_i \neq K \neq K_j$, it follows that K is not G-essential in K_s for $1 \leq s \leq m$. So, there exist uniform ideals U_s ($1 \leq s \leq m$) of G such that $U_s \subseteq K_s$ and $U_s \cap K = (0)$. From the proof of Lemma 5.1.28, we conclude that the sum $K + U_1 + U_2 + \ldots + U_m$ is direct.

We now show that $T = K + U_1 + U_2 + \ldots + U_m$ is essential in G. Let J be an ideal of G such that $T \cap J = (0)$. Then $T \cap (J + K) = (T \cap J) + K = K$ and $(T/K) \cap ((J + K)/K) = (0)$. Since K_i/K is uniform and $((U_i + K)/K)$ is a nonzero ideal of K_i/K, we have that $((U_i + K)/K)$ is (G/K)-essential in K_i/K, for $1 \leq i \leq m$.

By Corollary 3.4.10, $((U_1 + K)/K) + ((U_2 + K)/K) + \ldots + ((U_m + K)/K)$ is (G/K)-essential in $(K_1/K) + (K_2/K) + \ldots + (K_m/K)$. Since the sum $(K_1/K) + (K_2/K) + \ldots + (K_m/K)$ is essential in G/K, by Remark 3.4.2 (ii), $((U_1 + K)/K) + ((U_2 + K)/K) + \ldots + ((U_m + K)/K)$ is essential in G/K.

That is, T/K is essential in G/K.

Since $((J + K)/K) \cap (T/K) = (0)$, we have $((J + K)/K) = (0)$ and $J \subseteq K$, which implies $J = J \cap T = (0)$. Hence, the sum $T = K + U_1 + U_2 + \ldots + U_m$ is essential in G. Now

$$\begin{aligned}
\dim G &= \dim(K + U_1 + U_2 + \ldots + U_m) \\
&= \dim K + \dim U_1 + \dim U_2 + \ldots + \dim U_m \\
&= \dim K + m \\
&= \dim K + \dim(G/K).
\end{aligned}$$

The proof is complete.   ∎

As a consequence of this, we now obtain the following.

Corollary 5.1.31

If K is a complement and G has FGD, then G/K also has FGD, and dim $G =$ dim $K +$ dim(G/K).

Proof

If K is essential, then $K = G$, and the result is trivially true. So, assume that K is not essential. Since every complement ideal is E-irreducible, the result follows from Theorem 5.1.30. ∎

Corollary 5.1.32 (Goldie, 1972)

Let R be a ring and M be an R-module satisfying, $Rm = 0, m \in M \Rightarrow m = 0$. If K is a complement submodule of M, and M has FGD, then M/K also has FGD, and dim $M =$ dim $K +$ dim(M/K).

In the remaining part of this section, we prove the converse of Theorem 5.1.30.

Lemma 5.1.33

Let G be an N-group with FGD. If K is a proper ideal of G such that G/K has FGD and dim $G =$ dim $K +$ dim(G/K), then K is not essential in G.

Proof

If K is essential, then by Proposition 5.1.4 (ii), dim $G =$ dim K. Then by hypothesis, dim$(G/K) = 0$ and so $G/K = (0)$, that is, $G = K$. This contradicts our hypothesis. ∎

Theorem 5.1.34 (Theorem 2.2.12 of Satyanarayana, 1984)

Let G be an N-group with FGD and K an ideal of G such that dim $G =$ dim $K +$ dim(G/K). Then K has no proper G-essential extensions in G.

Proof

If $K = G$, there is nothing to prove. So it suffices to show that if $K \neq G$, then K has no proper G-essential extensions. Suppose K has a proper G-essential extension K^1. That is, K is G-essential in K^1 and $K \neq K^1$. So dim $K =$ dim K^1 and

$\dim(K^1/K) \geq 1$. Let K_1 be a complement of K^1. Then the sum $K^1 + K_1$ is direct and essential in G. By Proposition 5.1.7 and Theorem 5.1.8, we have

$$\dim G = \dim(K^1 + K_1) = \dim K^1 + \dim K_1.$$

Since $K_1 \cap K = (0)$, $(K_1 + K)/K$ is isomorphic with K_1, and hence dim $((K_1 + K)/K) = \dim K_1$. Since $K^1 \cap (K_1 + K) = (K^1 \cap K_1) + K$ and $(K^1 \cap K_1) + K = K$, the sum $(K^1/K) + ((K_1 + K)/K)$ is direct. Therefore, by Theorem 5.1.8, we have

$$\begin{aligned}
\dim(G/K) &\geq \dim((K^1/K) + ((K_1 + K)/K)) \\
&= \dim(K^1/K) + \dim((K_1 + K)/K) \\
&\geq 1 + \dim((K_1 + K)/K) \\
&= 1 + \dim K_1 \\
&= 1 + \dim G - \dim K^1 \\
&= 1 + \dim G - \dim K \\
&= 1 + \dim(G/K).
\end{aligned}$$

This is a contradiction. Therefore, K has no proper G-essential extensions. ∎

Remark 5.1.35

Every proper ideal of G that has no proper G-essential extensions is E-irreducible and nonessential.

Combining Theorems 5.1.30 and 5.1.34, we have the following.

Theorem 5.1.36

Let G be an N-group with FGD. Then a proper ideal K of G is nonessential and E-irreducible if and only if G/K has FGD and $\dim G/K = \dim G - \dim K$.

Now we present some equivalent conditions for an ideal to be a complement ideal.

We start with the following.

Lemma 5.1.37

Let K be an E-irreducible ideal of G. Then either (i) K is essential or (ii) K has no proper G-essential extensions.

Proof

It suffices to show that if K has a proper G-essential extension K^1 in G, then K is essential in G. Now suppose K^1 is a proper G-essential extension of K in G. To show K^1 is an essential ideal of G, let I be an ideal of G such that $K^1 \cap I = (0)$. Now $K^1 \cap (K + I) = K + (I \cap K^1) = K$. Since K is E-irreducible and $K \neq K^1$, K is essential in K^1, we have $K = K + I$, which implies $I \subseteq K^1$. So, $I = (0)$ and K^1 is essential in G. By Remark 3.4.2 (ii), K is essential in G. ∎

Lemma 5.1.38

Suppose K is an E-irreducible ideal of G. Then either (i) K is an essential ideal of G or (ii) K is a complement.

Proof

Suppose K is not essential. Then by Lemma 5.1.37, K has no proper G-essential extensions. By Proposition 5.1.14, K is a complement ideal. ∎

Theorem 5.1.39 (Theorem 2.3.3 of Satyanarayana, 1984)

Let K be an ideal of an N-group G and $\pi: G \to G/K$ be the canonical epimorphism. Then the following conditions are equivalent:

(i) $K = G$ or K is nonessential and E-irreducible.
(ii) K has no proper G-essential extensions.
(iii) K is a complement.
(iv) For any ideal K^1 of G containing K, K^1 is a complement in G if and only if $\pi(K^1)$ is a complement in G/K.
(v) For any essential ideal S of G, $\pi(S)$ is essential in G/K.
   Moreover, if G has FGD, then these conditions are equivalent to:
(vi) G/K has FGD and $\dim G/K = \dim G - \dim K$.

Proof

(i) $\Rightarrow$ (ii): Follows from Lemma 5.1.37.
   (ii) $\Rightarrow$ (iii): Proposition 5.1.14.
   (iii) $\Rightarrow$ (ii): Proposition 5.1.14.
   (ii) $\Rightarrow$ (i): is trivial.
   (iii) $\Leftrightarrow$ (iv): Follows from Theorem 5.1.15.

(iii) $\Leftrightarrow$ (v): Follows from Theorem 5.1.17.
(i) $\Leftrightarrow$ (vi): Follows from Proposition 5.1.21.   ∎
Combining Results 5.1.11 and 5.1.39, we have the following.

Corollary 5.1.40

Let G be an N-group. Then the following are equivalent.

(i) Every ideal of G is a complement.
(ii) Every proper ideal of G is nonessential and E-irreducible.
(iii) Every proper ideal of G has no proper G-essential extensions.
(iv) For an ideal K of G with canonical epimorphism
    $\pi: G \to G/K$, it follows that K^1 is a complement in G containing K if
    and only if $\pi(K^1)$ is a complement in G/K.
(v) For any ideal K of G with canonical epimorphism $\pi: G \to G/K$, we
    have S is an essential ideal of G implies that $\pi(S)$ is essential in G/K.
    Moreover, if G has FGD, then the preceding conditions are equiva-
    lent to each of the following.
(vi) G is completely reducible and G has descending chain condition
    (DCC) on ideals.
(vii) For any ideal K of G, G/K has FGD and dim G/K = dim G – dim K.

5.2 Linearly Independent Elements in N-Groups

Result 5.2.1 (Result 0.3 of Satyanarayana and Prasad, 2005)

Let U be an ideal of an N-group G; then the following two conditions are
equivalent:

(i) U is uniform.
(ii) $0 \neq x \in U$ and $0 \neq y \in U \Rightarrow <x> \cap <y> \neq (0)$.

Proof

(i) $\Rightarrow$ (ii): Suppose $0 \neq x \in U$ and $0 \neq y \in U$. Let us suppose that $<x> \cap$
    $<y> = (0)$. Since U is uniform, $<x> \subseteq U$, $<y> \subseteq U$, we have either
    $<x> = (0)$ or $<y> = (0) \Rightarrow x = 0$ or $y = 0$, which is a contradiction.
    Therefore, $<x> \cap <y> \neq (0)$.
(ii) $\Rightarrow$ (i): Suppose U is an ideal with property (ii). To show U is uniform,
    let I, J be ideals of G such that $I \subseteq U, J \subseteq U, I \cap J = (0)$. Now we have

to show that $I = (0)$ or $J = (0)$. Let us assume that $I \neq (0)$, $J \neq (0) \Rightarrow$ there exist $0 \neq a \in I$ and $0 \neq b \in J$. Now $(0) \neq <a> \subseteq I$ and $(0) \neq <b> \subseteq J$. By (ii), we have $(0) \neq <a> \cap <b> \subseteq I \cap J \Rightarrow I \cap J \neq (0)$, which is a contradiction. Therefore, $I = (0)$ or $J = (0)$. This shows that U is uniform. ∎

Lemma 5.2.2 (Satyanarayana and Prasad, 2005)

Suppose $f: G \rightarrow G^1$ is an N-group isomorphism and I_i, $1 \leq i \leq n$ are ideals of G. Then

(i) The sum of ideals I_i, $1 \leq i \leq n$ of G is direct in G if and only if the sum of ideals $f(I_i)$, $1 \leq i \leq n$ of G^1 is direct in G^1.

(ii) $K_1 \leq_e K_2$, if and only if $f(K_1) \leq_e f(K_2)$.

Proof

(i) Suppose the sum of ideals I_i, $1 \leq i \leq n$ is direct. Then we have $I_i \cap (I_1 + \ldots + I_{i-1} + I_{i+1} + \ldots + I_n) = (0)$. Now we show that $f(I_i) \cap [f(I_1) + \ldots + f(I_{i-1}) + f(I_{i+1}) + \ldots + f(I_n)] = (0)$.

Take $y \in f(I_i) \cap [f(I_1) + \ldots + f(I_{i-1}) + f(I_{i+1}) + \ldots + f(I_n)]$

$\Rightarrow y = f(x_i) = f(x_1) + \ldots + f(x_{i-1}) + f(x_{i+1}) + \ldots + f(x_n)$ with $x_i \in I_i$ for $1 \leq i \leq n$.

$\Rightarrow f(x_i) = f(x_1 + \ldots + x_{i-1} + x_{i+1} + \ldots + x_n)$

$\Rightarrow x_i = x_1 + \ldots + x_{i-1} + x_{i+1} + \ldots + x_n$ (since f is one–one)

$\Rightarrow x_i \in I_i \cap (I_1 + \ldots + I_{i-1} + I_{i+1} + \ldots + I_n) = (0)$

$\Rightarrow x_i = 0 \Rightarrow y = f(x_i) = f(0) = 0$ (since f is one–one).

This shows that $f(I_i) \cap [f(I_1) + \ldots + f(I_{i-1}) + f(I_{i+1}) + \ldots + f(I_n)] = (0)$, for $1 \leq i \leq n$. Hence, the sum of ideals $f(I_1), \ldots, f(I_n)$ is direct.

Converse

Suppose $f(I_i)$, $1 \leq i \leq n$ is direct.
Take $x \in I_i \cap (I_1 + \ldots + I_{i-1} + I_{i+1} + \ldots + I_n)$

$\Rightarrow f(x) \in f(I_i) \cap f(I_1 + \ldots + I_{i-1} + I_{i+1} + \ldots + I_n)$

$= f(I_i) \cap [f(I_1) + \ldots + f(I_{i-1}) + f(I_{i+1}) + \ldots + f(I_n)] = (0)$

$\Rightarrow f(x) = 0 \Rightarrow x = 0$ (since f is one–one).

Thus, we proved that $I_i \cap (I_1 + \ldots + I_{i-1} + I_{i+1} + \ldots + I_n) = (0)$, for $1 \leq i \leq n$. Hence, the sum of ideals I_i, $1 \leq i \leq n$ is direct.

(ii) Suppose $K_1 \leq_e K_2$. Let us suppose $f(K_1)$ is not essential in $f(K_2)$. Then there exists an ideal I of G^1 such that $f(K_1) \cap I = (0)$, $I \neq (0)$, and $I \subseteq f(K_2)$. Since $I \neq (0)$, it follows that $K = f^{-1}(I)$ is a nonzero ideal in G. Since the sum of $f(K_1)$ and $f(K) = I$ is direct in G^1, by (i), the sum of K_1 and K is direct in G.

Now $K_1 \cap K = (0)$, $K = f^{-1}(I) \subseteq f^{-1}(f(K_2)) = K_2$. This shows that K_1 is not essential in K_2, which is a contradiction.

The converse is a straightforward verification as in the earlier part. ∎

Lemma 5.2.3 (Lemma 0.4 of Satyanarayana and Prasad, 2005)

Suppose $f: G \to G^1$ is an epimorphism. Then for any $x \in G$, we have $f(<x>) = <f(x)>$.

Proof

We know that $<x> = \bigcup_{i=0}^{\infty} A_i$, where $A_{i+1} = A_i^* \cup A_i^\circ \cup A_i^+$ for all $i \geq 0$, and

$$A_i^* = \{g + x - g \mid x \in A_i, g \in G\},$$

$$A_i^\circ = \{a - b \mid a, b \in A_i\},$$

$$A_i^+ = \{n(g + a) - ng \mid a \in A_i, n \in N, g \in G\} \text{ with } A_0 = \{x\}.$$

Also, $<f(x)> = \bigcup_{i=0}^{\infty} B_i$, where $B_{i+1} = B_i^* \cup B_i^\circ \cup B_i^+$ with $B_0 = \{f(x)\}$.

We verify that $B_0 = f(A_0), \ldots, B_i = f(A_i)$ for all $i \geq 0$. Now $B_0 = \{f(x)\} = f(A_0)$, since $A_0 = \{x\}$. Suppose the induction hypothesis. That is, $B_k = f(A_k)$. Now we have to verify that $B_{k+1} = f(A_{k+1})$.

Part 1: Take $y \in B_{k+1} = B_k^* \cup B_k^\circ \cup B_k^+$.

Case (i): Suppose $y \in B_k^*$. Then $y = g + b - g$ for some $b \in B_k$ and $g \in G^1$. Now $b \in B_k = f(A_k) \Rightarrow b = f(a)$ for some $a \in A_k$. Since f is onto, there exists $g_1 \in G$ such that $f(g_1) = g$. Now

$$y = g + b - g = f(g_1) + f(a) - f(g_1) = f(g_1 + a - g_1) \in f(A_k^*).$$

Case (ii): Suppose $y \in B_k^\circ$. Then $y = b_1 - b_2$ for some $b_1, b_2 \in B_k$. Since $B_k = f(A_k)$, we have $b_1 = f(a_1)$, $b_2 = f(a_2)$ for some $a_1, a_2 \in A_k$. Now $y = b_1 - b_2 = f(a_1) - f(a_2) = f(a_1 - a_2) \in f(A_k^\circ)$.

Case (iii): Suppose $y \in B_k^+$. Then $y = n(g + b) - ng$ for some $b \in B_k$, $n \in N$, $g \in G^1$. Since f is onto, there exists $g_1 \in G$ such that $f(g_1) = g$. Since $b \in B_k = f(A_k)$, we have $b = f(a)$ for some $a \in A_k$. Now

$y = n(f(g_1) + f(a)) - nf(g_1) = f(n(g_1 + a) - ng_1) \in f(A_k^+)$. Hence, $B_{k+1} = B_k^* \cup B_k^\circ \cup B_k^+ \subseteq f(A_k^*) \cup f(A_k^\circ) \cup f(A_k^+) \subseteq f(A_{k+1})$.

Part 2: Let $z \in A_k^*$. Then $z = g + a - g$ for some $a \in A_k$, $g \in G \Rightarrow f(z) = f(g + a - g) = f(g) + f(a) - f(g) \in B_k^*$ (since $f(a) \in f(A_k) = B_k$). Therefore, $f(A_k^*) \subseteq B_k^*$.

Similarly, we can show that $f(A_k^\circ) \subseteq B_k^\circ$, $f(A_k^+) \subseteq B_k^+$. Finally, it follows that $f(A_{k+1}) \subseteq B_{k+1}$.

From Parts 1 and 2, we have $f(A_{k+1}) = B_{k+1}$.

By mathematical induction, we conclude that $f(A_i) = B_i$ for all $i = 1, 2, \dots$.

Hence, $<f(x)> = \bigcup_{i=0}^{\infty} = \bigcup_{i=0}^{\infty} f(A_i) = f\left(\bigcup_{i=0}^{\infty} A_i\right) = f(<x>)$. ■

Definition 5.2.4

Let X be a subset of G. X is said to be a **linearly independent** (l.i.) set if the sum $\Sigma_{a \in X} <a>$ is direct. If $\{a_i / 1 \le i \le n\}$ is an l.i. set, then we say that the elements a_i, $1 \le i \le n$ are l.i. If X is not an l.i. set, then we say that X is a **linearly dependent** (l.d.) set.

Definition 5.2.5

An element $0 \ne u \in U$ is said to be a **uniform element** (u-**element**) if $<u>$ is a uniform ideal of G.

Remark 5.2.6 (Remark 1.3 of Satyanarayana and Prasad, 2005)

(i) Let G be an N-group. Then every ideal contained in a uniform ideal of G is uniform.

Verification

Suppose U is a uniform ideal of G. Let I be an ideal of G such that $I \subseteq U$. We show that I is uniform. Let A, B be ideals of G such that $A, B \subseteq I$, $A \cap B = (0)$. Now $A, B \subseteq I \subseteq U$, $A \cap B = (0)$ and U is uniform; it follows that $A = (0)$ or $B = (0)$. Therefore, U is uniform.

(ii) Suppose G has FGD. If H is a nonzero ideal of G, then H contains a u-element.

Verification

Let H be a nonzero ideal of G. Then by Lemma 5.1.1, H contains a uniform ideal U. Since $U \neq (0)$, there exists $0 \neq u \in U$. By (i), $<u>$ is uniform (since $<u> \subseteq U$). Hence, u is a u-element.

Result 5.2.7 (Result 1.4 of Satyanarayana and Prasad, 2005)

(i) If a_i, $1 \leq i \leq m$ are l.i. elements in G, then $m \leq n$, where $n = \dim G$.
(ii) $\dim G$ is equal to the least upper bound of the set A, where $A = \{m \mid m$ is a natural number and there exist $a_i \subset G$, $1 \leq i \leq m$ such that a_i, $1 \leq I \leq m$ are l.i.$\}$.
(iii) If $m = \dim G$ and a_i, $1 \leq i \leq n$ are l.i. elements, then each $<a_i>$ is a uniform ideal (in other words, each a_i is a u-element).

Proof

(i) Since a_i, $1 \leq i \leq m$ are l.i. elements, the sum $\sum_{i=1}^{m} <a_i>$ is direct. By Corollary 5.1.4, we have $m \leq n$.
(ii) By (i), it is clear that $\dim G$ is an upper bound for A. Write $n = \dim G$. By Theorem 5.1.2, there exist uniform ideals U_i, $1 \leq i \leq n$ whose sum is direct and essential in G. So, $n \in A$. Thus, n is the least upper bound for A.
(iii) Suppose $<a_k>$ is not uniform for some $1 \leq k \leq n$. Then $<a_k>$ contains two nonzero ideals A and B such that $A \cap B = (0)$. So, there exist nonzero uniform ideals U and V such that $U \subseteq A$ and $V \subseteq B$. Let $0 \neq u \in U$ and $0 \neq v \in V$. Now the sum of

$$<a_1>, <a_2>, \ldots, <a_{k-1}>, <u>, <v>, <a_{k+1}>, \ldots, <a_n>$$

is direct. The number of summands in this sum is $n + 1$. By (ii), $n + 1 \leq \dim G = n$, which is a contradiction. The proof is complete. ■

Definition 5.2.8

If $n = \dim G$ and a_i, $1 \leq i \leq n$ are l.i. elements, then $\{a_i \mid 1 \leq i \leq n\}$ is called an **essential basis** for G.

A straightforward verification gives the following note.

Note 5.2.9

(i) G has FGD if and only if G contains no direct sum of infinite number of nonzero ideals, if and only if every l.i. subset X of G is a finite set.

(ii) G has no FGD if and only if G contains an infinite l.i. subset.

(iii) Suppose dim $G = n$ and X is a l.i. set. Then $|X| = n$ (here $|X|$ denotes the cardinality of X) $\Rightarrow X$ is a maximal l.i. set (in this case, we say that X is an essential basis for G).

(Example for the converse is not true: Take $G = \mathbb{Z}_2 \oplus \mathbb{Z}_6$. Then G is an N-group, where $N = \mathbb{Z}$. Since G is a finite N-group, it has FGD. Write $u = (1, 0)$. Then $2u = (0, 0)$. So $<u> = \{(1, 0), (0, 0)\}$ is an ideal of the N-group G. Also, u is a u-element. Write $v = (0, 2)$. Then $<v> = \{(0, 2), (0, 4), (0, 0)\}$ is an ideal of an N-group G. Also, v is a u-element.

Write $w = (0, 3)$. Then $<w> = \{(0, 3), (0, 0)\}$ is an ideal of an N-group G. Also, w is a u-element. Now it is easy to verify that $G = <u>\oplus<v>\oplus<w>$. Thus, $\dim(G) = 3$. Write $X = \{(1, 0), (0, 1)\}$. Now X is an l.i. set. Thus, there is no $Y \subseteq G$ such that $X \subsetneq Y$ and Y is l.i. Thus, X is a maximal l.i. set with cardinality of X (i.e., $|X|) = 2 \neq 3 = \dim(G)$.)

Lemma 5.2.10 (Lemma 1.7 (i) of Satyanarayana and Prasad, 2005)

Let $f: G \to G^1$ be an isomorphism and $x_i \in G$, $1 \le i \le k$. Then $x_1, x_2, \ldots, x_k$ are l.i. elements in G if and only if $f(x_1), f(x_2), \ldots, f(x_k)$ are l.i. elements in G^1.

Proof

$x_1, x_2, \ldots, x_k$ are l.i. elements
$\Leftrightarrow$ the sum of $<x_1>, <x_2>, \ldots, <x_k>$ is direct.
$\Leftrightarrow$ the sum of $f(<x_1>), f(<x_2>), \ldots, f(<x_k>)$ is direct (by Lemma 5.2.2)
$\Leftrightarrow$ the sum of $<f(x_1)>, <f(x_2)>, \ldots, <f(x_k)>$ is direct (by Lemma 5.2.3)
$\Leftrightarrow f(x_1), f(x_2), \ldots, f(x_k)$ are l.i. elements. ∎

Lemma 5.2.11 (Lemma 1.7 (ii) of Satyanarayana and Prasad, 2005)

Let $f: G \to G^1$ be an N-group isomorphism. Then u is a u-element in G, if and only if $f(u)$ is a u-element in G^1.

Proof

Suppose u is a u-element in G.

Let us suppose $f(u)$ is not a u-element.

Since $<f(u)>$ is not uniform, there exist $w_1, w_2 \in <f(u)>$ such that $<w> \cap <w_2> = (0)$.

By Lemma 5.2.3, $<f(u)> = f(<u>)$, so there exist $u_1, u_2 \in <u>$ such that $w_1 = f(u_1)$, $w_2 = f(u_2)$.

Since $<u>$ is uniform, we have $<u_1> \cap <u_2> \neq (0) \Rightarrow u_1, u_2$ are not l.i. $\Rightarrow f(u_1), f(u_2)$ are not l.i. (by Lemma 5.2.10)

$\Rightarrow w_1, w_2$ are not l.i. $\Rightarrow <w_1> \cap <w_2> \neq (0)$, which is a contradiction. Thus, we have proved that u is a u-element implies $f(u)$ is a u-element.

Converse

Suppose $f(u)$ is a u-element.

Let us suppose u is not a u-element

$\Rightarrow$ there exist $u_1, u_2 \in <u>$ such that $<u_1> \cap <u_2> = (0)$

$\Rightarrow u_1, u_2$ are l.i. $\Rightarrow f(u_1), f(u_2)$ are l.i. (by Lemma 5.2.10)

$\Rightarrow <f(u_1)> \cap <f(u_2)> = (0)$ and $<f(u_1)> \subseteq <f(u)>$ and $<f(u_2)> \subseteq <f(u)>$.

Since $f(u)$ is uniform, we have $<f(u_1)> = (0)$ or $<f(u_2)> = (0)$, which is a contradiction. The proof is complete.   ∎

Now we generalize the concept of "essentially spanned" given by Anderson and Fuller (1974) to N-groups.

Definition 5.2.12

Let H be an ideal of G and $X \subseteq H$. We say that

(i) H is **essentially spanned** by a collection of ideals $\{I_\alpha\}_{\alpha \in \Delta}$ (here Δ is an index set) of G if $\Sigma_{\alpha \in \Delta} I_\alpha$ is essential in H (or $\{I_\alpha\}_{\alpha \in \Delta}$ spans H essentially).

(ii) H is **spanned** by a collection of ideals $\{I_\alpha\}_{\alpha \in \Delta}$ of G if $\Sigma_{\alpha \in \Delta} I_\alpha$ $H\left(\{I_\alpha\}_{\alpha \in \Delta}$ spans $H\right)$.

(iii) H is **essentially spanned** by X if $\Sigma_{x \in X} <x>$ is essential in H (X spans H essentially (or X is an essentially spanning set for H)).

(iv) H is **spanned** by X if $\Sigma_{x \in X} <x> = H$. (In this case, we also say that X spans H or X is a spanning set for H.)

From the preceding definitions, the following is clear.

Note 5.2.13

(i) $\{I_\alpha\}_{\alpha \in \Delta}$ spans $H \Rightarrow \{I_\alpha\}_{\alpha \in \Delta}$ spans H essentially. The converse is not true.

(ii) X spans $H \Rightarrow X$ spans H essentially. The converse is not true.

(iii) If the N-group G is a module, then the two concepts: "$\{I_\alpha\}_{\alpha \in \Delta}$ spans H" in modules and "$\{I_\alpha\}_{\alpha \in \Delta}$ spans H" in N-groups (see Definition 5.2.12) are equivalent.

Consider the following examples.

Example 5.2.14

Let $N = \mathbb{Z}$ be the near ring of integers, and $G = \mathbb{Z}$ be the additive group of integers. Now G is an N-group.

(i) Consider $I = 2\mathbb{Z}$. Clearly, the ideal I is essential in G. Therefore, I spans G essentially. Since $I \neq G$, it follows that I does not span G.

(ii) Write $X = \{2\}$. Clearly, $\sum_{x \in X} <x> = 2\mathbb{Z} = I$ is essential in G. So, X spans G essentially. Since $\sum_{x \in X} <x> \neq G$, we have that X does not span G.

Definition 5.2.15

(i) An ideal H of G with a finite spanning set X is said to be a **finitely spanned ideal**.

(ii) An ideal H of G with a finite essential spanning set X is said to be **finitely essentially spanned**.

(iii) If $X = \{x\}$ and X essentially spans H, then we say that H is an **essentially cyclic ideal**.

Note 5.2.16

(i) If U is a uniform ideal, then U is an essentially cyclic ideal.

(ii) If $n = \dim G$, then there exist essentially cyclic ideals $U_i, 1 \leq i \leq n$ such that the sum $\sum_{i=1}^n U_i$ is direct and essential in G.

Remark 5.2.17

Every essentially cyclic ideal need not be uniform.

For example, write $N = Z$, the near ring of integers; and $G = Z_6$, the group of integers modulo 6. Now G is an N-group. Since $G = <1>$, G is essentially cyclic. Since $\{0, 2, 4\}$ and $\{0, 3\}$ are two ideals whose intersection is zero, it follows that G cannot be uniform.

Result 5.2.18 (Result 1.13 of Satyanarayana and Prasad, 2005)

Suppose G is a semisimple N-group with FGD. Then

(i) Let I be an ideal of G. Then "I is N-simple $\Rightarrow I$ is simple $\Rightarrow I$ is uniform."
(ii) There exist simple ideals $H_1, H_2, \ldots, H_n$ such that $H_1 \oplus H_2 \oplus \ldots \oplus H_n = G$.
(iii) There exist uniform ideals U_i, $1 \leq i \leq n$ such that $G = U_1 \oplus U_2 \oplus \ldots \oplus U_n$.

(Note that an ideal K of G is said to be an **N-simple** ideal if K contains no nonzero proper N-subgroups; an N-group G is said to be **semisimple** if it is a sum of simple ideals; equivalently, G is equal to the direct sum of a collection of simple ideals of G.)

Proof

(i) This is clear from the definitions of simple, N-simple, uniform ideals.
(ii) Let us suppose that G cannot be expressed as a sum of a finite number of simple ideals. Let H_1 be a simple ideal. Clearly, $H_1 \neq G$. So, there exists a simple ideal H_2 such that $H_1 \neq H_2$. Now $H_1 \cap H_2 = (0)$, and so $H_1 + H_2$ is a direct sum.

Since $H_1 + H_2 \neq G$, there exists a simple ideal H_3 of G such that $H_1 + H_2 + H_3 \neq H_1 + H_2$.

If $H_3 \cap (H_1 + H_2) \neq (0)$, then $H_3 \subseteq H_1 + H_2$ (since H_3 is simple) $\Rightarrow H_1 + H_2 + H_3 = H_1 + H_2$, which is a contradiction.

Therefore, $H_3 \cap (H_1 + H_2) = (0)$, and so the sum $H_1 + H_2 + H_3$ is direct. Now $H_1 + H_2 + H_3 \neq G$. If we continue this process up to infinite steps, we get an infinite chain $H_1 + H_1 \oplus H_2 + H_1 \oplus H_2 \oplus H_3 + \ldots$ such that for each m, $H_1 \oplus H_2 \oplus \ldots \oplus H_m$ is not essential in $H_1 \oplus H_2 \oplus \ldots \oplus H_m \oplus H_{m+1}$, which is a contradiction, since G has FGD. Hence, there exists n such that $G = H_1 \oplus H_2 \oplus \ldots \oplus H_n$.

(iii) Follows from (i) and (ii). ∎

Definition 5.2.19

A subset X of G is said to be a **u-linearly independent** (u.l.i.) set if every element of X is a u-element and X is an l.i. set. Let $a_i \in G$ for $1 \leq i \leq n$. a_i, $1 \leq i \leq n$ are said to be **l.i.u-elements** (or u.l.i. elements) if $\{a_i / 1 \leq i \leq n\}$ is a u.l.i. set.

Result 5.2.20 (Result 2.2 of Satyanarayana and Prasad, 2005)

Suppose $n = \dim G$ and a_i, $1 \leq i \leq n$ are l.i. elements. Then

(i) a_i, $1 \leq i \leq n$ are l.i.u-elements.
(ii) $\{a_i / 1 \leq i \leq n\}$ forms an essential basis for G.
(iii) The conditions (i) and (ii) are equivalent.

Proof

(i) Follows from Result 5.2.7 (iii).
(ii) Follows from (i) and Definition 5.2.8.
(iii) Straightforward. ∎

Definition 5.2.21

A u.l.i. set X is said to be a **maximal** u.l.i. set if $X \cup \{b\}$ is a u-l.d. set for each $0 \neq b \in G \backslash X$.

Result 5.2.22 (Result 2.3 of Satyanarayana and Prasad, 2005)

Suppose G has FGD. Then

(i) If b_i, $1 \leq i \leq k$ are l.i. elements, then there exist u-elements $a_i \in \langle b_i \rangle$, $1 \leq I \leq k$ such that a_i, $1 \leq i \leq k$ are l.i.u-elements.
(ii) If H is a nonzero ideal of G, then there exists a u.l.i. set $X = \{a_i \mid 1 \leq i \leq k\}$ such that $\langle X \rangle = \Sigma_{i=1}^{k} \langle a_i \rangle \leq_e H$. Moreover, dim $H = k$.

Proof

(i) By Remark 5.2.6, there exist u-elements $0 \neq a_i \in \langle b_i \rangle$, $1 \leq i \leq k$. Since $\langle b_1 \rangle + \langle b_2 \rangle + \ldots + \langle b_k \rangle$ is direct, it follows that $\langle a_1 \rangle + \langle a_2 \rangle + \ldots + \langle a_k \rangle$ is direct. Hence, a_i, $1 \leq i \leq k$ are l.i.u-elements.

(ii) Since G has FGD and H is a nonzero ideal of G, by Theorem 5.1.2, H has FGD. Suppose dim $H = k$. Then there exist uniform ideals U_i, $1 \leq i \leq k$ whose sum is direct and essential in H. Take $0 \neq a_i \in U_i$. Now each a_i is a u-element and $<a_i> \leq_e U_i$. So by Corollary 3.4.5, it follows that

$$\sum_{i=1}^{k} <a_i> \leq_e \sum_{i=1}^{k} U_i.$$

Since $<a_1> + \ldots + <a_k>$ is direct, the set $X = \{a_i / 1 \leq i \leq k\}$ is a u.l.i. set. Also, $<X> = \sum_{i=1}^{k} <a_i> \leq_e \sum_{i=1}^{k} U_i \leq_e H$. The proof is complete. ∎

Theorem 5.2.23 (Lemma 1.7 of Satyanarayana and Prasad, 2005)

Let f: $G \to G^1$ be an N-group isomorphism. Then $x_1, x_2, \ldots, x_n$ are l.i.u-elements in G if and only if $f(x_1), f(x_2), \ldots, f(x_n)$ are l.i.u-elements in G^1.

Proof

Combination of Lemmas 5.2.10 and 5.2.11. ∎

Theorem 5.2.24

If G has FGD and $n = $ dim G, then G contains l.i.u-elements $u_1, u_2, \ldots, u_n$ such that $\{u_1, u_2, \ldots, u_n\}$ spans G essentially.

Proof

Suppose dim $G = n$. Then by Theorem 5.1.2, there exist uniform ideals $U_1, U_2, \ldots, U_n$ in G whose sum is direct and essential in G.

Let $0 \neq u_i \in U_i$ for $1 \leq i \leq n$. Now $(0) \neq <u_i> \subseteq U_i$ and U_i is uniform, and by Remark 5.2.6, it follows that $<u_i>$ is uniform, and so u_i is a u-element. Since U_i is uniform, it follows that $<u_i> \leq_e U_i$. By Corollary 3.4.5, $<u_1> + \ldots + <u_n> \leq_e U_1 + \ldots + U_n$. So the sum $<u_1> + \ldots + <u_n>$ is direct and essential in G. This shows that $u_1, u_2, \ldots, u_n$ are l.i.u-elements that span G essentially. ∎

5.3 Primary Decomposition in Near Rings

Lemma 5.3.1 (Lemma 5.1.1 of Satyanarayana, 1984)

(i) If I is a completely semiprime ideal of a near ring N, then for any $a, b \in N$, $ab \in I \Rightarrow ba \in I$.
(ii) If $a \in N$ and I is a semiprime ideal of N, then $(I:a) = \{x \in N \mid xa \in I\}$ is an ideal of N.

Proof

(i) Suppose $a, b \in N$ such that $ab \in I$. Then $(ba)^2 = (ba)(ba) = b(ab)a \in I$. Since I is completely semiprime and $(ba)^2 \in I$, we have $ba \in I$.
(ii) It is obvious that $(I:a)$ is a left ideal. We now show that $(I:a)$ is a right ideal of N. Let $x \in (I:a)$ and $n \in N$. Since $x \in (I:a)$, $xa \in I$. By (i), $ax \in I$. Since I is a left ideal, $nax \in I$, and by (i), $xna \in I$. Thus $(I:a)$ is a right ideal. Hence $(I:a)$ is an ideal of N. ∎

Theorem 5.3.2 (Theorem 5.1.2 of Satyanarayana, 1984)

(a) If S is a semiprime ideal of N, then the following are equivalent:
   (i)   If $x^2 \in S$, then $<x>^2 \subseteq S$.
   (ii) S is completely semiprime.
   (iii) If $xy \in S$, then $<x><y> \subseteq S$.
(b) An ideal P of N is prime and completely semiprime if and only if it is completely prime.

Proof

(a)
   (i) $\Rightarrow$ (ii) is obvious.
   (ii) $\Rightarrow$ (iii): Let $x, y \in N$ such that $xy \in S$. By Lemma 5.3.1 (ii), $(I:y)$ is an ideal. So $<x> \subseteq (I:y)$, which implies $<x>y \subseteq I$. By Lemma 5.3.1 (i), $y<x> \subseteq I$. Again, using the same lemma we get $<y><x> \subseteq I$, and hence we have $<x><y> \subseteq I$.
   (iii) $\Rightarrow$ (i) is clear.
(b) follows from (a). ∎

Notation 5.3.3

We recall the following

P-rad(N) = The prime radical of N (i.e., the intersection of all prime ideals of N).

C-rad(N) = The completely prime radical of N (i.e., the intersection of all completely prime ideals of N).

L = The set of all nilpotent elements of N.

sn(N) = The sum of all nilpotent ideals of N.

Remark 5.3.4

(i) P-rad(N) $\subseteq$ C-rad(N).

(ii) sn(N) $\subseteq$ L $\subseteq$ S, for any completely semiprime ideal S of N.

Before proving the main theorem of this section, we need the following three lemmas.

Lemma 5.3.5 (Lemma 5.1.5 of Satyanarayana, 1984)

If M is an m-system and I is a completely semiprime ideal of N such that $M \cap I = \varphi$, then the subsemigroup of (N) generated by M is contained in

$$N \backslash I = \{n \in N \mid n \notin I\}.$$

Proof

It suffices to show that if $a_1, a_2, \ldots, a_k \in M$, then $a_1 a_2 a_3 \ldots a_k \in N \backslash I$. Let $a_1, a_2, \ldots, a_k \in M$. Since $a_1, a_2 \in M$ and M is an m-system, there exist $a_2^1 \in <a_1>$ and $a_2^* \in <a_2>$ such that $a_2^1 a_2^* \in M$. Since $a_2^1 a_2^* \in M$ and $a_3 \in M$, there exist $a_3^1 \in <a_2^1 a_2^*>$, $a_3^* \in <a_3>$ such that $a_3^1 a_3^* \in M$.

If we continue this up to k steps, we have elements

$a_2^1, a_3^1, \ldots, a_k^1, a_2^*, a_3^*, \ldots, a_k^*$ in N such that $a_i^1 a_i^* \in M$ for $2 \leq i \leq k$ and $a_i^1 \in <a_{i-1}^1 a_{i-1}^*>$, $a_i^* \in <a_i>$.

If $a_1 a_2 \ldots a_k \notin (N \backslash I)$, then $a_1 a_2 \ldots a_k \in I$. Now

$$a_1 a_2 \ldots a_k \in I \Rightarrow a_2^{\ 1} a_2 a_3 \ldots a_k \in I \text{ (by Lemma 5.3.1 (ii))}$$

$$\Rightarrow a_2 a_3 \ldots a_k a_2^{\ 1} \in I \text{ (by Lemma 5.3.1 (i))}$$

$$\Rightarrow a_2^{\ *} a_3 \ldots a_k a_2^{\ 1} \in I$$

$$\Rightarrow a_2^{\ 1} a_2^{\ *} a_3 \ldots a_k \in I$$

$$\Rightarrow a_3^{\ 1} a_3 \ldots a_k \in I.$$

If we continue this process up to k steps, we get $a_k^{\ 1} a_k^{\ *} \in I$. Hence, $a_k^{\ 1} a_k^{\ *} \in M \cap I = \varphi$, which is a contradiction.

Therefore, $a_1, a_2, \ldots, a_n \in N \backslash I$.

Recall that an ideal P is called a minimal prime (minimal completely prime) ideal of I if it is minimal in the set of all prime (completely prime) ideals containing I. ∎

Lemma 5.3.6 (Lemma 5.1.6 of Satyanarayana, 1984)

If P is a minimal prime ideal of I, then $N \backslash P$ is an m-system maximal relative to the property that $(N \backslash P) \cap I = \varnothing$.

Proof

If $M = N \backslash P$ is not a maximal m-system relative to the property $M \cap I = \varnothing$, then by Theorem 4.2.10, there exists an M-system M^* maximal relative to the properties $M \subseteq M^*$ and $M^* \cap I = \varnothing$. Again by Theorem 4.2.10, $N \backslash M^*$ is a minimal prime ideal of I in N. Now $N \backslash P = M \subseteq M^*$, and hence $P = (N \backslash M) \supseteq (N \backslash M^*)$. Since $(N \backslash M^*)$ is a minimal prime ideal of I, we have $P = N \backslash M^*$. Now it follows that $M = M^*$. This completes the proof of the lemma. ∎

Lemma 5.3.7 (Lemma 5.1.7 of Satyanarayana, 1984)

Every minimal prime ideal P of a completely semiprime ideal I is completely prime. Moreover, P is a minimal completely prime ideal of I.

Proof

Suppose $P \neq N$. Then $N \backslash P$ is an m-system such that $M \cap I = \emptyset$. By Lemma 5.3.5, the multiplicative subsemigroup S generated by M is contained in $N \backslash I$. Since P is a minimal prime ideal of I, by Lemma 5.3.6, M is a maximal m-system relative to having an empty intersection with I. Since S is a subsemigroup, it is an m-system and hence $M = S$. Since $S = (N \backslash P)$ is a subsemigroup of N, it follows that P is a completely prime ideal containing I. ∎

Theorem 5.3.8 (Theorem 5.1.8 of Satyanarayana, 1984)

Let I be a completely semiprime ideal of N. Then I is the intersection of all minimal completely prime ideals of I.

Proof

If I is a completely semiprime ideal of N, then I is a semiprime ideal. By Theorem 4.2.10, I is the intersection of all minimal prime ideals of I. Since I is completely semiprime, by Lemma 5.3.7, every minimal prime ideal of I is a minimal completely prime ideal of I, and hence I is the intersection of some collection of minimal completely prime ideals of I. Therefore, I is the intersection of all minimal completely prime ideals of I. The proof is complete. ∎

Theorem 5.3.9 (Theorem 5.1.9 of Satyanarayana, 1984)

If P is a prime ideal and I is a completely semiprime ideal, then P is a minimal prime of I if and only if P is a minimal completely prime of I.

Proof

Suppose P is a minimal completely prime ideal of I. If P is not a minimal prime ideal of I, then there exists a prime ideal P^1 containing I and $P \neq P^1$. Now

$\wp = \{J \mid J$ is a prime ideal such that $P \supseteq J \supset I\}$ is nonempty, and by Zorn's lemma $\wp$ contains a minimal element, say P^*. Now P^* is a minimal prime of I. Now by Lemma 5.3.7, P^* is a completely prime of I. Since $P \supseteq P^* \supset I$ and P is a minimal completely prime of I, we have $P = P^*$, and so P is a minimal prime, which is a contradiction. The rest follows from Lemma 5.3.7. ∎

TABLE 5.1

Multiplication Table

.	0	a	b	c
0	0	0	0	0
a	a	a	a	a
b	0	a	b	c
c	a	0	c	b

Corollary 5.3.10

If I is a completely semiprime ideal of N, then I is the intersection of all completely prime ideals of N containing I.

The following example shows that Theorem 5.3.8 fails if N is not a zero-symmetric near ring.

Example 5.3.11

Let $(G, +)$ be the Klein four group and $G = \{0, a, b, c\}$. Define multiplication on G as follows (Table 5.1):

Now $(G, +, .)$ is a near ring that is not zero-symmetric since $a0 = a \neq 0$. In this near ring $\{0, a\}$ is only the nontrivial ideal, and also it is completely prime. (0) is completely semiprime, but not completely prime (since $c \cdot a = 0$ and $a \neq 0 \neq c$). Hence, the completely semiprime ideal (0) cannot be written as the intersection of its minimal completely prime ideals.

The following example shows that even in rings $P\text{-rad}(N)$, $C\text{-rad}(N)$, and L are not equal.

Example 5.3.12

Let N be the set of all 3×3 matrices over the ring of integers. If

$$x = \begin{bmatrix} 0 & 0 & 0 \\ 1 & 0 & 0 \\ 1 & 1 & 0 \end{bmatrix}, y = \begin{bmatrix} 0 & 1 & 1 \\ 0 & 0 & 1 \\ 0 & 0 & 0 \end{bmatrix},$$ then x and y are nilpotent elements of N

(since $x^4 = y^4 = 0$). Consider $x + y = \begin{bmatrix} 0 & 1 & 1 \\ 1 & 0 & 1 \\ 1 & 1 & 0 \end{bmatrix}, (x+y)^2 = \begin{bmatrix} 2 & 1 & 1 \\ 1 & 2 & 1 \\ 1 & 1 & 2 \end{bmatrix}.$

$(x + y)^2$ is not nilpotent. Hence, L is not an ideal. So $L \neq P$-rad(N) and $L \neq$ C-rad(N).

By Corollary 4.2.18, P-rad$(N) \subseteq L$. Since C-rad(N) is completely semiprime, every nilpotent element belongs to C-rad(N), and hence $L \subseteq C$-rad(N). Thus, we have

P-rad$(N) + L + C$-rad(N).

Theorem 5.3.13 (Lemma 5.1.14 of Satyanarayana, 1984)

If N is a near ring such that P-rad(N) is completely semiprime, then P-rad$(N) =$ $L = C$-rad(N).

Proof

Since P-rad(N) is completely semiprime, by Theorem 5.3.8, P-rad$(N) = \cap P$, P ranges over all completely semiprime ideals containing P-rad(N), and hence P-rad$(N) \supseteq C$-rad(N). Already we have P-rad$(N) \subseteq L$, $D \subseteq C$-rad(N).

Thus, P-rad$(N) = L = C$-rad(N). ■

Theorem 5.3.14 (Theorem 5.1.15 of Satyanarayana, 1984)

If N has DCC on N-subgroups of N, then P-rad(N) is completely semiprime if and only if L is an ideal. In this case, P-rad$(N) = L = C$-rad(N).

Proof

If L is an ideal, then by Theorem 4.2.10, L is nilpotent and since P-rad(N) is semiprime, we have $L \subseteq P$-rad(N). Since L is the set of all nilpotent elements, L is completely semiprime. By Theorem 5.3.8, $L = C$-rad(N). So, $L = P$-rad$(N) =$ C-rad(N), which shows that P-rad(N) is a completely semiprime ideal. The converse follows from Theorem 5.3.13. ■

In the following, we obtain "prime-essential decomposition" for ideals of near rings.

Lemma 5.3.15 (Lemma 5.2.1 of Satyanarayana, 1984)

If N is a semiprime near ring, then every complement ideal is semiprime.

Proof

Let K be a complement of a nonzero ideal B of N. Let us suppose that K is not a semiprime ideal. Then there exists an ideal A of N such that $A^2 \subseteq K$ and $A \nsubseteq K$. Since $A \nsubseteq K$, $(K + A) \cap B \neq (0)$. Let $0 \neq a \in (K + A) \cap B$. Now $<a>^2 \subseteq (K + A)^2 \subseteq (A^2 + K) \subseteq K$, and since $a \in B$, $<a>^2 \subseteq B$. So $<a>^2 \subseteq (K \cap B) = (0)$. Since N is a semiprime near ring, $a = 0$, which is a contradiction. Hence, K is a semiprime ideal. ∎

Lemma 5.3.16 (Lemma 5.2.2 of Satyanarayana, 1984)

If N is a semiprime near ring, then every proper ideal I of N is essential in its prime radical P-rad(I).

Proof

By Result 5.1.11, there exist two ideals K and K^* such that K is a complement of I, K^* is a complement of K containing I, and $I \oplus K$, $K^* \oplus K$ are essential ideals. Now, by Theorem 3.4.11, I is essential in K^*. By Lemma 5.3.15, K^* is a semiprime ideal and so $I \subseteq (P\text{-rad}(I)) \subseteq K^*$, which shows that I is essential in P-rad(I). ∎

Theorem 5.3.17 (Theorem 5.2.3 of Satyanarayana, 1984)

Let I be a proper ideal of a semiprime near ring N. Then there exists an essential ideal J of N such that $I = J \cap (P\text{-rad}(I))$.

Proof

If I is essential in N, then $I = I \cap (P\text{-rad}(I))$. Suppose I is not an essential ideal. Let K be a complement of I.

Now $J = I \oplus K$ is essential in N and by modular law,

$$(P\text{-rad}(I)) \cap J = (K \cap (P\text{-rad}(I))) + I.$$

Now $K \cap (P\text{-rad}(I))$ is an ideal of N contained in P-rad(I), and $I \cap (K \cap (P\text{-rad}(I)) = (0)$. Since I is essential in P-rad(I), by Lemma 5.3.16, it follows that $K \cap (P\text{-rad}(I)) = (0)$.
From this, it follows that $I = J \cap (P\text{-rad}(I))$. ∎

Definition 5.3.18

A decomposition $I = J \cap (P\text{-rad}(I))$ of an ideal I, where J is an essential ideal is said to be **reduced** if $I = J_1 \cap (P\text{-rad}(I))$, J_1 is an essential ideal such that $J_1 \subseteq J$, then $J_1 = J$.

Theorem 5.3.19 (Theorem 5.2.5 of Satyanarayana, 1984)

Let N be a near ring with DCC on ideals. Then every ideal I of N has a reduced decomposition.

Proof

By Theorem 5.3.17, there exists an essential ideal J such that $I = J \cap (P\text{-rad}(I))$. Write $\Omega = \{J^1$ is an ideal of $N/I = J^1 \cap (P\text{-rad}(I))$ and J^1 is essential$\}$.

Since N has DCC on ideals, Ω contains a minimal element, say J^*. Now $I = J^* \cap (P\text{-rad}(I))$ is a reduced decomposition. ∎

Theorem 5.3.20 (Theorem 5.2.6 of Satyanarayana, 1984)

If $I = J_1 \cap (P\text{-rad}(I)) = J_2 \cap (P\text{-rad}(I))$ are two reduced decompositions, then $J_1 = J_2$, and hence the reduced decomposition is unique.

Proof

Follows from the definition of the reduced decomposition. ∎

Definition 5.3.21

A decomposition $I = J \cap P_1 \cap \dots \cap P_k$, where J is an essential ideal and P_i ($1 \leq i \leq k$) are distinct minimal primes of I, is said to be **reduced** if $I = J \cap (P\text{-rad}(I))$ is reduced and $P\text{-rad}(I) = P_1 \cap P_2 \cap \dots \cap P_k$.

Theorem 5.3.22 (Theorem 5.2.8 of Satyanarayana, 1984)

Let N be a semiprime near ring having DCC on ideals and I an ideal of N. Then there exists an essential ideal J and distinct minimal prime ideals P_1, P_2, ..., P_k of I such that the decomposition $I = J_1 \cap P_1 \cap \dots \cap P_k$ is reduced.

Moreover, if there exists an essential ideal J_2 and distinct minimal prime ideals $P_1^1, P_2^1, \ldots, P_m^1$ of I such that the decomposition
$I = J_2 \cap P_1^1 \cap P_2^1 \cap \ldots \cap P_m^1$ is reduced, then $J_1 = J_2$, $k = m$ and

$$\{P_1, P_2, \ldots, P_k\} = \{P_1^1, P_2^1, \ldots, P_m^1\}.$$

Proof

By Theorem 5.3.19, there exists an essential ideal J_1 such that the decomposition $I = J_1 \cap (P\text{-rad}(I))$ is reduced. Since N is Artinian (DCC on ideals), the set

$$\Omega = \left\{ \bigcap_{i=1}^{t} P_i \mid P_i\ (1 \le i \le t) \text{ are minimal prime of } I \right\}$$

contains a minimal element, say $\bigcap_{i=1}^{k} P_i$. We may assume that none of the P_i's can be deleted from $\bigcap_{i=1}^{k} P_i$. Now it can be easily verified that every minimal prime ideal of I contains $\bigcap_{i=1}^{k} P_i$, and hence $P\text{-rad}(I) = \bigcap_{i=1}^{k} P_i$.

Hence, the decomposition $I = J_1 \cap P_1 \cap \ldots \cap P_k$ is reduced.

Suppose there exist an essential ideal J_2 and minimal prime ideals $P_1^1, P_2^1, \ldots, P_m^1$ of I such that the decomposition $I = J_2 \cap P_1^1 \cap \ldots \cap P_m^1$ is reduced. By Theorem 5.3.20, $J_1 = J_2$. Since

$P\text{-rad}(I) = P_1^1 \cap P_2^1 \cap \ldots \cap P_m^1 = P_1 \cap P_2 \cap \ldots \cap P_k$, we have
$P_1 P_2 \ldots P_k \subseteq P_1^1 \cap P_2^1 \cap \ldots \cap P_k = P_1^1 \cap \ldots \cap P_m^1 \subseteq P_j^1$ for $1 \le j \le m$.

Fix j $(1 \le j \le m)$. Then $P_1 P_2 \ldots P_k \subseteq P_j^1$ implies that $P_i \subseteq P_j^1$ for some $1 \le i \le k$. Since P_j^1 is a minimal prime of I, $P_i = P_j^1$. Thus $\{P_1^1, P_2^1, \ldots, P_m^1\} \subseteq \{P_1, P_2, \ldots, P_k\}$. Therefore, $m \le k$.

Similarly, we can show that $\{P_1, P_2, \ldots, P_k\} \subseteq \{P_1^1, P_2^1, \ldots, P_m^1\}$ and $k \le m$. Hence, we have that $m = k$ and $\{P_1^1, P_2^1, \ldots, P_m^1\} = \{P_1, P_2, \ldots, P_k\}$. ∎

Definition 5.3.23

An ideal Q of a near ring N is said to be **primary** if for any two ideals I, J of N, $IJ \subseteq Q$, which implies that $I \subseteq Q$ or $J \subseteq (P\text{-rad}(Q))$.

Theorem 5.3.24 (Theorem 5.3.1 of Satyanarayana, 1984)

For an ideal Q of N, the following conditions are equivalent:

(i) Q is primary.
(ii) $<a><b> \subseteq Q$, $a, b \in N \Rightarrow a \in Q$ or $b \in P\text{-rad}(Q)$.
(iii) $a, b \in N$, $a \notin Q$, $b \notin P\text{-rad}(Q) \Rightarrow <a><b> \not\subseteq Q$.

(iv) I, J are ideals of N, such that $I \not\subseteq Q$.
      $J \not\subseteq (P\text{-rad}(Q))$, which implies $IJ \not\subseteq Q$.
The proof is straightforward.

Theorem 5.3.25 (Theorem 5.3.2 of Satyanarayana, 1984)

Let Q_1 and Q_2 be primary ideals such that $C = P\text{-rad}(Q_1) = P\text{-rad}(Q_2)$.
Then $Q_1 \cap Q_2$ is primary with $P\text{-rad}(Q_1 \cap Q_2) = C$.

Proof

$P\text{-rad}(Q_1 \cap Q_2) = P\text{-rad}(Q_1) \cap P\text{-rad}(Q_2) = C$. To show that $Q_1 \cap Q_2$ is primary, let I, J be two ideals of N such that $IJ \subseteq (Q_1 \cap Q_2)$ and $I \not\subseteq (Q_1 \cap Q_2)$. Then, either $I \not\subseteq Q_1$ or $I \not\subseteq Q_2$. If $I \not\subseteq Q_1$, then

$$J \subseteq (P\text{-rad}(Q_1)) = C. \text{ If } I \not\subseteq Q_2, \text{ then}$$

$$J \subseteq (P\text{-rad}(Q_2)) = C. \text{ Hence, if } I \not\subseteq (Q_1 \cap Q_2), \text{ then}$$

$$J \subseteq (P\text{-rad}(Q_1 \cap Q_2)). \text{ The proof is complete.} \quad \blacksquare$$

By induction, we have the following.

Corollary 5.3.26

Let Q_i, $1 \leq i \leq n$, be primary ideals with radical C.
Then $Q_1 \cap Q_2 \cap \ldots \cap Q_n$ is primary with the same radical C.
In the rest of this section, we assume that N is a near ring with ACC on ideals.

Theorem 5.3.27 (Theorem 5.3.4 of Satyanarayana, 1984)

Any ideal of N contains the product of a finite number of its minimal primes.

Proof

Let us suppose that N contains an ideal that does not contain the product of a finite number of its minimal primes. Since N has ACC on ideal, the set of all ideals not meeting the stated condition contains a maximal element, say I. Then I is not prime itself, and hence there exist ideals B and C properly containing I such that $BC \subseteq I$. Then each B and C must satisfy the stated condition, and hence I contains a product of a finite number of prime ideals containing I. Since every prime ideal containing I contains a minimal prime ideal of I, it follows that there exists a finite number or minimal prime ideals of I whose product is contained in I, which is a contradiction. This completes the proof. ∎

Corollary 5.3.28

Any ideal I of N has only a finite number of minimal prime ideals.

Proof

Let $P_1, P_2, ..., P_n$ be minimal prime ideals of I such that $P_1, P_2, ..., P_n \subseteq I$. Let P be a minimal prime ideal of I. Then $P_1 P_2, ..., P_n \subseteq I \subseteq P$. Since P is prime, $P_i \subseteq P$ for some i, $1 \le i \le n$, and since P is minimal prime, we have $P_i = P$. ∎

Corollary 5.3.29

If I is an ideal of N, then $(P\text{-rad}(I))^n \subseteq I$ for some integer n.

Proof

Suppose $P_1 P_2, ..., P_n \subseteq I$ for some minimal prime ideals $P_1, P_2, ..., P_n$ of I. Since each P_i contains $P\text{-rad}(I)$, we have $(P\text{-rad}(I))^n \subseteq P_1 P_2, ..., P_n \subseteq I$. ∎

Theorem 5.3.30 (Theorem 5.3.7 of Satyanarayana, 1984)

If Q is primary, then $P\text{-rad}(Q) = P$ is a prime ideal.

Proof

Suppose B, C are ideals such that $BC \subseteq P$. Also suppose that $B \not\subseteq P$. Then for some n (by Corollary 5.3.29), $P^n \subseteq Q$. Thus $(BC)^n \subseteq Q$. Let k be the least positive integer such that $(BC)^k \subseteq Q$. Now $(BC)^{k-1}B \subseteq (Q:C)$, and hence $<(BC)^{k-1}B>C \subseteq Q$. Since Q is primary, either $<(BC)^{k-1}B> \subseteq Q$ or $C \subseteq P$. If $(BC)^{k-1}B \subseteq Q$, then since $(BC)^{k-1} \not\subseteq Q$, we have $B \subseteq P$, which is a contradiction. Thus, $C \subseteq P$, and hence $P = P\text{-rad}(Q)$ is a prime ideal. ∎

Corollary 5.3.31

If Q is a primary ideal, then Q has a unique minimal prime ideal $P = P\text{-rad}(Q)$.

Theorem 5.3.32 (Theorem 5.3.9 of Satyanarayana, 1984)

For any ideal Q of N, the following conditions are equivalent:

(i) Q is primary.
(ii) A, B are ideals of N such that $AB \subseteq Q$ implies that $A \subseteq Q$ or $B^n \subseteq Q$ for some integer n.
(iii) $a, b \in N$ and $<a><b> \subseteq Q$ imply $<a> \subseteq Q$ or $<b>^n \subseteq Q$ for some integer n.

Proof

(i) $\Rightarrow$ (ii): Follows from Corollary 5.3.29.
(ii) $\Rightarrow$ (iii) is clear.
(iii) $\Rightarrow$ (i): By Theorem 5.3.24, it suffices to prove that $<a><b> \subseteq Q$, which implies that $a \in Q$ or $b \in P\text{-rad}(Q)$. Suppose $<a><b> \subseteq Q$. By (iii), $a \in Q$ or $<b>^n \subseteq Q$. If $<b>^n \subseteq Q$, then $<b>^n \subseteq (P\text{-rad}(Q))$. Since $P\text{-rad}(Q)$ is semiprime, we have $b \in (P\text{-rad}(Q))$. ∎

Definition 5.3.33

Let Q be an ideal of N. Then Q is called **P-primary** if Q is primary and $P = P\text{-rad}(Q)$.

Theorem 5.3.34 (Theorem 5.3.11 of Satyanarayana, 1984)

Let Q and P be ideals of N. Then Q is P-primary if and only if the following conditions hold:

(i) $Q \subseteq P \subseteq (P\text{-rad}(Q))$.
(ii) If A and B are ideals such that $AB \subseteq Q$ and $A \not\subseteq Q$, then $B \subseteq P$.

Proof

If Q is P-primary, then clearly (i) and (ii) hold. To prove the converse, we first show that P is a semiprime ideal. Let A be an ideal of N such that

$A^2 \subseteq P$. By Corollary 5.3.29, there exists an integer n such that $(A^2)^n \subseteq P^n \subseteq (P\text{-rad}(Q))^n \subseteq Q$. Let k be the least positive integer such that $A^k \subseteq Q$. Since $<A^{k-1}>A \subseteq Q$ and $A^{k-1} \not\subseteq Q$, by condition (ii), $A \subseteq P$. Thus, P is a semiprime ideal, and since $Q \subseteq P$, we have $(P\text{-rad}(Q)) \subseteq P$. Hence, $P = P\text{-rad}(Q)$ and by condition (ii), Q is -primary. ∎

Theorem 5.3.35 (Theorem 5.3.12 of Satyanarayana, 1984)

An ideal Q of N is prime if and only if it is semiprime and primary.

Proof

Suppose Q is semiprime and primary. Since Q is semiprime, by Theorem 4.2.10, $Q = P\text{-rad}(Q)$. Let A, B be two ideals of N such that $AB \subseteq Q$. Since Q is primary, either $A \subseteq Q$ or $B \subseteq (P\text{-rad}(Q))$. Thus, either $A \subseteq Q$ or $B \subseteq Q$, which shows that Q is a prime ideal. The rest is clear. ∎

Now we prove the existence theorem of primary decomposition for ideals in a class of near rings.

Theorem 5.3.36 (Theorem 2.1 of Satyanarayana, 1984)

Let N be a near ring satisfying

   (i) N has ACC on ideals.
  (ii) $a \in Na$ for all $a \in N$.
 (iii) Every left N-subgroup of N is an ideal.
    Then every ideal of N has a primary decomposition.

Proof

Suppose there exists an ideal that has no primary decomposition. Then there exists a maximal element 0 in the set of all ideals that have no primary decomposition. Clearly, Q is not primary. So there exist elements a, b in N such that $<a><b> \subseteq Q$ and $a \notin Q$, $b \notin P\text{-rad}(Q)$. Since $b \notin P\text{-rad}(Q)$, by Theorem 4.2.19, there exists a sequence $\{b_0, b_1, \ldots\}$ of elements from N such that $b_0 = b$, and for $n \geq 1$, $b_n = b_{n-1}b^*_{n-1}$ for some

$$b^*_{n-1} \in <b_{n-1}> \text{ and } b_n \notin Q. \text{ Write } B_n = \{x \in N \mid xb_n \in Q\}.$$

For each n, B_n is a left ideal and hence it is ideal by (iii). Since

$B_1 \subseteq B_2 \subseteq B_3 \subseteq \ldots$, by (i), there exists a least positive integer k such that $B_k = B_i$ for all $i \geq k$. Write $B^* = \{x \in N / xb_k^* \in Q\}$ and $D = Nb_k + Q$. B^* and D are ideals by (iii). By (ii) we have $b_k \in Nb_k$, and hence D contains Q properly.

Since $ab_k^* \in a<b_k> \subseteq a<b> \subseteq Q$, we have $a \in B^*$ and hence B^* contains Q properly.

To show $Q = B^* \cap D$, let $z \in B^* \cap D$. Then $z = mb_k + q$ for some $m \in N, q \in Q$, and $zb_k^* \in Q$.

Now $mb_k b_k^* + qb_k^* = zb_k^* \in Q$

$$\Rightarrow mb_{k+1} = mb_k b_k^* \in Q \text{ (since } qb_k^* \in Q)$$
$$\Rightarrow m \in B_{k+1} = B_k$$
$$\Rightarrow mb_k \in Q \Rightarrow z \in Q.$$

Hence, we have $Q = B^* \cap D$. Since B^* and D are ideals properly containing Q, D and B^* have primary representation, then $D \cap B^* = Q$ has primary representation, which is a contradiction. The proof is complete. ∎

Definition 5.3.37 (Santha Kumari, 1982)

A near ring N is called a Q-near ring if N contains a multiplicatively closed subset Q satisfying the following properties:

(i) $a \in Q$ implies Na is a left ideal of N.
(ii) $aN = Na$ for all a in Q.
(iii) For all ideals A, B of N such that $A \subset B$ (properly), B contains an element of Q that is not in A.

Corollary 5.3.38 (Corollary 0.5.4 of Satyanarayana, 1984)

Let N be a near ring with identity satisfying ACC on ideals. Suppose N is a Q-near ring with $N = Q$. Then every ideal of N can be represented as the intersection of a finite number of primary ideals.

Remark 5.3.39 (Remark 2.2 of Satyanarayana, 1984)

The class of all near rings N satisfy the following:

(i) $a \in Na$ for all $a \in N$ and
(ii) Every left N-subgroup of N is an ideal contains

(a) The class of all Boolean near rings (i.e., every element of N is an idempotent).

(b) The class of all strongly regular near rings (i.e., for every $a \in N$, there exists $x \in N$ such that $a = xa^2$).

(iii) The class of all left duo rings with left identity (i.e., rings in which every one-sided ideal is two-sided).

Theorem 5.3.40 (Krull Intersection Theorem) (Theorem 2.3 of Satyanarayana, 1984)

Let N be a near ring satisfying conditions (i)–(iii) of Theorem 5.3.36. Let I be an ideal of N.

$$\text{If } B = \bigcap_{n=1}^{\infty} <I^n>, \text{ then } B = <BI>.$$

Proof

If $<BI> = N$, then $N = <BI> \subseteq B$, and so $B = N = <BI>$. So, assume that $<BI> \neq N$. We now show that $B \subseteq <BI>$. By Theorem 5.3.36, $<BI>$ has a primary decomposition $<BI> = Q_1 \cap Q_2 \cap \ldots \cap Q_m$, where $P_i = P\text{-rad}(Q_i)$, $1 \leq i \leq m$. Let us fix some j ($1 \leq j \leq m$). If $I \subseteq P_j$, then $I^k \subseteq P_j^k \subseteq Q_j$ for some integer k (by Corollary 5.3.29). So,

$$B = \bigcap_{n=1}^{\infty} <I^n> \subseteq <I^k> \subseteq Q_i.$$

If $I \not\subseteq P_j$, then since $BI \subseteq Q_j$ and Q_j is P_j-primary, we have $B \subseteq Q_j$. Hence, in either case $B \subseteq Q_j$. Thus, $B \subseteq Q_1 \cap Q_2 \cap \ldots \cap Q_m = <BI>$. ∎

5.4 Tertiary Decomposition in N-Groups

In this section, in analogy with the "tertiary decomposition" for modules over rings, we present the tertiary decomposition for Noetherian (ACC on ideals) N-groups.

Throughout this section, K stands for an ideal of G. We start this section with the following definitions.

Definition 5.4.1

(i) An element $a \in N$ is said to be an **annihilating element** for G if $aK = \{0\}$ for some nonzero ideal K of G.

(ii) An element $a \in N$ is said to be an **annihilating element** for K if there exists a nonzero ideal K^1 of G such that $K^1 \subseteq K$ and $aK^1 = \{0\}$.

Definition 5.4.2

The tertiary radical of K, written as t-rad(K), is the set of all elements of N that are annihilating elements for all nonzero ideals of G/K.

Theorem 5.4.3 (Theorem 4.1.3 of Satyanarayana, 1984)

If G is completely reducible, then for any ideal K of G, t-rad(K) = $(0:(G/K))$.

(Here, an N-group G is completely reducible if G is a sum of simple ideals; equivalently, G is equal to the direct sum of a collection of simple ideals of G.)

Proof

Let $x \in (0:(G/K))$. Then for all $g \in G$, $x\langle g\rangle \subseteq K$ implies $x \in$ t-rad(K). Therefore, t-rad(K) $\supseteq (0:(G/K))$. To show $(0:(G/K)) \supseteq$ t-rad(K), let $a \in$ t-rad(K). Since G is completely reducible, $G = \Sigma_{i\in\Delta} H_i$ (direct sum), where each H_i is a simple ideal. If $H_i \subseteq K$, then $aH_i \subseteq aK \subseteq K$. Now suppose $H_i \not\subseteq K$.

Then $H_i \cap K = (0)$, and hence $(H_i \oplus K)/K$ is a nonzero ideal. Since $a \in$ t-rad(K), a is an annihilating element for $(H_i \oplus K)/K$. Since $(H_i \oplus K)/K$ is simple, it follows that $a((H_i \oplus K)/K = \{0\}$, and so $aH_i \subseteq K$. Thus, in either case we have $aH_i \subseteq K$. Now by Theorem 3.2.26, $aG = a(\Sigma H_i) \subseteq (\Sigma aH_i) \subseteq K$, and so $a \neq (0:(G/K))$. ∎

Theorem 5.4.4 (Theorem 4.1.4 of Satyanarayana, 1984)

If G is completely reducible and satisfies the condition $0 \neq a \in N$, $0 \neq g \in G$ implies that $ag \neq 0$, then for any ideal K of G, we have t-rad(K) = (0).

Proof

If $a \in$ t-rad(K), then a is an annihilating element for G/K. So there exists a nonzero ideal K^1/K of G/K such that $a \in (0:(K^1/K))$. Let K^* be an ideal of G such that $K \oplus K^* = K^1(K)$. Now $K^* \neq (0)$. Since $(K^1/K) \cong K^*$, we have $(0:(K^1/K)) = (0:K^*)$.

Hence, $a \in (0:K^*)$. Since $K^* \neq (0)$, by hypothesis, $a = 0$. ∎

Now we present an interesting characterization of t-rad(K).

Lemma 5.4.5 (Lemma 1.3 of Satyanarayana, 1982)

t-rad$(K) = \{a \in N \mid \alpha \notin K \Rightarrow$ there exists $\beta \in <\alpha>$ such that $\beta \notin K$ and $a <\beta> \subseteq K\}$.

Proof

Write $X = \{a \in N \mid \alpha \notin K \Rightarrow$ there exists $\alpha \in <\alpha>$ such that $\beta \notin K$ and $a <\beta> \subseteq K\}$. Let $x \in$ t-rad(K) and $\alpha \notin K$. Then $(<\alpha> + K)/K$ is a nonzero ideal of G/K, and hence x is an annihilating element for $(<\alpha> + K)/K$. Therefore, there exists a nonzero ideal K^1/K of G/K such that $(K^1/K) \subseteq ((<\alpha> + K)/K)$ and $x(K^1/K) = \{0\}$, which implies $xK^1 \subseteq K$. Let $y \in K^1$ such that $y \notin K$. Then $y + K = \beta + K$ for some $\beta \in <a>$.

Now $\beta \notin K$ and $\beta \in K^1$. Therefore, $x<\beta> \subseteq xK^1 \subseteq K$, which implies that $x \in X$.

Conversely, let $x \in X$, and K^1/K be a nonzero ideal of G/K. Then there exists $\alpha \in K^1$ such that $\alpha \notin K$. Since $x \in X$, there exists $\beta \in <\alpha>$ such that $\beta \notin K$ and $x <\beta> \subseteq K$.

Therefore, $x((<\beta> + K)/K) = (0)$. Hence, $x \in$ t-rad(K). ∎

Lemma 5.4.6 (Lemma 1.4 of Satyanarayana, 1982)

Let $a_1, a_2, ..., a_n \in$ t-rad(K) and $\alpha \notin K$; then there is an element $\beta \in <\alpha>$ such that $\beta \notin K$ and $a_i<\beta> \subseteq K$ for $1 \leq i \leq n$.

Proof

Since $a_1 \in$ t-rad(K), by Lemma 5.4.5, there exists

$\beta_1 \in <\alpha>$ such that $\beta_1 \notin K$ and $a_1<\beta_1> \subseteq K$. After choosing

$\beta_1, \beta_2, ..., \beta_{n-1}$ such that $\beta_{k+1} \in <\beta_k>$, $\beta_{k+1} \notin K$, and

$a_{k+1}<\beta_{k+1}> \subseteq K$ for $1 \leq k \leq n - 2$, choose β_n such that $\beta_n \in <\beta_{n-1}>$, $\beta_n \notin K$, and $a_n<\beta_n> \subseteq K$.

Write $\beta = \beta_n$. Show $a_i<\beta> \subseteq a_i<\beta_i> \subseteq K$ for $1 \leq i \leq n$. ∎

Corollary 5.4.7 (Corollary 1.5 of Satyanarayana, 1982)

t-rad(K) is a two-sided ideal of N.

Proof

Let $a_1, a_2 \in$ t-rad(K) and $\alpha \notin K$. Then by Lemma 5.4.6, there exists $\beta \in <\alpha>$ such that $\beta \notin K$ and $a_1<\beta> \subseteq K$, $a_2<\beta> \subseteq K$.

Now $(a_1 - a_2) <\beta> \subseteq a_1 <\beta> -a_2<\beta> \subseteq K$.

Therefore, $a_1 - a_2 \in$ t-rad(K).

Let $n \in N$. Since $a_1<\beta> \subseteq K$, we have $(n + a_1 - n) <\beta> \subseteq \{nx + a_1x - nx \mid x \in <\beta>\} \subseteq K$.

Hence $(n + a_1 - n) \in$ t-rad(K). Let $n_1, n_2 \in N$.

Then $(n_1(n_2 + a_1) - n_1n_2) <\beta> \subseteq \{n_1(n_2x + a_1x) - n_1n_2x/x \in <\beta>\}$, and hence $(n_1(n_2 + a_1) - n_1n_2) <\beta> \subseteq K$. Thus, t-rad($K$) is a left ideal.

Also $(a_1n)<\beta> \subseteq a_1(n<\beta>) \subseteq a_1<\beta> \subseteq K$.

Hence, t-rad(K) is an ideal of N.   ∎

Corollary 5.4.8 (Corollary 1.6 of Satyanarayana, 1982)

If N satisfies ACC on ideals and $I =$ t-rad(K), then for any $\alpha \notin K$, there exists a $\beta \in <\alpha>$ such that $I <\beta> \subseteq K$ and $\beta \notin K$.

Proof

N has ACC on ideals, which implies that $I = <a_1> + <a_2> + \ldots + <a_n>$ for some $a_i \in N$, $1 \leq i \leq n$. Since $\alpha \notin K$, by Lemma 5.4.6, there exists $\beta \in <\alpha>$ such that $\beta \notin K$ and $a_i <\beta> \subseteq K$ for $1 \leq i \leq n$. Then $<a_i><\beta> \subseteq K$ for $1 \leq i \leq n$. Hence, $I <\beta> \subseteq K$.   ∎

Notation 5.4.9

For any $a \in N$, we write

$$X_a = \{g \in G/a <g> \subseteq K\}.$$

We note that $K \subseteq X_a$. With this notation, we have the following.

Proposition 5.4.10 (Proposition 4.1.10 of Satyanarayana, 1984)

If $a \in$ t-rad(K), then $X_a \cap A \neq (0)$ for every nonzero ideal A of G.

Proof

If $A \subseteq K$, then $aA \subseteq aK \subseteq K$, and hence $A \subseteq X_a$. Suppose $A \nsubseteq K$; then there exists $\alpha \in A$ such that $\alpha \notin K$. By Lemma 5.4.5, there exists $\beta \in <\alpha>$ such that $\beta \notin K$ and $a<\beta> \subseteq K$. Now $0 \neq \beta \in A \cap X_a$. So, in either case we get that $X_a \cap A \neq (0)$.   ∎

Corollary 5.4.11 (Corollary 4.1.11 of Satyanarayana, 1984)

If $a \in$ t-rad(K), then the ideal generated by X_a is essential in G.

Proposition 5.4.12 (Proposition 4.1.12 of Satyanarayana, 1984)

$(X_a \cap A)\backslash K \neq \emptyset$ for all ideals A of G with $A \not\subseteq K$ if and only if $a \in$ t-rad(K).

Proof

If $a \in$ t-rad(K) and $A \not\subseteq K$, then there exists $\alpha \in A\backslash K$. By Lemma 5.4.5, there exists $\beta \in <\alpha>\backslash K$ such that $a<\beta> \subseteq K$, that is, $\beta \in X_a$. Moreover, $\beta \in A$ and $\beta \notin K$. Hence, $\beta \in (X_a \cap A)\backslash K$.

Converse

Suppose $(X_a \cap A)\backslash K \neq \emptyset$ for all ideals A of G not contained in K. Let $\alpha \notin K$. Then $<\alpha> \not\subseteq K$, and so $(X_a \cap <\alpha>)\backslash K \neq \emptyset$. Let $\beta \in (X_a \cap <\alpha>)\backslash K$.

Now $\beta \in <\alpha>\backslash K$ and $a <\beta> \subseteq K$, since $\beta \in X_a$. Hence, by Lemma 5.4.5, $a \in$ t-rad(K). ∎

Corollary 5.4.13 (Corollary 4.1.13 of Satyanarayana, 1982)

If $a \in$ t-rad(K), then $<X_a/K>$ is essential in G/K.

Proof

Let A be an ideal of G containing K properly. Then $A \not\subseteq K$, and by Proposition 5.4.12, $(X_a \cap A)\backslash K \neq \emptyset$. If $\beta \in (x_a \cap A)\backslash K$, then

$$(0) \neq ((<\beta> + K)/K) \subseteq (<x_a/K> \cap (A/K)).$$

Hence, $<X_a/K>$ is essential in G/K. ∎

Proposition 5.4.14 (Proposition 4.1.14 of Satyanarayana, 1982)

If there exists an ideal K^1 of G such that $K \subseteq K^1 \subseteq X_a$ and K^1/K is essential in G/K, then $a \in$ t-rad(K).

Proof

Since K^1/K is essential in G/K, for any ideal A of G such that $A \not\subseteq K$, we have $((A + K)/K) \cap (K^1/K) \neq (0)$.

So $(A \cap K^1) \not\subseteq K$ and $(X_a \cap A) \not\subseteq K$. Therefore $(X_a \cap A)\backslash K \neq \varnothing$ for all ideals A of G not contained in K. Hence, by Proposition 5.4.12, $a \in$ t-rad(K). ∎

Combining Proposition 5.4.14 and Corollary 5.4.13 yields the following.

Corollary 5.4.15 (Corollary 4.1.15 of Satyanarayana, 1984)

Let $a \in N$ and suppose X_a is an ideal of G. Then $a \in$ t-rad(K) $\Leftrightarrow x_a/K$ is an ideal of G/K.

Now we present the primary radical of a given ideal K and prove that for a class of N-groups, the primary radical of a given ideal K contains the tertiary radicals of K.

Definition 5.4.16

The **primary radical** of K, written Pri-rad(K), is the intersection of all prime ideals of N that contain $(0:(G/K))$, that is, Pri-rad(K) = P-rad($0:(G/K)$).

Lemma 5.4.17 (Lemma 2.2 of Satyanarayana, 1982)

If N satisfies ACC on ideals and if G satisfies the property that Ia is an ideal of G, for any ideal I of N and $a \in G$, then t-rad(K) is a semiprime ideal.

Proof

By Remark 4.2.4, it suffices to show that $a{<}a{>} \subseteq$ t-rad(K) and $a \in N$, which implies that $a \in$ t-rad(K). Suppose $a{<}a{>} \subseteq$ t-rad(K) and $\alpha \notin K$. By Corollary 5.4.8, there exists $\beta \in {<}\alpha{>}$ such that $\beta \notin K$ and $a{<}a{>}{<}\beta{>} \subseteq K$. If $a{<}\beta{>} \not\subseteq K$, then there exists $\beta^* \in {<}\beta{>}$ such that $a\beta^* \notin K$. Since ${<}a{>}\beta^*$ is an ideal of G, we have ${<}a\beta^*{>} \subseteq {<}a{>}\beta^*$. Now $a{<}a\beta^*{>} \subseteq a({<}a{>}\beta^*) \subseteq a{<}a{>}{<}\beta{>} \subseteq K$. Hence, given $\alpha \notin K$, β or $a\beta^*$ is in ${<}\alpha{>}$, not in K such that $a{<}\beta{>} \subseteq K$ or $a {<}a\beta^*{>} \subseteq K$. By Lemma 5.4.5, we have $a \in$ t-rad(K). ∎

Theorem 5.4.18 (Theorem 2.3 of Satyanarayana, 1982)

Suppose N and G are as in Lemma 5.4.17. Then Pri-rad(K) $\subseteq$ t-rad(K).

Proof

Since $(0:(G/K)) \subseteq$ t-rad(K) and by Lemma 5.4.17, t-rad(K) is a semiprime ideal, we have Pri-rad(K) = P-rad$(0:(G/K)) \subseteq$ t-rad(K). ∎

Definition 5.4.19

An ideal K of G is said to be a **tertiary** ideal of G if the annihilating elements for G/K are all in t-rad(K).

If K is a tertiary ideal, then t-rad(K) has the following characterization.

Lemma 5.4.20 (Lemma 3.2 of Satyanarayana, 1982)

K is a tertiary ideal of G if and only if t-rad(K) = $\{x \in N \mid x <\beta> \subseteq K$ for some $\beta \notin K\}$.

Proof

Write $X = \{x \in N/x <\beta> \subseteq K$ for some $\beta \notin K\}$. Suppose K is tertiary.

Let $x \in$ t-rad(K) and $\alpha \notin K$. Then there exists $\beta \in <\alpha>$ such that $\beta \notin K$ and $a<\beta> \subseteq K$, which implies that $x \in X$. To show $X \subseteq$ t-rad(K), let $y \in X$. Then there exists $\beta \in G$ such that $\beta \notin K$ and $y <\beta> \subseteq K$. $y<\beta> \subseteq K$, which implies that $y((<\beta> + K)/K) = (0)$, and so y is an annihilating element for G/K. Since K is tertiary, by definition, $y \in$ t-rad(K).

Converse

Suppose that t-rad(K) = X. To show K is tertiary, let z be an annihilating element for G/K. Then there exists an ideal K^1 of G properly containing K such that $z(K^1/K) = \{0\}$, and so $zK^1 \subseteq K$. Since K^1 contains K properly, there exists an element $u \in K^1$ such that $u \notin K$. Now $z<u> \subseteq zK^1 \subseteq K$ and $u \notin K$, which imply that $z \in X =$ t-rad(K). The proof is complete. ∎

Theorem 5.4.21 (Theorem 3.3 of Satyanarayana, 1982)

Suppose N has ACC on ideals and G has the property that Ia is an ideal of G, for any ideal I of N and $a \in G$. If K is a tertiary ideal of G, then $P =$ t-rad(K) is a prime ideal of N.

Proof

By Remark 4.2.4, it suffices to show that $a<b> \subseteq P$. $a, b \in N$ implies either $a \in P$ or $b \in P$. Suppose $a<b> \subseteq P$ and $\alpha \notin K$. Then by Corollary 5.4.8, there exists

$\beta \in <\alpha>$ such that $\beta \notin K$ and $a<b><\beta> \subseteq K$. If $b<\beta> \subseteq K$, then by Lemma 5.4.20, $b \in$ t-rad$(K) = P$. If not, there exists an element $\beta^* \in <\beta>$ such that $b\beta^* \notin K$. Now $a<b\beta^*> \subseteq a<b>\beta^* \subseteq a<b><\beta> \subseteq K$.

Since $a<b\beta^*> \subseteq K$, $b\beta^* \notin K$, and K is tertiary, by Lemma 5.4.20, we have $a \in P =$ t-rad(K).   ∎

Now we prove the existence and uniqueness of the tertiary decomposition of ideals in N-groups.

Lemma 5.4.22 (Lemma 3.5 of Satyanarayana, 1982)

Every meet-irreducible ideal of G is tertiary.

Proof

Let K be an irreducible ideal. Suppose K is not tertiary. Then there exists $a \in N$ and $\alpha \in G$ a such that $a \notin$ t-rad(K), $\alpha \notin K$, and $a<\alpha> \subseteq K$. Since $a \notin$ t-rad(K), by Lemma 5.4.5, there exists $\beta \notin K$ with $a<\beta^1> \subseteq K$ and $\beta^1 \in <\beta>$, which implies that $\beta^1 \in K$. Write $K^* = (K + <\alpha>) \cap (K + <\beta>)$.

Clearly, $K^* \supseteq K$. If $x \in K^*$, then $x = y_1 + \alpha_1 = y_2 + \beta_1$ for some $y_1, y_2 \in K$ and $\alpha_1 \in <\alpha>$, $\beta_1 \in <\beta>$. Therefore, $\beta_1 = -y_2 + y_1 + \alpha_1$, and hence $<\beta_1> \subseteq (<-y_2 + y_1> + <\alpha_1>)$.

To show $a <\beta_1> \subseteq K$, let $\beta^* \in <\beta_1>$. Then $\beta^* = y^* + \alpha^*$ for some $y^* \in <-y_2 + y_1>$ and $\alpha^* \in <\alpha_1>$. By Proposition 3.2.28, we have

$a\beta^* \equiv ay^* + a\alpha^* \bmod(K \cap <\alpha_1>)$, which implies

$a\beta^* \equiv a(y^* + \alpha^*) \equiv ay^* + a\alpha^* \bmod(K \cap <\alpha_1>)$, which implies

$a\beta^* \equiv ay^* + a\alpha^* \bmod(K)$. Since $y^* \in K$ and $a\alpha^* \in K$, $a\beta^* \in K$. Thus, $a <\beta_1> \subseteq K$.

Hence, $\beta_1 \in K$, which implies that $x = y_2 + \beta_1 \in K$.

Therefore, $K = K^*$ and K is not a meet-irreducible ideal, which is a contradiction. The proof is complete.   ∎

Definition 5.4.23

The decomposition $K = K_1 \cap K_2 \cap \ldots \cap K_r$ of K by tertiary ideal K_i $(1 \leq I \leq r)$ is called irredundant if no K_i can be omitted.

Lemma 5.4.24 (Lemma 3.7 of Satyanarayana, 1982)

If $K = K_1 \cap K_2 \cap \ldots \cap K_r$ is an irredundant decomposition by tertiary ideals K_i $(1 \leq I \leq r)$, then t-rad$(K) = P_1 \cap P_2 \cap \ldots \cap P_r$, where $P_i =$ t-rad(K_i).

Proof

Let $a \in P_1 \cap P_2 \cap \ldots \cap P_r$ and $\alpha \notin K$. Then there exists i such that $\alpha \notin K_i$, say $\alpha \notin K_1$. Then there exists $\beta_1 \in \langle\alpha\rangle$ such that $a \langle\beta_1\rangle \subseteq K_1$ and $\beta_1 \notin K_1$. If $\beta_1 \in K_2$, write $\beta_2 = \beta_1$. If not, there exists $\beta^* \in \langle\beta_1\rangle$ such that $\beta^* \notin K_2$ and $a\langle\beta^*\rangle \subseteq K_2$. Write $\beta_2 = \beta^*$. After choosing $\beta_1, \beta_2, \ldots, \beta_{r-1}$ with the properties $a\langle\beta_i\rangle \subseteq K_i$ and $\beta_i \in \langle\beta_{i-1}\rangle$ for $2 \le i \le r-1$, choose β_r as follows. If $\beta_{r-1} \in K_r$, then write $\beta_r = \beta_{r-1}$ and if $\beta_{r-1} \notin K_r$, then there exists $\beta_1^* \in \langle\beta_{r-1}\rangle$ such that $a\langle\beta_1^*\rangle \subseteq K_r$ and $\beta_1^* \notin K_r$. Then write $\beta_r = \beta_1^*$. Now we get $a\langle\beta_r\rangle \subseteq a\langle\beta_i\rangle \subseteq K_i$ for $1 \le i \le r$ and $\beta_r \notin K$. This implies that $a\langle\beta_r\rangle \subseteq K$ and $\beta_r \notin K$.

Hence, $a \in \text{t-rad}(K)$.

Converse

Let $a \in \text{t-rad}(K)$. Fix some i $(1 \le i \le r)$. Since $K = K_1 \cap K_2 \cap \ldots \cap K_r$ is irredundant, there exists $\alpha \in G_i$ such that $\alpha \in K_j$ for every $j \ne i$ and $\alpha \notin K_i$. $\alpha \notin K$ implies that there exists $\beta \in \langle\alpha\rangle$ such that $\beta \notin K$ and $a\langle\beta\rangle \subseteq K$. Since $\beta \in \langle\alpha\rangle \subseteq K_j$ for $j \ne i$ and $\beta \notin K$, we have $\beta \notin K_i$. Since $K \subseteq K_i$, we have $a\langle\beta\rangle \subseteq K_i$. By Lemma 5.4.20, we get $a \in \text{t-rad}(K_i)$. ∎

Lemma 5.4.25 (Lemma 3.8 of Satyanarayana, 1982)

If K_1 and K_2 are tertiary ideals of G such that $\text{t-rad}(K_1) = \text{t-rad}(K_2)$, then $K_1 \cap K_2$ is tertiary and $\text{t-rad}(K_1 \cap K_2) = \text{t-rad}(K_1) = \text{t-rad}(K_2)$.

Proof

Write $K = K_1 \cap K_2$. By Lemma 5.4.24, $\text{t-rad}(K) = (\text{t-rad}(K_1) \cap \text{t-rad}(K_2)) = \text{t-rad}(K_1) = \text{t-rad}(K_2)$. To show that K is tertiary, let $a \in N$ and $\beta \in G\backslash K$ such that $a\langle\beta\rangle \subseteq K$. Since $\beta \notin K = K_1 \cap K_2$, we have $\beta \notin K_1$ or $\beta \notin K_2$. If $\beta \notin K_1$, then $a \in \text{t-rad}(K_1) = \text{t-rad}(K)$. If $\beta \notin K_2$, then $a \in \text{t-rad}(K_2) = \text{t-rad}(K)$.

Hence, by Lemma 5.4.20, we have that K is a tertiary ideal of G. ∎

Definition 5.4.26

A decomposition $K = K_1 \cap K_2 \cap \ldots \cap K_r$ of an ideal K by tertiary ideals K_i is called **reduced** if it is irredundant and the ideals $\text{t-rad}(K_i)$s are distinct.

Theorem 5.4.27 (Theorem 3.10 of Satyanarayana, 1982)

If G has ACC on ideals, then every ideal of G has a reduced decomposition by tertiary ideals.

Proof

Let $\mathcal{R}$ be the set of all ideals of G that do not have a reduced decomposition. If $\mathcal{R} \neq \varnothing$ (since G has ACC on ideals), then $\mathcal{R}$ contains a maximal element K^1 (say). Since K^1 is maximal in $\mathcal{R}$, we have K^1 is meet-irreducible (if not K^1 is an intersection of two ideals with reduced decomposition, and hence K^1 has a decomposition by tertiary ideals. By Lemma 5.4.25, K^1 has a reduced decomposition, which is a contradiction since $K^1 \in \mathcal{R}$). Since K^1 is meet-irreducible, by Lemma 5.4.22, we have K^1 is tertiary, which is a contradiction. Hence, $\mathcal{R} = \varphi$. ∎

To prove the uniqueness theorem of tertiary decomposition, we need the following lemma.

Lemma 5.4.28 (Lemma 3.11 of Satyanarayana, 1982)

Let A, B, A^1, B^1 be ideals of G such that A, A^1 are tertiary, t-rad(A) $\neq$ t-rad(A^1), and $K = A \cap B = A^1 \cap B^1$. Then $K = B = B^1$.

Proof

Let $\alpha \in B \cap B^1$. Since t-rad(A) $\neq$ t-rad(A^1), there is an element in one but not in the other, say $a \in$ t-rad(A) such that $a \notin$ t-rad(A^1). Suppose $\alpha \notin K$.
Since $\alpha \in B$, it follows that $\alpha \notin A$. By Lemma 5.4.5, there exists $\beta \in <\alpha>$ such that $a <\beta> \subseteq A$ and $\beta \notin A$. Also, $a<\beta> \subseteq a<\alpha> \subseteq <\alpha> \subseteq B$. Therefore, $a<\beta> \subseteq K \subseteq A^1$. Since $\beta \notin A$, we have $\beta \notin K$ and hence $\beta \notin A^1$. Since A^1 is a tertiary ideal, by Lemma 5.4.20, $a \in$ t-rad(A^1). This is a contradiction. Hence, $B \cap B^1 \subseteq K$. ∎

Theorem 5.4.29 (Theorem 3.12 of Satyanarayana, 1982)

If the ideal K of G has two reduced decompositions $K = K_1 \cap K_2 \cap \ldots \cap K_r = K_1^1 \cap K_2^1 \cap \ldots \cap K_s^1$, then $r = s$, and the set of ideals t-rad(K_i) ($1 \leq i \leq r$) coincides with the set of ideals t-rad(K_j^1) ($1 \leq j \leq s$).

Proof

Suppose t-rad(K_1) $\neq$ t-rad(K_j^1) for all $1 \leq j \leq s$. Since $K = K_1 \cap K_2 \cap \ldots \cap K_r = K_1^1 \cap K_2^1 \cap \ldots \cap K_s^1$, by Lemma 5.4.28, we have

$K = K_2 \cap K_3 \cap \ldots \cap K_r \cap K_2^1 \cap K_3^1 \cap \ldots \cap K_s^1$.

Now K has two decompositions, $K = K_1 \cap K_2 \cap \ldots \cap K_r$ and $K = K_2 \cap K_3 \cap \ldots \cap K_r \cap K_2^1 \cap K_3^1 \cap \ldots \cap K_s^1$. Since t-rad($K_1$) and t-rad($K_2^1$) are distinct, we have $K = K_2 \cap K_3 \cap \ldots \cap K_r \cap K_3^1 \cap \ldots \cap K_s^1$.

If we continue up to s steps, finally we get $K = K_2 \cap K_3 \cap \ldots \cap K_r$, which is a contradiction (since $K = K_1 \cap K_2 \cap \ldots \cap K_r$ is a reduced decomposition).

Hence, t-rad(K_1) = t-rad(K_j^1) for some j $(1 \leq j \leq s)$. In the same way, given i $(1 \leq i \leq r)$, there exists j $(1 \leq j \leq s)$ such that t-rad(K_i) = t-rad(K_j^1), which implies that $r \leq s$.

A similar argument shows that $s \leq r$.

Hence, $s = r$, and the set of ideals t-rad(K_i), $1 \leq i \leq r$ coincides with the set of ideals t-rad(K_j^1), $1 \leq j \leq s$. ∎

6

Matrix Near Rings

6.1 Preliminary Results

In this chapter, we present investigations related to the relationship of two-sided ideals of a near ring N to those of a matrix near ring $M_n(N)$.

We recall the notation for a Cartesian product of n copies of a set: if L is a subset of N, then L^n denotes the Cartesian product of n copies of L (n is a positive integer).

Result 6.1.1

(i) (**Proposition 4.1 of Meldrum and Van der Walt, 1986**): If L is a left ideal of N, then L^n is an ideal of the $M_n(N)$-group N^n.

(ii) (**Corollary 4.2 of Meldrum and Van der Walt, 1986**): If L is a left ideal of N, then $(L^n : N^n) = \{A \in M_n(N) \mid A\rho \in L^n \text{ for all } \rho \in N^n\}$ is a two-sided ideal of $M_n(N)$.

Proof

(i) First, we have to show that L^n is a normal subgroup of N^n.

Let $\alpha = (a_1, \ldots, a_n), \beta = (b_1, \ldots, b_n) \in L^n$.

Now $\alpha - \beta = (a_1 - b_1, \ldots, a_n - b_n) \in L^n$ (since L is a left ideal of N).

Let $r = (r_1, \ldots, r_n) \in N^n$. Now $r + \alpha - r = (r_1 + a_1 - r_1, \ldots, r_1 + a_1 - r_1) \in L^n$.

Therefore, $r + \alpha - r \in L^n$ for all $\alpha \in L^n$ and $r \in N^n$. Thus, L^n is a normal subgroup of N^n.

Suppose $\alpha = (a_1, \ldots, a_n) \in L^n$ and $r = (r_1, \ldots, r_n) \in N^n$. Now we have to show that for any matrix $A \in M_n(N)$, there is $\alpha^1 \in L^n$ such that $A(r + \alpha) = Ar + \alpha^1$. We will show this by induction on $w(A)$, the weight of A.

Let $w(A) = 1$ and $A = f_{ij}^s$.

Now $A(r+\alpha) = f_{kl}^s(r+\alpha)$

$$= (0, \ldots, s(r_j + a_j), \ldots, 0)$$

$$= (0, \ldots, sr_j - sr_j + s(r_j + a_j) - sr_j + sr_j, \ldots, 0)$$

$$= (0, \ldots, sr_j + a_j^1, \ldots, 0) \text{ (where } a_j^1 \in L)$$

$$= (0, \ldots, sr_j, \ldots, 0) + (0, \ldots, a_j^1, \ldots, 0)$$

$$= \left(f_{ij}^s\right)(r_1, \ldots, r_n) + \alpha^1 \text{ (where } \alpha^1 \in L^n)$$

$$= Ar + \alpha^1$$

Suppose the preceding statement is true for all the matrices with weight $< m$ ($m \geq 2$).
Let $w(m) = m$ and $A = B + C$ or $A = BC$, where $w(B), w(C) < m$.
Now $A(r+\alpha) = (B+C)(r+\alpha)$

$$= B(r+\alpha) + C(r+\alpha)$$

$$= B(r+\alpha^1) + C(r+\alpha^{11}), \text{ where } \alpha^1, \alpha^{11} \in L^n$$

$$= Br + Cr - Cr + \alpha^1 + Cr + \alpha^{11} \in L^n$$

$$= (B+C)\, r + \alpha^{111}, \text{ where } \alpha^{111} = -Cr + 1 + Cr + \alpha^{11} \in L^n$$

$$= Ar + \alpha^{111}.$$

Let $A = BC$.
Then $A(r+\alpha) = (BC)(r+\alpha)$

$$= B[C(r+\alpha)]$$

$$= B[Cr + \alpha^{11}] \text{ (where } \alpha^{11} \in L^n)$$

$$= B(Cr) + \alpha^{111} \text{ (where } \alpha^{111} \in L^n)$$

$$= (BC)r + \alpha^{111} \text{ (where } \alpha^{111} \in L^n)$$

$$= Ar + \alpha^{111} \text{ (where } \alpha^{111} \in L^n).$$

Let $A \in M_n(N)$, $r \in N^n$, and $\alpha \in L^n$.
Now $A(r + \alpha) - Ar = Ar + \alpha^1 - Ar \in L^n$ (since $\alpha^1 \in L^n$ and $Ar \in N^n$).
Therefore, L^n is an ideal of the $M_n(N)$-group N^n. The proof is complete. ■

(ii) Let $A, B \in (L^n : N^n)$ and $\rho \in N^n$. Now

$$(A - B)\rho = (A + (-B))\rho = A\rho + (-B)\rho = A\rho - B\rho \in L^n.$$

This means that $(A - B) \in (L^n : N^n)$. Therefore $(L^n : N^n)$ is a subgroup of $M_n(N)$.
Take $A \in (L^n : N^n)$ and $C \in M_n(N)$. Then $(C + A - C)\rho = C\rho + A\rho - C\rho \in L^n$ (since $A\rho \in L^n$ and L^n is a normal subgroup of N^n). Therefore $(L^n : N^n)$ is a normal subgroup of $M_n(N)$. For $A \in (L^n : N^n)$ and $C \in M_n(N)$, we have $(AC)\rho = A(C\rho) \in L^n$. Therefore, L^n is a right ideal of N^n. Finally, for $A \in (L^n : N^n)$, $C, C^1 \in M_n(N)$, we have

$$(C(C^1 + A) - CC^1)\rho = (C(C^1 + A))\rho - CC^1\rho = C(C^1\rho + A\rho) - CC^1\rho \in L^n.$$

Hence $(L^n : N^n)$ is a two-sided ideal of $M_n(N)$.

Proposition 6.1.2 (Proposition 4.3 of Meldrum and Van der Walt, 1986)

Suppose I_1 and I_2 are two-sided ideals of N. Then $I_1 \neq I_2$ if and only if $(I_1^n : N^n) \neq (I_2^n : N^n)$.

Proof

Suppose $I_1 \neq I_2$. Take $a \in I_2 \backslash I_1$. Since I_2 is a right ideal, we have $f_{11}^a \in (I_2^n : N^n)$. Further, since $f_{11}^a (1, 0, \ldots, 0) \notin I_1^n$, we have that $f_{11}^a \notin (I_1^n : N^n)$. Thus, we proved that $(I_1^n : N^n) \neq (I_2^n : N^n)$.

Converse

Suppose that $(I_1^n : N^n) \neq (I_2^n : N^n)$. Take $A \in (I_1^n : N^n) \neq (I_2^n : N^n)$. Let $\rho \in N^n$ such that $A\rho \notin I_2^n$. Suppose that $A\rho = (a_1, a_2, \ldots, a_n)$ with $a_k \notin I_2$. Then

$$E_{1k} A (f_{11}^{a_1} + \ldots + f_{n1}^{a_n})(1, 0, \ldots, 0) = (a_k, 0, \ldots, 0).$$

By Corollary 2.4.22, it follows that the matrix on the left is equal to $f_{11}^{a_k}$. Now, since E_{1k} is zero-symmetric, we have $f_{11}^{a_k} \in (I_1^n : N^n)$. But $f_{11}^{a_k} \notin (I_2^n : N^n)$. Therefore, $a_k \in I_1 \backslash I_2$. The proof is complete. ∎

Definition 6.1.3

(i) Let I be an ideal of $M_n(N)$; then $I_* = \{x \in N \mid x \in \text{im}(\pi_j A)$ for some $A \in I$ and $j, 1 \leq j \leq n\}$.

(ii) If I is an ideal of N, then the related ideal I^* in $M_n(N)$ is defined as

$$I^* = \{A \in M_n(N) \mid A\rho \in I^n \text{ for all } \rho \in N^n\}.$$

Result 6.1.4 (Lemma 4.4 of Meldrum and Van der Walt, 1986)

If I is a two-sided ideal of $M_n(N)$, then $a \in I_*$, if and only if $f_{11}^a \in I$.

Proof

Suppose $a \in I_*$. This implies that there exist $A \in I$ and $\rho \in N^n$ and $j \in \{1, 2, \ldots, n\}$ such that $\pi_j(A\rho) = a$.

Since $A \in I$ and I is a right ideal, there exists $B \in I$ such that $A\rho = Be_1$.

Now, $Be_1 = B\left(f_{11}^1 + f_{21}^0 + \ldots + f_{n1}^0\right)e_1$

$$= \left(f_{11}^{a_1} + f_{21}^{a_2} + \ldots + f_{n1}^{a_n}\right)e_1 \qquad \text{(by Theorem 2.4.21)}$$

Therefore, $a = \pi_j(A\rho) = \pi_j(Be_1)$

$$= \pi_j\left(f_{11}^{a_1} + \ldots + f_{n1}^{a_n}\right)e_1$$

$$= \pi_j(a_1, \ldots, a_n)$$

$$= a_j.$$

Now $E_{1j}B\left(f_{11}^1 + f_{21}^0 + \ldots + f_{n1}^0\right)(\alpha_1, \ldots, \alpha_n) = E_{1j}\left(f_{11}^{a_1} + f_{21}^{a_2} + \ldots + f_{n1}^{a_n}\right)(\alpha_1, \ldots, \alpha_n)$

$$= E_{1j}(a_1\alpha_1, a_2\alpha_1, \ldots, a_n\alpha_1)$$

$$= (a_j\alpha_1, 0, \ldots, 0)$$

$$= (a\alpha_1, 0, \ldots, 0).$$

Since $f_{11}^1(\alpha_1, \ldots, \alpha_n) = (a\alpha_1, 0, \ldots, 0)$, we have

$$E_{1j}B\left(f_{11}^1 + f_{21}^0 + \ldots + f_{n1}^0\right)(\alpha_1, \ldots, \alpha_n) = f_{11}^0(\alpha_1, \ldots, \alpha_n).$$

Therefore, $E_{1j}B\left(f_{11}^1 + f_{21}^0 + \ldots + f_{n1}^0\right) = f_{11}^a \in I$.

Converse

Suppose that $f_{11}^a \in I$. Take $A = f_{11}^a$.

$$\pi_1 A(1, 0, \ldots, 0) = \pi_1(a, 0, \ldots, 0) = a.$$

Therefore, $a \in I_*$.   ∎

Result 6.1.5 (Corollary 4.5 of Meldrum and Van der Walt, 1986)

If I is a two-sided ideal of $M_n(N)$, then $a \in I_*$ if and only if $f_{ij}^a \in I$ for all $1 \le i \le n$, $1 \le j \le n$.

Proof

Suppose $a \in I_*$. Now, by Result 6.1.4, we have $f_{11}^a \in I$.

$\Rightarrow f_{11}^a f_{1j}^1 \in I$ (since I is an ideal of $M_n(N)$ and E_{1j} is zero symmetric)

$\Rightarrow f_{1j}^a \in I$ $\quad \Rightarrow f_{i1}^1 f_{1j}^a \in I$ $\quad \Rightarrow f_{ij}^a \in I.$

Converse

Suppose that $f_{ij}^a \in I$ for all $1 \le i, j \le n$. This implies that $f_{11}^a \in I$, and so $a \in I_*$. ∎

Theorem 6.1.6 (Proposition 4.6 of Meldrum and Van der Walt, 1986)

If I is a two-sided ideal in $M_n(N)$, then I_* is a two-sided ideal in N.

Proof

Suppose I is a two-sided ideal in $M_n(N)$. We will show that I_* is a two-sided ideal in N.

Let $a, b \in I_*$. This implies $f_{11}^a, f_{11}^b \in I$.

Now $f_{11}^{a-b} = f_{11}^a - f_{11}^b \in I$. This implies $a - b \in I_*$.

Let $a \in I_*$ and $r \in N$. Now $f_{11}^{r+a-r} = f_{11}^r + f_{11}^a - f_{11}^r \in I$. This implies $r + a - r \in I_*$.

Also $f_{11}^{ar} = f_{11}^a \, f_{11}^r \in I$. This implies $ar \in I_*$. For any $r, r^1 \in N$, and $a \in I_*$,

$$f_{11}^{r(r^1+a)-rr^1} = f_{11}^{r(r^1+a)} + f_{11}^{rr^1} = f_{11}^r \left[f_{11}^{r^1} + f_{11}^a \right] - f_{11}^{rr^1} \in I.$$

Therefore, $r(r^1 + a) - rr^1 \in I_*$.

Hence, I_* is a two-sided ideal of N.

Now we state certain results from Meldrum and Van der Walt (1985) without proofs. ∎

Proposition 6.1.7 (Proposition 4.7 of Meldrum and Van der Walt, 1986)

For two-sided ideals I^* in $M_n(N)$ and I in N, we have

(i) $(I_*)^* \supseteq I$ and (ii) $(I^*)_* = I$

Proposition 6.1.8 (Proposition 4.8 of Meldrum and Van der Walt, 1986)

There is a bijection between the set of ideals of N and the set of full ideals of $M_n(N)$ given by $I \to I^*$ and $I \to I_*$ such that $(I^*)_* = I$ and $(I_*)^* = I$ for an ideal I of N and a full ideal I^* of $M_n(N)$. Moreover, 0 and $M_n(N)$ are full ideals of $M_n(N)$.

Definitions 6.1.9

(i) An element $A \in M_n(N)$ is said to be **nilpotent** if there exists a positive integer k such that $A^k = 0$.

(ii) The matrix near ring $M_n(N)$ is said to be **reduced** if $M_n(N)$ has no nonzero nilpotent elements.

Theorem 6.1.10 (Theorem 1.3 of Satyanarayana, Syam Prasad, and Pradeep Kumar, 2004)

Consider $M_n(N)$ as an $M_n(N)$-group.

(i) If $N^2 = 0$, then $M_n(N) = 0$.

(ii) If $M_n(N)$ has insertion of factors property (IFP), then $N^4 = 0$, and hence, $N^k = 0$ for all $k \geq 4$.

(iii) If $N^k \neq 0$ for some $k \geq 4$, then $M_n(N)$ has no IFP.

(iv) If N is a near ring with identity 1 and $N^2 \neq 0$, then $M_n(N)$ has no IFP.

Proof

(i) Suppose N is a near ring such that $N^2 = 0$.

Let $a \in N$ and consider f_{ij}^a. For any $(x_1, x_2, \ldots, x_n) \in N^n$, we have

$$f_{ij}^a(x_1, x_2, \ldots, x_n) = (0, \ldots, ax_j, 0, \ldots, 0) = (0, 0, \ldots, 0) = 0 \text{ (since } ax_j \in N^2 = 0\text{)}.$$

Now $\{ f_{ij}^a \mid a \in N \} = \{0\}$. Therefore, $M_n(N) = (0)$.

(ii) Let us suppose that $N^4 \neq 0$.

Then there exist $a, b, c, d \in N$ such that $abcd \neq 0$. Now

$$f_{12}^\alpha f_{34}^c(x_1, x_2, x_3, x_4, x_5, \ldots, x_n) = f_{12}^a(0, 0, cx_4, 0, \ldots, 0)$$
$$= (a0, 0, \ldots, 0)$$
$$= (0, 0, \ldots, 0) \text{ (since } N \text{ is zero-symmetric)}.$$

Therefore, $f_{12}^a f_{34}^c = O$ (here $O = (0, 0, \ldots, 0)$).

Since $M_n(N)$ has IFP, we get that $f_{12}^a A f_{34}^c = O$ for all $A \in M_n(N)$.

Take $A = f_{23}^b$ Then $f_{12}^a f_{23}^b f_{34}^c = O$.

Now $(0, 0, \ldots, 0) = f_{12}^a f_{23}^b f_{34}^c(x_1, x_2, x_3, d, 0, \ldots, 0)$
$$= f_{12}^a f_{23}^b(0, 0, cd, 0, \ldots)$$
$$= f_{12}^a(0, bcd, 0, 0, \ldots, 0)$$
$$= (abcd, 0, \ldots, 0).$$

Therefore, $abcd = 0$, which is a contradiction.

Thus, we have proved that if $M_n(N)$ has IFP, then $N^4 = 0$.

(iii) Follows from (ii).

(iv) Since $N^2 \neq 0$, there exist $a, b \in N$ such that $ab \neq 0$.

Let us suppose $M_n(N)$ has IFP. Now

$$f_{12}^a f_{34}^1(x_1, x_2, x_3, x_4, x_5, \ldots, x_n) = f_{12}^a(0, 0, x_4, 0, \ldots, 0) = (0, 0, \ldots, 0).$$

Since $M_n(N)$ has IFP, we have that $f_{12}^a A f_{34}^1 = O$ for any $A \in M_n(N)$.

Therefore, $f_{12}^a f_{23}^b f_{34}^1 = O$.

Now $(0, \ldots, 0) = f_{12}^a f_{23}^b f_{34}^1 (1, 1, \ldots, 1)$
$$= f_{12}^a f_{23}^b (0, 0, 1, 0, \ldots, 0)$$
$$= f_{12}^a (0, b, \ldots, 0)$$
$$= (ab, 0, \ldots, 0).$$

Therefore, $ab = 0$, which is a contradiction. Hence, $M_n(N)$ has no IFP. ∎

Proposition 6.1.11 (Proposition 1.1 of Syam Prasad and Satyanarayana, 1999)

If $M_n(N)$ is reduced, then N has IFP.

Proof

Suppose $M_n(N)$ is reduced.

Let $a, b, n \in N$ such that $ab = 0$. Since N is zero-symmetric, we have that $f_{11}^{ab} = O$.

Consider $\left(f_{11}^{ba}\right)^2$.

Now $\left(f_{11}^{ba}\right)^2 = f_{11}^{ba} f_{11}^{ba} = f_{11}^{baba}$ (by Lemma 2.4.14)

$\qquad\qquad = f_{11}^{boa}$ (since $ab = 0$)

$\qquad\qquad = f_{11}^{bo} = O$ (since N is zero-symmetric)

Therefore $\left(f_{11}^{ba}\right)^2 = O$. Since $M_n(N)$ is reduced, we have that $f_{11}^{ba} = O$.

Now, $f_{11}^{ba} = O \Rightarrow \left(f_{11}^{ba}\right)(1, 1, \ldots, 1) = (0, \ldots, 0)$

$\qquad\qquad\qquad \Rightarrow (ba, \ldots, 0) = (0, 0, \ldots, 0)$

$\qquad\qquad\qquad \Rightarrow ba = 0.$

Consider $\left(f_{11}^{anb}\right)^2$.

Now $\left(f_{11}^{anb}\right)^2 = f_{11}^{anb} f_{11}^{anb} = f_{11}^{anbanb}$ (by Lemma 2.4.14)

$\qquad\qquad = f_{11}^{anonb}$ (since $ba = 0$)

$\qquad\qquad = f_{11}^{ano} = O$ (since N is zero-symmetric).

Therefore $\left(f_{11}^{anb}\right)^2 = O$. Since $M_n(N)$ is reduced, we have that $f_{11}^{anb} = O$.

So $\left(f_{11}^{anb}\right)(1, 1, \ldots, 1) = O$. This means that $(anb, 0, \ldots, 0) = (0, 0, \ldots, 0)$.

Therefore, $anb = 0$. Hence, N has IFP. ∎

Proposition 6.1.12 (Proposition 1.2 of Syam Prasad and Satyanarayana, 1999)

Let I be an ideal of $M_n(N)$. If I has IFP, then I_* has IFP in N.

Proof

Suppose I has IFP. Let $a, b, c \in N$ such that $ab \in I_*$.

It is enough to show that $acb \in I_*$.

Since $ab \in I_*$, by Result 6.1.5, $f_{ij}^{ab} \in I$. So, by Lemma 2.4.14, we have $f_{i1}^a f_{1j}^b = f_{ij}^{ab} \in I$. Since I has IFP, we have that $f_{i1}^a A f_{1j}^b \in I$ for all $A \in M_n(N)$. Put $A = f_{11}^c$.

Then $f_{ij}^{acb} = f_{i1}^a f_{11}^c f_{1j}^b \in I$. Therefore, $acb \in I_*$. ∎

Proposition 6.1.13 (Proposition 4.9 of Meldrum and Van der Walt, 1986)

N is simple if and only if $M_n(N)$ is simple.

Proof

Suppose N is simple and let I be any nonzero ideal of $M_n(N)$. Since I_* is a non-zero ideal and N is simple, it follows that $I_* = N$. Now by Result 6.1.4, we have that $E_{ij} \in I$ for all $1 \leq i, j \leq n$. This means that $I = M_n(N)$. Hence, $M_n(N)$ is simple.

The other part follows from Proposition 6.1.2.   ■

6.2 Prime Ideals in Matrix Near Rings

We begin this section with the following result.

Proposition 6.2.1 (Proposition 4.11 of Meldrum and Van der Walt, 1986)

If an ideal I of N is a prime (semiprime, respectively) ideal, then I^* is a prime ideal (semiprime, respectively) in $M_n(N)$.

Proof

Let I be a prime ideal in N. To show I^* is prime in $M_n(N)$, suppose $\mathring{A}$ and $\mathcal{B}$ are two ideals of $M_n(N)$ such that $\mathring{A}\mathcal{B} \subseteq I^*$. We have to show that $\mathring{A} \subseteq I^*$ or $\mathcal{B} \subseteq I^*$. Let us suppose that $\mathring{A} \nsubseteq I^*$ and $\mathcal{B} \nsubseteq I^*$. Then there exist $A \in \mathring{A}$ and $B \in \mathcal{B}$ such that $A, B \notin I^*$. This means there exist $\rho, \sigma \in N^n$ such that $A\rho \notin I^n$ and $B\sigma \in I^n$. Now write $A\rho = (a_1, a_2, \ldots, a_n)$. Since $A \in <A>$ and $<A>$ is a right ideal of $M_n(N)$, we have $C \in <A>$ such that $A\rho = Ce_1$. This implies that

$$E_{1k}C\left[f_{11}^1 + f_{21}^0 + \ldots + f_{n1}^0 \right](\alpha_1, \ldots, \alpha_n) = E_{1k}\left[f_{11}^{a_1} + f_{21}^{a_2} + \ldots + f_{n1}^{a_n} \right](\alpha_1, \ldots, \alpha_n)$$
$$= f_{1k}^1 (a_1\alpha_1, \ldots, a_n\alpha_n) = (a_k\alpha_1, 0, \ldots, 0)$$
$$= f_{11}^{a_k}(\alpha_1, \ldots, \alpha_n).$$

This means $f_{11}^{a_k} \in <A>$.

So, $f_{11}^{a_k} \in <A> \backslash I^*$. Therefore, there exists $a = a_k \notin I$ such that $f_{11}^a \in <A> \backslash I^*$. Similarly, since $B\sigma \notin I^n$, we have that there exists $b \notin I$ such that $f_{11}^b \in <B> \backslash I^*$. Since $a \notin I, b \notin I$, we have $<a> \nsubseteq I$ and $<b> \nsubseteq I$. Again, since $<a><b> \nsubseteq I$, there exist $c \in <a>$ and $d \in <b>$ such that $cd \notin I$.

This implies

$$f_{11}^{cb} \in \Gamma. \tag{6.1}$$

Now $c \in <a>$. Write $X = \{a\}$, using $<a> = \bigcup_{i=0}^{\infty} X_i$. To prove $f_{11}^c \in <f_{11}^a>$, we use induction on m, where $c \in X_m$.

Suppose $c \in X_m$ and $m = 0$. Then $c \in X_0 = X = \{a\}$. In this case, $f_{11}^c = f_{11}^a \in <f_{11}^a>$. Suppose $m = 1$. Then $c \in X_1 = X_0^* \cup X_0^0 \cup X_0^+ \cup X_0^1$.

If $c \in X_0^*$, then $c = n + a - n$. Then $f_{11}^c = f_{11}^{n+a-n} = f_{11}^n + f_{11}^a - f_{11}^n \in <f_{11}^a>$. If $c \in X_0^0$, then $c = a - a = 0$. Now $f_{11}^c = f_{11}^0 = 0 \in <f_{11}^a>$. If $c \in X_0^+$, then $c = n(n^1 + a) - nn^1$. Now $f_{11}^c = f_{11}^{n(n^1+a) - nn^1} = f_{11}^n(f_{11}^{n^1} + f_{11}^a) - f_{11}^n f_{11}^{n^1} \in <f_{11}^a>$. If $c \in X_0^1$, then $c = an$ for some $n \in N$. Now $f_{11}^c = f_{11}^a f_{11}^n \in <f_{11}^a>$. Therefore, $c \in <a>$. Thus, $f_{11}^c \in <f_{11}^a>$ for $m = 1$.

Induction Hypothesis: Suppose $f_{11}^c \in <f_{11}^a>$ for all $c \in X_{k-1}$.

Suppose $c \in X_k = X_{k-1}^* \cup X_{k-1}^0 \cup X_{k-1}^+ \cup X_{k-1}^1$. If $c \in X_{k-1}^*$, then $c = n + x - n$ for some $x \in X_{k-1}$. Now $f_{11}^c = f_{11}^{n+x-n} = f_{11}^n + f_{11}^x - f_{11}^n \in <f_{11}^a>$. If $c \in X_{k-1}^0$, then $c = x - y$ for some $x, y \in X_{k-1}$. Now $f_{11}^c = f_{11}^{x-y} = f_{11}^x - f_{11}^y \in <f_{11}^a>$.

In a similar way, we can prove that if $c \in X_{k-1}^+$ or $c \in X_{k-1}^1$, then $f_{11}^c \in <f_{11}^a>$.

Therefore, $c \in <a>$ implies $f_{11}^c \in <f_{11}^a> \subseteq <A>$. Similarly, $d \in <b>$ implies $f_{11}^d \in <f_{11}^b> \subseteq <B>$. Thus, $f_{11}^{cd} = f_{11}^c f_{11}^d \in <A><B> \subseteq \mathring{A}\mathcal{B} \subseteq \Gamma$.

Therefore,

$$f_{11}^{cd} \in \Gamma. \tag{6.2}$$

Thus, from Equations 6.1 and 6.2, we have a contradiction.

So, either $\mathring{A} \subseteq \Gamma$ or $\mathcal{B} \subseteq \Gamma$. Hence, Γ is a prime ideal in $M_n(N)$. The proof is complete. ∎

The other part of the proof is similar.

A straightforward verification provides the following note.

Note 6.2.2

$(I_1)^* \subsetneq (I_2)^*$ for any two ideals I_1 and I_2 of N with $I_1 \subsetneq I_2$.

Definition 6.2.3 (Booth and Groenewald, 1991)

(i) A near ring N is said to be **equiprime** if for all $0 \neq a \in N$ and $x, y \in N$, $arx = ary$ for all $r \in N$ implies $x = y$.

(ii) An ideal P of N is called an **equiprime** ideal of N if N/P is an equiprime near ring.

(iii) A near ring N is called **strongly equiprime** if for all $0 \neq a \in N$, there exists a finite subset F of N such that $x, y \in N$, $afx = afy$ for all $y \in F$ implies $x = y$.

Definition 6.2.4

An ideal P of N is said to be a **3-prime ideal** if $a, b \in N$, $aNb \subseteq P$ implies $a \in P$ or $b \in P$. If (0) is a 3-prime ideal in N, then N is called a **3-prime near ring**.

Definition 6.2.5

An element $A \in M_n(N)$ is called a **diagonal matrix** if $A = f_{11}^{r_1} + f_{22}^{r_2} + \ldots + f_{nn}^{r_n}$, where $r_i \in N$ for $1 \leq i \leq n$.

Now we present the following notions:

(i) "A is jth row equivalent to B" for two elements $A, B \in M_n(N)$
(ii) jth row strictly equiprime near ring
(iii) jth column scalar r-matrix

We prove that (i) for any $A, B \in M_n(N)$, $A = B$ if and only if A is jth row equivalent to B for all $1 \leq j \leq n$; and (ii) every equiprime ideal is a 3-prime ideal.

Definition 6.2.6

(i) Let $A, B \in M_n(N)$. Then we say that A and B are **jth row equivalent matrices** (or A is jth row equivalent to B) if $(f_{1j}^1 + \ldots + f_{nj}^1)A = (f_{1j}^1 + \ldots + f_{nj}^1)B$.
(ii) Take j such that $1 \leq j \leq n$. Then $M_n(N)$ is said to be **jth row strictly equiprime**, if for any $0 \neq A \in M_n(N)$ there exist $\varphi \in M_n(N)$ such that $A \varphi U = A \varphi V$, implying U and V are jth row equivalent.
(iii) N is said to be **strictly equiprime**, if for any $0 \neq a \in N$ there corresponds an element $x_a \in N$ such that $u, v \in N$ and $ax_a u = a x_a v \Rightarrow u = v$.

The following note is straightforward.

Note 6.2.7

(i) jth row equivalence is an equivalence relation.
(ii) Every strictly equiprime near ring is strongly equiprime.

Definition 6.2.8

Let $1 \leq j \leq n$ and $A \in M_n(N)$. Then

(i) A is said to be a *jth column matrix* if there exists $r_i \in N, 1 \le i \le n$ such that $A = (f_{1j}^n + \dots + f_{nj}^{r_n})$.

(ii) A is said to be a *jth column scalar matrix* if there exists $r \in N$ such that $A = (f_{1j}^{r_1} + \dots + f_{nj}^{r_n})$. This matrix is also called a *jth column scalar r-matrix*.

Note 6.2.9

Let $U, V \in M_n(N)$. Then $(f_{1j}^1 + \dots + f_{nj}^1)U = (f_{1j}^1 + \dots + f_{nj}^1)V$

$\Leftrightarrow [(f_{1j}^1 + \dots + f_{nj}^1)U](s_1, \dots, s_n) = [(f_{1j}^1 + \dots + f_{nj}^1)V](s_1, \dots, s_n)$

for all $(s_1, \dots, s_n) \in N^n$.

$\Leftrightarrow (f_{1j}^1 + \dots + f_{nj}^1)U(s_1, \dots, s_n) = (f_{1j}^1 + \dots + f_{nj}^1)V(s_1, \dots, s_n)$

for all $(s_1, \dots, s_n) \in N^n$.

$\Leftrightarrow [U(s_1, \dots, s_n)]_j, \dots, [U(s_1, \dots, s_n)]_j = [V(s_1, \dots, s_n)]_j, \dots, [V(s_1, \dots, s_n)]_j$

for all $(s_1, \dots, s_n) \in N^n$.

$\Leftrightarrow [U(s_1, \dots, s_n)]_j = [V(s_1, \dots, s_n)]_j$ for all $(s_1, \dots, s_n) \in N^n$.

A straightforward verification results in the following Remark 6.2.10 and Proposition 6.2.11. An immediate consequence of Proposition 6.2.11 is Corollary 6.2.12.

Remark 6.2.10 (Remark 2.5 of Satyanarayana, Rao, and Syam Prasad, 1996)

$A = B$ if and only if A and B are *jth row equivalent* for all j with $1 \le j \le n$.

Proof

Suppose that A and B are *jth row equivalent* for all j ($1 \le j \le n$). Let $(r_1, \dots, r_n) \in N^n$. We have to show that $A(r_1, \dots, r_n) = B(r_1, \dots, r_n)$. Let us suppose that $A(r_1, \dots, r_n) \ne B(r_1, \dots, r_n)$. Then $[A(r_1, \dots, r_n)]_j \ne [B(r_1, \dots, r_n)]_j$ for some j.

This implies that $(f_{1j}^1 + \dots + f_{nj}^1)A \ne (f_{1j}^1 + \dots + f_{nj}^1)B$ (by Note 6.2.9). This implies that A and B are not *jth row equivalent*, which is a contradiction. Hence $A = B$.

The converse part is clear. ∎

Proposition 6.2.11 (Proposition 2.6 of Satyanarayana, Rao, and Syam Prasad, 1996)

In a zero-symmetric near ring N, every equiprime ideal is a 3-prime ideal.

Proof

Suppose P is an equiprime ideal of N. To show P is a 3-prime ideal, take $a, b \in N$
such that $aNb \subseteq P$. We have to show that either $a \in P$ or $b \in P$. Suppose $a \notin P$.
  Then $a + P \neq 0 + P$ and $(a + P)N(b + P) = 0 + P$.
  This implies $a + P \neq 0 + P$ and $(a + P)(n + P)(b + P) = (0 + P)$ for all $n \in N$.
  Since N is zero-symmetric, for all $n \in N$, we have that $a + P \neq 0 + P$ and
$(a + P)(n + P)(b + P) = (a + P)(n + P)(0 + P)$ for all $n \in N$.
  This implies $b + P = 0 + P \Rightarrow b \in P$.
  Hence, P is a 3-prime ideal.   ∎

Corollary 6.2.12 (Corollary 2.7 of Satyanarayana, Rao, and Syam Prasad, 1996)

Let N be a zero-symmetric near ring. If N is equiprime, then N is 3-prime.

Proof

Suppose N is an equiprime near ring. Then by definition, <0> is an equiprime ideal. By Proposition 6.2.11, <0> is a 3-prime ideal. This implies that N
is a 3-prime near ring.
  Now we consider near rings N with 1.   ∎

Proposition 6.2.13 (Proposition 3.1 of Satyanarayana, Rao, and Syam Prasad, 1996)

Let N be a strictly equiprime near ring with 1. Then $M_n(N)$ is jth row strictly
equiprime for $1 \leq j \leq n$. More precisely, if $A \in M_n(N)$ such that

(i) $A \neq 0$ with $[A(r_1, r_2, ..., r_n)]_j \neq 0$, then there exists a diagonal matrix B,
and

(ii) for each j ($1 \leq j \leq n$), there is a jth column scalar r_a-matrix C_j such that
$U, V \in M_n(N)$ and $A(BC_j)U = A(BC_j)V \Rightarrow U$ is the jth row equivalent
to V, for $1 \leq j \leq n$.

Proof

Given that $A \in M_n(N)$ such that (i) $A \neq 0$ with $[A(r_1, r_2, ..., r_n)]_l \neq 0$. Since N is
strictly equiprime, there exists $x_a \in N$ such that $ax_a u = ax_a v \Rightarrow u = v$. Consider
the diagonal matrix $B = f_{11}^{r_1} + f_{22}^{r_2} + ... + f_{nn}^{r_n}$ and jth column scalar r-matrices
$C_j = f_{1j}^{r_1} + ... + f_{nj}^{r_n}$ for $1 \leq j \leq n$, where $r = x_a$. Write $D = BC_j \in M_n(N)$.

(ii) Now suppose that $ADU = ADV$ (that is, $ABC_jU = ABC_jV$). Now we have to show that U is jth row equivalent to V.

Let us suppose that U is not jth row equivalent to V. This means
$(f_{1j}^1 + \ldots + f_{nj}^1)U \neq (f_{1j}^1 + \ldots + f_{nj}^1)V$

$\Rightarrow [(f_{1j}^1 + \ldots + f_{nj}^1)U](s_1, \ldots, s_n) \neq [(f_{1j}^1 + \ldots + f_{nj}^1)V](s_1, \ldots, s_n)$

$$\text{for some } (s_1, \ldots, s_n) \in N^n$$

$\Rightarrow [U(s_1, \ldots, s_n)]_j \neq [V(s_1, \ldots, s_n)]_j$ $\qquad$ (by Note 6.2.9)

$\Rightarrow a\, x_a[U(s_1, \ldots, s_n)]_j \neq a\, x_a[V(s_1, \ldots, s_n)]_j$

$\Rightarrow [A(r_1, r_2, \ldots, r_n)]_i\, x_a[U(s_1, \ldots, s_n)]_j \neq [A(r_1, r_2, \ldots, r_n)]_i\, x_a[V(s_1, \ldots, s_n)]_j$

$\Rightarrow [A(r_1, r_2, \ldots, r_n)]\, [x_a(U(s_1, \ldots, s_n)_j] \neq [A(r_1, r_2, \ldots, r_n)]\, [x_a(V(s_1, \ldots, s_n)_j]$

$\Rightarrow A[(f_{11}^{r_1} + f_{22}^{r_2} + \ldots + f_{nn}^{r_n})\, (f_{1j}^{x_a} + f_{2j}^{x_a} + \ldots + f_{nj}^{x_a})\, U(s_1, \ldots, s_n)]$

$\qquad \neq A[(f_{11}^{r_1} + f_{22}^{r_2} + \ldots + f_{nn}^{r_n})\, (f_{1j}^{x_a} + f_{2j}^{x_a} + \ldots + f_{nj}^{x_a})\, V(s_1, \ldots, s_n)]$

$\Rightarrow A[(f_{11}^{r_1} + f_{22}^{r_2} + \ldots + f_{nn}^{r_n})\, (f_{1j}^{r} + f_{2j}^{r} + \ldots + f_{nj}^{r})\, U(s_1, \ldots, s_n)]$

$\qquad \neq A[(f_{11}^{r_1} + f_{22}^{r_2} + \ldots + f_{nn}^{r_n})\, (f_{1j}^{r} + f_{2j}^{r} + \ldots + f_{nj}^{r})\, V(s_1, \ldots, s_n)]$ $\quad$ (since $r = x_a$)

$\Rightarrow ABC_jU \neq ABC_jV$, a contradiction to our supposition.

Hence, U and V are jth row equivalent for $1 \leq j \leq n$. $\qquad\blacksquare$

Corollary 6.2.14 (Satyanarayana, Rao, and Syam Prasad, 1996)

Suppose N is a near ring with 1. If N is a strictly equiprime near ring, then $M_n(N)$ is strongly equiprime.

Proof

Suppose N is a strictly equiprime near ring. Let $0 \neq A \in M_n(N)$. Since $M_n(N)$ is a strictly equiprime near ring for each $1 \leq j \leq n$, there exists $D_j \in M_n(N)$ such that $AD_jU = AD_jV \Rightarrow U$ and V are jth row equivalent.

Now, write $F = \{D_j / 1 \leq j \leq n\}$. Suppose $AfU = AfV$ for all $f \in F \Rightarrow AD_jU = AD_jV$ for all j ($1 \leq j \leq n$) $\Rightarrow U$ and V are jth row equivalent for all j ($1 \leq j \leq n$) $\Rightarrow U = V$ (by Remark 6.2.10). $\qquad\blacksquare$

Note 6.2.15

If $\mathcal{R} \subseteq M_n(N)$ is right invariant, then $\mathcal{R}N^n = \mathcal{R}e_1$ (a subset I of N is called right invariant in N if $IN \subseteq I$).

Verification

Suppose $\mathcal{R}$ is right invariant in $M_n(N)$.

Then

$$\mathcal{R}M_n(N) \subseteq \mathcal{R} \qquad (6.3)$$

Let $A\alpha \in \mathcal{R}N^n$, where $A \in \mathcal{R}$ and $\alpha \in N^n$. Since $\alpha \in N^n$, there exists $B \in M_n(N)$ such that $\alpha = Be_1$. So $A\alpha = A(Be_1) = (AB)e_1 \in \mathcal{R}e_1$ (by Condition 6.3).

Therefore, $A\alpha \in \mathcal{R}e_1$. This implies that $\mathcal{R}N^n \subseteq \mathcal{R}e_1$. Also, clearly, $\mathcal{R}e_1 \subseteq \mathcal{R}N^n$. Hence, $\mathcal{R}N^n = \mathcal{R}e_1$.

Theorem 6.2.16 (Theorem 3.3 of Satyanarayana, Rao, and Syam Prasad, 1996)

If I is a prime left ideal of N, then I^* is a prime left ideal in $M_n(N)$.

Proof

Suppose I is a prime left ideal in N. Since I is a left ideal in N, by Result 6.1.1 (ii), we have that I^* is an ideal in $M_n(N)$.

Now we have to show that I^* is a prime left ideal. Let us suppose that I^* is not a prime left ideal. Then there exist two left ideals L_1, L_2 in $M_n(N)$ such that $L_1, L_2 \subseteq I^*$, $L_1 \not\subset I^*$ $L_2 \not\subset I^*$. Then there exist $A \in L_1 \setminus I^*$, $B \in L_1 \setminus I^*$ which imply that there exist $p, q \in Nn$ such that $Ap \notin I^n$ and $Bq \notin I^n$. Write $Ap = (a_1, a_2, ..., a_n)$ and $Bq = (c_1, c_2, ..., c_n)$.

Since $Ap \notin I^n$ and $Bq \notin I^n$, there exist k, s such that $a_k \notin I$ and $C_s \notin I$. $AL_2 \subseteq L_1L_2 \subseteq I^*$, and I^* is an ideal. By Theorem 4.1.19, we get $<A>L_2 \subseteq I^*$

$$\Rightarrow <A><B>_1 \subseteq <A> L_2 \subseteq I^* \text{ (since } B \subseteq L_2)$$

$$\Rightarrow <A><B>_1 \subseteq I^* \quad \Rightarrow <A><B>_1 M_n(N) \subseteq I^* \text{ (since } I^* \text{ is an ideal)}. \quad (6.4)$$

Since $<A>$ is right invariant, by Note 6.2.15, $<A>N^n = <A> e_1$, where $e_1 = (1, 0, ..., 0)$.

Now $Ap \in <A>N^n = <A> e_1 \Rightarrow$ there exists $C \in <A>$ such that $Ap = Ce_1$. Since $Ap = (a_1, a_2, ..., a_n)$, we have

$$(a_1, a_2, ..., a_n) = Ap = Ce_1$$
$$= C[(f_{11}^1 + f_{22}^0 + ... + f_{nn}^0)]e_1$$
$$= [(f_{11}^{b_1}1 + f_{22}^{b_2} + ... + f_{nn}^{b_n})]e_1 \text{ (by Theorem 2.4.21)}$$
$$= (b_1, b_2, ..., b_n) \Rightarrow a_i = b_i \text{ for all } 1 \le i \le n.$$
$$C[(f_{11}^1 + f_{22}^0 + ... + f_{nn}^0)]e_1 = [(f_{11}^{b_1}1 + f_{22}^{b_2} + ... + f_{nn}^{b_n})] = [f_{11}^{a_1}1 + f_{22}^{a_2} + ... + f_{nn}^{a_n}].$$

Now consider $E_{1k}C[f_{11}^1 + f_{22}^0 + ... + f_{nn}^0)] (\alpha_1, \alpha_2, ..., \alpha_n)$

$$= E_{1k}[f_{11}^{a_1}1 + f_{22}^{a_2} + ... + f_{nn}^{a_n})] (\alpha_1, \alpha_2, ..., \alpha_n) = f_{11}^{a_k} (\alpha_1, \alpha_2, ..., \alpha_n).$$

Therefore, $E_{1k}C[(f_{11}^1 + f_{22}^0 + ... + f_{nn}^0)] = f_{11}^{a_k} \Rightarrow f_{11}^{a_k} \in <A>$ (since $C \in <A>$ and E_{1k} is zero-symmetric).

Therefore, $f_{11}^{a_k} \in <A> \backslash I^*$ (since $a_k \notin I$, $f_{11}^{a_k} \notin I^*$). Suppose $q = (x_1, x_2, ..., x_n)$. By Theorem 2.4.21, there exists y_i, $1 \leq i \leq n$ in N such that

$$B[f_{111}^{x_1} + f_{21}^{x_2} + ... + f_{n1}^{x_n}] = [f_{11}^{y_1} + f_{21}^{y_2} + ... + f_{n1}^{y_n}].$$

Consider $(c_1, c_2, ..., c_n) = Bq = B[f_{11}^{x_1}1 + f_{21}^{x_2} + ... + f_{n1}^{x_n}] (1, 0, ..., 0)$

$$= [f_{11}^{y_1}1 + f_{21}^{y_2} + ... + f_{n1}^{y_n}] (1, 0, ..., 0)$$

$$= (y_1, y_2, ..., y_n)$$

$$\Rightarrow c_i = y_i \text{ for } 1 \leq i \leq n.$$

Now $B[f_{11}^{x_1}1 + f_{22}^{x_2} + ... + f_{n1}^{x_n}] = [f_{11}^{y_1}1 + f_{22}^{y_2} + ... + f_{n1}^{y_n}] = [f_{11}^{c_1}1 + f_{22}^{c_2} + ... + f_{n1}^{c_n}]$.

Consider $E_{1s}B[f_{11}^{x_1}1 + f_{21}^{x_2} + ... + f_{n1}^{x_n}](\alpha_1, \alpha_2, ..., \alpha_n)$

$$= E_{1s}[f_{11}^{c_1}1 + f_{21}^{c_2} + ... + f_{n1}^{c_n}](\alpha_1, \alpha_2, ..., \alpha_n) = f_{n1}^{c_s}(\alpha_1, \alpha_2, ..., \alpha_n).$$

Therefore, $f_{n1}^{c_s} = E_{1s}B[f_{111}^{x_1} + f_{21}^{x_2} + ... + f_{n1}^{x_n}] \in <B>_1 M_n(N)$.

Hence, $f_{11}^{a_k} \in <A>$ and $f_{n1}^{c_s} \in <B>_1 M_n(N)$. This implies that $f_{11}^{a_k c_k} = f_{11}^{a_k} f_{n1}^{c_s} \in <A><B>_1 M_n(N) \subseteq I^*$ (by Condition 6.4)

$\Rightarrow a_k c_k \in I$, which is a contradiction.

Hence, I^* is a prime left ideal of $M_n(N)$. ∎

Proposition 6.2.17 (Syam Prasad, 2000)

If I is a completely prime ideal in $M_n(N)$, then I_* is a completely prime ideal in N.

Proof

Suppose I is a completely prime ideal in $M_n(N)$. By Theorem 6.1.6, I_* is an ideal of N. Let $a, b \in N$ such that $ab \in I_*$. Now $ab \in I_*$. By Result 6.1.4, we have $f_{11}^{ab} \in I$. Again, by Result 2.2.14, we have $f_{11}^a f_{11}^b \in I$. Since I is completely prime, we get that $f_{11}^a \in I$ or $f_{11}^b \in I$.

Thus, by Result 6.1.4, we have $a \in I_*$ or $b \in I_*$. Therefore, I_* is a completely prime ideal in N. ∎

6.3 Finite Goldie Dimension in $M_n(N)$-Group N^n

In this section, we present some results on linearly independent elements (l.i-elements) and linearly independent uniform elements (l.i.u-elements) in N-groups N, and $M_n(N)$-group N^n, and finally we present the result that the

Goldie dimension of an N-group N is equal to that of an $M_n(N)$-group N^n. We use the notation N^n to denote the $M_n(N)$-group N^n.

We first present some important results on finite dimension in N-groups, which are useful for proving dimension results related to the $M_n(N)$-group N^n. We start with the following theorem.

Theorem 6.3.1 (Theorem 2.5 of Satyanarayana and Syam Prasad, 2005)

If G has finite Goldie dimension (FGD), then K is a complement ideal of G if and only if there exist l.i.u-elements $u_1 + K$, $u_2 + K$, ..., $u_m + K$ in G/K, which span G/K essentially with $m = \dim G - \dim K$.

Proof

Suppose that K is a complement ideal of G. Since K is a complement, by Corollary 5.1.31, we have that $\dim (G/K) = \dim G - \dim K$. So, $\dim (G/K) = m$. Hence, by Theorem 5.1.2, G/K contains m uniform ideals whose sum is direct and essential in G/K. We select one and only one nonzero element from each of these uniform ideals. Suppose these elements are $u_i + K$, $1 \le i \le m$.

Since $u_i + K$ is a nonzero element of a uniform ideal, we have that $u_i + K$ is a u-element in G/K. Now $u_i + K$, $1 \le i \le m$ are l.i. and $<u_1 + K> \oplus ... \oplus <u_m + K>$ is essential in G/K.

Converse

Suppose that there exist l.i.u-elements $u_1 + K$, ..., $u_m + K$ in G/K, which span G/K essentially. Then $<u_1 + K> \oplus ... \oplus <u_m + K> \le_e G/K$. This shows that dim $(G/K) = m$. Therefore, dim $(G/K) = m = \dim G - \dim K$. Now, by Corollary 5.1.31, we have that K is a complement ideal of G. ∎

Theorem 6.3.2 (Theorem 2.6 of Satyanarayana and Syam Prasad, 2005)

Suppose G has FGD, dim $G = n$, $k < n$. If $u_1, u_2, ..., u_k$ are l.i.u-elements of G, then there exist $u_{k+1}, ..., u_n$ in G such that $u_1, u_2, ..., u_k, u_{k+1}, ..., u_n$ are elements of G that span G essentially.

Proof

Given that u_i, $1 \le i \le k$ are l.i.u-elements. Write $H = <u_1> \oplus ... \oplus <u_k>$. Now, dim $H = k$. Since dim $H = k < n = \dim G$, by Corollary 5.1.4, we have that H is not essential in G. Since H is not essential in G, there exists a nonzero ideal H^1 of G such that $H \cap H^1 = (0)$. By Zorn's lemma,

$$B = \{I/I \text{ is a nonzero ideal of } G \text{ such that } H \cap I = (0)\}$$

contains a maximal element, say J. By Result 5.1.11, we have that $H \oplus J$ is an essential ideal in G. Now $n = \dim G = \dim (H \oplus J) = \dim H + \dim J = k + \dim J$.

This implies that $\dim J = n - k$. Since $\dim J = n - k$, there exist l.i.u-elements v_1, $v_2, ..., v_{n-k}$ in J such that the sum of $<v_i>$, $1 \le i \le n - k$ is direct and essential in J.

Since $H \cap J = (0)$, $<u_1> \oplus ... \oplus <u_k> = H$, $<v_1> \oplus ... \oplus <v_{n-k}>$ is essential in J, by Corollary 3.4.5, we have that $<u_1> \oplus ... \oplus <u_k> \oplus <v_1> \oplus ... \oplus <v_{n-k}>$ is essential in $H \oplus J$.

Since $H \oplus J$ is essential in G, we have that $<u_1> \oplus ... \oplus <u_k> \oplus <v_1> \oplus ... \oplus <v_{n-k}>$ is essential in G.

This shows that $u_1, u_2, ..., u_k, v_1, ..., v_{n-k}$, are l.i.u-elements that span G essentially. ∎

Theorem 6.3.3 (Theorem 2.7 of Satyanarayana and Syam Prasad, 2005)

If G has FGD, then the following are equivalent:

- (i) $\dim G = n$
- (ii) there exist n uniform ideals U_i, $1 \le i \le n$ whose sum is direct and essential in G
- (iii) the maximum number of l.i.u-elements in G is n
- (iv) n is maximum with respect to the property that for any given $\{x_1, x_2, ..., x_k\}$ of l.i.u-elements with $k < n$, there exist $x_{k+1}, ..., x_n$ such that $\{x_1, x_2, ..., x_n\}$ are l.i.u-elements
- (v) the maximum number of l.i-elements that can span G essentially is n and
- (vi) the minimum number of l.i.u-elements that can span G essentially is n

Proof

(i) ⇔ (ii): Follows from Theorem 5.1.2.

(i) ⇒ (iii): Follows from Result 5.2.7.

(iii) ⇒ (ii): From (iii), there exist l.i.u-elements u_i, $1 \le i \le n$. This means $<u_i>$, $1 \le i \le n$ are uniform ideals and their sum is direct. If $<u_1> \oplus ... \oplus <u_n>$ is not essential in G, then G contains a nonzero ideal H such that the sum of $<u_1> \oplus ... \oplus <u_n>$ and H is direct. Since H is nonzero, by Remark 5.2.6 (ii), there exists a u-element $u \in H$. Now $u_1, u_2, ..., u_n$, u are l.i.u-elements, and they are $n + 1$ in number, which is a contradiction. Thus, $<u_1> \oplus ... \oplus <u_n> \le_e G$.

(i) ⇒ (iv): Follows from Theorem 6.3.1 and Result 5.2.7.

(iv) ⇒ (iii): Straightforward.

(i) ⇔ (v): Follows from Result 5.2.7.

(i) ⇒ (vi): Let us suppose that there exist l.i.u-elements u_i, $1 \le i \le k$, and $k < n$ such that u_i, $1 \le i \le k$ span G essentially. This means

$\langle u_1 \rangle \oplus \dots \oplus \langle u_n \rangle \leq_e G$. By Theorem 6.3.1, there exist l.i.u-elements $u_{k+1}, \dots, u_n$ such that $u_1, u_2, \dots, u_n$ are l.i.u-elements

$\Rightarrow (\langle u_1 \rangle \oplus \dots \oplus \langle u_k \rangle) \cap (\langle u_{k+1} \rangle) = (0)$. Since $\langle u_1 \rangle \oplus \dots \oplus \langle u_k \rangle \leq_e G$, we have that $\langle u_{k+1} \rangle = (0) \Rightarrow u_{k+1} = 0$, which is a contradiction.

(vi) $\Rightarrow$ (ii): Suppose there exist l.i.u-elements $u_i, 1 \leq i \leq n$ in G that span G essentially. This means $\langle ui \rangle, 1 \leq i \leq n$ are uniform ideals whose sum is direct and essential in G. ∎

We now present the notation relating to ideals of $M_n(N)$-group N^n.

Notation 6.3.4

For any ideal I of N^n, we write

$$I_{**} = \{x \in N/x = \pi_j A \text{ for some } A \in I, 1 \leq j \leq n\},$$

where π_j is the jth projection map from N^n to N.

Lemma 6.3.5 (Lemma 1.2 of Satyanarayana and Syam Prasad, 2005)

Let I be an ideal of N^n. Then

$$I_{**} = \{x \in N/(x, 0, \dots, 0) \in I\}.$$

Proof

Let $x \in \text{RHS} \Rightarrow (x, 0, \dots, 0) \in I$

$\Rightarrow \pi_1(x, 0, \dots, 0) = x \in I_*$

Take $x \in I_{**} \Rightarrow x = \pi_j A$ for some $A = (x_1, \dots, x_n) \in I \Rightarrow x = x_j$. Since $f_{1j}^1 \in M_n(R)$ and $A \in I$, by Theorem 3.2.16, we have that $(x_j, 0, \dots, 0) = f_{1j}^1 A \in I$, which implies $x_j \in \text{RHS}$. The proof is complete. ∎

Lemma 6.3.6 (Lemma 1.3 of Satyanarayana and Syam Prasad, 2005)

I_{**} is a left ideal of N.

Proof

Let $n \in N$ and $x \in I_{**}$. Consider $(n + x - n, 0, \dots, 0)$. Now $(n + x - n, 0, \dots, 0) = (n, 0, \dots, 0) + (x, 0, \dots, 0) - (n, 0, \dots, 0) \in I$ (since I is an ideal of N^n). Therefore, $n + x - n \in I_{**}$. This shows that I_{**} is a normal subgroup of N.

For $n_1, n_2 \in N$ and $x \in I_{**}$, we have

$$(n_1(n_2 + x) - n_1 n_2, 0, \ldots, 0) = f_{11}^{m_1}((n_2, 0, \ldots, 0) + (x, 0, \ldots, 0)) - f_{11}^{m_1}(n_2, 0, \ldots, 0) \in I,$$

which implies that $n_1(n_2 + x) - n_1 n_2 \in I_{**}$. Hence, I_{**} is a left ideal of N. ∎

Proposition 6.3.7 (Theorem 1.4 (i) of Satyanarayana and Syam Prasad, 2005)

Suppose L is a subset of N. Then L is an ideal of N if and only if L^n is an ideal of N^n.

Proof

Suppose L^n is an ideal of N^n. Now

$$X \in L \Leftrightarrow (x, 0, \ldots, 0) \in L^n$$
$$\Leftrightarrow x \in (L^n)_{**} \qquad \text{(by Lemma 6.3.5)}.$$

Therefore, $L = (L^n)_{**}$. By Lemma 6.3.6, we have that $L = (L^n)_{**}$ is an ideal of N. The converse follows from Result 6.1.1.

Lemma 6.3.8 (Lemma 1.5 (i) of Satyanarayana and Syam Prasad, 2005)

If I is an ideal of N^n, then $(I_{**})^n = I$.

Proof

Let $(x_1, x_2, \ldots, x_n) \in (I_{**})^n$
$\Rightarrow x_i \in I_{**}$ for $1 \leq i \leq n$
$\Rightarrow (x_i, 0, \ldots, 0) \in I$ (by Lemma 6.3.5)
$\Rightarrow f_{i1}^1(x_i, 0, \ldots, 0) \in I$ (since I is an ideal of N^n)
$\Rightarrow (0, \ldots, \underset{ith}{x_i}, \ldots, 0) \in I$ for $1 \leq i \leq n$.
Therefore $(x_1, x_2, \ldots, x_n) = \Sigma_i(0, \ldots, x_i, \ldots, 0) \in I$. Hence $(I_{**})^n \subseteq I$.
The other part is clear from Notation 6.3.4. ∎

Remark 6.3.9 (Remark 1.6 of Satyanarayana and Syam Prasad, 2005)

Suppose I, J are ideals of N. Then

(i) $(I \cap J)^n = I^n \cap J^n$ and
(ii) $I \cap J = (0) \Leftrightarrow (I \cap J)^n = (0) \Leftrightarrow I^n \cap J^n = (0)$.

Proof

(i) Clearly, $I \cap J \subseteq I$, $I \cap J \subseteq J$. So $(I \cap J)^n \subseteq I^n$, $(I \cap J)^n \subseteq J^n$.
Therefore $(I \cap J)^n \subseteq I^n \cap J^n$. Now
$(x_1, x_2, \ldots, x_n) \in I^n \cap J^n$
$\Rightarrow (x_1, \ldots, x_n) \in I^n$ and $(x_1, \ldots, x_n) \in J^n$
$\Rightarrow x_i \in I$, $x_i \in J$ for $1 \leq i \leq n$.
$\Rightarrow x_i \in I \cap J$ for $1 \leq i \leq n$
$\Rightarrow (x_1, x_2, \ldots, x_n) \in (I \cap J)^n$.
Therefore, $I^n \cap J^n \subseteq (I \cap J)^n$. Hence $(I \cap J)^n = I^n \cap J^n$.

(ii) $I \cap J = (0) \Leftrightarrow (I \cap J)^n = (0)$, and the rest follows from (i). ∎

Note 6.3.10

(i) If u is an element of N, then $<(0, \ldots, \underset{ith,}{u}, \ldots, 0)> = <(u, 0, \ldots, 0)>$.

(ii) If U is an ideal of N and I is an ideal of N^n such that $I \subseteq U^n$, then $I_{**} \subseteq U$.

(iii) If $A \leq_e N$, then $A^n \leq_e N^n$.

Verification

(i) Suppose u is an element of N.
Then $(u, 0, \ldots, 0) \in N^n$. Now

$$(u, 0, \ldots, 0) = f_{1i}^1 (0, \ldots, \underset{ith,}{u}, 0, \ldots, 0) \in <(0, \ldots, \underset{ith,}{u}, \ldots, 0)>.$$

Therefore, $<(u, 0, \ldots, 0)> \subseteq <(0, \ldots, \underset{ith,}{u}, 0, \ldots, 0)>$.
Consider $(0, \ldots, \underset{ith,}{u}, \ldots, 0)$. Now

$$(0, \ldots, \underset{ith,}{u}, \ldots, 0) = f_{i1}^1 (u, 0, \ldots, 0) \in <(u, 0, \ldots, 0)>.$$

Therefore, $<(0, \ldots, \underset{ith,}{u}, \ldots, 0)> \subseteq <(u, 0, \ldots, 0)>$.
Thus, we verified that $<(0, \ldots, \underset{ith,}{u}, \ldots, 0)> = <(u, 0, \ldots, 0)>$.

(ii) Suppose U is an ideal of N and I is an ideal of N^n such that $I \subseteq U^n$.
Now $x \in I_{**} \Rightarrow (x, 0, 0, \ldots, 0) \in I$ (by Lemma 6.3.5)
$\Rightarrow (x, 0, 0, \ldots, 0) \in U^n$ (since $I \subseteq U^n$)
$\Rightarrow x \in U$. Therefore, $I_{**} \subseteq U$.

(iii) Suppose $A^n \cap I = (0)$ and I is an ideal of N^n. Now
$I = (I_{**})^n$ (by Lemma 6.3.8). So, $A^n \cap (I_{**})^n = (0)$.
$A^n \cap (I_{**})^n = (0) \Rightarrow (A \cap I_{**})^n = (0)$ (by Remark 6.3.9)
$\Rightarrow A \cap I_{**} = (0)$ (by Remark 6.3.9)
$\Rightarrow I_{**} = (0)$ (since $A \leq_e N$)
$\Rightarrow I = (0)$.

Lemma 6.3.11 (Lemma 1.8 of Satyanarayana and Syam Prasad, 2005)

If x, u are elements of N and $x \in <u>$, then $(x, 0, \ldots, 0) \in <(u, 0, \ldots, 0)>$.

Proof

Following the notation, we have $<u> = \bigcup_{i=1}^{\infty} A_{i+1}$, where $A_{i+1} = A_i^* \cup A_i^\circ \cup A_i^+$ with $A_o = \{u\}$. Here

$$A_i^* = \{m + y - m \mid m \in N,\ y \in A_i\};$$
$$A_i^\circ = \{a - b \mid a,\ b \in A_i\};\ \text{and}$$
$$A_i^+ = \{n_1(n_2 + a) - n_1 n_2 \mid n_1,\ n_2 \in N \text{ and } a \in A_i\}.$$

Since $x \in <u>$, we have $x \in A_k$ for some k. Now we show (by induction on k) that $x \in A_k \Rightarrow (x, 0, \ldots, 0) \in <(u, 0, \ldots, 0)>$.

Suppose $k = 0$. Then $x \in A_k = A_o = \{u\} \Rightarrow x = u$.

Therefore, $(x, 0, \ldots, 0) = (u, 0, \ldots, 0) \in <(u, 0, \ldots, 0)>$. This shows that the statement is true for $k = 0$.

Suppose the induction hypothesis for $k = i$, that is,

$$x \in A_i \Rightarrow (x, 0, \ldots, 0) \in <(u, 0, \ldots, 0)>.$$

Let $x \in A_{i+1} = A_i^* \cup A_i^\circ \cup A_i^+$.

Case (i): Suppose $x \in A_i^*$.

$X \in A_i^* \Rightarrow x = m + y - m$ for some $m \in N, y \in A_i$. Now

$$y \in A_i \Rightarrow (y, 0, \ldots, 0) \in <(u, 0, \ldots, 0)>,$$

and so

$$(x, 0, \ldots, 0) = (m, 0, \ldots, 0) + (y, 0, \ldots, 0) - (m, 0, \ldots, 0) \in <(u, 0, \ldots, 0)>.$$

Case (ii): Suppose $x \in A_i^\circ$.

$x \in A_i^\circ \Rightarrow x = a - b$ for some $a, b \in A_i$. Now $a, b \in A_i$

$\Rightarrow (a, 0, \ldots, 0), (b, 0, \ldots, 0) \in <(u, 0, \ldots, 0)>$, and so

$$(x, 0, \ldots, 0) = (a - b, 0, \ldots, 0) = (a, 0, \ldots, 0) - (b, 0, \ldots, 0) \in <(u, 0, \ldots, 0)>$$

(by the induction hypothesis).

Case (iii): Suppose $x \in A_i^+$.

$x \in A_i^+ \Rightarrow x = n_1(n_2 + a) - n_1 n_2$ for some $n_1, n_2 \in N$, and $a \in A_i$. Now

$$(x, 0, \ldots, 0) = f_{11}^{n_1}((n_2, 0, \ldots, 0) + (a, 0, \ldots, 0)) - f_{11}^{n_1}(n_2, 0, \ldots, 0) \in <(u, 0, \ldots, 0)>.$$

Therefore, $(x, 0, \ldots, 0) \in \langle (u, 0, \ldots, 0) \rangle$.

Thus, by mathematical induction, we have that $(x, 0, \ldots, 0) \in \langle (u, 0, \ldots, 0) \rangle$ for any $x \in \langle u \rangle$. The proof is complete. ∎

Theorem 6.3.12 (Theorem 1.9 of Satyanarayana and Syam Prasad, 2005)

If u is an element of N, then $\langle u \rangle^n = \langle (u, 0, \ldots, 0) \rangle$.

Proof

Since $(u, 0, \ldots, 0) \in \langle u \rangle^n$, we have $\langle (u, 0, \ldots, 0) \rangle \subseteq \langle u \rangle^n$.

Take $(x_1, x_2, \ldots, x_n) \in \langle u \rangle^n$

$\Rightarrow x_i \in \langle u \rangle$ for $1 \leq i \leq n$.

By Lemma 6.3.11, we have that $(x_i, 0, \ldots, 0) \in \langle (u, 0, \ldots, 0) \rangle$ for all i. Now

$$(x_1, x_2, \ldots, x_n) = (x_1, 0, \ldots, 0) + f_{21}^1 (x_2, 0, \ldots, 0) + \ldots + f_{n1}^1 (x_n, 0, \ldots, 0) \in \langle (u, 0, \ldots, 0) \rangle$$

(by Theorem 3.2.16). Therefore, $\langle u \rangle^n \subseteq \langle (u, 0, \ldots, 0) \rangle$. The proof is complete. ∎

Lemma 6.3.13 (Lemma 2.1 of Satyanarayana and Syam Prasad, 2005)

For any $x_1, x_2, \ldots, x_n \in N$, we have

$$(\langle x_1 \rangle + \langle x_2 \rangle + \ldots + \langle x_n \rangle)^n = \langle x_1 \rangle^n + \langle x_2 \rangle^n + \ldots + \langle x_n \rangle^n.$$

Proof

Since $\langle x_i \rangle \subseteq \langle x_1 \rangle + \ldots + \langle x_n \rangle$, we have

$\langle x_i \rangle^n \subseteq (\langle x_1 \rangle + \ldots + \langle x_n \rangle)^n$ for $1 \leq i \leq n$.

Therefore, $\langle x_1 \rangle^n + \ldots + \langle x_n \rangle^n \subseteq (\langle x_1 \rangle + \ldots + \langle x_n \rangle)^n$.

Let $(a_1, a_2, \ldots, a_n) \in (\langle x_1 \rangle + \ldots + \langle x_n \rangle)^n$

$\Rightarrow a_i \in \langle x_1 \rangle + \ldots + \langle x_n \rangle$ for $1 \leq i \leq n$

$\Rightarrow a_i = a_{i_1} + \ldots + a_{i_n}$, where $a_{i_j} \in \langle x_j \rangle$ for $1 \leq i \leq n, 1 \leq j \leq n$. Now

$$(a_1, a_2, \ldots, a_n) = (a_{1_1} + \ldots + a_{1_n}, a_{2_1} + \ldots + a_{2_2}, \ldots, a_{n_1} + \ldots + a_{n_n})$$
$$= (a_{1_1}, a_{2_1}, \ldots, a_{n_1}) + (a_{1_2}, a_{2_2}, \ldots, a_{n_2}) + \ldots + (a_{1_n}, a_{2_n}, \ldots, a_{n_n})$$

$\in \langle x_1 \rangle^n + \langle x_2 \rangle^n + \ldots + \langle x_n \rangle^n.$

Therefore $(\langle x_1 \rangle + \langle x_2 \rangle + \ldots + \langle x_n \rangle)^n \subseteq \langle x_1 \rangle^n + \langle x_2 \rangle^n + \ldots + \langle x_n \rangle^n.$

The proof is complete. ∎

Theorem 6.3.14 (Theorem 2.2 of Satyanarayana and Syam Prasad, 2005)

$x_1, x_2, \ldots, x_n$ are linearly independent in N if and only if $(x_1, 0, \ldots, 0)$, $(x_2, 0, \ldots, 0)$, $\ldots (x_n, 0, \ldots, 0)$ are linearly independent in N^n.

Proof

$x_1, x_2, \ldots, x_n$ are linearly independent in N

$\Leftrightarrow$ the sum $<x_1> + \ldots + <x_n>$ is direct

$\Leftrightarrow <x_i> \cap (<x_1> + \ldots + <x_{i-1}> + <x_{i+1}> + \ldots + <x_n>) = 0$ for $1 \le i \le n$.

$\Leftrightarrow [<x_i> \cap (<x_1> + \ldots + <x_{i-1}> + <x_{i+1}> + \ldots + <x_n>)]^n = 0$ for $1 \le i \le n$

(by Remark 6.3.9)

$\Leftrightarrow <x_i>^n \cap [(<x_1> + \ldots + <x_{i-1}> + <x_{i+1}> + \ldots + <x_n>)]^n = 0$ for $1 \le i \le n$

(by Remark 6.3.9)

$\Leftrightarrow <x_i>^n \cap [<x_1>^n + \ldots + <x_{i-1}>^n + <x_{i+1}>^n + \ldots + <x_n>^n] = 0$ for $1 \le i \le n$

(by Lemma 6.3.13)

$\Leftrightarrow <x_1>^n + <x_2>^n + \ldots + <x_n>^n$ is direct

$\Leftrightarrow <(x_1, 0, \ldots, 0)> + <(x_2, 0, \ldots, 0)> + \ldots + <(x_n, 0, \ldots, 0)>$ is direct

(by Lemma 6.3.13)

$\Leftrightarrow (x_1, 0, \ldots, 0), (x_2, 0, \ldots, 0), \ldots, (x_n, 0, \ldots, 0)$ are linearly independent in N^n.

The proof is complete. ∎

Lemma 6.3.15 (Lemma 3.2 of Satyanarayana and Syam Prasad, 2005)

Let $u \in N$. Then u is a uniform element in N if and only if $(u, 0, \ldots, 0)$ is a uniform element in N^n.

Proof

Suppose u is a uniform element in N.

Let us suppose $(u, 0, \ldots, 0)$ is not a uniform element in N^n. Then, $<(u, 0, \ldots, 0)>$ is not a uniform ideal in N^n. So, there exist two nonzero ideals I and J of N^n, which are contained in $<(u, 0, \ldots, 0)>$ such that $I \cap J = (0)$.

Since I, J are ideals of N^n, by Lemma 6.3.8, we have that $I = (I_{**})^n$ and $J = (J_{**})^n$.

Now $I \cap J = (0) \Rightarrow (I_{**})^n \cap (J_{**})^n = (0)$

$\Rightarrow (I_{**} \cap J_{**})^n = (0)$ (by Remark 6.3.9)

$\Rightarrow I_{**} \cap J_{**} = (0)$.

Also, by Note 6.3.10 (ii), we have $I_{**} \subseteq <u>$ and $J_{**} \subseteq <u>$. So, we have that I_{**} and J_{**} are two nonzero ideals of N whose intersection is zero and they are contained in $<u>$. This is a contradiction to the fact that $<u>$ is a uniform ideal of N.

Converse

Suppose $(u, 0, \ldots, 0)$ is uniform in N^n.

This means $<(u, 0, \ldots, 0)>$ is a uniform ideal of N^n

$\Rightarrow <u>^n$ is a uniform ideal of N^n (by Theorem 6.3.12).

If $<u>$ is not uniform, then there exist two nonzero ideals A and B of N, contained in $<u>$ such that $A \cap B = (0)$. Now A^n and B^n are nonzero ideals of N^n, which are contained in $<u>^n$. Since $A \cap B = (0)$, we have $(A \cap B)^n = (0)$, and by Remark 6.3.9, $A^n \cap B^n = (0)$, a contradiction to the fact that $<u>^n$ is uniform. The proof is complete.   ∎

Theorem 6.3.16 (Lemma 3.3 of Satyanarayana and Syam Prasad, 2005)

Suppose $x_1, x_2, \ldots, x_n \in N$. Then $x_1, x_2, \ldots, x_n$ are l.i.u-elements in N if and only if $(x_1, 0, \ldots, 0), \ldots, (x_n, 0, \ldots, 0)$ are l.i.u-elements in N^n.

Proof

By using Theorem 6.3.14 and Lemma 6.3.15, we get the following:

$x_1, x_2, \ldots, x_n$ are l.i.u-elements

$\Leftrightarrow x_1, x_2, \ldots, x_n$ are l.i. and each x_i is uniform

$\Leftrightarrow (x_1, 0, \ldots, 0), (x_2, 0, \ldots, 0), \ldots, (x_n, 0, \ldots, 0)$ are l.i. and each $(x_i, 0, \ldots, 0)$ is uniform, $1 \leq i \leq n$.

$\Leftrightarrow (x_1, 0, \ldots, 0), (x_2, 0, \ldots, 0), \ldots, (x_n, 0, \ldots, 0)$ are l.i.u-elements in N^n.

The proof is complete.   ∎

We now present a theorem relating to the dimension of an N-group N and that of an $M_n(N)$-group N^n.

Theorem 6.3.17 (Theorem 3.4 of Satyanarayana and Syam Prasad, 2005)

$\dim N = \dim N^n$.

Proof

Suppose $\dim N = n$. By Theorem 5.2.24, there exist l.i.u-elements $x_1, x_2, \ldots, x_n$ in N, which span N essentially

$\Rightarrow x_1, x_2, \ldots, x_n$ are l.i.u-elements and $<x_1> + <x_2> + \ldots + <x_n> \leq_e N$

$\Rightarrow (x_1, 0, \ldots, 0), (x_2, 0, \ldots, 0), \ldots, (x_n, 0, \ldots, 0)$ are l.i.u-elements in N^n and $(<x_1> + <x_2> + \ldots + <x_n>)^n \leq_e N^n$ (by Theorem 6.3.16 and Note 6.3.10 (iii))

$\Rightarrow (x_1, 0, \ldots, 0), (x_2, 0, \ldots, 0), \ldots, (x_n, 0, \ldots, 0)$ are l.i.u-elements in N^n and $<x_1>^n + <x_2>^n + \ldots + <x_n>^n \leq_e N^n$ (by Lemma 6.3.13)

$\Rightarrow (x_1, 0, \ldots, 0), (x_2, 0, \ldots, 0), \ldots, (x_n, 0, \ldots, 0)$ are l.i.u-elements in N^n and $<(x_1, 0, \ldots, 0)> + <(x_2, 0, \ldots, 0)> + \ldots + <(x_n, 0, \ldots, 0)> \leq_e N^n$ (by Theorem 6.3.12)

$\Rightarrow \dim N^n = n$ (by Theorem 5.1.2). This shows that $\dim N = \dim N^n$. ∎

7

Gamma Near Rings

The concept of "gamma near rings" (defined by Satyanarayana in 1984) is a generalization of both the concepts "gamma ring" and "near ring." Subsequently, several mathematicians, for example, Booth, Babu Prasad, Cho, Jun, Kim, Ozturk, Kown, Park, Sapanci, Sreenadh, Eswaraiahsettey, George, Groenewald, Godloza, Nagaraju, Pradeep Kumar, Satyanarayana, Selvaraj, Subba Rao, Syam Prasad, Venkatachalam, and Vijayakumari, studied different concepts in gamma near rings, including ideals, prime ideals, semiprime ideals, fuzzy ideals, and fuzzy prime ideals.

7.1 Definition and Examples

In this section, we introduce the notion a Γ-near ring. We present basic definitions and results. Before going for a formal definition of a gamma near ring, let us consider a natural example.

Let $(G, +)$ be a group and X a nonempty set. Let M be the set of all mappings of X into G. Then M is a group under pointwise addition. If G is non-Abelian, then $(M, +)$ is also non-Abelian. To see this, let $a, b \in G$ such that $a + b \neq b + a$. Now define f_a, f_b from X to G by $f_a(x) = a$, $f_b(x) = b$ for all $x \in X$. Then $f_a, f_b \in M$ and $f_a + f_b \neq f_b + f_a$. Thus, if G is non-Abelian, then M is also non-Abelian.

Let Γ be the set of all mappings of G into X. If $f_1, f_2 \in M$ and $g \in \Gamma$, then obviously, $f_1 g f_2 \in M$. For all $f_1, f_2, f_3 \in M$ and $g_1, g_2 \in \Gamma$, it is clear that

(i) $(f_1 g_1 f_2) g_2 f_3 = f_1 g_1 (f_2 g_2 f_3)$ and
(ii) $(f_1 + f_2) g_1 f_3 = f_1 g_1 f_3 + f_2 g_1 f_3$

But $f_1 g_1 (f_2 + f_3)$ need not be equal to $f_1 g_1 f_2 + f_1 g_1 f_3$. To see this, fix $0 \neq z \in G$ and $u \in X$. Define $g_u : G \to X$ by $g_u(x) = u$ for all $x \in G$ and $f_z : X \to G$ by $f_z(x) = z$ for all $x \in X$. Now for any two elements $f_2, f_3 \in M$, consider $f_z g_u (f_2 + f_3)$ and $f_z g_u f_2 + f_z g_u f_3$.

For all $x \in X$, $f_z g_u (f_2 + f_3))(x) = f_z(g_u((f_2(x) + f_3(x))) = f_z(u) = z$, and $(f_z g_u f_2 + f_z g_u f_3)(x) = f_z g_u f_2(x) + f_z g_u f_3(x) = f_z(u) + f_z(u) = z + z$.

Since $z \neq 0$, $z \neq z + z$, and hence $f_z g_u (f_2 + f_3) \neq f_z g_u f_2 + f_z g_u f_3$.

Now we have the following:

If $(G, +)$ is a non-Abelian group and X is a nonempty set, then $M = \{f \mid f: X \to G\}$ is a non-Abelian group under pointwise addition and there exists a mapping $M \times \Gamma \times M \to M$, where $\Gamma = \{g \mid g: G \to X\}$ satisfies the following conditions:

(i) $(f_1 g_1 f_2) g_2 f_3 = f_1 g_1 (f_2 g_2 f_3)$ and

(ii) $(f_1 + f_2) g_1 f_3 = f_1 g_1 f_3 + f_2 g_1 f_3$

for all $f_1, f_2, f_3 \in M$ and for all $g_1, g_2 \in \Gamma$.

Keeping these axioms in mind, we define the concept of a "gamma near ring."

Definition 7.1.1

Let $(M, +)$ be a group (not necessarily Abelian) and Γ be a nonempty set. Then M is said to be a Γ-**near ring** if there exists a mapping $M \times \Gamma \times M \to M$ (denote the image of (m_1, α_1, m_2) by $m_1 \alpha_1 m_2$ for $m_1, m_2 \in M$ and $\alpha_1 \in \Gamma$) satisfying the following conditions:

(i) $(m_1 + m_2)\alpha_1 m_3 = m_1 \alpha_1 m_3 + m_2 \alpha_1 m_3$ and

(ii) $(m_1 \alpha_1 m_2)\alpha_2 m_3 = m_1 \alpha_1 (m_2 \alpha_2 m_3)$

for all $m_1, m_2, m_3 \in M$ and for all $\alpha_1, \alpha_2 \in \Gamma$.

Furthermore, M is said to be a **zero-symmetric Γ-near ring** if $m\alpha 0 = 0$ for all $m \in M, \alpha \in \Gamma$ (where "0" is an additive identity in M).

Remark 7.1.2

If M is a Γ-near ring, then for $\alpha \in \Gamma$, define a binary operation $*\alpha$ on M by $m_1 *\alpha m_2 = m_1 \alpha m_2$ for all $m_1, m_2 \in M$. Then $(M, +, *\alpha)$ is a near ring. Conversely, if $(M, +)$ is a group and Γ is a set of binary operations on M such that

(i) $(M, +, *)$ is a near ring for all $* \in \Gamma$ and

(ii) $(m_1 *_1 m_2) *_2 m_3 = m_1 *_1 (m_2 *_2 m_3)$ for all $*_1, *_2 \in \Gamma$, and for all $m_1, m_2, m_3 \in M$, then M is a Γ-near ring.

Example 7.1.3

Let $M = \mathbb{Z}_6$. Take $\Gamma = \{\alpha, \beta\}$, where α, β are given by scheme 1: $(0, 1, 0, 0, 0, 0)$ and scheme 2: $(0, 0, 1, 0, 0, 0)$ (Tables 7.1 and 7.2) (refer to page 409 of Pliz, 1983). Then M is a Γ-near ring.

TABLE 7.1

Operation Table with α

α	0	1	2	3	4	5
0	0	0	0	0	0	0
1	0	1	0	0	0	0
2	0	2	0	0	0	0
3	0	3	0	0	0	0
4	0	4	0	0	0	0
5	0	5	0	0	0	0

TABLE 7.2

Operation Table with β

β	0	1	2	3	4	5
0	0	0	0	0	0	0
1	0	0	1	0	0	0
2	0	0	2	0	0	0
3	0	0	3	0	0	0
4	0	0	4	0	0	0
5	0	0	5	0	0	0

7.2 Ideals of Gamma Near Rings

In this section, we present the definitions of substructures of gamma near rings like ideals.

Definition 7.2.1 (Definition 6.1.3 of Satyanarayana, 1984)

Let M be a Γ-near ring. Then a normal subgroup $(I, +)$ of $(M, +)$ is called

(i) **a left** (respectively **right**) **ideal** if $m_1\alpha(m_2 + i) - m_1\alpha m_2 \in I$ (respectively $i\alpha m_1 \in I$) for all $m_1, m_2 \in M$, $\alpha \in \Gamma$, and $i \in I$

(ii) an **ideal** if it is both a left and a right ideal

Remark 7.2.2 (Remark 6.1.4 of Satyanarayana, 1984)

(i) If I is an ideal of a zero-symmetric Γ-near ring M, then $m\alpha i \in I$ for all $m \in M$, $\alpha \in \Gamma$, $i \in I$.

(ii) A subset I of a Γ-near ring M is an ideal if and only if I is an ideal of the near ring $(M, +, {}^*\alpha)$ for every $\alpha \in \Gamma$.

(iii) If M is a Γ-near ring, then M is a Γ^1-near ring for every $\emptyset \neq \Gamma^1 \subseteq \Gamma$.

(iv) If I is an ideal of a Γ-near ring M, then I is an ideal of a Γ^1-near ring M, for every nonempty subset Γ^1 of Γ.

Example 7.2.3 (Example 6.1.5 of Satyanarayana, 1984)

Consider $\mathbb{Z}_8 = \{0, 1, 2, 3, \ldots, 7\}$, the group of integers modulo 8, and a set $X = \{a, b\}$. Write $M = \{f \mid f\colon X \to \mathbb{Z}_8 \text{ and } f(a) = 0\}$.

Then $M = \{f_0, f_1, f_2, \ldots, f_7\}$, where f_i is defined by $f_i(a) = 0$ and $f_i(b) = i$ for $0 \le i \le 7$.

Define two mappings $g_0, g_1\colon \mathbb{Z}_8 \to X$ by setting $g_0(i) = a$ for all $i \in \mathbb{Z}_8$ and $g_1(i) = a$ if $i \notin \{3, 7\}$, $g_1(i) = b$ if $i \in \{3, 7\}$. Write $\Gamma = \{g_0, g_1\}$, $\Gamma^* = \{g_0\}$.

Then M is a Γ-near ring and a Γ^*-near ring. Now $Y = \{f_0, f_2, f_4, f_6\}$ is an ideal of the Γ^*-near ring M, but not an ideal of the Γ-near ring M (since $f_2 \in Y$ and $(f_3 g_1(f_1 + f_2) - f_3 g_1 f_1) = f_3 \notin Y$).

Let I be a subgroup of M and define a congruence relation on M by $a \equiv b \pmod{I}$ if and only if $a - b \in I$. Then the following holds.

Proposition 7.2.4 (Proposition 6.1.6 of Satyanarayana, 1984)

Let M be a Γ-near ring and $(I, +)$ a subgroup $(M, +)$. Then the following two conditions are equivalent.

(i) I is an ideal of M and
(ii) $m_1 \equiv m_1{}^1 \pmod{I}$ and $m_2 \equiv m_2{}^1 \pmod{I}$
$\Rightarrow m_1 + m_2 \equiv m_1{}^1 + m_2{}^1 \pmod{I}$ and $m_1 m_2 \equiv m_1{}^1 m_2{}^1 \pmod{I}$ are equivalent for all $m_1, m_1{}^1, m_2, m_2{}^1 \in M$, $\alpha \in \Gamma$

Verification

(i) $\Rightarrow$ (ii): Suppose that I is an ideal of M.

Also suppose that $m_1 \equiv m_1{}^1 \pmod{I}$ and $m_2 \equiv m_2{}^1 \pmod{I} \Rightarrow m_1 - m_1{}^1 \in I$, $m_2 - m_2{}^1 \in I$.

Now we have to show that $m_1 + m_2 \equiv m_1{}^1 + m_2{}^1 \pmod{I}$ and $m_1 m_2 \equiv m_1{}^1 m_2{}^1 \pmod{I}$.

Now $(m_1 + m_2) - (m_1^1 + m_2^1) = m_1 + (m_2 - m_2^1) - m_1^1 = m_1 + (m_2 - m_2^1) + m_1 - m_1 - m_1^1$

$$= [m_1 + (m_2 - m_2^1) + m_1] - (m_1^1 - m_1) \in I$$

$$[\text{since } m_2 - m_2^1 \in I \text{ and } I \text{ is a normal subgroup of } M]$$

$$\Rightarrow (m_1 + m_2) \equiv (m_1^1 + m_2^1) \pmod{I}.$$

Now

$$m_1 \alpha m_2 - m_1^1 \alpha m_2^1 = m_1 \alpha (m_2 - m_2^1 + m_2^1) - m_1 \alpha m_2^1 + m_1 \alpha m_2^1 - m_1^1 \alpha m_2^1$$

$$= m_1 \alpha [(m_2 - m_2^1) + m_2^1] - m_1 \alpha m_2^1 + (m_1 - m_1^1) \alpha m_2^1 \in I + I = I.$$

The proof of (i) $\Rightarrow$ (ii) is complete.   ■

(ii) $\Rightarrow$ (i): Suppose that $m_1 \equiv m_1^1 \pmod{I}$ and $m_2 \equiv m_2^1 \pmod{I}$.
Now we show that I is an ideal of M.
Let $m \in M$ and $I \in I$.
We know that $m \equiv m \pmod{I}$ and $i \equiv 0 \pmod{I}$
$\Rightarrow m + i \equiv m + 0 \pmod{I}$
$\Rightarrow m + i \equiv m \pmod{I}$
$\Rightarrow m + i - m \in I$.
This shows that I is a normal subgroup of M.
Now we show that I is a right ideal.
Let $i \in I$ and $m \in M$.
We know $i \equiv 0 \pmod{I}$ and $m \equiv m \pmod{I}$
$\Rightarrow i\alpha m \equiv 0\alpha m \pmod{I}$ (by the assumption (ii))
$\Rightarrow i\alpha m \equiv 0 \pmod{I}$
$\Rightarrow i\alpha m - 0 \in I$
$\Rightarrow i\alpha m \in I$.
This is true for all $i \in I$ and $m \in M$.
To show that I is a left ideal, take $m, m^1 \in M$ and $i \in I$.
We know $m \equiv m \pmod{I}$, and $m^1 + i \equiv m^1 \pmod{I}$
$\Rightarrow m\alpha(m^1 + i) \equiv m\alpha m^1 \pmod{I}$ (by the assumption (ii))
$\Rightarrow m\alpha(m^1 + i) - m\alpha m^1 \in I$.
So, I is a left ideal of M and hence I is an ideal of M.
The proof is complete. ∎

Definition 7.2.5

Let M_1 and M_2 be Γ-near rings. A group homomorphism θ of $(M_1, +)$ into $(M_2, +)$ is said to be a Γ-**homomorphism** if $\theta(x\alpha y) = (\theta x)\alpha(\theta y)$ for all $x, y \in M$, $\alpha \in \Gamma$. We say that θ is a Γ-**isomorphism** (or simply isomorphism) if θ is one–one and onto.

For an ideal I of a Γ-near ring, we define the quotient Γ-near ring M/I, similar to the quotient near ring N/I, where I is an ideal of the near ring N (refer to Remark 2.2.11).

Theorem 7.2.6 (Theorem 6.1.8 of Satyanarayana, 1984)

Let I be an ideal of M and θ the canonical group epimorphism of M onto M/I. Then θ is a Γ-homomorphism of M onto M/I with kernel I. Conversely, if θ is a Γ-homomorphism of M_1 onto M_2 and I is the kernel of θ, then M_1/I is isomorphic to M_2.

Proof

Define the canonical mapping $\pi: M \to M/I$ by $\pi(m) = m + I$. We show that π is an epimorphism. Take $m_1, m_2 \in M$.

Now $\pi(m_1 + m_2) = (m_1 + m_2) + I = (m_1 + I) + (m_2 + I) = \pi(m_1) + \pi(m_2)$.

And $\pi(m_1 \alpha m_2) = (m_1 \alpha m_2) + I = (m_1 + I)\alpha(m_2 + I) = \pi(m_1)\alpha\pi(m_2)$.

Therefore, π is a Γ-near ring homomorphism.

Let $x \in M/I$. Then there exists some $m_1 \in M$ such that $x = m_1 + I$.

Now $\pi(m_1) = m_1 + I = x$, and so π is onto. Therefore, $\pi: M \to M/I$ is a near ring epimorphism. Hence, M/I is a homomorphic image of M.

(ii) Suppose $h: M \to M^1$ is an epimorphism.

Part 1: Now we show that ker h is an ideal of M, and $M/\ker h \cong M^1$.

First we show that ker h is a normal subgroup of M.

Clearly, $0 \in \ker h$ and so ker $h \neq \varnothing$.

Let $m \in M$ and $x \in \ker h$.

Now $h(m + x - m) = h(m) + h(x) - h(m)$ (since h is a homomorphism)

$$= h(m) + 0 - h(m) \qquad (since \ h(x) = 0)$$

$$= h(m) - h(m) = 0.$$

This shows that $m + x - m \in \ker h$.

Part 2: We show that ker h is a right ideal of M.

Let $m \in M$, $x \in \ker h$, and $\alpha \in \Gamma$.

Now, $h(x\alpha m) = h(x)\alpha h(m) = 0\alpha h(m) = 0$. Therefore, $x\alpha m \in \ker h$.

Part 3: In this part, we show that ker h is a left ideal of M.

Let $m, m^1 \in M$ and $x \in \ker h$.

Now, $h(m\alpha(m^1 + x) - m\alpha m^1) = h(m)\alpha h(m^1 + x) - h(m)\alpha h(m^1)$

$$= h(m)\alpha(h(m^1) + h(x)) - h(m)\alpha h(m^1)$$

$$= h(m)\alpha(h(m^1) + 0) - h(m)\alpha h(m^1)$$

$$= h(m)\alpha h(m^1) - h(m)\alpha h(m^1) = 0.$$

Therefore, $m\alpha(m^1 + x) - m\alpha m^1 \in \ker h$. Hence, ker h is an ideal of M.

Part 4: Define $\varphi: M/\ker h \to M^1$ as $\varphi(m + \ker h) = h(m)$ for all $m + \ker h \in M/\ker h$.

We show that φ is a near ring isomorphism. Take $m_1 + \ker h, m_2 + \ker h \in M/\ker h$.

Now, $\varphi(m_1 + \ker h) = \varphi(m_2 + \ker h)$

if and only if $h(m_1) = h(m_2)$

if and only if $h(m_1) - h(m_2) = 0$

if and only if $h(m_1 - m_2) = 0$

if and only if $(m_1 - m_2) \in \ker h$

if and only if $m_1 + \ker h = m_2 + \ker h$.

Therefore, φ is one–one and well-defined.

Let $m^1 \in M^1$. Since h is onto, there exists $m \in M$ such that $h(m) = m^1$.

Now, $\varphi(m + \ker h) = h(m) = m^1$, and so φ is onto.

Part 5: Let $m_1 + \ker h, m_2 + \ker h \in M/\ker h$.
Now $\varphi((m_1 + \ker h) + (m_2 + \ker h)) = \varphi((m_1 + m_2) + \ker h)$
$$= h(m_1 + m_2)$$
$$= h(m_1) + h(m_2) \text{ (since } h \text{ is epimorphism)}$$
$$= \varphi(m_1 + \ker h) + \varphi(m_2 + \ker h).$$
Also, $\varphi((m_1 + \ker h)\alpha(m_2 + \ker h)) = \varphi((m_1\alpha m_2) + \ker h)$
$$= h(m_1\alpha m_2)$$
$$= h(m_1)\alpha h(m_2)$$
$$= \varphi(m_1 + \ker h)\alpha\varphi(m_2 + \ker h).$$

Therefore, φ is a near ring homomorphism. Hence, $\varphi: M/\ker h \to M^1$ is a Γ-near ring isomorphism. ∎

The proofs of the following theorems are analogous to the proofs provided for near rings (refer to Theorems 2.3.2 and 2.3.5), and hence omitted.

Theorem 7.2.7 (Theorem 6.1.9 of Satyanarayana, 1984)

Let θ be a Γ-homomorphism of M_1 onto M_2 with kernel I and J^1 a nonempty subset of M_2. Then J^1 is an ideal of M_2 if and only if $\theta^{-1}(J^1) = J$ is an ideal of M_1 containing I. In this case, we have that $M_1/J, M_2/J^1$, and $(M_1/J)/(J/I)$ are Γ-isomorphic.

Theorem 7.2.8 (Theorem 6.1.10 of Satyanarayana, 1984)

Let I and J be two ideals of a Γ-near ring M. Then $(I + J)/J$ is isomorphic to $I/(I \cap J)$.

Notation 7.2.9

The ideal generated by a subset A of M will be denoted by $<A>$, and the ideal generated by an element $a \in M$ will be denoted by $<a>$. If $A_1, A_2, \ldots, A_{k+1}$ are subsets of M, then we write
$$A_1\Gamma A_2\Gamma A_3\Gamma \ldots \Gamma A_k\Gamma A_{k+1} = \{a_1\alpha_1 a_2\alpha_2 \ldots a_k\alpha_k a_{k+1} \mid a_i \in A_i, \alpha_j \in \Gamma \text{ for } 1 \leq i \leq k+1$$
and $1 \leq j \leq k\}$.
If $A_i = A$ for $1 \leq i \leq k+1$, then $A_1\Gamma A_2\Gamma \ldots \Gamma A_{k+1}$ is denoted by A^{k+1}.

Example 7.2.10

Let G be a nontrivial group and X a nonempty set. If M is the set of all mappings from X into G and the set of all mappings from G into X, then M is a Γ-near ring. Let y be a nonzero fixed element of G. Define $\varphi: X \to G$ by

$\varphi(x) = y$ for every $x \in X$. Then $0 \neq \varphi \in M$, where "0" is an additive identity in M and $\varphi g0 = \varphi \neq 0$ for any $g \in \Gamma$. Therefore, M is a Γ-near ring that is not a zero-symmetric Γ-near ring.

Henceforth, we consider zero-symmetric Γ-near rings only. Throughout, M stands for a Γ-near ring.

7.3 Prime Ideals and Nilpotent Ideals in Gamma Near Rings

In this section, we present definitions of prime ideals, m-system, f-prime and f-s-prime ideals, and nilpotent ideals of gamma near rings.

Suppose f is a mapping from M into the set of all ideals of M, satisfying the following conditions:

(i) $a \in f(a)$ and
(ii) $x \in f(a) + A$, A is an ideal $\Rightarrow f(x) \subseteq f(a) + A$

Such a mapping is called an "ideal mapping."
Now we provide a natural example for this.

Example 7.3.1

Let M be a Γ-near ring and Q a subset of M. For each $a \in M$, define $f(a) = <(\{a\} \cup Q)>$, the ideal generated by $\{a\} \cup Q$; then $f(a)$ satisfies the two previous conditions.

Remark 7.3.2

The following are equivalent

(i) for any element $a \in M$, $f(a) = <a>$
(ii) $f(0) = <0>$
(iii) for any ideal A of M, $x \in A \Rightarrow f(x) \subseteq A$
(iv) for any element $a \in M$, $x \in <a> \Rightarrow f(x) \subseteq <a>$

Definition 7.3.3 (Definition 2.1 of Satyanarayana, 1999)

A subset H of M is said to be

(i) a **multiplicative** set if for any h_1, h_2 in H, there corresponds an element $\alpha \in \Gamma$ such that $h_1 \alpha h_2 \in H$
(ii) an **m-system** if for every h_1, $h_2 \in H$, there exists $h_1^1 \in <h_1>$, $h_2^1 \in <h_2>$ $\alpha \in \Gamma$ such that $h_1^1 \alpha h_2^1 \in H$

(iii) an **s-system** if H contains a multiplicative set H^* called a kernel of H, such that for every $h \in H$, $\langle h \rangle \cap H^* \neq \emptyset$

(iv) an **f-system** if H contains an m-system H^* called a kernel of H, such that for every $h \in H$, $f(h) \cap H^* \neq \emptyset$

(v) an **f-s-system** if H contains a multiplicative set H^* called a kernel of H, such that for every $h \in H$, $f(h) \cap H^* \neq \emptyset$

Definition 7.3.4 (Definition 2.2 of Satyanarayana, 1999)

An ideal A of M is said to be

(i) **prime** if $B\Gamma C \subseteq A$, B, and C are ideals of M, implying either $B \subseteq A$ or $C \subseteq A$

(ii) **completely prime** if $a\Gamma b \subseteq A$, $a, b \in M$, implying either $a \in A$ or $b \in A$

(iii) **s-prime** if $M \backslash A$ is an s-system

(iv) **f-prime** if $M \backslash A$ is an f-system

(v) **f-s-prime** if $M \backslash A$ is an f-s-system

Here, we write $H(H^*)$ for either s-system, f-system, or f-s-system H with kernel H^*.

Theorem 7.3.5 (Note 2.3 of Satyanarayana, 1999)

For an ideal P of M, the following hold:

(i) P is prime if and only if $M \backslash P$ is an m-system.

(ii) P is completely prime if and only if $M \backslash P$ is a multiplicative set.

(iii) P is s-prime $\Rightarrow P$ is prime $\Rightarrow P$ is f-prime.

(iv) P is completely prime $\Rightarrow P$ is s-prime $\Rightarrow P$ is f-s-prime.

(v) If P is f-prime, then $f(a)f(b) \subseteq P$, $a, b \in M$, which implies $a \in P$ or $b \in P$.

Proof

It is a straightforward verification similar to some earlier results (refer to Theorems 4.1.32 and 4.1.44). ∎

Example 7.3.6 (Example 2.4 of Satyanarayana, 1999)

Suppose N is a zero-symmetric near ring such that N contains a direct sum of nonzero ideals I_1, I_2, I_3 of itself. Take $M = N$ and $\Gamma = \{0\}$. Then M is a Γ-near ring. Let $0 \neq x \in I_1$. Define for any $a \in M$, $f(a) = \langle\{a, x\}\rangle$, and write $S^* = \{x, x^2, x^3, \ldots\}$. Now S^* is an m-system and $S^* \subseteq I_1$.

Now $M\backslash I_2$ contains S^*, and $M\backslash I_2$ is an f-system with kernel S^*. Therefore, I_2 is an f-prime ideal. But I_2 is not a prime ideal, since $I_1 \not\subseteq I_2$, $I_3 \not\subseteq I_2$, and $I_1 I_3 \subseteq I_2$. Hence, in general, an f-prime ideal need not be a prime.

Definition 7.3.7 (Definition 2.5 of Satyanarayana, 1999)

(i) An element $a \in M$ is said to be **nilpotent** if there exists an integer $k \geq 2$ such that $\{a\}^k = \{0\}$ (refer to Notation 7.2.9).

(ii) A subset H of M is said to be **nil** if every element of H is nilpotent; H is said to be **nilpotent** if there exists an integer $n \geq 2$ such that $H^n = \{0\}$.

(iii) An element $a \in M$ is said to be **f-nilpotent** (**f-nil**) if $f(a)$ is nilpotent (nil).

(iv) A subset H of M is said to be **f-nil** if every element of H is f-nilpotent. H is said to be **f-nilpotent** if there exists an integer $k \geq 2$ such that

$$f(a_1)\Gamma f(a_2)\Gamma \ldots \Gamma f(a_{k-1})\Gamma f(a_k) = \{0\} \text{ for all } a_1, a_2, \ldots, a_k \in H.$$

With these definitions, we have the following.

Remark 7.3.8 (Remark 2.6 of Satyanarayana, 1999)

Let $a \in M$ and $H \subseteq M$.

(i) a is f-nilpotent $\Rightarrow a$ is f-nil $\Rightarrow a$ is nilpotent.
(ii) H is f-nilpotent $\Rightarrow H$ is f-nil $\Rightarrow H$ is nil.
(iii) H is f-nilpotent $\Rightarrow H$ is nilpotent $\Rightarrow H$ is nil.

Examples 7.3.9

Let N be a near ring. If we write $M = N$ and $\Gamma = \{0\}$, then M is a Γ-near ring. Now we have the following:

(i) Suppose x and y are two elements of N such that x is nilpotent and y is not nilpotent. Define $f(a) = <\{a, y\}>$ for all $a \in M$. Now $y \in f(a)$ and $f(a)$ is not nil. Hence, x is not f-nil, but it is evident that x is nilpotent.

(ii) Suppose Q is an ideal of N that is nil, but not nilpotent. Define $f(a) = <(\{a\} \cup Q)>$ for all $a \in M$. Now for any $q \in Q$, $f(q) = Q$, and hence q is f-nil, but not f-nilpotent. Further, Q is f-nil, but not f-nilpotent.

(iii) Suppose K is a nonzero nilpotent ideal of N. Write for all $a \in M$, $f(a) = <(\{a\} \cup K)>$.

Then $x \in K$, $f(x) = K$, and every element of K is f-nilpotent. Hence, K is f-nilpotent.

Lemma 7.3.10 (Lemma 3.1 of Satyanarayana, 1999)

Let P be an ideal of M. Then for any two subsets A and B of M, we have $(A + P)\Gamma(B + P) \subseteq A\Gamma B + P$.

Proof

Let $a \in A$, $b \in B$, $p_1, p_2 \in P$, and $\alpha \in \Gamma$. Consider

$$(a + p_1)\alpha(b + p_2) = a\alpha\,(b + p_2) + p_1\alpha(b + p_2)$$
$$= a\alpha b - a\alpha b + a\alpha(b + p_2) + p_1\alpha(b + p_2).$$

Since $a\alpha b \in A\Gamma B$ and $p_1\alpha(b + p_2)$, $-a\alpha b + a\alpha(b + p_2) \in P$, we have $(a + p_1)\alpha(b + p_2) \in A\Gamma B + P$. The proof is complete. ∎

Lemma 7.3.11 (Lemma 3.2 of Satyanarayana, 1999)

Let $S(S^*)$ be an f-system in M and let A be an ideal in M that does not meet S. Then A is contained in a maximal ideal P that does not meet S. The ideal P is necessarily an f-prime ideal.

Proof

Suppose S is nonempty. Then the existence of P follows from Zorn's lemma. We now show that $M \backslash P$ is an f-system with kernel $S^* + P$. For any element a in $M \backslash P$, by the maximality of P, $P + f(a)$ contains an element s of S, and we can choose an element $s^* \in f(s) \cap S^*$. Since $f(s) \subseteq f(a) + P$, we have $s^* = a^1 + P$ for some $a^1 \in f(a)$ and $p \in P$.

Now $a^1 = s^* - p \in f(a) \cap (S^* + P)$. Now, it remains to be shown that $S^* + P$ is an m-system.

Let $x, y \in S^* + P$. Then $x = s_1 + p_1$, $y = s_2 + p_2$ for some $s_1, s_2 \in S^*$, $p_1, p_2 \in P$, and $s_1 = x - p_1$, $s_2 = y - p_2$.

So, $\langle s_1 \rangle \Gamma \langle s_2 \rangle \subseteq \langle x - p_1 \rangle \Gamma \langle y - p_2 \rangle$

$$\subseteq (\langle x \rangle + P)\,\Gamma(\langle y \rangle + P)$$
$$\subseteq \langle x \rangle \Gamma \langle y \rangle + P.$$

Since $s_1, s_2 \in S^*$ and S^* is an m-system, there exists $s_1^* \in \langle s_1 \rangle$, $s_2^* \in \langle s_2 \rangle$, $\alpha \in \Gamma$ such that $s_1^* \alpha s_2^* \in S^*$. Now $s_1^* \alpha s_2^* = x_1\beta y_1 + p_3$ for some $x_1 \in \langle x \rangle$, $y_1 \in \langle y \rangle$, $\beta \in \Gamma$, $p_3 \in P$, and hence $s_1^* \alpha s_2^* - p_3 = x_1\beta y_1 \in \langle x \rangle \Gamma \langle y \rangle \cap (S^* + P)$.

Hence, $S^* + P$ is an m-system contained in $M \backslash P$. Hence, $M \backslash P$ is an f-system with kernel $S^* + P$. If $S = \varnothing$, then $P = M$. ∎

Definition 7.3.12

The **f-radical** of an ideal A (denoted by **f-rad(A)**) will be defined to be the set of all elements a of M with the property that every f-system that contains a, contains an element of A.

Theorem 7.3.13 (Theorem 3.4 of Satyanarayana, 1999)

The f-radical of an ideal A is the intersection of all the f-prime ideals containing A.

Proof

Suppose f-rad$(A) \not\subseteq P$, for some f-prime ideal P that contains A. Let $x \in$ f-rad$(A)\backslash P$. Since $M\backslash P$ is an f-system, by the definition of f-rad(A), $(M\backslash P) \cap A \neq \emptyset$. But this contradicts the fact that P contains A. So f-rad$(A) \subseteq P$ for all f-prime ideals P containing A.

Let $a \notin$ f-rad(A). Then there exists an f-system $S(S^*)$ such that $a \in S$ and $S \cap A = \emptyset$. By Lemma 7.3.11, there exists an f-prime ideal P^1 containing A such that $P^1 \cap S = \emptyset$. Therefore, $a \notin P^1$. Hence, f-rad(A) = the intersection of all f-prime ideals containing A.   ∎

Definition 7.3.14

Let A be an ideal of M. Then we define the **prime radical** of A, denoted by P-rad(A), as the set of all elements x of M for which every m-system that contains x must contain an element of A.

It is evident that if we define f by $f(a) = <a>$, then the notions f-rad(A) and f-prime coincide with P-rad(A) and prime, respectively. Hence, Theorem 7.3.13, yields the following.

Corollary 7.3.15

For any ideal A of M, P-rad(A) = the intersection of all prime ideals containing A.

Definition 7.3.16

Let A be an ideal of M. An **f-prime ideal P** is said to be a minimal f-prime ideal of A if P contains A, and there exists a kernel S^* for the f-system $M\backslash P$ such that S^* is a maximal m-system that does not meet A.

We now try for the existence of a minimal f-prime ideal of a given ideal A. For this, we start with the following.

Let $S(S^*)$ be an f-system in M and let A be an ideal not meeting S. Since S^* is an m-system such that $A \cap S^* = \varnothing$, by Zorn's lemma, the set

$$\{K \mid K \text{ is an m-system such that } K \cap A = \varnothing \text{ and } K \supseteq S^*\}$$

contains a maximal element, say S_1^*. Now write $S_1 = \{x \in M/f(x) \cap S_1^* \neq \varnothing\} \cap (M \backslash A)$.

Then S_1 is an f-system with kernel S_1^*. Again, by Zorn's lemma, the set $\{J \mid J$ is an ideal, $J \supseteq A$ and $J \cap S_1 = \varnothing\}$ contains a maximal element, say J^*, and by Lemma 7.3.11, J^* is an f-prime ideal. Since $M \backslash J^*$ is an f-system with kernel $S_1^* + J^*$ (for this, see the proof of Lemma 7.3.11), we have $S_1^* = S_1^* + J^*$, and so $M \backslash J = S_1$. Thus, J^* is a minimal f-prime ideal of A. Hence, we have the following.

Lemma 7.3.17 (Lemma 6.4.2 of Satyanarayana, 1984)

Let $S(S^*)$ be an f-system and A be an ideal that does not meet S. Then there exists an f-system S_1 containing S with kernel S_1^*, and S_1^* is maximal with respect to the properties $S_1^* \cap A = \varnothing$, $S_1^* \supseteq S^*$. Moreover, $M \backslash S$ is a minimal f-prime ideal of A.

It is easy to verify the following.

Corollary 7.3.18

Every f-prime ideal P of an ideal A contains a minimal f-prime ideal of A.

Theorem 7.3.19 (6.4.4 of Satyanarayana, 1984)

Let A be an ideal of M. Then
f-rad(A) = the intersection of all minimal f-prime ideals of A.

Proof

Let Y be the intersection of all minimal f-prime ideals of A. Clearly, Y contains the intersection of all f-prime ideals that contain A, and hence by Theorem 7.3.13, Y contains f-rad(A).

On the other hand, if P is an f-prime ideal containing A, then by Corollary 7.3.18, there exists a minimal f-prime P^* that is contained in P. Now $P \supseteq P^* \supseteq Y$, and hence $\bigcap_{p \text{ is } f\text{-prime}} P \supseteq Y$.

Thus, f-rad(A) = Y. ∎

Definitions 7.3.20

Let A be an ideal of M. An element a in M is said to be **strongly nilpotent modulo** A, if for every sequence $x_1, x_2, \ldots$, of elements of M satisfying the conditions $x_1 = a$ and $x_i = x_{i-1}^1 \alpha_{i-1} x_{i-1}^*$ for some $\alpha_{i-1} \in \Gamma$ and $x_{i-1}^1, x_{i-1}^* \in <x_{i-1}>$, there exists an integer k such that $x_s \in A$ for all $s \geq k$.

An element $a \in M$ is said to be **strongly nilpotent** if it is strongly nilpotent modulo (0) (the zero ideal of M).

An element $x \in M$ is said to be **f-strongly nilpotent modulo** A if every element of $f(x)$ is strongly nilpotent modulo A. An element x is said to be **f-strongly nilpotent** if every element of $f(x)$ is strongly nilpotent.

Since $x \in f(x)$, every f-strongly nilpotent element is strongly nilpotent. The following example establishes that the converse is not true.

Example 7.3.21

Let N be a near ring such that $(0) \neq \text{P-rad}(N)$ and $\text{P-rad}(N) \neq N$.

Let $x \in N \backslash (\text{P-rad}(N))$. We can consider N as a Γ-near ring, where $\Gamma = \{0\}$.

In this case, the definition of strong nilpotency coincides with the corresponding Definition 7.3.20. Write $f(a) = <\{a, x\}>$ for every element $a \in N$. Since x is not strongly nilpotent, we have that a is not f-strongly nilpotent for every $a \in N$. So, N contains no f-strongly nilpotent elements, while all elements of $\text{P-rad}(N)$ are strongly nilpotent.

Definition 7.3.22

We define f-rad(0) to be the **f-prime radical** of M, and it is also often denoted by f-rad(M).

Now we characterize the f-prime radical of M in terms of f-strongly nilpotent elements.

Lemma 7.3.23 (Lemma 4.3 of Satyanarayana, 1999)

Let $a_1, a_2, \ldots$, be a sequence of elements of M with a_i, where $a_i = a_{i-1}^1 \alpha_{i-1} a_{i-1}^*$, for some $\alpha_{i-1} \in \Gamma$ and $a_{i-1}^1, a_{i-1}^* \in <a_{i-1}>$. Then $\{a_i \mid i \geq 1\}$ is an m-sequence.

Proof

Let $y_1, y_2 \in \{a_i \mid i \geq 1\}$. Then $y_1 = a_{s_1}, y_2 = a_{s_2}$ for some integers $s_1, s_2 \geq 1$. Suppose $s_1 \leq s_2$. Since $a_i \in <a_{i-1}>$ for all i, we have $<a_1> \supseteq <a_2> \supseteq <a_3> \supseteq \cdots$. Now $<a_{s_1}> \supseteq <a_{s_2}> \supseteq \ldots$, and so $a_{s_2+1} = a_{s_2}^1 \alpha_{s_2} a_{s_2}^*$ for some $a_{s_2}^1, a_{s_2}^* \in <a_{s_2}> \subseteq <a_{s_1}>$.

Therefore, $a_{s_2}^1 \in \langle \alpha_{s_1} \rangle$, $\alpha_{s_2} \in \Gamma$ such that $a_{s_2}^1 \alpha_{s_2} a_{s_2}^* \in \{a_i \mid i \geq 1\}$.
The proof is complete. ■

Theorem 7.3.24 (Theorem 4.4 of Satyanarayana, 1999)

f-rad $(M) = \{x \in M \mid x$ is f-strongly nilpotent$\} \cup \{0\}$.

Proof

Let $0 \neq x \in$ f-rad (M). If x is not f-strongly nilpotent, then there exists $x_1 \in f(x)$ such that x_1 is not strongly nilpotent. Therefore, there exists a sequence $a_1, a_2, a_3, \ldots$, such that $a_1 = x_1$ and $a_i = a_{i-1}^1 \alpha_{i-1} a_{i-1}^*$ for some $\alpha_{i-1} \in \Gamma$, a_{i-1}^1, $a_{i-1}^* \in \langle a_{i-1} \rangle$ and $a_i \neq 0$ for all i. By Lemma 7.3.23, we have that $S = \{a_i \mid i \geq 1\} \cup \{x\}$ is an f-system with kernel $S^* = S \backslash \{x\}$. Since $S \cap \{0\} = \emptyset$, by Lemma 7.3.11, there exists an f-prime ideal P such that $P \cap S = \emptyset$.
Now $x \notin P$, and so $x \notin$ f-rad(M), which is a contradiction.
Thus, x is f-strongly nilpotent.

Converse

Suppose that x is an f-strongly nilpotent element. If $x \notin$ f-rad (M), then there exists an f-prime ideal P such that $x \notin P$. Suppose $M \backslash P$ is an f-system with kernel Q^*. Then $x \in M \backslash P$, and hence there exists an element x_1 in $f(x) \cap Q^*$. Since Q^* is an m-system, there exists a sequence $x_1, x_2, \ldots$ of elements from Q^* such that $x_i = x_{i-1}^1 \alpha_{i-1} x_{i-1}^*$ for some $\alpha_{i-1} \in \Gamma$. $x_{i-1}^1, x_{i-1}^* \in \langle x_{i-1} \rangle$. Since x is f-strongly nilpotent, there exists an integer $t \geq 1$ such that $x_s = 0$ for all $s \geq t$. Thus, $0 = x_s \in Q^*$, and this is a contradiction since $Q^* \subseteq (M \backslash P)$ and $0 \in P$.
The proof is complete. ■
In a similar approach, we can obtain the following theorem.

Theorem 7.3.25 (Theorem 4.5 of Satyanarayana, 1999)

If A is an ideal of M, then

f-rad$(A) = \{x \in M \mid x$ is f-strongly nilpotent modulo $A\} \cup A$.

Theorems 7.3.24 and 7.3.25 yield the following theorem.

Theorem 7.3.26 (Theorem 6.5.7 of Satyanarayana, 1984)

(i) P-rad $(M) = \{x \in M \mid x$ is strongly nilpotent$\}$
(ii) For any ideal A of M, P-rad$(A) = \{x \in M \mid x$ is strongly nilpotent modulo $A\}$.

Theorem 7.3.27 (Theorem 6.5.8 of Satyanarayana, 1984)

(a) If $f(0) \subseteq$ P-rad (M), then
   (i) "0" is f-strongly nilpotent.
   (ii) f-rad(M) = {x ∈ M/x is f-strongly nilpotent}.
   (iii) Let P be an f-prime ideal. Then P is prime if and only if $f(x) \subseteq P$ for all x ∈ P.
   (iv) An ideal P is prime if and only if P is f-prime and $f(x) \subseteq P$ for all x ∈ P.
(b) $f(0) \subseteq$ P-rad(M) if and only if f-rad(M) = the set of all f-strongly nilpotent elements of M.

Proof

(a)   (i) Since $f(0) \subseteq$ P-rad(M) and P-rad(M) is the set of all strongly nilpotent elements, we have that the element "0" is f-strongly nilpotent.
   (ii) Follows from (i) and Theorem 7.3.24.
   (iii) Let P be an f-prime ideal. Suppose P is prime.

Let x ∈ P, x ∈ <x> ⊆ (f(0) + <x>), which imply $f(x) \subseteq (f(0) + <x>) \subseteq$ (P-rad(M) + P) ⊆ P.

Converse

Suppose that $f(x) \subseteq P$ for all x ∈ P. Let A, B be ideals of M such that $A\Gamma B \subseteq P$. If $A \not\subseteq P$ and $B \not\subseteq P$, then there exist a ∈ A\P, b ∈ B\P. Now we have

$$f(a)\Gamma f(b) \subseteq (f(0) + <a>)\Gamma(f(0) + <b>)$$
$$\subseteq (P + <a>)\Gamma(P + <b>)$$
$$\subseteq P + <a>\Gamma<b> \subseteq P + A\Gamma B \subseteq P.$$

Since P is f-prime, we have either $f(a) \subseteq P$ or $f(b) \subseteq P$, which is a contradiction. Therefore, P is prime.
(iv) Follows from (iii).
   (b) Suppose $f(0) \subseteq$ P-rad (M). Then by (a) (i), "0" is f-strongly nilpotent.

Now by Theorem 7.3.24, f-rad(M) = {x ∈ M/x is f-strongly nilpotent}.

Converse

Suppose that f-rad(M) = {x ∈ M/x is f-strongly nilpotent}. Then "0" is f-strongly nilpotent. So, f(0) is strongly nilpotent, and by Theorem 7.3.26 (i), $f(0) \subseteq$ P-rad(M). ∎

Definitions 7.3.28

A subset S of M is said to be an **sp-system** if $a \in S$ implies $(<a>\Gamma<a>) \cap S \neq \emptyset$.

A nonempty subset K of M is called an **f-sp-system** if K contains an sp-system K^* such that $f(x) \cap K^* \neq \emptyset$ for every $x \in K$. In this case, K^* is called a kernel of K.

An ideal I of M is said to be **f-semiprime** if $M\backslash I$ is an f-sp-system.

Remark 7.3.29

Every m-system is an sp-system, and so every f-system is an f-sp-system. But, the converse may not be true.

Example 7.3.30

Consider $M = \mathbb{Z}$, a ring of integers, as a Γ-near ring with $\Gamma = \mathbb{Z}$. Let p, q be two distinct prime numbers. Define f by $f(a) = <\{a, pq\}>$. Now $f(pq) = <pq>$ and $M\backslash<pq>$ is an sp-system. Thus, $M\backslash<pq>$ is an f-sp-system.

Now $f(p)f(q) = <\{p, pq\}><\{q, pq\}> \subseteq <pq>$, but $p \notin <pq>$ and $q \notin <pq>$. By Theorem 7.3.5 (v), $<pq>$ is not f-prime. Hence, $<pq>$ is f-semiprime, but not f-prime.

Let K be a subset of M with the property that for any element a in K, there exists an sp-system $S \subseteq K$ such that $f(a) \cap S \neq \emptyset$. Let X be the union of all sp-systems that are contained in K. One can easily verify that K is an f-sp-system with kernel X. Hence, we have the following.

Remark 7.3.31

A subset K of M is an f-sp-system if and only if K satisfies the condition: $a \in K \Rightarrow$ there exists an sp-system $S \subseteq K$ such that $f(a) \cap S \neq \emptyset$.

Lemma 7.3.32 (Lemma 6.6.5 of Satyanarayana, 1984)

If S is an sp-system and $x \in S$, then there exists an m-system X such that $x \in X \subseteq S$.

Proof

Let $x \in S$ and S be an sp-system. Then we can choose an element x_1 in $(<x>\Gamma<x>) \cap S$. Again, since S is an sp-system, we can choose x_2 in $(<x_1>\Gamma<x_1>) \cap S$.

If we continue this process, we get a sequence $\{x_i\}$ of elements in S such that $x_0 = x$ and $x_i \in (<x_{i-1}>\Gamma<x_{i-1}>) \cap S$ for $i \geq 1$. Now $x_i \in (<x_{i-1}>\Gamma<x_{i-1}>) \subseteq <x_{i-1}>$ for each i, so that $<x_0> \supseteq <x_1> \supseteq <x_2> \supseteq \ldots$.

Now it is easy to verify that $X = \{x_0, x_1, x_2, \ldots\}$ is an m-system and $x = x_0 \in X \subseteq S$.   ∎

Theorem 7.3.33 (Theorem 6.6.6 of Satyanarayana, 1984)

Let A be an ideal of M. Then A is f-semiprime if and only if $A = $ f-rad (A).

Proof

Suppose A is f-semiprime. Clearly, $A \subseteq$ f-rad(A). To show f-rad$(A) \subseteq A$, let $a \in$ rad(A). Suppose that $a \notin A$. Since $M\backslash A$ is an f-sp-system, there exists an element x and an sp-system K such that $x \in f(a)$ and $x \in K \subseteq (M\backslash A)$.

Now by Lemma 7.3.32, there exists an m-system K^* such that $x \in K^* \subseteq K$. Write

$$Q = \{y \in M\backslash A \mid f(y) \, K^* \neq \emptyset\}.$$

Clearly, $K^* \subseteq Q \subseteq (M\backslash A)$, $a \in Q$, and Q is an f-system with kernel K^*. By Lemma 7.3.11, there exists an f-prime ideal P such that $P \cap Q = \emptyset$ and $P \supseteq A$.

Now $a \notin P$, and $P \supseteq$ f-rad(A), which is a contradiction.

Converse

Suppose that $A = $ f-rad(A). To show A is semiprime, we have to show $M\backslash A$ is an f-sp-system. We now show that $M\backslash A$ is an f-sp-system with kernel $(M\backslash P$-rad$(A))$. Since P-rad(A) is semiprime (because the intersection of any collection of prime ideals is a semiprime ideal), $(M\backslash P$-rad$(A))$ is an sp-system.

Let $x \in M\backslash A$. Now $x \notin A = $ f-rad(A), and so there is an f-system Y with kernel X such that $x \in Y$ and $Y \cap A = \emptyset$. Since $x \in Y$ and X is a kernel of Y, there exists an element z in $f(x) \cap X$. Now $z \in X$, X is an m-system, and $X \cap A = \emptyset$.

So, there is a prime ideal P containing A such that $z \notin P$, and hence $z \notin$ P-rad(A). Since $z \notin$ P-rad(A) and $z \in f(x)$, $f(x) \cap (M\backslash(P$-rad$(A))) \neq \emptyset$. The proof is complete.   ∎

Corollary 7.3.34 (Corollary 6.6.7 of Satyanarayana, 1984)

The intersection of any collection of f-prime ideals of M is an f-semiprime ideal, and every f-semiprime ideal is the intersection of all f-prime ideals containing it.

Corollary 7.3.35 (Corollary 6.6.8 of Satyanarayana, 1984)

Let A be an ideal of M. Then A is semiprime if and only if $A = $ P-rad(A).

7.4 Modules over Gamma Near Rings

In this section, we present the notion of a "module over a gamma near ring" and its substructures like $M\Gamma$-group, ideal (or $M\Gamma$-submodule). We also present some results related to these concepts.

Definition 7.4.1

Let M be a Γ-near ring. An additive group G is said to be a **Γ-near ring-module** (or **$M\Gamma$-module**) if there exists a mapping $M \times \Gamma \times G \to G$ (denote the image of (m, α, g) by $m\alpha g$ for $m \in M, \alpha \in \Gamma, g \in G$) satisfying the conditions

(i) $(m_1 + m_2)\alpha_1 g = m_1\alpha_1 g + m_2\alpha_1 g$ and
(ii) $(m_1\alpha_1 m_2)\alpha_2 g = m_1\alpha_1 (m_2\alpha_2 g)$

for all $m_1, m_2 \in M, \alpha_1, \alpha_2 \in \Gamma,$ and $g \in G$.
Throughout this section, G stands for an $M\Gamma$-module.

Definition 7.4.2

(i) An additive subgroup H of G is said to be a **$M\Gamma$-subgroup** if $m\alpha h \in H$ for all $m \in M, \alpha \in \Gamma,$ and $h \in H$. (Note that (0) and G are the trivial $M\Gamma$-subgroups.)
(ii) A normal subgroup H of G is said to be a **submodule** (or ideal) of G if $m\alpha(g + h) - m\alpha g \in H$ for $m \in M, \alpha \in \Gamma, g \in G,$ and $h \in H$.
(iii) For two $M\Gamma$-modules G_1 and G_2, a group homomorphism $\theta: G_1 \to G_2$ is said to be a **module homomorphism** (or **Γ-module homomorphism**) if $\theta(m\alpha g) = m\alpha(\theta g)$ for $m \in M, \alpha \in \Gamma,$ and $g \in G_1$.

Remark 7.4.3 (Remark 1.3 of Satyanarayana, 2004)

If H is a submodule of G, then the factor group G/H becomes an $M\Gamma$-module with respect to the operation $m\alpha(g + H) = m\alpha g + H$, and the mapping $g \to (g + H)$ is a module epimorphism (say a natural epimorphism) of G onto G/H.

Definition 7.4.4

By an **irreducible (simple**, respectively) $M\Gamma$-module G, we mean a non-zero $M\Gamma$-module G containing no nontrivial $M\Gamma$-subgroups (submodules, respectively) of G. A nonzero submodule of G is said to be **irreducible (simple**, respectively) if it is an irreducible (simple, respectively) $M\Gamma$-module.

Since we are considering zero-symmetric Γ-near rings, every submodule is an $M\Gamma$-subgroup, and hence every irreducible $M\Gamma$-module (submodule, respectively) is a simple $M\Gamma$-module (submodule, respectively). But, the converse may not be true.

Example 7.4.5

Consider $G = \mathbb{Z}_4 = \{0, 1, 2, 3\}$, the ring of integers modulo 4, and $X = \{a, b\}$. Write $M = \{g \mid g: X \to G, g(a) = 0\} = \{g_0, g_1, g_2, g_3\}$, where $g_i(a) = 0$, $g_i(b) = i$ for $0 \le i \le 3$. Let $\Gamma = \{f_1, f_2, f_3, f_4\}$, where each $f_i: G \to X$ defined by $f_1(i) = a$ $(0 \le i \le 3)$, $f_2(i) = a$ $(0 \le i \le 2)$, $f_2(3) = b$, $f_3(i) = a$ for $i \in \{0, 2, 3\}$, $f_3(1) = b$, $f_4(i) = a$ if $i \in \{0, 2\}$, and $f_4(i) = b$ if $i \notin \{0, 2\}$. For $g \in M, f \in \Gamma, x \in G$, write $gfx = g(f(x))$. Now G becomes a $M\Gamma$-module.

Further, $Y = \{0, 2\}$ is only the nontrivial subgroup and also the $M\Gamma$-subgroup, but not a submodule of G (since $3 \notin Y$ and $g_3 f_2 (1 + 2) - g_3 f_2(1) = 3$). Hence, G is simple, but not irreducible.

Definition 7.4.6

A proper submodule of G is said to be **maximal (strictly maximal)** if it is maximal in the set of all proper submodules ($M\Gamma$-subgroups) of G.

If H is a strictly maximal submodule of G, then it is maximal. The converse may not be true.

Example 7.4.7

Consider $\mathbb{Z}_8$, the ring of integers modulo 8 and $X = \{a, b\}$.

Write $M = \{f \mid f: X \to \mathbb{Z}_8, f(a) = 0\} = \{f_0, f_1, \ldots, f_7\}$, where $f_i(a) = 0$, $f_i(b) = i$ for $1 \le i \le 7$, and $\Gamma = \{g_0, g_1\}$, where $g_0, g_1: \mathbb{Z}_8 \to X$ defined by $g_0(i) = a$ for all i and $g_1(i) = a$ if $i \notin \{3, 7\}$, $g_1(i) = b$ if $i \in \{3, 7\}$. $\mathbb{Z}_8$ becomes an $M\Gamma$-module and all its proper subgroups (0), $\{0, 4\}$, $\{0, 2, 4, 6\}$ are $M\Gamma$-subgroups, but $Y = \{0, 2, 4, 6\}$ is not a submodule (since $f_5 g_1 (1 + 2) - f_5 g_1(1) = 5 \notin Y$). Hence $\{0, 4\}$ is maximal, but not strictly maximal.

Definition 7.4.8 (Definition 2.1 of Satyanarayana, 2004)

Let $\mathscr{S}$ denote the set of all strictly maximal submodules of G. We define the submodule $J(G) = \bigcap_{B \in \mathscr{S}} B$ to be the **radical** of G. It is understood that if $\mathscr{S}$ is empty, then G is its own radical, in which case, we say C is a radical $M\Gamma$-module.

Example 7.4.9

(i) In Example 7.4.7, write $\Gamma^1 = \{g_0\}$. Then $\mathbb{Z}_8$ is an $M\Gamma^1$-module and every subgroup of $\mathbb{Z}_8$ is a submodule. So $\{0, 2, 4, 6\}$ is only the strictly maximal submodule and $J(\mathbb{Z}_8) = \{0, 2, 4, 6\}$, and hence $\mathbb{Z}_8$ is not a radical $M\Gamma^1$-module.

(ii) In Example 7.4.7, $\mathbb{Z}_8$ contains no strictly maximal submodules, and hence it is a radical $M\Gamma$-module.

(iii) In Example 7.4.5, write $\Gamma^1 = \{f_1\}$. Then $\mathbb{Z}_4$ is a $M\Gamma^1$-module with its unique submodule $\{0, 2\}$. So, its radical is $\{0, 2\}$ and hence it is not a radical $M\Gamma^1$-module.

The proofs of Lemma 7.4.10 and Corollary 7.4.11 stated in the following are straightforward verifications.

Lemma 7.4.10 (Note 2.3 (i) of Satyanarayana, 2004)

A submodule B of G is maximal (strictly maximal) if and only if G/B is a simple (irreducible) $M\Gamma$-module.

Corollary 7.4.11 (Note 2.3 (ii) of Satyanarayana, 2004)

(i) If G is a direct sum of irreducible submodules, then $J(G) = (0)$, and hence G is not a radical module.

(ii) $J(G/J(G)) = (0)$.

(iii) If H is a proper submodule and $J(G/H) = (0)$, then $J(G) \subseteq H$.

Definition 7.4.12

A submodule H of G is said to be **small** (**strictly small**, respectively) if $G = B$ for each submodule ($M\Gamma$-subgroup, respectively) B of G such that $G = H + B$.

Remark 7.4.13 (Remark 2.5 of Satyanarayana, 2004)

(i) Every strictly small submodule is small, but the converse may not be true.
(ii) If A, B are small (strictly small), then so is $A + B$.
(iii) Every small (strictly small) submodule is contained in every maximal submodule (maximal $M\Gamma$-subgroup). If A is small, then $A \subseteq J(G)$.

Verification

Let A be a small submodule and B be a maximal submodule of G. Now $B \subseteq A + B \subseteq G$ and B is maximal; we have $A + B = B$ or $A + B = G$. This implies $A \subseteq B$ or $A + B = G$. Suppose $A + B = G$. Since A is small we have $B = G$. So $A \subseteq B$. Also, every small submodule is contained in any strictly maximal submodule. So, every small submodule is contained in $J(G)$.

(iv) $J(G)$ contains the sum of all small submodules.
(v) If $J(G)$ is small, then $J(G)$ = sum of all small submodules.
(vi) $J(G)$ contains the sum of all strictly small submodules (this follows since every strictly small submodule is a small submodule). Moreover, if $J(G)$ is strictly small, then $J(G)$ is equal to the sum of all strictly small submodules.

Example 7.4.14

Consider S_3, the group of permutations on $\{1, 2, 3\}$. Write $e = (1)$, $\alpha = (12)$, and $\beta = (123)$. For any two elements x, y of S_3, define $x + y$ to be the composition of x and y. Then

$$S_3 = \{e, \alpha, \beta, \beta + \beta, \alpha + \beta, \beta + \alpha\}.$$

Write $X = \{a, b\}$ and $M = \{f \mid f: X \to S_3$ such that $f(a) = e\}$. Now $M = \{f_x / x \in S_3\}$, where f_x is defined as the functions $f_x(a) = e$ and $f_x(b) = x$. It is clear that M is a group under the addition of mappings. Write $\Gamma = \{g\}$, where $g: S_3 \to X$ defined by $g(x) = a$ for all $x \in S_3$. Now M is a Γ-near ring, and if we define $fgd = f(g(d))$ for all $f \in M$, $d \in S_3$, then S_3 becomes an $M\Gamma$-module.

In S_3, $A = \{e, \beta, \beta + \beta\}$ is a unique nontrivial normal subgroup, and the other subgroups of G are not normal. From the construction, A is a submodule of S_3, and the other subgroups of G are only $M\Gamma$-subgroups. Write $B = \{e, \alpha\}$. Then $A + B = G$ and $B \neq G$, which implies A is small, but not strictly small.

Let $\mathscr{L}$ ($\mathscr{L}_1$, respectively) denote the collection of maximal submodules (maximal $M\Gamma$-subgroups, respectively) of G.

Theorem 7.4.15 (Theorem 2.7 (i) of Satyanarayana, 2004)

If $J(G)$ is small (strictly small, respectively), then $J(G) = \bigcap_{B \in L} B (= \bigcap_{B \in L_1} B$, respectively).

Proof

Suppose that $J(G)$ is small. Since every strictly maximal ideal is a maximal ideal, we have that $\cap \{B \mid B$ is a maximal ideal of $G\} \subseteq J(G)$.

Since $J(G)$ is small, by Remark 7.4.13 (iii), we have that $J(G) \subseteq B$ for any maximal ideal B of G. Hence, $J(G) = \cap \{B \mid B$ is a maximal ideal of $G\}$.

The proof for the part related to "strictly small" is similar.

The proof of the following corollary is trivial. ∎

Corollary 7.4.16

If $J(G)$ is strictly small, then

$$J(G) = \bigcap_{B \in L} B \ (= \bigcap_{B \in L_1} B).$$

Theorem 7.4.17 (Theorem 2.7 (iii) of Satyanarayana, 2004)

Suppose G has ascending chain condition (ACC) on its submodules. Then $J(G)$ is small if and only if $J(G) = \bigcap_{B \in L} B$.

Proof

Suppose $J(G) = \bigcap_{B \in L} B$. Let us suppose $J(G)$ is not small. Since $J(G)$ is not small, the set $N = \{A \mid A$ is a proper submodule of G and $J(G) + A = G\}$ is nonempty. Since G has ACC, there exists a (proper) maximal submodule B of G such that $G = J(G) + B$. By Remark 7.4.13 (iii), we have $J(G) \subseteq B$. This shows that $G = B$, a contradiction. ∎

The rest follows from Theorem 7.4.15 and Corollary 7.4.16.

By using a similar argument, we can obtain the following theorem.

Theorem 7.4.18 (Theorem 6.4.18 of Satyanarayana, 1984)

Suppose G has ACC on M-subgroups. Then $J(G)$ is strictly small if and only if $J(G) = \bigcap_{B \in L} B$.

The proof of the following corollary is a straightforward verification.

Corollary 7.4.19 (Theorem 6.9.19 of Satyanarayana, 1984)

   (i) $J(G)$ contains the sum of all small submodules.
  (ii) If $J(G)$ is small (strictly small), then $J(G)$ is the sum of all small (strictly small) submodules.

Theorem 7.4.20 (Theorem 6.9.20 of Satyanarayana, 1984)

$J(G)$ is the sum of all small submodules if and only if every submodule B generated by a finite subset of $J(G)$ is small.

Proof

A straightforward verification (use Remark 7.4.13 and Corollary 7.4.19).
   A similar argument shows us the following.   ■

Theorem 7.4.21 (Theorem 6.9.21 of Satyanarayana, 1984)

$J(G)$ is the sum of all strictly small submodules if and only if every submodule B generated by a finite subset of $J(G)$ is strictly small.
   Henceforth, we consider G to be an $M\Gamma$-module that satisfies descending chain condition (DCC) on submodules. By using Lemma 7.4.10, we obtain the following.

Lemma 7.4.22 (Lemma 6.9.22 of Satyanarayana, 1984)

If there exist finitely many strictly maximal submodules $I_1, I_2, ..., I_n$ such that $\bigcap_{j=1}^{n} I_j = (0)$, then G is a direct sum of finite number of irreducible submodules.

Lemma 7.4.23 (Theorem 3.1 of Satyanarayana, 2004)

If $J(G) = (0)$, then there exists a finite number of strictly maximal submodules $I_1, I_2, ..., I_n$ of G such that $\bigcap_{k=1}^{n} I_k = (0)$.

Proof

Let D be the collection of all strictly maximal submodules of G.
   Since G has DCC on submodules, the set
   $\zeta = \{\bigcap_{i=1}^{k} J_i / J_i \in D, 1 \leq i \leq k\}$ has a minimal element, say $\bigcap_{i=1}^{n} I_i$.
   Now $J \in D \Rightarrow J \cap (\bigcap_{i=1}^{n} I_i) \in \zeta \Rightarrow \bigcap_{i=1}^{n} I_i \subseteq J$. Hence, $\bigcap_{i=1}^{n} I_i \subseteq J(G) = (0)$.   ■

Now combining Lemmas 7.4.22 and 7.4.23 and Corollary 7.4.11 (i), we get the following.

Theorem 7.4.24 (Theorem 6.9.24 of Satyanarayana, 1984)

Suppose G has DCC on submodules. Then $J(G) = (0)$ if and only if G can be expressible as a sum of a finite number of minimal submodules.

Corollary 7.4.25 (Corollary 6.9.25 of Satyanarayana, 1984)

If $J(G) = (0)$ and A is a submodule of G, then there exists a submodule B of G such that $G = A \oplus B$.

Proof

A straightforward verification. ∎

Lemma 7.4.26 (Note 3.2 of Satyanarayana, 2004)

If $J_1, J_2, ..., J_n$ are finite numbers of submodules, $m \in M$ and $\alpha \in \Gamma$, then for all $a_i \in J_i$, we have

$$m\alpha(a_1 + a_2 + ... + a_n) = m\alpha a_1 + m\alpha a_2 + ... + m\alpha a_n \left(\bmod \left(\bigcap_{i=1}^{n} J_i \right) \right).$$

Proof

Refer to Proposition 7.2.4 and use mathematical induction. ∎

Theorem 7.4.27 (Theorem 6.9.27 of Satyanarayana, 1984)

If $J(G) = (0)$, then every simple submodule A is irreducible.

Proof

By Theorem 7.4.24, we have that $G = I_1 \oplus I_2 \oplus ... \oplus I_n$ for some irreducible submodules $I_1, I_2, ..., I_n$. Let $0 \neq a \in A$. Then $a = a_1 + a_2 + ... + a_n$ for some $a_i \in I_i$, $1 \leq i \leq n$, and $a_k \neq 0$ for some $1 \leq k \leq n$. Let f_k be the map that carries each element of A into its component in I_k. Then f_k is a nonzero module homomorphism from A into I_k. Since I_k is irreducible and A is simple, it follows that the image of f_k is I_k and the kernel of f_k is (0). Therefore, f_k is a module isomorphism, and hence A is irreducible. ■

Theorem 7.4.28 (Theorem 3.3 of Satyanarayana, 2004)

If $J(G) = (0)$, then every maximal submodule is strictly maximal.

Proof

Let A be a maximal submodule. By Corollary 7.4.25, there exists a submodule B such that $G = A \oplus B$. Now B is $M\Gamma$-isomorphic to G/A. By Lemma 7.4.10, we have that B is simple and by Theorem 7.4.27, we have that B is irreducible, and hence A is strictly maximal by Lemma 7.4.10. ■

Theorem 7.4.29 (Theorem 3.5 (i) of Satyanarayana, 2004)

If A is a maximal submodule that contains $J(G)$, then A is strictly maximal.
The proof is a straightforward verification (use Corollary 7.4.11 and Theorem 7.4.28).

Corollary 7.4.30 (Theorem 3.5 (ii) of Satyanarayana, 2004)

If $J(G)$ is small, then every maximal submodule is strictly maximal.

Proof

If $J(G)$ is small, then $J(G) \subseteq A$ for any maximal submodule and hence follows from the previous theorem. ■
Combining Theorem 7.4.17 and Corollary 7.4.30, we have the following theorem.

Theorem 7.4.31 (Theorem 3.5 (iii) of Satyanarayana, 2004)

If G has ACC and DCC on submodules, then $J(G)$ is small if and only if every maximal submodule is strictly maximal.

8

Fuzzy Aspects in Near Rings and Gamma Near Rings

8.1 Fuzzy Ideals and Prime Ideals in Gamma Near Rings

In this chapter, M stands for a zero-symmetric gamma near ring. Proofs for some of the results of this chapter are either parallel or minor modifications of the proofs of the related results in ring theory or in near ring theory. For the sake of completeness, we provide the proofs in detail. We begin this section with the following definition.

Definition 8.1.1

Let σ and τ be two fuzzy subsets of M. Then the fuzzy subset $\sigma \circ \tau$ (the composition of σ and τ) of M is defined by

$$(\sigma \circ \tau)(x) = \sup_{x=y\alpha z}\{\min(\sigma(y), \tau(z))\} \text{ if } x \text{ is expressible as a product } x = y\alpha z$$

$$\text{for some } \alpha \in \Gamma.$$

$$= 0, \text{ otherwise, for all } x, y, z \in M.$$

Definition 8.1.2

Let $\mu: M \to [0, 1]$. Then μ is said to be a **fuzzy ideal** of M if it satisfies the following conditions:

(i) $\mu(x + y) \geq \min\{\mu(x), \mu(y)\}$
(ii) $\mu(-x) = \mu(x)$
(iii) $\mu(x) = \mu(y + x - y)$
(iv) $\mu(x\alpha y) \geq \mu(x)$ and
(v) $\mu\{(x\alpha(y + z) - x\alpha y\} \geq \mu(z)$ for all $x, y, z \in M$ and $\alpha \in \Gamma$.

Proposition 8.1.3 (Theorem 3.5 of Jun, Sapanci, and Ozturk, 1998)

Let μ be a fuzzy subset of M. Then the level subsets

$$\mu_t = \{x \in M \mid \mu(x) \geq t\}, \, t \in \text{Im } \mu,$$

are ideals of M if and only if μ is a fuzzy ideal of M.

Proof

Suppose μ is a fuzzy ideal of M. To show μ_t is an ideal of M, let $x, y \in \mu_t$.
Then $\mu(x) \geq t, \mu(y) \geq t$.
Consider $\mu(x + y)$. Now

$\mu(x + y) \geq \min \{\mu(x), \mu(y)\}$ (since μ is a fuzzy ideal of M) $\geq \min \{t, t\} = t$.

Therefore, $x + y \in \mu_t$.
Since μ is a fuzzy ideal of M, $\mu(-x) = \mu(x) \geq t$. Therefore, $-x \in \mu_t$. Hence, μ_t is a subgroup of M.
Consider $\mu(y + x - y)$, where $x, y \in \mu_t$. Now

$\mu(y + x - y) = \mu(x)$ (since μ is a fuzzy ideal of M) $\geq t$ (since $x \in \mu_t$).

Therefore, $y + x - y \in \mu_t$. Hence, μ_t is a normal subgroup of M.
Let $i \in \mu_t, x, y \in M$, and $\alpha \in \Gamma$. Consider $\mu\{x\alpha(y + i) - x\alpha y\}$. Now

$\mu\{x\alpha(y + i) - x\alpha y\} \geq \mu(i)$ (since μ is a fuzzy ideal of M) $\geq t$ (since $i \in \mu_t$).

Therefore, $\mu\{x\alpha(y + i) - x\alpha y\} \in \mu_t$. Hence, μ_t is a left ideal of M.
Consider $\mu(i\alpha x)$. Now

$\mu(i\alpha x) \geq \mu(i)$ (since μ is a fuzzy ideal of M) $\geq t$ (since $i \in \mu_t$).

Therefore, $i\alpha x \in \mu_t$.
Thus, we proved that μ_t is a right ideal of M. Hence, μ_t is an ideal of M.

Converse

Suppose that μ_t is an ideal of M for all $t \in \text{Im } \mu$. Now we show that μ is a fuzzy ideal of M.

(i) Let $x, y \in M$. Without loss of generality, we may assume that $\mu(x) < \mu(y)$, and $\mu(x) = t$. Now
$x, y \in \mu_t \Rightarrow x + y \in \mu_t$ (since μ_t is a subgroup)
$\Rightarrow \mu(x + y) \geq t = \min \{\mu(x), \mu(y)\}$.
Therefore, $\mu(x + y) \geq \min \{\mu(x), \mu(y)\}$.

(ii) Suppose $\mu(x) = t$. Then

$$x \in \mu_t \Rightarrow -x \in \mu_t \text{ (since } \mu_t \text{ is a subgroup)} \Rightarrow \mu(-x) = t.$$

(iii) Suppose $\mu(x) = t$. Then

$$x \in \mu_t \Rightarrow y + x - y \in \mu_t \text{ (since } \mu_t \text{ is a normal subgroup)}$$
$$\Rightarrow \mu(y + x - y) \geq t = \mu(x) \tag{8.1}$$

In Equation 8.1, by taking $-y + x + y$ in place of x, we get
$\mu(y - y + x + y - y) \geq \mu(-y + x + y)$
$\Rightarrow \mu(x) \geq \mu(-y + x + y)$.
Now take y in place of $-y$, to get

$$\mu(x) \geq \mu(y + x - y) \tag{8.2}$$

From Equations 8.1 and 8.2, we see that $\mu(x) = \mu(y + x - y)$.

(iv) Take $x, y \in M$, $\alpha \in \Gamma$. Put $\mu(x) = t$. Then $x \in \mu_t$. Since μ_t is a right ideal, it follows that $x\alpha y \in \mu_t$. This implies that $\mu(x\alpha y) \geq t = \mu(x)$.

(v) Let $x, y, z \in M$ and $\alpha \in \Gamma$. Put $\mu(z) = t$. Then $z \in \mu_t$. Since μ_t is a left ideal, it follows that $x\alpha(y + z) - x\alpha y \in \mu_t \Rightarrow \mu\{(x\alpha(y + z) - x\alpha y\} \geq t = \mu(z)$. Therefore, μ is a fuzzy left ideal of M. Thus, we proved that μ is a fuzzy ideal of M. ∎

Lemma 8.1.4 (Theorem 1.1 (i), (ii) of Prasad and Satyanarayana, 2005)

(i) If μ is a fuzzy ideal of M, then $\mu(x + y) = \mu(y + x)$ for all $x, y \in M$.

(ii) If μ is a fuzzy ideal of M, then $\mu(0) \geq \mu(x)$ for all $x \in M$.

Verification

(i) Put $z = x + y$. Now

$$\mu(x + y) = \mu(z) = \mu(-x + z + x) \text{ (since } \mu \text{ is a fuzzy ideal)}$$
$$= \mu(-x + x + y + x) = \mu(y + x).$$

(ii) Clearly, $0 = 0\alpha x$ for all $\alpha \in \Gamma$ and $x \in M$. This implies $\mu(0) = \mu(0\alpha x)$. Consider $\mu(0)$. Now

$$\mu(0) = \mu\{0\alpha(0 + x) - 0\alpha0\} \geq \mu(x) \text{ (since } \mu \text{ is a fuzzy ideal of } M).$$

Therefore, $\mu(0) \geq \mu(x)$ for all $x \in M$.

Lemma 8.1.5 (Theorem 1.1 (iii) of Prasad and Satyanarayana, 2005)

Let μ be a fuzzy ideal of M. If $\mu(x - y) = \mu(0)$, then $\mu(x) = \mu(y)$ for all $x, y \in M$.

Proof

Suppose $\mu(x - y) = \mu(0)$. Now
$\mu(y - x) = \mu(-(y - x))$ (since μ is a fuzzy ideal of M) $= \mu(x - y) = \mu(0)$. Now

$$\mu(x) = \mu(x - y + y) \geq \min \{\mu(x - y), \mu(y)\}$$
$$= \min \{\mu(0), \mu(y)\} \text{ (since } \mu(x - y) = \mu(0)) = \mu(y) \text{ (by Lemma 8.1.4 (ii))}.$$

Therefore, $\mu(x) \geq \mu(y)$.
On the other hand, take $\mu(y)$. Now

$$\mu(y) = \mu(y - x + x) \geq \min \{\mu(y - x), \mu(x)\} \quad \text{(since } \mu \text{ is a fuzzy ideal)}$$
$$\geq \min \{\mu(0), \mu(x)\} \quad \text{(since } \mu(y - x) = 0)$$
$$= \mu(x) \quad \text{(by Lemma 8.1.4 (ii))}.$$

Therefore, $\mu(y) \geq \mu(x)$. The proof is complete. ∎

Proposition 8.1.6 (Theorem 3.3 of Jun, Sapanci, and Ozturk, 1998)

Let I be an ideal of a gamma near ring M and $t < s \neq 0$ in [0,1]. Then the fuzzy subset μ defined by

$$\mu(x) = \begin{cases} s & \text{if } x \in I \\ t & \text{otherwise} \end{cases} \quad \text{is a fuzzy ideal of } M.$$

Proof

Let $x, y \in M$.

(i) Suppose $\mu(x) = t$, $\mu(y) = t$. Now
$\mu(x + y) \geq t$ (by the definition of μ) $= \min \{\mu(x), \mu(y)\}$.
Suppose $\mu(x) = s$, $\mu(y) = s$. Then $x, y \in I$. So, $x + y \in I$ (since I is an ideal of M). Now $\mu(x + y) = s = \min \{\mu(x), \mu(y)\}$. Therefore,

$$\mu(x + y) \geq \min \{\mu(x), \mu(y)\}.$$

Suppose $\mu(x) = t$, $\mu(y) = s$. Then $x \notin I$ and $y \in I$. Now
$\mu(x + y) \geq t = \min \{t, s\} = \min \{\mu(x), \mu(y)\}$.

(ii) Take $x \in M$. Then $\mu(x) = t$ or $\mu(x) = s$. Suppose $\mu(x) = t$. Then $x \notin I$, and so $-x \notin I$. Therefore, $\mu(-x) = t$. Suppose $\mu(x) = s$. Then $x \in I$, and so $-x \in I$ (since I is a subgroup of M). Therefore, $\mu(-x) = s$. Hence, $\mu(x) = \mu(-x)$.

(iii) Take $x, y \in M$. To show that $\mu(x) = \mu(y + x - y)$, suppose $\mu(x) = s$. Then $x \in I$. Since I is normal, we have $y + x - y \in I$. Therefore, $\mu(y + x - y) = s$. So, $\mu(x) = \mu(y + x - y)$.
Suppose $\mu(x) = t$. Then $x \notin I$, and so $y + x - y \notin I$. Therefore, $\mu(y + x - y) = t$. Hence, $\mu(x) = \mu(y + x - y)$.

(iv) Take $x, y \in M$ and $\alpha \in \Gamma$. Suppose $\mu(x) = s$. Then $x \in I$. Since I is an ideal of M, it follows that $x\alpha y \in I$. Therefore, $\mu(x\alpha y) = s = \mu(x)$.
If $\mu(x) = t$, then $\mu(x\alpha y) \geq t = \mu(x)$. Hence, in all cases, we have $\mu(x\alpha y) \geq \mu(x)$.

(v) Let $x, y, z \in M$, $\alpha \in \Gamma$. If $\mu(z) = s$, then $z \in I$. Since I is an ideal of M, it follows that

$$x\alpha(y + z) - x\alpha y \in I, \text{ and so } \mu\{x\alpha(y + z) - x\alpha y\} = s = \mu(z).$$

If $\mu(z) = t$, then $\mu\{x\alpha(y + z) - x\alpha y\} \geq t = \mu(z)$.
Hence, we conclude that μ is a fuzzy ideal of M. ∎

Lemma 8.1.7 (Lemma 7.1.9 of Prasad, 2000)

Let M and M^1 be two Γ-near rings and $f: M \to M^1$ be a gamma near ring homomorphism. If f is surjective and μ is a fuzzy ideal of M, then so is $f(\mu)$. If σ is a fuzzy ideal of M^1, then $f^{-1}(\sigma)$ is a fuzzy ideal of M.

Proof

Part (i): Assume that μ is a fuzzy ideal of M. Let $u, v \in M^1$.

(i) $(f(\mu))(u + v) = \sup\limits_{f(x) = u + v} \mu(x)$

(since $u + v \in M^1$ and f is onto, there exists $x \in M$ such that $f(x) = u + v$)

$$\geq \min\left\{ \sup\limits_{f(y) = u} \mu(y), \sup\limits_{f(z) = v} \mu(z)\right\} \overset{\infty}{\underset{k=0}{U}} X_i$$

$$= \min \{(f(\mu))(u), (f(\mu))(v)\}.$$

(ii) $(f(\mu))(-u) = \sup\limits_{f(z) = -u} \mu(z) = \sup\limits_{f(-z) = u} \mu(-z) = (f(\mu)(u))$

(since $f(z) = -u, f(-z) = f(0-z) = f(0) - f(z) = 0 - (-u) = u$, and μ is a fuzzy ideal, it follows that $\mu(z) = \mu(-z)$).

(iii) $(f(\mu)) (v + u - v) = \sup\limits_{f(z) = v + u - v} \mu(z)$

$$\geq \mu(y + x - y)$$

(since f is onto, there exist $x, y \in M$ such that $f(x) = u$, $f(y) = v$, and
$f(y + x - y) = f(y) + f(x) - f(y) = v + u - v$).
Hence $(f(\mu))(v + u - v) \geq \sup\limits_{f(x) = u} \mu(x) = (f(\mu))(u)$.

(iv) Take $m_1{}^1, m_2{}^1 \in M^1$. Since f is onto, there exist $m_1, m_2 \in M$ such that
$f(m_1) = m_1{}^1, f(m_2) = m_2{}^1$. Now

$$(f(\mu))(m_1{}^1 \alpha m_2{}^1) = \sup\limits_{f(z) = m_1{}^1 \alpha m_2{}^1} \mu(z)$$
$$\geq \mu(m_1 \alpha m_2) \text{ (since } f(m_1 \alpha m_2) = f(m_1) \alpha f(m_2) = m_1{}^1 \alpha m_2{}^1)$$
$$\geq \mu(m_1) \text{ (since } \mu \text{ is a fuzzy ideal of } M).$$

Therefore $(f(\mu))(m_1{}^1 \alpha m_2{}^1) \geq \sup\limits_{f(m_1) = m_1{}^1} \mu(m_1) = (f(\mu))(m_1{}^1)$.

(v) Let $m_1{}^1, m_2{}^1, m_3{}^1 \in M^1$ and $\alpha \in \Gamma$. Now

$$(f(\mu))(m_1{}^1 \alpha(m_2{}^1 + m_3{}^1) - m_1{}^1 \alpha m_2{}^1) = \sup\limits_{f(z) = (m_1{}^1 \alpha(m_2{}^1 + m_3{}^1) - m_1{}^1 \alpha m_2{}^1)} \mu(z)$$
$$\geq \mu(m_1 \alpha(m_2 + m_3) - m_1 \alpha m_2)$$

[since $m_1{}^1, m_2{}^1, m_3{}^1 \in M^1$ and f is onto, there exist $m_1, m_2, m_3 \in \mu$ such
that $f(m_1) = m_1{}^1, f(m_2) = m_2{}^1, f(m_3) = m_3{}^1$; and

$$f(m_1 \alpha(m_2 + m_3) - m_1 \alpha m_2) = f(m_1 \alpha(m_2 + m_3)) - f(m_1 \alpha m_2)$$
$$= f(m_1) \alpha(f(m_2) + f(m_3)) - f(m_1) \alpha f(m_2)$$
$$= (m_1{}^1 \alpha(m_2{}^1 + m_3{}^1) - m_1{}^1 \alpha m_2{}^1)$$
$$\geq \mu(m_1) \text{ (since } \mu \text{ is a fuzzy ideal of } M).$$

Therefore $(f(\mu)(m_1{}^1 \alpha(m_2{}^1 + m_3{}^1) - m_1{}^1 \alpha m_2{}^1) \geq \sup\limits_{f(m_1) = m_1{}^1} \mu(m_1) = (f(\mu))(m_1{}^1)$.
Hence, $f(\mu)$ is a fuzzy ideal of M^1.

Part (ii): Suppose σ is a fuzzy ideal of M^1. Now we show that $f^{-1}(\sigma)$ is a fuzzy
ideal of M. Take $x, y \in M$.

(i) $(f^{-1}(\sigma))(x + y) = \sigma(f(x + y))$ (by definition)
                  $= \sigma(f(x)) + (f(y))$ (since σ is a homomorphism)
                  $\geq \min \{\sigma(f(x)), \sigma(f(y))\}$ (since σ is a fuzzy ideal of M^1)
                  $= \min \{f^{-1}(\sigma(x)), f^{-1}(\sigma(y))\}$ (by the definition of $f^{-1}(\sigma)$).

(ii) $(f^{-1}(\sigma))(-x) = \sigma(f(-x))$
              $= \sigma(-f(-x))$ (since σ is a fuzzy ideal)
              $= \sigma(f(x))$ (since f is a homomorphism)
              $= (f^{-1}(\sigma))(x)$.

(iii) $(f^{-1}(\sigma))(x) = \sigma(f(x))$
              $= \sigma(f(y) + f(x) - f(y))$ (since σ is a fuzzy ideal)
              $= \sigma(f(y + x) - f(y))$ (since f is a homomorphism)
              $= \sigma(f(y + x - y))$ (since f is a homomorphism)
              $= f^{-1}(\sigma)(y + x - y)$ for all $x, y \in M$.

(iv) $(f^{-1}(\sigma))(x\alpha y) = \sigma(f(x\alpha y))$

$$= \sigma(f(x)\alpha f(y)) \qquad \text{(since } f \text{ is a homomorphism)}$$
$$\geq \sigma(f(x)) \qquad \text{(since } \sigma \text{ is a fuzzy ideal)}$$
$$= (f^{-1}(\sigma))(x) \text{ for all } x, y \in M \text{ and } \alpha \in \Gamma.$$

(v) Let $m_1, m_2, m_3 \in M$ and $\alpha \in \Gamma$. Now

$$(f^{-1}(\sigma)) \{m_1\alpha(m_2 + m_3) - m_1\alpha m_2\}$$
$$= \sigma(f(m_1\alpha(m_2 + m_3) - m_1\alpha m_2)) \text{ (by the definition of } f^{-1})$$
$$= \sigma(f(m_1\alpha(m_2 + m_3)) - f(m_1\alpha m_2)) \text{ (since } f \text{ is a homomorphism)}$$
$$= \sigma[(f(m_1)\alpha f(m_2 + m_3)) - f(m_1)\alpha f(m_2))]$$
$$\geq \sigma(f(m_3)) \text{ (since } \sigma \text{ is a fuzzy ideal)}$$
$$= (f^{-1}(\sigma))(m_3).$$

Hence, $f^{-1}(\sigma)$ is a fuzzy ideal of M. ∎

Proposition 8.1.8 (Proposition 7.1.10 of Prasad, 2000)

Let M and M^1 be two gamma near rings, $h: M \rightarrow M^1$ be a gamma epimorphism and μ, σ be fuzzy ideals of M and M^1, respectively; then

(i) $h(h^{-1}(\sigma)) = \sigma$

(ii) $h^{-1}(h(\mu)) \supseteq \mu$

(iii) $h^{-1}(h(\mu)) = \mu$ if μ is a constant on ker h.

Proof

(i) Let $m_1^1 \in M^1$. Since h is onto, there exists $m_1 \in M$ such that $h(m_1) = m_1^1$. Since $m_1^1 \in M^1$, $\sigma(m_1^1) \in [0, 1]$. Put $t = \sigma(m_1^1)$. Now, we show that $h(h-^{-1}(\sigma))(m_1^1) = t$.

$$h^{-1}(\sigma)(m_1) = \sigma(h(m_1))$$
$$= \sigma(m_1^1) = t.$$

Therefore, $h^{-1}(\sigma(m_1)) = t$ for all $t \in h^{-1}(m_1^1)$. Hence, $h(h^{-1}(\sigma))(m_1^1) = \sup\limits_{m_1 \in h^{-1}(m_1^1)} (h^{-1}(\sigma))(m_1) = \sup \{t\} = t$. Thus, $h(h^{-1}(\sigma)) = \sigma$.

(ii) Consider $h^{-1}(h(\mu))(x)$. Now

$$h^{-1}(h(\mu)) (x) = (h(\mu)) (h(x))$$
$$= (h(\mu))(y)$$
$$= \sup\limits_{h(z) = y} \mu(z)$$
$$\geq \mu(x) \text{ (since } h(x) = y)$$

Therefore, $h^{-1}(h(\mu)) \supseteq \mu$.

(iii) Let $x \in M$. We have to show that $h^{-1}(h(\mu)) (x) = \mu(x)$. Write $\mu(x) = t$ and $h(x) = y$.

Let $x_1, x_2 \in h^{-1}(y) \Rightarrow h(x_1) = h(x_2) \Rightarrow x_1 - x_2 \in$ ker h
$\Rightarrow \mu(x_1 - x_2) = \mu(0)$ (since $0, x_1 - x_2 \in$ ker h and μ is constant on ker h)

Since $x \in h^{-1}(y)$, we have $\mu(x_1) = \mu(x_2) = \mu(x) = t$. Therefore,

$$h^{-1}(h(\mu))(x) = h(\mu)(h(x)) = h(\mu)(y) = \sup_{x \in h^{-1}(y)} \mu(z) = \sup \{t\} = t = \mu(x).$$

Hence, $h^{-1}(h(\mu)) = \mu$.   ∎

Definition 8.1.9

A fuzzy ideal μ of M is said to be a **fuzzy prime ideal** of M if μ is not a constant function; and for any two fuzzy ideals σ and τ of M, $\sigma \circ \tau \subseteq \mu$ implies that either $\sigma \subseteq \mu$ or $\tau \subseteq \mu$.

Theorem 8.1.10 (Theorem 2.3 of Prasad and Satyanarayana, 2005)

If μ is a fuzzy prime ideal of M, then $M_\mu = \{x \in M \mid \mu(x) = \mu(0)\}$ is a prime ideal of M.

Proof

Put $t = \mu(0)$ in Proposition 8.1.3. Then we see that M_μ is an ideal of M (here $M_\mu = \mu_{\mu(0)}$ is an ideal of M).

Now to show that M_μ is a prime ideal of M,

let A and B be two ideals of μ such that $A\Gamma B \subseteq M_\mu$. Define the fuzzy subsets σ and τ as

$$\sigma(x) = \begin{cases} \mu(0) & \text{if } x \in A \\ 0 & \text{if } x \notin A \end{cases}, \quad \tau(x) = \begin{cases} \mu(0) & \text{if } x \in B \\ 0 & \text{if } x \notin B \end{cases}$$

Now we show that $\sigma(x)$ is a fuzzy ideal of M. Let $x, y \in M$. If $x, y \in A$, then $x + y \in A$.

So, $\sigma(x + y) = \mu(0) = \min \{\mu(0), \mu(0)\} = \min \{\sigma(x), \sigma(y)\}$ (since $x, y \in A$).

$\sigma(-x) = \mu(0)$ (since $x \in A$, we have $-x \in A$)

   $= \sigma(x)$.

$\sigma(y + x - y) = \mu(0)$ (since $y + x - y \in I$) $= \sigma(x)$.

$\sigma(x\alpha y) = \mu(0)$ (since $x, y \in A$ and A is an ideal of M, we have $x\alpha y \in A$).

   $= \sigma(x)$.

Let $z \in A$. $\sigma\{x\alpha(y + z) - x\alpha y\} = \mu(0)$ (since $x\alpha(y + z) - x\alpha y \in A$)

   $= \sigma(z)$.

Suppose $x, y \notin A$. If $x + y \notin A$, then $\sigma(x + y) = 0 = \min \{\sigma(x), \sigma(y)\}$.

If $x + y \in A$, then $\sigma(x + y) \geq 0 = \min \{\sigma(x), \sigma(y)\}$.

Also, it can be easily verified that $\sigma(-x) = \sigma(x)$, $\sigma(y + x - y) = \sigma(x)$, $\sigma(x\alpha y) \geq \sigma(x)$, and $\sigma\{x\alpha(y + z) - x\alpha y\} \geq \sigma(z)$.

Hence, σ is a fuzzy ideal of M.

Similarly, we can verify that τ is a fuzzy ideal of M.

Next we show that $\sigma \circ \tau \subseteq \mu$.

If $(\sigma \circ \tau)(x) = 0$ for all $x \in M$, then there is nothing to prove (since $(\sigma \circ \tau)(x) = 0 \leq \mu(x)$ for all $x \in M$).

Otherwise, $(\sigma \circ \tau)(x) = \sup_{x = y\alpha z} \{\min (\sigma(y), \tau(z))\}$. So, we have to consider only the cases where $\min \{\sigma(y), \tau(z)\} > 0$. For all these cases, $\sigma(y) = \tau(z) = \mu(0)$. Therefore, $y \in A$ and $z \in B$. So $x = y\alpha z \in A\Gamma B \subseteq M_\mu \Rightarrow \mu(x) = \mu(0)$. Hence $(\sigma \circ \tau)(x) \leq \mu(x)$ for all x. Thus, $\sigma \circ \tau \subseteq \mu$.

Since μ is fuzzy prime, it follows that $\sigma \subseteq \mu$ or $\tau \subseteq \mu$. Suppose $\sigma \subseteq \mu$.

If $A \not\subseteq M_\mu$, then there exists $a \in A$ such that $a \notin M_\mu$.

This means $\mu(a) \neq \mu(0)$.

Now $\mu(0) = \mu(0\alpha a)$

$$= \mu\{0\alpha(0 + a) - 0\alpha 0\}$$

$$\geq \mu(a) \text{ (since } \mu \text{ is an ideal)}.$$

Therefore, $\mu(a) < \mu(0)$. Hence, $\sigma(a) = \mu(0) > \mu(a)$, which is a contradiction to the fact that $\sigma \subseteq \mu$.

Thus, we proved that if $\sigma \subseteq \mu$, then $A \subseteq M_\mu$.

Similarly, if $\tau \subseteq \mu$, we can show that $B \subseteq M_\mu$. Hence, M_μ is a prime ideal of M. ∎

Lemma 8.1.11 (Theorem 1.3 of Prasad and Satyanarayana, 2005)

If μ is a fuzzy ideal of M, and $a \in M$, then $\mu(x) \geq \mu(a)$ for all $x \in <a>$.

Proof

Let $a \in M$ and $x \in <a>$. We show that $\mu(x) \geq \mu(a)$. Clearly, $<a> = \bigcup_{i=0}^{\infty} \mu_i$.

Here,

$$\mu_{k+1} = \mu_k^* \cup \mu_k^+ \cup \mu_k^\circ \cup \mu_k^{++},$$

where $\mu_k^* = \{n + x - n \mid n \in M, x \in \mu_k\}$,

$\mu_k^\circ = \{x - y \mid x, y \in \mu_k\}$,

$\mu_k^+ = \{n_1\alpha(n_2 + a) - n_1\alpha n_2 \mid n_1, n_2, \in M, a \in \mu_k \text{ and } \alpha \in \Gamma\}$, and

$\mu_k^{++} = \{x\alpha m \mid x \in \mu_k, \alpha \in \Gamma, m \in M\}$.

We prove that $\mu(u) \geq \mu(a)$ for all $u \in \mu_m$. To prove this, we use the principle of induction on m. The verification is clear in the case $m = 0$. Suppose the induction hypothesis for k. That is, $\mu(x) \geq \mu(a)$ for all $x \in \mu_k$. Now let $v \in \mu_{k+1} = \mu_k^* \cup \mu_k^\circ \cup \mu_k^+ \cup \mu_k^{++}$. If $v \in \mu_k^*$ or μ_k°, then the proof is clearly $\mu(v) \geq \mu(a)$. Suppose that $v \in \mu_k^+$. Then $v = n_1\alpha(n_2 + a) - n_1\alpha n_2$ for some $n_1, n_2 \in M, x \in \mu_k$, and $\alpha \in \Gamma$. Now $\mu(v) = \mu(n_1\alpha(n_2 + x) - n_1\alpha n_2) \geq \mu(x)$ (since μ is a fuzzy ideal of M) $\geq \mu(a)$ (by induction hypothesis for k).

Suppose $v \in \mu_k^{++}$. Then $v = x\alpha m$ for some $x \in \mu_k$, $\alpha \in \Gamma$, and $m \in M$.
Now $\mu(v) = \mu(x\alpha m) \geq \mu(x)$ (since μ is a fuzzy ideal of M)
$\qquad\qquad\quad \geq \mu(a)$ (by induction hypothesis for k).

Thus, in all cases, we proved that $\mu(v) \geq \mu(a)$ for all $v \in \mu_{k+1}$. Hence by mathematical induction, we have $v \in \mu_m$, which implies that $\mu(v) \geq \mu(a)$ for every positive integer m. Thus, we conclude that $\mu(x) \geq \mu(a)$ for all $x \in <a>$. The proof is complete. ∎

Proposition 8.1.12 (Theorem 2.8 of Prasad and Satyanarayana, 2005)

Let I be an ideal of M and $\alpha \in [0, 1)$. Let μ be a fuzzy subset of M, defined by

$$\mu(x) = \begin{cases} 1 & \text{if } x \in I \\ s & \text{otherwise} \end{cases}.$$

Then μ is a fuzzy prime ideal of M if I is a prime ideal of M.

Proof

Suppose I is a prime ideal of M. By Proposition 8.1.6, μ is a nonconstant fuzzy ideal of M.

Let σ and τ be two fuzzy ideals of M such that $\sigma o \tau \subseteq \mu$, $\sigma \subseteq \mu$, $\tau \subseteq \mu$. Then there exist $x, y \in M$ such that $\sigma(x) > \mu(x)$ and $\tau(y) > \mu(y)$.

$\mu(x) = \mu(y) = s$ (by the definition of μ).

So, $x, y \notin I$. Since I is a prime ideal, it follows that $<x>\Gamma<y> \not\subseteq I$. Therefore, there exists $a \in <x>\Gamma<y>$ such that $a \notin I$. So $\mu(a) = s$.

$$\text{Hence } (\sigma o \tau)(a) \leq \mu(a) = s \qquad\qquad (8.3)$$

Let $a = c\alpha d$, where $c \in <x>$, $d \in <y>$, and $\alpha \in \Gamma$. Now
$\sigma o \tau(a) = \sup_{x = p\alpha q} \{\min (\sigma(p), \tau(q))\}$
$\qquad\quad \geq \min \{\sigma(c), \tau(d)\}$
$\qquad\quad \geq \min \{\sigma(x), \tau(y)\}$ (by Lemma 8.1.11)
$\qquad\quad \geq \min \{\mu(x), \mu(y)\} = s$.
Therefore, $\sigma o \tau(a) > s$, which is a contradiction to Equation 8.3.
Hence, μ is a fuzzy prime ideal of M. The proof is complete. ∎

Corollary 8.1.13 (Corollary 2.10 of Prasad and Satyanarayana, 2005)

Let I be an ideal of M. Then λ_I is a fuzzy prime ideal of M if and only if I is a prime ideal of M.

Proof

By Proposition 8.1.6, I is an ideal of M if and only if λ_I is a fuzzy ideal of M. Suppose I is prime. By taking $s = 0$ in Proposition 8.1.12, we have

$$\lambda_I(x) = \begin{cases} 1 & \text{if } x \in I \\ 0 & \text{otherwise} \end{cases}$$

is a fuzzy prime ideal of M.

Converse

Suppose that λ_I is a fuzzy prime ideal of M. We show that I is a prime ideal of M. Let A, B be two ideals of M such that $A\Gamma B \subseteq I$. Now we show that $A \subseteq I$ or $B \subseteq I$. Consider $(\lambda_A \circ \lambda_B)(x)$. Now

$$(\lambda_A \circ \lambda_B)(x) = \sup_{x = p\alpha q} \{\min (\lambda_A(p), \lambda_B(q)\} \text{ for some } \alpha \in \Gamma.$$
$$(\lambda_A \circ \lambda_B)(x) = 1 = \sup_{x = p\alpha q} \{\min (\lambda_A(p), \lambda_B(q)\} \text{ for some } \alpha \in \Gamma$$
$$\Rightarrow \text{ there exist } p, q \text{ such that } \min \{\lambda_A(p), \lambda_B(q)\} = 1$$
$$\Rightarrow \lambda_A(p) = 1, \lambda_B(q) = 1 \Rightarrow p \in A, q \in B$$
$$\Rightarrow p\alpha q \in A\Gamma B \subseteq I \Rightarrow \lambda_I(p\alpha q) = 1 \Rightarrow \lambda_I(x) = 1.$$

Therefore, $\lambda_A \circ \lambda_B \subseteq \lambda_I$. Since λ_I is fuzzy prime, it follows that $\lambda_A \subseteq \lambda_I$ or $\lambda_B \subseteq \lambda_I$. Suppose $\lambda_A \subseteq \lambda_I$. Then $x \in A \Rightarrow \lambda_A(x) = 1 \leq \lambda_I(x) \leq 1 \Rightarrow x \in I$. Therefore, $A \subseteq I$.
Similarly, if $\lambda_B \subseteq \lambda_I$, then $B \subseteq I$. Thus, I is a prime ideal of M. ∎

Lemma 8.1.14 (Lemma 2.6 of Prasad and Satyanarayana, 2005)

If μ is a fuzzy prime ideal of M, then $\mu(0) = 1$.

Proof

Suppose μ is a fuzzy prime ideal of M. Clearly, $\mu(0) > 1$. Suppose $\mu(0) < 1$. Since μ is nonconstant, there exists $a \in M$ such that $\mu(a) < \mu(0)$. Define fuzzy subsets σ and θ of M as

$$\sigma(x) = \begin{cases} 1 & \text{if } \mu(x) = \mu(0) \\ 0 & \text{otherwise} \end{cases}$$

and $\theta(x) = \mu(0)$ for all $x \in M$.
  By Proposition 8.1.6, it follows that σ and θ are fuzzy ideals.
  Since $\sigma(0) = 1 > \mu(0)$ and $\theta(a) = \mu(0) > \mu(a)$, it follows that $\sigma \not\subseteq \mu$, $\theta \not\subseteq \mu$.
  Now for any $b \in M$, $(\sigma \circ \theta)(b) = \sup_{b = x\alpha y} \{\min (\sigma(x), \theta(y))\}$ for some $\alpha \in \Gamma$.

Case (i): If $\sigma(x) = 0$, then $\mu(x) < \mu(0)$. Now

$$\min \{\sigma(x), \theta(y)\} = \min \{0, \mu(0)\} \leq \mu(x\alpha y) = \mu(b).$$

Therefore, $\sup_{b = x\alpha y} \{\min (\sigma(x), \theta(y))\} \leq \mu(b).$

Hence $(\sigma \circ \theta)(b) \leq \mu(b)$. This is true for all $b \in M$. Thus, $\sigma \circ \theta \subseteq \mu$.
Case (ii): Suppose $\sigma(x) = 1$. Then $\mu(x) = \mu(0)$. Now

$$\min\{\sigma(x), \theta(y)\} = \min \{\mu(0), \mu(0)\} = \mu(0) = \mu(x) \leq \mu(x\alpha y)$$
$$(\text{since } \mu \text{ is a fuzzy ideal of } M) = \mu(b).$$

Therefore, $(\sigma \circ \theta)(b) \leq \mu(b)$ for all $b \in M$.

Hence, $\sigma \circ \theta \subseteq \mu$. Since μ is fuzzy prime, it follows that $\sigma \subseteq \mu$ or $\theta \subseteq \mu$, which is a contradiction.

Therefore, $\mu(0) = 1$. ∎

Proposition 8.1.15 (Theorem 2.7 of Prasad and Satyanarayana, 2005)

If μ is a fuzzy prime ideal of M, then $|\text{Im } \mu| = 2$.

Proof

Suppose μ is a fuzzy prime ideal of M. We show that Im μ contains exactly two values. Let a, b be elements of M such that $\mu(a) < 1$ and $\mu(b) < 1$. It is enough to show that $\mu(a) = \mu(b)$.
Part (i): Define fuzzy subsets σ and θ as follows:

$$\sigma(x) = \begin{cases} 1 & \text{if } x \in <a> \\ 0 & \text{otherwise} \end{cases}$$

and $\theta(x) = \mu(a)$ for all $x \in M$. By Proposition 8.1.6, it follows that σ and θ are fuzzy ideals of M. Since $a \in <a>$, it follows that $\sigma(a) = 1 > \mu(a)$.

Therefore, $\sigma \not\subseteq \mu$. Let $z \in M$. Consider $(\sigma \circ \theta)(z)$.

Now $(\sigma \circ \theta)(z) = \sup_{z = x\alpha y} \{\min (\sigma(x), \theta(y))\}$ for some $\alpha \in \Gamma$. Suppose $\sigma(x) = 0$.

Then $\min \{\sigma(x), \theta(y)\} = \min \{0, \mu(a)\} = 0 \leq \mu(x\alpha y) = \mu(z).$
This is true for all $z \in M$ such that $z = x\alpha y$ for some $\alpha \in \Gamma, x, y \in M$.

Therefore, $\sup_{z = x\alpha y} \{\min (\sigma(x), \theta(y))\} \leq \mu(z).$

Hence $(\sigma \circ \theta)(z) \leq \mu(z)$. Thus, $\sigma \circ \theta \subseteq \mu$.
Suppose $\sigma(x) = 1$. Then

$$\min \{\sigma(x), \theta(y)\} = \min \{1, \mu(a)\}$$
$$= \mu(a) \leq \mu(x) \text{ (since } x \in <a>, \text{ by Lemma 8.1.11)}$$
$$\leq \mu(x\alpha y) \text{ (since } \mu \text{ is a fuzzy ideal of } M) = \mu(z).$$

Therefore, $\sigma \circ \theta \subseteq \mu$. Since μ is fuzzy prime, it follows that $\sigma \subseteq \mu$ or $\theta \subseteq \mu$. Since $\sigma \not\subseteq \mu$, it follows that $\theta \subseteq \mu$. Now

$$\mu(b) \geq \theta(b) = \mu(a) \tag{8.4}$$

Part (ii): We construct fuzzy ideals τ and δ as follows:

$$\tau(x) = \begin{cases} 1 & \text{if } x \in <b> \\ 0 & \text{otherwise} \end{cases}$$

and $\delta(x) = \mu(b)$ for all $x \in M$.

By Proposition 8.1.6, τ and δ are fuzzy ideals of M. Now $\tau(b) = 1 > \mu(b)$. Therefore, $\tau \not\subseteq \mu$.

Consider $\tau \circ \delta$.

Now $(\tau \circ \delta)(z) = \sup_{z=x\alpha y} \{\min (\tau(x), \delta(y))\}$. If $\tau(x) = 0$, then clearly $\tau \circ \delta \subseteq \mu$.

Suppose $\tau(x) = 1$. Then $x \in <b>$.

Now $\mu(x\alpha y) \geq \mu(x)$ (since μ is a fuzzy ideal of M)

$\geq \mu(b)$ (since $x \in <b>$)

$= \delta(y)$ (by the definition of δ)

$= \min \{1, \delta(y)\} = \min \{\tau(x), \delta(y)\}$.

This is true for all x, y such that $z = x\alpha y$.

Therefore, $\mu(z) = \mu(x\alpha y) \geq \sup_{z=x\alpha y} \{\min (\tau(x), \delta(y))\} = (\tau \circ \delta)(z)$ for all $z \in M$.

Hence, $\tau \circ \delta \subseteq \mu$. Since μ is a fuzzy prime ideal of M, it follows that $\tau \subseteq \mu$ or $\delta \subseteq \mu$. Since $\tau \not\subseteq \mu$, it follows that $\delta \subseteq \mu$. Therefore, $\mu(a) \geq \delta(a) = \mu(b)$. Hence, $\mu(a) \geq \mu(b)$.

Thus, from Parts (i) and (ii), we can conclude that $\mu(a) = \mu(b)$. ∎

Proposition 8.1.16

If μ is a fuzzy subset of M, then μ is a fuzzy prime ideal of M, if and only if Im $\mu = \{1, \alpha\}$, where $\alpha \in (0, 1)$.

Proof

Suppose Im $\mu = \{1, \alpha\}$. Then by Proposition 8.1.12, μ is a fuzzy prime ideal of M.

Converse

Suppose μ is a fuzzy prime ideal of M. Then by Proposition 8.1.15, $|\text{Im } \mu| = 2$, that is, Im $\mu = \{1, \alpha\}$, where $\alpha \in (0, 1)$. ∎

8.2 Fuzzy Cosets of Gamma Near Rings

In this section, we define the concept of a fuzzy coset of a fuzzy ideal of a gamma near ring M. We present certain fundamental results.

Definition 8.2.1

Let μ be a fuzzy ideal of a gamma near ring M and $m \in M$. Then, a fuzzy subset $m + \mu$ defined by

$$(m + \mu)(m^1) = \mu(m^1 - m) \text{ for all } m^1 \in M$$

is called a **fuzzy coset** of the fuzzy ideal μ.

Proposition 8.2.2 (Lemma 2.2 (i) of Satyanarayana and Prasad, 2005)

If μ is a fuzzy ideal of M, then $x + \mu = y + \mu$ if and only if $\mu(x - y) = \mu(0)$.

Proof

Suppose μ is a fuzzy ideal of M and $x + \mu = y + \mu$

$$
\begin{aligned}
&\Rightarrow (x + \mu)(y) = (y + \mu)(y) \\
&\Rightarrow \mu(y - x) = \mu(y - y) &&\text{(by the definition of a fuzzy coset)} \\
&\Rightarrow \mu(y - x) = \mu(0) \\
&\Rightarrow \mu(x - y) = \mu(0) &&\text{(since } \mu \text{ is a fuzzy ideal of } M).
\end{aligned}
$$

Converse

Suppose that $\mu(x - y) = \mu(0)$. Now we show that $x + \mu = y + \mu$. Let $z \in M$. Consider $(x + \mu)(z)$. Now

$$
\begin{aligned}
(x + \mu)(z) &= \mu(z - x) \\
&= \mu(-y + z - x + y) &&\text{(since } \mu \text{ is a fuzzy ideal of } M) \\
&= \mu((-y + z) + (-x + y)) \\
&\geq \min \{\mu(-y + z), \mu(-x + y)\} &&\text{(since } \mu \text{ is a fuzzy ideal of } M) \\
&= \min \{\mu(-y + z - y + y), \mu(-(-x + y))\} \\
&= \min \{\mu(z - y), \mu(x - y)\} &&\text{(since } \mu \text{ is a fuzzy ideal of } M) \\
&= \min \{\mu(z - y), \mu(0)\} &&\text{(since } \mu(x - y) = \mu(0)) \\
&= \mu(z - y) &&\text{(by Lemma 8.1.4)} \\
&= (y + \mu)(z) &&\text{(by the definition of a coset)}
\end{aligned}
$$

Therefore $(x + \mu)(z) \geq (y + \mu)(z)$ for all $z \in M$. Hence $(x + \mu) \supseteq (y + \mu)$.

Similarly, by interchanging y and x in the proof, we can show that $(y + \mu) \supseteq (x + \mu)$. ∎

Corollary 8.2.3 (Lemma 2.2 (ii) of Satyanarayana and Prasad, 2005)

If $x + \mu = y + \mu$, then $\mu(x) = \mu(y)$.

Proof

Suppose $x + \mu = y + \mu$. By Proposition 8.2.2, $\mu(x - y) = \mu(0)$. Now by Lemma 8.1.4, we see that $\mu(x) = \mu(y)$. ∎

Proposition 8.2.4 (Lemma 2.2 (v) of Satyanarayana and Prasad, 2005)

Every fuzzy coset of a fuzzy ideal μ of M is constant on every coset of an ordinary ideal M_μ, where $M_\mu = \{x \in M \mid \mu(x) = \mu(0)\}$.

Proof

Let $y, z \in M_\mu$. We show that $(x + \mu)(y) = (x + \mu)(z)$. Since $y, z \in M_\mu$, it follows that $\mu(y) = \mu(0)$ and $\mu(z) = \mu(0)$. Since M_μ is an ideal, it follows that $y - z \in M_\mu$. So $\mu(y - z) = \mu(0)$. Consider $(x + \mu)(y)$.

Now
$$
\begin{aligned}
(x + \mu)(y) &= \mu(x - y) \\
&= \mu(-(x - y)) && \text{(since } \mu \text{ is a fuzzy ideal of } M) \\
&= \mu(y - x) \\
&= \mu(-z + y - x + z) \\
&\geq \min\{\mu(-z + y), \mu(x - z)\} \\
&= \min\{\mu(y - z), \mu(x - z)\} \\
&= \min\{\mu(0), \mu(x - z)\} && \text{(since } \mu(y - z) = \mu(0)) \\
&= \mu(y - z) \\
&= (x + \mu)(z).
\end{aligned}
$$

Therefore $(x + \mu)(y) \geq (x + \mu)(z)$.

Similarly, by interchanging y and z in the proof, we can show that $(x + \mu)(z) \geq (x + \mu)(y)$. Hence $(x + \mu)(y) = (x + \mu)(z)$ for all $y, z \in M_\mu$. ∎

Corollary 8.2.5 (Lemma 2.2 (v) of Satyanarayana and Prasad, 2005)

If $z \in M_\mu$, then $(x + \mu)(z) = \mu(x)$.

Proof

Let $z \in M_\mu$. Then $\mu(z) = \mu(0)$. Now $z, 0 \in M_\mu$; we have

$(x + \mu)(z) = (x + \mu)(0)$ (by Proposition 8.2.4)

$\qquad \Rightarrow \mu(z - x) = \mu(0 - x) = \mu(-x) = \mu(x)$ (since μ is a fuzzy ideal of M).

Therefore, $\mu(z - x) = \mu(x)$. Hence $(x + \mu)(z) = \mu(x)$. ∎

Theorem 8.2.6 (Theorem 2.4 of Satyanarayana and Prasad, 2005)

Let μ be a fuzzy ideal of M. Then the set $M/\mu = \{x + \mu \mid x \in M\}$ of fuzzy cosets of μ is a gamma near ring with respect to the operations defined by

$$(x + \mu) + (y + \mu) = (x + y) + \mu; \text{ and}$$
$$(x + \mu)\alpha(y + \mu) = x\alpha y + \mu \text{ for all } x, y \in M \text{ and } \alpha \in \Gamma.$$

Proof

Suppose $x + \mu = u + \mu$, $y + \mu = v + \mu$.

By Proposition 8.2.2, $\mu(x - u) = \mu(y - v) = \mu(0)$. Now

$$
\begin{aligned}
\mu\{(x + y) - (u + v)\} &= \mu(x + y - v - u) \\
&= \mu\{(x + y - v) - u\} \\
&= \mu\{-u + (x + y - v)\} \\
&= \mu\{(-u + x) + (y - v)\} \\
&\geq \min\{\mu(-u + x), \mu(y - v)\} \text{ (since } \mu \text{ is a fuzzy ideal of } M) \\
&= \min\{\mu(x - \mu), \mu(y - v)\} \\
&= \mu(0) \qquad\qquad\qquad \text{(since } \mu(x - u) = \mu(y - v) = \mu(0)).
\end{aligned}
$$

Clearly, by Lemma 8.1.4 (ii), it follows that $\mu(0) \geq \mu\{(x + y) - (u + v)\}$.

Therefore, $\mu\{(x + y) - (u + v)\} = \mu(0)$. Hence, by Proposition 8.2.2, we have $x + y + \mu = u + v + \mu$.

Next we show that $x\alpha y + \mu = u\alpha v + \mu$. Consider $\mu(x\alpha y - u\alpha y)$. Now

$$
\begin{aligned}
\mu(u\alpha v - x\alpha y) &= \mu(u\alpha v - x\alpha v + x\alpha v - x\alpha y) \\
&= \mu((u - x)\alpha v + x\alpha(y + (-y + v)) - x\alpha y) \\
&\geq \min\{\mu(u - x), \mu(-y + v)\} \text{ (since } \mu \text{ is a fuzzy ideal of } M) \\
&= \min\{\mu(0), \mu(0)\} = \mu(0).
\end{aligned}
$$

Therefore, $\mu(x\alpha y - u\alpha y) = \mu(0) \Rightarrow x\alpha y + \mu = u\alpha v + \mu$ (by Proposition 8.2.2).

To verify that $M/\mu = \{x + \mu \mid x \in M\}$ is a gamma near ring with the preceding operations:

(i) Take $x + \mu, y + \mu \in M/\mu$. Now $(x + \mu) + (y + \mu) = (x + y) + \mu \in M/\mu$, since $x + y \in M$.

(ii) For every $x + \mu \in M/\mu$, $0 + \mu \in M/\mu$, $(x + \mu) + (0 + \mu) = (x + 0) + \mu = x + \mu$.

(iii) Let $x + \mu, y + \mu, z + \mu \in M/\mu$. Then

$$
\begin{aligned}
(x + \mu) + ((y + \mu) + (z + \mu)) &= (x + \mu) + ((y + z) + \mu) \\
&= (x + (y + z) + \mu) \\
&= ((x + y) + z + \mu) \text{ (since } (M, +) \text{ is a group)} \\
&= ((x + y) + \mu) + (z + \mu) \\
&= ((x + \mu) + (y + \mu)) + (z + \mu).
\end{aligned}
$$

(iv) To each $x + \mu \in M/\mu$, $(-x) + \mu \in M/\mu$ such that $(x + \mu) (-x + \mu) = 0 + \mu$.
Therefore $(M/\mu, +)$ is a group.
Let $x, y, z \in M$; then $x + \mu, y + \mu, z + \mu \in M/\mu$, and $\alpha, \beta \in \Gamma$.

$$
\begin{aligned}
((x + \mu) + (y + \mu))\alpha(z + \mu) & \\
&= ((x + y) + \mu)\alpha(z + \mu) \\
&= ((x + y)\alpha z) + \mu && \text{(by the definition of multiplication)} \\
&= ((x\alpha z) + (y\alpha z)) + \mu && \text{(by right distributivity in } M) \\
&= (x\alpha z) + \mu + (y\alpha z) + \mu \\
&= ((x + \mu)\alpha(z + \mu)) + ((y + \mu)\alpha(z + \mu)).
\end{aligned}
$$

Also $((x + \mu)\alpha(y + \mu))\beta(z + \mu)$

$$
\begin{aligned}
&= ((x\alpha y) + \mu)\beta(z + \mu) && \text{(by the definition of addition)} \\
&= (x\alpha y)z + \mu \\
&= x\alpha(y\beta z) + \mu && \text{(by associativity in } M) \\
&= (x + \mu)\alpha(y\beta z + \mu) \\
&= (x + \mu)\alpha((y + \mu)\beta(z + \mu)).
\end{aligned}
$$

Hence, M/μ is a gamma near ring. ∎

Proposition 8.2.7 (Lemma 2.6 of Satyanarayana and Prasad, 2005)

Let μ be a fuzzy ideal of M; the fuzzy subset θ_μ of M/μ defined by $\theta_\mu(x + \mu) = \mu(x)$ for all $x \in M$ is a fuzzy ideal of M/μ.

Proof

Given that $\theta_\mu(x + \mu) = \mu(x)$.
To show θ_μ is well-defined, suppose $x + \mu = y + \mu$

$$
\begin{aligned}
&\Rightarrow \mu(x - y) = \mu(0) \\
&\Rightarrow \mu(x) = \mu(y) \\
&\Rightarrow \theta_\mu(x + \mu) = \theta_\mu(y + \mu).
\end{aligned}
$$

We show that θ_μ is a fuzzy ideal of M/μ.

(i) $\theta_\mu((x + \mu) + (y + \mu)) = \theta_\mu(x + y + \mu)$
$\qquad = \mu(x + y)$
$\qquad \geq \min \{\mu(x), \mu(y)\}$ (since μ is a fuzzy ideal of M)
$\qquad = \min \{\theta_\mu(x + \mu), \theta_\mu(y + \mu)\}.$

Therefore, $\theta_\mu((x + \mu) + (y + \mu)) \geq \min \{\theta_\mu(x + \mu), \theta_\mu(y + \mu)\}.$
(ii) $\theta_\mu(x + A) = \mu(x) = \mu(-x)$ $\qquad\qquad$ (since μ is a fuzzy ideal of M)
$\qquad = \theta_\mu(-x + \mu).$
(iii) $\theta_\mu((y + A) + (x + \mu) - (y + \mu))$
$\qquad\qquad = \theta_\mu((y + x - y) + \mu) = \mu(y + x - y) = \mu(x) = \theta_\mu(x + \mu).$
(iv) $\theta_\mu((x + \mu)\alpha(y + \mu))$
$\qquad\qquad = \theta_\mu(x\alpha y + \mu)$
$\qquad\qquad = \mu(x\alpha y)$
$\qquad\qquad \geq \mu(x)$ $\qquad\qquad$ (since μ is a fuzzy ideal of M)
$\qquad\qquad = \theta_\mu(x + \mu).$
(v) $\theta_\mu\{(x + \mu)\alpha((y + \mu) + (z + \mu)) - (x + \mu)\alpha(y + \mu)\}$
$\qquad\qquad = \theta_\mu\{(x + \mu)\alpha((y + z) + \mu) - (x + \mu)\alpha(y + \mu)\}$
$\qquad\qquad = \theta_\mu\{(x\alpha(y + z)) + \mu - (x\alpha y) + \mu\}$
$\qquad\qquad = \theta_\mu\{(x\alpha(y + z) - (x\alpha y)) + \mu)\}$
$\qquad\qquad = \mu\{x\alpha(y + z) - x\alpha y\}$
$\qquad\qquad \geq \mu(z)$ $\qquad\qquad$ (since μ is a fuzzy ideal of M)
$\qquad\qquad = \theta_\mu(z + \mu).$
Hence, θ_μ is a fuzzy ideal of M/μ. $\qquad\qquad\qquad\qquad\qquad\qquad\qquad\qquad$ ∎

Theorem 8.2.8

If μ is a fuzzy ideal of M, then the map $\theta: M \to M/\mu$ defined by $\theta(x) = x + \mu$, $x \in M$, is a gamma near ring homomorphism with kernel $M_\mu = \{x \in M \mid \mu(x) = \mu(0)\}.$

Proof

Given $\theta : M \to M/\mu$ by $\theta(x) = x + \mu$. Now

$\qquad\qquad \theta(x + y) = (x + y) + \mu$ $\qquad\qquad$ (by the definition of θ)
$\qquad\qquad\qquad = (x + \mu) + (y + \mu)$
$\qquad\qquad\qquad = \theta(x) + \theta(y)$ $\qquad\qquad$ (by the definition of θ).

Consider $\theta(x\alpha y)$. Now

$\qquad\qquad \theta(x\alpha y) = (x\alpha y) + \mu$ $\qquad\qquad$ (by the definition of θ)
$\qquad\qquad\qquad = (x + \mu)\alpha(y + \mu)$
$\qquad\qquad\qquad = \theta(x)\alpha\theta(y).$ Therefore, θ is a homomorphism.

To prove $\ker \theta = M_\mu$:

Take $x \in \ker \theta$ if and only if $\theta(x) = 0 = 0 + \mu$

if and only if $x + \mu = 0 + \mu$
if and only if $\mu(x - 0) = \mu(0)$
if and only if $\mu(x) = \mu(0)$
if and only if $x \in M_\mu$.

Therefore, $\ker \theta = M_\mu$. ∎

Theorem 8.2.9 (Theorem 3.3 of Satyanarayana and Prasad, 2005)

The gamma near ring M/μ is isomorphic to the gamma near ring M/M_μ. The isomorphic correspondence is given by $x + \mu :\to x + M_\mu$ (refer to Definition 7.2.5 for gamma near ring isomorphism).

Proof

Define $f\, M \to M/\mu$ by $f(x) = x + \mu$. We verify that f is a gamma near ring homomorphism. Let $x, y \in M$ and $\alpha \in \Gamma$. Now
$f(x + y) = (x + y + \mu) = (x + y) + \mu = (x + \mu) + (y + \mu) = f(x) + f(y)$, and
$f(x\alpha y) = x\alpha y + \mu = (x + \mu)\alpha(y + \mu) = f(x)\alpha f(y)$.
Take $y \in M/\mu$; then $y = x + \mu$ for some $x \in M$. Now $f(x) = x + \mu = y$.
Since M_μ is the kernel of f (by Theorem 8.2.8), it follows that $M/M_\mu \cong M/\mu$. ∎

Lemma 8.2.10 (Lemma 3.5 of Satyanarayana and Prasad, 2005)

Let μ and σ be two fuzzy ideals of M such that $\sigma \supseteq \mu$ and $\sigma(0) = \mu(0)$. Then the fuzzy subset θ_σ of M/μ defined by $\theta_\sigma(x + \mu) = \sigma(x)$ for all $x \in M$ is a fuzzy ideal of M/μ such that $\theta_\sigma \supseteq \theta_\mu$.

Proof

We show that θ_σ is well-defined.
Suppose $x + \mu = y + \mu$. Then $\mu(x - y) = \mu(0)$. Now $\sigma(x - y) \geq \mu(x - y) = \mu(0) = \sigma(0)$ (since $\mu(0) = \sigma(0)$). By Lemma 8.1.4 (ii), it follows that $\sigma(0) \geq \sigma(x - y)$.
Therefore, $\sigma(x - y) = \sigma(0)$, and so by Proposition 8.1.6, $\sigma(x) = \sigma(y)$.
Hence, θ_σ is well-defined.
Now we verify that θ_σ is a fuzzy ideal of M/μ.

(i) $\theta_\sigma((x + \mu) + (y + \mu)) = \theta_\sigma((x + y) + \mu)$
$\qquad\qquad\qquad = \sigma(x + y)$
$\qquad\qquad\qquad \geq \min \{\sigma(x), \sigma(y)\}$ (since σ is a fuzzy ideal of M)
$\qquad\qquad\qquad = \min \{\theta_\sigma(x + \mu), \theta_\sigma(y + \mu)\}$.

(ii) $\theta_\sigma((-x) + \mu) = \theta_\sigma(-x + \mu) = \sigma(-x) = \sigma(x)$ (since σ is a fuzzy ideal of M)
$$= \theta_\sigma(x + \mu).$$

(iii) $\theta_\sigma((x + \mu)\alpha(y + \mu)) = \theta_\sigma(x\alpha y + \mu)$
$$= \sigma(x\alpha y) \geq \sigma(x) \qquad \text{(since } \sigma \text{ is a fuzzy ideal of } M)$$
$$= \theta_\sigma(x + \mu).$$

(iv) $\theta_\sigma\{(y + \mu) + (x + \mu) - (y + \mu)\} = \theta_\sigma((y + x - y) + \mu)$
$$= \sigma(y + x - y)$$
$$= \sigma(x) \qquad\qquad\qquad \text{(since } \sigma \text{ is a fuzzy ideal)}$$
$$= \theta_\sigma(x + \mu).$$

(v) $\theta_\sigma\{(x + \mu)\alpha(y + \mu) + (z + \mu)) - (x + \mu)\alpha(y + \mu)\}$
$$= \theta_\sigma\{(x + \mu)\alpha((y + z) + \mu) - (x\alpha y + \mu)\}$$
$$= \theta_\sigma\{(x\alpha(y + z)) + \mu - (x\alpha y) + \mu)\}$$
$$= \theta_\sigma\{(x\alpha(y + z) - x\alpha y) + \mu\}$$
$$= \sigma(x\alpha(y + z) - x\alpha y) \geq \sigma(z) \qquad \text{(since } \sigma \text{ is a fuzzy ideal of } M)$$
$$= \theta_\sigma(z + \mu).$$

Now $\theta_\sigma(x + \mu) = \sigma(x) \geq \mu(x) = \theta_\mu(x + \mu)$. Hence, $\theta_\sigma \supseteq \theta_\mu$.   ∎

Notation 8.2.11

The fuzzy ideal θ_σ of M/μ is denoted by σ/μ.

Lemma 8.2.12 (Lemma 3.7 of Satyanarayana and Prasad, 2005)

Let μ be a fuzzy ideal of M and θ be a fuzzy ideal of M/μ such that $\theta \supseteq \theta_\mu$. Then the fuzzy subset σ_θ of M defined by $\sigma_\theta(x) = \theta(x + \mu)$ for all $x \in M$ is a fuzzy ideal of M such that $\sigma_\theta \supseteq \mu$.

Proof

Let μ be a fuzzy ideal of M and θ be a fuzzy ideal of M/μ such that $\theta \supseteq \theta_\mu$.
  We show that σ_θ defined by $\sigma_\theta(x) = \theta(x + \mu)$ is a fuzzy ideal of M.
  Let $x, y \in M$.

(i) $\sigma_\theta(x + y) = \theta(x + y + \mu)$
$$= \theta((x + \mu) + (y + \mu))$$
$$\geq \min\{\theta(x + \mu), \theta(y + \mu)\} \qquad \text{(since } \theta \text{ is a fuzzy ideal)}$$
$$= \min\{\sigma_\theta(x), \sigma_\theta(y)\} \qquad \text{(by the definition of } \sigma_\theta).$$

(ii) $\sigma_\theta(-x) = \theta(-x) + \mu) = \theta(x + y) = \sigma_\theta(x).$

(iii) $\sigma_\theta(x\alpha y) = \theta(x\alpha y + \mu)$
$$= \theta((x + \mu)\alpha(y + \mu))$$
$$\geq \theta(x + \mu) \qquad\qquad \text{(since } \theta \text{ is a fuzzy ideal of } M)$$
$$= \sigma_\theta(x).$$

Clearly, $\sigma_\theta(x) = \sigma_\theta(y + x - y)$. Consider $\sigma_\theta\{x\alpha(y + z) - x\alpha y\}$. Now

$$
\begin{aligned}
\sigma_\theta\{x\alpha(y + z) - x\alpha y\} &= \theta\{(x\alpha(y + z) - x\alpha y) + \mu\} \\
&= \theta\{(x + \mu)\alpha((y + \mu) + (z + \mu)) - ((x + \mu)\alpha(y + \mu))\} \\
&\geq \theta(x + \mu) \qquad \text{(since } \theta \text{ is a fuzzy ideal of } M) \\
&= \sigma_\theta(x).
\end{aligned}
$$

Hence, σ_θ is a fuzzy ideal of M.

$$\sigma_\theta(x) = \theta(x + \mu) \geq \theta_\mu(x + \mu) \text{ (since } \theta \supseteq \theta_\mu) = \mu(x). \text{ So } \sigma_\theta \supseteq \mu. \qquad \blacksquare$$

Theorem 8.2.13 (Theorem 3.9 of Satyanarayana and Prasad, 2005)

Let μ be a fuzzy ideal of M. There exists an order-preserving bijective correspondence between the set P of all fuzzy ideals of σ of M such that $\sigma \supseteq \mu$ and $\sigma(0) = \mu(0)$ and the set Q of all fuzzy ideals θ of M/μ such that $\theta \supseteq \theta_\mu$.

Proof

Write $P = \{\sigma/\sigma \text{ is a fuzzy ideal of } M, \sigma \supseteq \mu, \sigma(0) = \mu(0)\}$,

$\quad Q = \{\theta/\theta \text{ is a fuzzy ideal of } M/\mu, \theta \supseteq \theta_\mu\}$.

Define $\eta : P \rightarrow Q$ by $\eta(\sigma) = \theta_\sigma$.

By Lemma 8.2.10, it follows that $\eta(\sigma) = \theta_\sigma$ is a fuzzy ideal of M/μ containing θ_μ.

Converse

Define a mapping $\xi : Q \rightarrow P$ by $\xi(\theta) = \sigma_\theta$ as a fuzzy ideal of M containing μ.

Now $\xi\eta(\sigma) = \xi(\eta(\sigma)) = \xi(\theta_\sigma) = \sigma\theta_\sigma$ and $\sigma\theta_\sigma(x) = \theta_\sigma (x + \mu) = \sigma(x)$.

Therefore, $\xi\eta$ is an identity mapping.

Again, consider $\eta\xi(\theta)$. Now $\eta\xi(\theta) = \eta(\xi(\theta)) = \eta(\sigma_\theta) = \sigma_{\theta\sigma}$, and $\sigma_{\theta\sigma}(x + \mu) = \sigma_\theta(x) = \theta(x + \mu)$ for all $x + \mu \in M/\mu$.

Therefore, $\eta\xi$ is also an identity mapping.

Suppose $\sigma \subseteq \tau$. We show that $\theta_\sigma \subseteq \theta_\tau$. $\theta_\sigma(x + \mu) = \sigma(x) \leq \tau(x) = \theta_\tau(x + \mu)$ for all $x + \mu \in M/\mu$. Now $\eta(\sigma) = \theta_\sigma \subseteq \theta_\tau = \eta(\tau)$.

Hence, η is an order-preserving bijective correspondence. $\qquad \blacksquare$

Proposition 8.2.14 (Theorem 3.11 of Satyanarayana and Prasad, 2005)

Let $h : M \rightarrow M^1$ be an epimorphism, and σ a fuzzy ideal of M^1 such that $\mu = h^{-1}(\sigma)$. Then the mapping $\psi : M/\mu \rightarrow M^1/\sigma$ defined by $\psi(x + \mu) = h(x) + \sigma$ is a gamma near ring isomorphism.

Proof

We show that the mapping $\psi(x + \mu) = h(x) + \sigma$ is well-defined.

Take $x + \mu, y + \mu \in M/\mu$. Suppose $x + \mu = y + \mu$

$$\Rightarrow \mu(x - y) = \mu(0) \qquad \text{(by Proposition 8.2.2),}$$

that is, $h^{-1}(\sigma)(x - y) = h^{-1}(\sigma)(\theta) \qquad$ (since $\mu = h^{-1}(\sigma)$)

$$\Rightarrow \sigma[h(x - y)] = \sigma[h(0)]$$
$$\Rightarrow \sigma[h(x) - h(y)] = \sigma[h(0)] \quad \text{(since } h \text{ is a homomorphism)}$$
$$\Rightarrow h(x) + \sigma = h(y) + \sigma.$$

Now we show that $\psi : M/\mu \rightarrow M^1/\sigma$, defined by $\psi(x + \mu) = h(x) + \sigma$, is an isomorphism.

$$\psi((x + \mu) + (y + \mu)) = \psi((x + y) + \mu)$$
$$= h(x + y) + \sigma$$
$$= (h(x) + h(y)) + \sigma \qquad \text{(since } h \text{ is a homomorphism)}$$
$$= (h(x) + \sigma) + (h(y)) + \sigma)$$
$$= \psi(x + y) + \psi(y + \mu)$$
$$\psi((x + \mu)\alpha(y + \mu)) = \psi(x\alpha y + \mu)$$
$$= h(x\alpha y) + \sigma \qquad \text{(by the definition of } \psi\text{)}$$
$$= (h(x)\alpha h(y)) + \sigma$$
$$= (h(x) + \sigma)\alpha(h(y) + \sigma)$$
$$= \psi(x + \mu)\alpha\psi(y + \mu).$$

Therefore, ψ is a homomorphism.

To verify that ψ is one–one, suppose $h(x) + \sigma = h(y) + \sigma$.

Then $\sigma(h(x) - h(y)) = \sigma(h(0))$

$$\Rightarrow \sigma(h(x - y)) = \sigma(h(0)) \qquad \text{(since } h \text{ is homomorphism)}$$
$$\Rightarrow (h-^1(\sigma))(x - y) = (h-^1(\sigma)) \, (0)$$
$$\Rightarrow \mu(x - y) = \mu(0) \qquad \text{(since } \mu = h^{-1}(\sigma)\text{)}$$
$$\Rightarrow x + \mu = y + \mu.$$

Therefore, ψ is one–one.

Take $y \in M^1/\sigma$. Then $y = h(x) + \sigma$ for some $x \in M$. Now $\psi(x + \mu) = h(x) + \sigma = y$. Therefore, ψ is onto. Hence, $\psi : M/\mu \rightarrow M^1/\sigma$ is an isomorphism. ∎

As a consequence of Proposition 8.2.14, we have the following.

Corollary 8.2.15 (Corollary 3.12 of Satyanarayana and Prasad, 2005)

Let μ and σ be two fuzzy ideals of a gamma near ring M such that $\mu \subseteq \sigma$ and $\sigma(0) = \mu(0)$. Then M/σ is isomorphic to $(M/\mu)/(\sigma/\mu)$.

Proof

Define $\psi : M \to M/\mu$ by $\psi(x) = x + \mu$ for all $x \in m$.

By Theorem 8.2.9, it follows that ψ is a Γ-near ring epimorphism. From Notation 8.2.11, that is, $\theta_\sigma = \sigma/\mu$, and by Lemma 8.2.10, it follows that σ/μ is a fuzzy ideal of M/σ such that $\theta_\mu \subseteq \theta_\sigma = \sigma/\mu$ and $\theta_\mu(0) \subseteq \theta_\sigma(0)$. Now $\psi^{-1}(\sigma/\mu)$ is a fuzzy set in M, and for any $x \in M$ we have $(\psi^{-1}(\sigma/\mu))(x) = \psi^{-1}(\theta_\sigma)(x) = \theta_\sigma(\psi(x)) = \theta_\sigma(x + \mu) = \sigma(x)$. Therefore, $\psi^{-1}(\sigma/\mu)$ is a fuzzy ideal of M. Define $\psi^* : M/\sigma \to (M/\mu)/(\sigma/\mu)$ by $\psi^*(x + \sigma) = \psi(x) + (\sigma/\mu)$. By Proposition 8.2.14, it follows that ψ^* is a Γ-near ring isomorphism. The proof is complete.

8.3 Fuzzy Ideals of $M\Gamma$-Groups

We made an introductory study of $M\Gamma$-groups in Section 7.4. In the present section, we discuss some results on fuzzy ideals of $M\Gamma$-groups. We begin this section with the definition of a fuzzy $M\Gamma$-group.

Definition 8.3.1 (Definition 3 of Jun, Kwon, and Park, 1995)

A fuzzy set μ of G is called a fuzzy $M\Gamma$-**subgroup** of G if it satisfies the following two conditions:

(i) $\mu(x - y) \geq \min\{\mu(x), \mu(y)\}$ and
(ii) $\mu(a\alpha y) \geq \mu(y)$ for all $x, y \in G$, $a \in M$, and $\alpha \in \Gamma$.

In this section, M denotes a gamma near ring, and G stands for an $M\Gamma$-group.

Now we present several fundamental important results.

Note 8.3.2

Let $(N, +)$ be a near ring, B an N-group, and H be an ideal of B. Write $M = N$ and $\Gamma = \{\cdot\}$. Then M is a Γ-near ring, B is an $M\Gamma$-group, and H is an ideal of the $M\Gamma$-group B.

Note 8.3.3

Let $(M, +)$ be a Γ-near ring, G be a $M\Gamma$-group, and H be an ideal of the $M\Gamma$-group G. Fix $\gamma \in \Gamma$. We know that $N_\gamma = (M, +, *_\gamma)$ is a near ring, where the operation $*_\gamma$ is defined as $a *_\gamma b = a\gamma b$, for all $a, b \in M$. For $n \in M$ and $g \in G$, define $ng = n\gamma g$. Then

    (i) G becomes an N-group with $N = N_\gamma$.
   (ii) H is an ideal of the N-group G, where $N = N_\gamma$.
  (iii) The statements (i) and (ii) are true for all $\gamma \in \Gamma$.

Note 8.3.4

Suppose M is a Γ-near ring. Let $\varnothing \neq \Gamma^1 \subseteq \Gamma$. Then

    (i) M is also a Γ^1-near ring.
   (ii) If G is an $M\Gamma$-group, then G is also an $M\Gamma^1$-group.

Note 8.3.5

The converse of Note 8.3.4 (ii) is not true, that is, there exists a Γ-near ring M, $\varnothing \neq \Gamma^1 \subseteq \Gamma$, and $M\Gamma^1$-group G such that G is not an $M\Gamma$-group.
    For this, observe the following example.

Example 8.3.6

Consider $\mathbb{Z}_8 = \{0, 1, 2, 3, \ldots, 7\}$, the group of integers modulo 8, and $X = \{a, b\}$, $a \neq b$. Write $M = \{f \mid f: X \to \mathbb{Z}_8 \text{ and } f(a) = 0\}$. Then $M = \{f_0, f_1, f_2, \ldots, f_7\}$, where f_i is defined by $f_i(a) = 0$ and $f_i(b) = i$ for $0 \leq i \leq 7$. Define two mappings $g_0, g_1: \mathbb{Z}_8 \to X$ by setting $g_0(i) = a$ for all $i \in \mathbb{Z}_8$ and

$$g_1(i) = \begin{cases} a & \text{if } i \notin \{3,7\} \\ b & \text{if } i \in \{3,7\} \end{cases}.$$

    Write $\Gamma = \{g_0, g_1\}$, $\Gamma^* = \{g_0\}$.
    Then M is a Γ-near ring and also a Γ^*-near ring. Write $G = \{f_0, f_2, f_4, f_6\}$.
    Now G is an $M\Gamma^*$-group $f_3 \in M$, $g_1 \in \Gamma$, $f_1 \in M$, $f_2 \in G$, and $f_3 g_1(f_1 + f_2) - f_3 g_1 f_1 = f_3 \notin G$.
    This shows that G cannot be an $M\Gamma$-group.
    Hence we have the following situation.
    $\mathbb{Z}_8 = \{0, 1, 2, 3, \ldots, 7\}$, $X = \{a, b\}$, $a \neq b$, $M = \{f \mid f: X \to \mathbb{Z}_8 \text{ and } f(a) = 0\}$, $\Gamma = \{g_0, g_1\}$, $\Gamma^* = \{g_0\}$, where g_0, g_1 have been defined earlier, $\Gamma^* \subseteq \Gamma$, $G \subseteq M$, and G is an $M\Gamma^*$-group, but G is not an $M\Gamma$-group.

Definition 8.3.7

A fuzzy mapping $\mu\colon G \to [0, 1]$ is said to be a **fuzzy ideal** of G if it satisfies the following properties:

(i) $\mu(x + y) \geq \min\{\mu(x), \mu(y)\}$
(ii) $\mu(x + y - x) \geq \mu(y)$
(iii) $\mu(x) = \mu(-x)$
(iv) $\mu(n\gamma(a + x) - n\gamma a) \geq \mu(x)$ for all $n \in M$, $\gamma \in \Gamma$, and $a, x, y \in G$.

If μ satisfies (i), (ii), and (iii), then we say that μ is a **fuzzy normal $M\Gamma$-subgroup** of G.

Example 8.3.8

(i) Every constant fuzzy set $\mu : G \to [0, 1]$ is a fuzzy ideal of G.
(ii) Write $N = \mathbb{Z}$, the usual operations "+" and "·".

Then $(N, +, \cdot)$ forms a near ring. Write $\Gamma = \{\cdot\}$ and $M = N$. Then M is a Γ-near ring. Write $G = \mathbb{Z}$. Then $(G, +)$ where + is the usual addition of integers is a group. Now, G is an N-group. G is also an $M\Gamma$-group. Define $\mu : G \to [0, 1]$ by

$$\mu(x) = \begin{cases} 0.6 & \text{if } x = 4n \text{ for some } n \in \mathbb{Z} \\ 0.2 & \text{if } x = 2n \text{ for some } n \in \mathbb{Z} \text{ and } x \neq 4m \text{ for any } m \in \mathbb{Z}. \\ 0 & \text{if } x \text{ is odd} \end{cases}$$

Now, a straightforward verification shows that μ is a fuzzy ideal of the $M\Gamma$-group G.

Definition 8.3.9

For any family $\{\mu_i / i \in A\}$ of fuzzy subsets of a set S, we define the **intersection of the fuzzy sets** μ_i, $i \in A$ as $(\wedge_{i \in A} \mu_i)(s) = \inf\{\mu_i(s) \mid i \in A\}$ for any $s \in S$. We call $\wedge \mu_i$ the intersection of the fuzzy subsets μ_i, $i \in A$.

The proof of the following two results is a straightforward verification.

Result 8.3.10

Let G be an $M\Gamma$-group and $\{\mu_i \mid i \in A\}$ be a family of fuzzy ideals of G. Then the intersection $\wedge \mu_i$ is also a fuzzy ideal of the $M\Gamma$-group G.

Result 8.3.11

If μ is a fuzzy ideal of $M\Gamma$-group G, then

(i) $\mu(g - g^1) \geq \min\{\mu(g), \mu(g^1)\}$ for all $g, g^1 \in G$.
(ii) $\mu(g + g^1) = \mu(g^1 + g)$ for all $g, g^1 \in G$.

From Result 8.3.11 (i), we can understand that the condition $\mu(g + g^1) \geq \min\{\mu(g), \mu(g^1)\}$ in Definition 8.3.7 may be replaced by $\mu(g - g^1) \geq \min\{\mu(g), \mu(g^1)\}$.

Proposition 8.3.12 (Theorem 2.6 of Satyanarayana, Vijaya Kumari, Godloza, and Nagaraju, accepted for publication)

If μ is a fuzzy ideal of the $M\Gamma$-group G, $x, y \in G$ with $\mu(x) > \mu(y)$, then $\mu(x + y) = \mu(y)$. Moreover, if $x, y \in G$ and $\mu(x) \neq \mu(y)$, then $\mu(x + y) = \min\{\mu(x), \mu(y)\}$.

Proof

Suppose $\mu(x) > \mu(y)$. By definition,

$$\mu(x + y) \geq \min\{\mu(x), \mu(y)\} = \mu(y) \text{ (since } \mu(x) > \mu(y)) \tag{8.5}$$
$$\mu(y) = \mu(y + x - x) \geq \min\{\mu(y + x), \mu(-x)\} = \min\{\mu(y + x), \mu(x)\} \tag{8.6}$$

If $\mu(x + y) \geq \mu(x)$, then

$\mu(y) \geq \mu(x)$ (by Equation 8.6)
$> \mu(y)$ (by the given condition), which is a contradiction.

Therefore,

$$\mu(x + y) < \mu(x) \tag{8.7}$$

Now, $\mu(y) \geq \min\{\mu(y + x), \mu(x)\}$ (by Equation 8.6) $= \mu(y + x)$ (by Equation 8.7). Now, from Equation 8.5, we can conclude that $\mu(x + y) = \mu(y)$. ∎
The rest follows by using similar arguments.

Theorem 8.3.13 (Theorem 2.7 of Satyanarayana, Vijaya Kumari, Godloza, and Nagaraju, accepted for publication)

Let M be a zero-symmetric Γ-near ring and G a $M\Gamma$-group. If μ is a fuzzy ideal of G, then $\mu(m\gamma x) \geq \mu(x)$ for all $x \in G$, $m \in M$, and $\gamma \in \Gamma$.

Proof

Now $\mu(m\gamma x) = \mu(m\gamma(x + 0)) = \mu[m\gamma(0 + x) - 0]$
$\qquad\qquad = \mu[m\gamma(0 + x) - m\gamma 0]$ (since M is a zero-symmetric Γ-near ring)
$\qquad\qquad \geq \mu(x).$ ∎

Note 8.3.14

If $\mu : G \to [0, 1]$ is a fuzzy ideal, then

(i) $\mu(0) \geq \mu(g)$ for all $g \in G$ and
(ii) $\mu(0) = \sup\limits_{g \in G} \mu(g).$

(The verification is similar to that of Lemma 8.1.4.)
From Note 8.3.14, we understand that if $\mu : G \to [0, 1]$ is a fuzzy ideal of G, then the image of μ (denoted by Im μ) is a subset of $[0, \mu(0)]$.

Definition 8.3.15

Let X and Y be two sets such that $X \subseteq Y$. Suppose that α and β are two numbers in $[0, 1]$ such that $\alpha > \beta$. Define a function μ from X to $[0, 1]$ by

$$\mu(x) = \begin{cases} \alpha & \text{if } x \in Y \\ \beta & \text{otherwise} \end{cases}.$$

Then the function μ on Y is called the **generalized characteristic function** of X related to α, β.
Every characteristic function is a generalized characteristic function.

Theorem 8.3.16 (Theorem 2.10 of Satyanarayana, Vijaya Kumari, Godloza, and Nagaraju, accepted for publication)

Let $I \subseteq G$, $\alpha, \beta \in [0, 1]$ such that $\alpha > \beta$.
If μ is the generalized characteristic function of I related to α, β, then the following two conditions are equivalent:

(i) μ is a fuzzy ideal of the $M\Gamma$-group G and
(ii) I is an ideal of the $M\Gamma$-group G

The proof is a straightforward verification.
Taking $\alpha = 1$ and $\beta = 0$ in Theorem 8.3.16, we get the following corollary.

Corollary 8.3.17

Let $I \subseteq G$. If μ is the characteristic function of I, then the following conditions are equivalent:

(i) μ is a fuzzy ideal and
(ii) I is an ideal of G.

From Theorem 8.3.16 and Corollary 8.3.17, we get the following corollary.

Corollary 8.3.18

Let $I \subseteq G$, and $\alpha, \beta \in [0, 1]$ such that $\alpha > \beta$. Then the following three conditions are equivalent:

(i) I is an ideal of the $M\Gamma$-group G.
(ii)   The generalized characteristic function of I related to α, β is a fuzzy ideal of the $M\Gamma$-group G.
(iii) The characteristic function of I is a fuzzy ideal of the $M\Gamma$-group G.
       In the statement of Theorem 8.3.16, the condition $\alpha > \beta$ is necessary.
       For this, observe the following Notes 8.3.19 and 8.3.20.

Note 8.3.19

If we replace the condition $\alpha > \beta$ of the statement of Theorem 8.3.16 by $\alpha = \beta$, then the new statement is not true. If $\alpha = \beta$, then $\mu(x) = \alpha$ for all $x \in G$, and so $\mu : G \to [0, 1]$ is a constant fuzzy set, and hence μ is a fuzzy ideal of G. Let $G = Z_7 = \{0, 1, 2, 3, 4, 5, 6\}$, the additive group of integers modulo 7. Write $M = Z$, the additive group of integers, $\Gamma = Z$. For $m_1, m_2 \in M$, $\gamma \in \Gamma$, define $m_1 \gamma m_2$ as the usual product of integers m_1, γ, m_2. Then M is a Γ-near ring. For any $m \in M$, $\gamma \in \Gamma, g \in G$, define $m\gamma g$ as the product of integers $m, \gamma,$ and g. Now G becomes an $M\Gamma$-group. Write $I = \{1, 2, 3, 4\}$. We know that I cannot be an ideal of G (because $2, 3 \in I$ but $2 + 3 = 5 \notin I$). Define $\mu : G \to [0, 1]$ by

$$\mu(x) = \begin{cases} \alpha & \text{if } x \in I \\ \beta & \text{if } x \notin I \end{cases}.$$

If $\alpha = \beta$, then μ is a constant fuzzy set, and so μ is a fuzzy ideal of G. But, I is not an ideal. Thus, if $\alpha = \beta$, then the statement of Theorem 8.3.16 is not true.

Note 8.3.20

If we replace the condition $\alpha > \beta$ of the statement of Theorem 8.3.16, by $\beta > \alpha$, then the new statement is not true. Suppose $\beta > \alpha$.

Let I be an ideal of a given $M\Gamma$-group G with $I \neq G$. Define $\mu : G \to [0, 1]$ by

$$\mu(x) = \begin{cases} \alpha & \text{if } x \in I \\ \beta & \text{if } x \notin I \end{cases}.$$

If possible, suppose μ is an ideal of an $M\Gamma$-group G. Then by Note 8.3.14, it follows that $\mu(0) \geq \mu(x)$ for all $x \in G$.

Take $y \in G\backslash I$.

Now $\alpha = \mu(0)$ (since $0 \in I$) $\geq \mu(y)$ (by Note 8.3.14)

$\qquad = \beta$ (by the definition of μ)

$\qquad \Rightarrow \alpha \geq \beta$, a contradiction to the fact that $\beta > \alpha$.

This shows that μ is not a fuzzy ideal. Now I is an ideal but μ is not a fuzzy ideal.

So, if we replace the condition $\alpha > \beta$ of the statement of Theorem 8.3.16 by $\beta > \alpha$ then the new statement is not true.

Lemma 8.3.21

If μ is a fuzzy ideal of the $M\Gamma$-group G, and if $\mu(x - y) = \mu(0)$, then $\mu(x) = \mu(y)$ for all $x, y \in G$.

(The proof of Lemma 8.3.21 is parallel to that of Lemma 8.1.5.)

The converse of Lemma 8.3.21 is not true. That is, we can find an $M\Gamma$-group G and a fuzzy ideal μ of G such that $\mu(x) = \mu(y)$, but $\mu(x - y) \neq \mu(0)$. For this, observe the following example.

Example 8.3.22

Consider Example 8.3.8 (ii). In this example, $\mu(1) = 0$ and $\mu(7) = 0$. Write $x = 7$, $y = 1$. Then $\mu(x) = 0$, and $\mu(y) = 0$. So $\mu(x) = \mu(y)$. Now $\mu(x - y) = \mu(7 - 1) = \mu(6) = 0.2 \neq 0.6 = \mu(0)$. So there exist x and y in the $M\Gamma$-group G of Example 8.3.8 (ii) such that $\mu(x) = \mu(y)$, but $\mu(x - y) \neq \mu(0)$.

Notation 8.3.23

Let $a \in G$.

We write $A_1 = \{a\}$, $A_k^* = \{g + x - g \mid g \in G, x \in A_k\}$ (refer to Notation 3.4.6), $A_k^+ = \{m\alpha(g + x) - m\alpha g \mid m \in M, \alpha \in \Gamma, g \in G, x \in A\}$, $A_k^0 = \{x - y \mid x, y \in A_k\}$, $A_{k+1} = A_k^* \cup A_k^+ \cup A_k^0$.

A straightforward verification provides the following.

Lemma 8.3.24

$<a> = \bigcup_{i=1}^{\infty} A_i$ where $A_1 = \{a\}$, and A_is are defined as in Notation 8.3.23.

Theorem 8.3.25 (Theorem 2.19 of Satyanarayana, Vijaya Kumari, Godloza, and Nagaraju, accepted for publication)

If μ is a fuzzy ideal of an $M\Gamma$-group G and $a \in G$, then $\mu(x) \geq \mu(a)$ for all $x \in <a>$, where $<a>$ is the ideal of G generated by a.

Proof

Similar to the proof of Lemma 8.1.11. ∎

In this section, we obtain some results on "level ideals."

Definition 8.3.26

Let μ be any fuzzy ideal of the given $M\Gamma$-group G. For any $t \in [0, \mu(0)]$, the set $\mu_t = \{x \in G/\mu(x) \geq t\}$ is an ideal of G. This ideal μ_t is called a **level ideal** of μ.

In general, μ_t is called a **level set**. In particular, the level set μ_s (denoted by $G_\mu = \{x \in G \mid \mu(x) = \mu(0)\}$, where $s = \mu(0)$) is an ideal of G.

Example 8.3.27

(i) Consider M, G, μ as in Example 8.3.8 (ii). In this example, the following are distinct level ideals of μ: $\mu_0 = \{x \in G /\mu(x) \geq 0\} = G$; $\mu_{0\cdot1} = \{x \in G/\mu(x) \geq 0.1\} = 2\mathbb{Z}$;

$\mu_t = 2\mathbb{Z}$ for all t with $0 \leq t < 0.2$; $\mu_{0\cdot2} = \{x \in G/\mu(x) \geq 0.2\} = 2\mathbb{Z}$; $\mu_{0.25} = \{x \in G/\mu(x) \geq 0.25\} = 4\mathbb{Z}$; $\mu_{0.5} = \{x \in G/\mu(x) \geq 0.5\} = 4\mathbb{Z}$; and $\mu_{0.6} = \{x \in G/\mu(x) \geq 0.6\} = 4\mathbb{Z}$.

Theorem 8.3.28 (Theorem 3.3 of Satyanarayana, Vijaya Kumari, Godloza, and Nagaraju, accepted for publication)

Let G be an $M\Gamma$-group and $\mu: G \to [0, 1]$ a fuzzy subset of G. Then the following conditions are equivalent:

(i) μ is a fuzzy ideal of the $M\Gamma$-group G and
(ii) μ_t is an ideal of the $M\Gamma$-group G for all $t \in [0, \mu(0)]$.

Proof

(i) $\Rightarrow$ (ii): Let t with $0 \le t \le \mu(0)$.
Since $\mu(0) \ge t$, it follows that $0 \in \mu_t = \{x \mid \mu(x) \ge t\}$.
So, μ_t is a nonempty subset of G. Let $x, y \in \mu_t$.
Then $\mu(x) \ge t$ and

$$\mu(y) \ge t \Rightarrow \mu(x - y) \ge \min\{\mu(x), \mu(y)\} \text{ (by Result 8.3.11)}$$
$$\ge \min\{t, t\} = t$$
$$\Rightarrow x - y \in \mu_t.$$

So $(\mu_t, +)$ is a subgroup of $(G, +)$. Let $g \in G$.
Now $\mu(g + x - g) \ge \mu(x) \ge t \Rightarrow g + x - g \in \mu_t$. So $(\mu_t, +)$ is a normal subgroup of $(G, +)$.
Let $m \in M, \gamma \in \Gamma$. Now

$$\mu[m\gamma(g + x) - m\gamma g] \ge \mu(x) \ge t \Rightarrow m\gamma(g + x) - m\gamma g \in \mu_t.$$

So, μ_t is an ideal of the $M\Gamma$-group G.
(ii) $\Rightarrow$ (i): Let $x, y \in G$, and write $t = \min\{\mu(x), \mu(y)\}$.
Now $\mu(x) \ge t, \mu(y) \ge t \Rightarrow x, y \in \mu_t \Rightarrow x - y \in \mu_t$ (since μ_t is an ideal)
$$\Rightarrow \mu(x - y) \ge t \text{ (by the definition of } \mu_t).$$
Therefore, $\mu(x - y) \ge \min\{\mu(x), \mu(y)\}$.
Write $t = \mu(x) \Rightarrow x \in \mu_t \Rightarrow g + x - g \in \mu_t$ for any $g \in G$ (since μ_t is a normal subgroup)
$$\Rightarrow \mu(g + x - g) \ge t = \mu(x) \Rightarrow \mu(g + x - g) \ge \mu(x).$$
Write $t = \mu(x) \Rightarrow x \in \mu_t \Rightarrow -x \in \mu_t$ (since μ_t is an ideal) $\Rightarrow \mu(-x) \ge t$.
So, $\mu(-x) \ge t = \mu(x) \Rightarrow \mu(-x) \ge \mu(x)$. Therefore, $\mu(x) = \mu(-x)$.
Since $x \in \mu_t$, and μ_t is an ideal of G, it follows that $m\gamma(g + x) - m\gamma g \in \mu_t$
for any $m \in M, \gamma \in \Gamma, g \in G$.
So, $\mu[m\gamma(g + x) - m\gamma g] \ge t = \mu(x)$ for $m \in M, \gamma \in \Gamma, x, g \in G$.
We have proved that μ is a fuzzy ideal of the $M\Gamma$-group G. ∎

Proposition 8.3.29 (Proposition 3.4 of Satyanarayana, Vijaya Kumari, Godloza, and Nagaraju, accepted for publication)

Let μ be a fuzzy ideal of the $M\Gamma$-group G, and μ_t, μ_s (with $t < s$) be two level ideals of μ. Then the following two conditions are equivalent:

(i) $\mu_t = \mu_s$ and
(ii) there is no $x \in G$ such that $t \leq \mu(x) < s$.

Proof

(i) $\Rightarrow$ (ii): Let us suppose that there exists an element $x \in G$ such that $t \leq \mu(x) < s$.
   Then $x \in \mu_t$ and $x \notin \mu_s$, and so $\mu_t \neq \mu_s$, which is a contradiction.
(ii) $\Rightarrow$ (i): Let $x \in \mu_s \Rightarrow \mu(x) \geq s \Rightarrow \mu(x) \geq s > t \Rightarrow \mu(x) > t \Rightarrow x \in \mu_t$.
   Therefore, $\mu_s \subseteq \mu_t$. Let $x \in \mu_t \Rightarrow \mu(x) \geq t$.
   By the given condition (ii), there is no y such that $s > \mu(y) \geq t$, and so $\mu(x) \geq s$, which implies $x \in \mu_s$. Hence, $\mu_t = \mu_s$.   ∎

8.4 Homomorphisms and Fuzzy Cosets of $M\Gamma$-Groups

In Section 8.2, we presented the concept of fuzzy cosets in gamma near rings. In this section, we present several important results related to homomorphisms and fuzzy cosets in $M\Gamma$-groups. In particular, we establish a one–one correspondence between the set of all h-invariant (where $h : G \to G^1$ is an onto homomorphism) fuzzy ideals of the $M\Gamma$-group G and the set of all fuzzy ideals of the $M\Gamma$-group G^1.

Notation 8.4.1

Let G be a $M\Gamma$-group and $x \in G$. Then $FI(x)$ denotes the family of all ideals of G that contain x.

Theorem 8.4.2 (Theorem 4.2 of Satyanarayana, Vijaya Kumari, Godloza, and Nagaraju, accepted for publication)

Let μ be a fuzzy ideal of the $M\Gamma$-group G.

(i) If $FI(x) \subset FI(y)$, then $<y> \subseteq <x>$ and $\mu(x) \leq \mu(y)$.
(ii) If $FI(x) = FI(y)$, then $<x> = <y>$ and $\mu(x) = \mu(y)$

Proof

(i) Suppose $FI(x) \subseteq FI(y)$.
Now, $<x> \in FI(x) \subseteq FI(y) \Rightarrow y \in <x> \Rightarrow <x> \supseteq <y>$.
Since $y \in <x>$, by Theorem 8.3.25, it follows that $\mu(y) \geq \mu(x)$.

(ii) $FI(x) = FI(y) \Rightarrow <x> = <y> \Rightarrow \mu(x) = \mu(y)$ (by (i)). ∎

Remark 8.4.3

The intersection of any family of fuzzy ideals $\{\mu_i\}$ of MΓ-group G is also a fuzzy ideal of G.

The proof is a straightforward verification.

Definition 8.4.4

Let A and B be two sets, μ a fuzzy subset of A, and $h : A \to B$, a function. Then μ is called ***h*-invariant** if $h(x) = h(y)$ implies $\mu(x) = \mu(y)$ for all $x, y \in A$.

Note 8.4.5

Let $h : G \to G^1$ be a MΓ-group homomorphism, and μ be a fuzzy ideal of G. Then the following are true:

(i) If μ is h-invariant, then $h^{-1}(h(\mu)) = \mu$.
(ii) If h is onto, then $h(h^{-1}(\gamma)) = \gamma$, where γ is a fuzzy ideal of G^1.

Verification

(i) Let $x \in G$.
Consider $h^{-1}(h(\mu))(x) = h(\mu)(h(x))$
$$= \sup_{h(x^1)=h(x)} \mu(x^1)$$
$= \mu(x)$ (since μ is h-invariant).
Therefore, $h^{-1}(h(\mu)) = \mu$.

(ii) Let $y \in G^1$.
Consider $h(h^{-1}(\gamma))(y) = \sup_{h(x)=y} h^{-1}(\gamma)(x)$
$$= \sup_{h(x)=y} \gamma(h(x)) \text{ (by Definition 1.7.19 of } h^{-1})$$
$= \sup \gamma(y) = \gamma(y)$.
Hence, $h(h^{-1}(\gamma)) = \gamma$.

In the following theorems, h denotes an $M\Gamma$-homomorphism from G onto G^1, where G and G^1 are $M\Gamma$-groups.

Theorem 8.4.6 (Theorem 4.7 of Satyanarayana, Vijaya Kumari, Godloza, and Nagaraju, accepted for publication)

 (i) If μ is a fuzzy ideal of the $M\Gamma$-group G, then $h(\mu)$ is a fuzzy ideal of the $M\Gamma$-group G^1, and
 (ii) If ν is a fuzzy ideal of the $M\Gamma$-group G^1, then $h^{-1}(\nu)$ is a fuzzy ideal of the $M\Gamma$-group G
 (iii) $h^{-1}(\nu)$ is constant on ker h.

Proof

Part (a): Assume that μ is a fuzzy ideal of the $M\Gamma$-group G.
 We prove that $h(\mu)$ is a fuzzy ideal of the $M\Gamma$-group G^1.

 (i) $h(\mu)(x + y) = \sup_{h(z)=x+y} \mu(z)$

 [Reason: Since $x + y \in G^1$ and h is onto, there exists $z \in G$ such that $h(z) = x + y$]

$$\geq \min \left\{ \sup_{h(x^1)=x} \mu(x^1), \sup_{h(y^1)=y} \mu(y^1) \right\} = \min\{h(\mu)(x), h(\mu)(y)\}.$$

 (ii) $h(\mu)(x + y - x) = \sup_{h(z) = x+y-x} \mu(z)$
$$\geq \mu(x^1 + y^1 - x^1)$$

 [Reason: Since h is onto, there exists $x^1, y^1 \in G$ such that $h(x^1) = x$, $h(y^1) = y$, and

$$h(x^1 + y^1 - x^1) = h(x^1) + h(y^1) - h(x^1) = x + y - x$$
$$\geq \mu(y^1) \text{ for all } y^1 \text{ such that } h(y^1) = y.]$$

 Therefore,
$$h(\mu)(x + y - x) \geq \sup_{h(y^1)=y} \mu(y^1)$$
$$= h(\mu)(y) \text{ (by the definition of } h(\mu)).$$

 (iii) $h(\mu)(-x) = \sup_{h(z) = -x} \mu(z)$ (by the definition of $h(\mu)$)
$$= \sup_{h(-z) = x} \mu(-z) = h(\mu)(x) \text{ (by the definition of } h(\mu)).$$

 (iv) $h(\mu)(n\gamma(a + x) - n\gamma a) = \sup_{h(z) = n\gamma(a+x)-n\gamma a} \mu(z)$ (by the definition of $h(\mu)$)
$$\geq \mu(n^1\gamma(a^1 + x^1) - n^1\gamma a^1)$$

[Reason: Since h is onto, there exist n^1, a^1, x^1 such that $h(n^1) = n$, $h(a^1) = a$, $h(x^1) = x$, and so

$$h(n^1\gamma(a^1 + x^1) - n^1\gamma a^1) = h(n^1)\gamma(h(a^1) + h(x^1)) - h(n^1)\gamma h(a^1)]$$
$$= n\gamma(a + x) - n\gamma a)$$
$$\geq \mu(x^1) \text{ (since } \mu \text{ is a fuzzy ideal, condition (iv) of Definition 8.3.7).}$$

Therefore,

$$h(\mu)(n\gamma(a + x) - n\gamma a) \geq \sup_{h(x^1)=x} \mu(x^1) = h(\mu)(x).$$

Hence, $h(\mu)$ is a fuzzy ideal of the $M\Gamma$-group G^1.

Part (b): Suppose that v is a fuzzy ideal of the $M\Gamma$-group G^1.
In the following, we verify that $h^{-1}(v)$ is a fuzzy ideal of the $M\Gamma$-group G.

(i) $h^{-1}(v)(x + y) = v(h(x + y))$ (by the definition of $h^{-1}(v)$)
$= v(h(x) + h(y))$ (since h is a homomorphism)
$\geq \min\{v(h(x)), v(h(y))\}$ (since v is a fuzzy ideal)
$= \min\{h^{-1}(v)(x), h^{-1}(v)(y)\}$ (by the definition of $h^{-1}(v)$).

(ii) $h^{-1}(v)(x + y - x) = v(h(x + y - x))$ (by the definition of $h^{-1}(v)$)
$= v(h(x) + h(y) - h(x))$ (since h is a homomorphism)
$\geq v(h(y))$ (since v is a fuzzy ideal)
$= h^{-1}(v)(y)$.

(iii) $h^{-1}(v)(-x) = v(h(-x))$ (by the definition of $h^{-1}(v)$)
$= v(-h(x))$ (since h is a homomorphism)
$= v(h(x))$ (since $v(y) = v(-y)$ for all y by condition (iii) of Definition 8.3.7)
$= h^{-1}(v)(x)$ (by the definition of $h^{-1}(v)$).

(iv) $h^{-1}(v)(n\gamma(a + x) - n\gamma a) = v(h(n\gamma(a + x) - n\gamma a))$ (by the definition of $h^{-1}(v)$)
$= v(n\gamma h(a + x) - n\gamma h(a))$ (since h is a homomorphism)
$= v(n\gamma(h(a) + h(x)) - n\gamma h(a))$ (since h is a homomorphism)
$\geq v(h(x))$ (since v is a fuzzy ideal)
$= h^{-1}(v(x))$ (by the definition of $h^{-1}(v)$).

Therefore, $h^{-1}(v)$ is a fuzzy ideal of the $M\Gamma$-group G.

Part (c): For any $x \in \ker h$, it follows that

$$h^{-1}(v)(x) = v(h(x)) \text{ (by the definition of } h^{-1}(v))$$
$$= v(0) \text{ (since } x \in \ker h).$$

This is true for all $x \in \ker h$. This shows that $h^{-1}(v)$ is constant on $\ker h$.
The proof is complete. ∎

Lemma 8.4.7 (Lemma 4.8 of Satyanarayana, Vijaya Kumari, Godloza, and Nagaraju, accepted for publication)

If μ is a fuzzy ideal of G and $h : G \to G^1$ is an onto homomorphism, such that μ is constant on ker h, then μ is h-invariant.

Proof

Suppose $h(x) = h(y)$ for some $x, y \in G$.
    Then $h(x - y) = 0$, and so $x - y \in$ ker h.
    Since $0, x - y \in$ ker h, and μ is constant on ker h, it follows that $\mu(0) = \mu(x - y)$.
    By Lemma 8.3.21, it follows that $\mu(x) = \mu(y)$.   ∎

Theorem 8.4.8 (Theorem 4.9 of Satyanarayana, Vijaya Kumari, Godloza, and Nagaraju, accepted for publication)

The mapping $\mu \to h(\mu)$ defines a one–one correspondence between the set of all h-invariant fuzzy ideals of the $M\Gamma$-group G and the set of all fuzzy ideals of the $M\Gamma$-group G^1, where $h : G \to G^1$ is an onto homomorphism.

Proof

Write $\mathcal{B}$ = the set of all h-invariant fuzzy ideals of the $M\Gamma$-group G, and
    $\mathcal{H}$ = the set of all fuzzy ideals of the $M\Gamma$-group G^1.
    Let $\mu \in \mathcal{B}$. Define $\varphi : \mathcal{B} \to \mathcal{H}$ by $\varphi(\mu) = h(\mu)$. Suppose $\varphi(\mu_1) = \varphi(\mu_2)$.
    Then $h(\mu_1) = h(\mu_2)$
$$\Rightarrow \mu_1 = h^{-1}(h(\mu_1)) = h^{-1}(h(\mu_2)) = \mu_2 \text{ (by using Note 8.4.5).}$$
    This shows that h is one–one.
    Let $v \in \mathcal{H}$.
    Then by Theorem 8.4.6, $h^{-1}(v)$ is a fuzzy ideal of G that is constant on ker h.
    By Lemma 8.4.7, it follows that $h^{-1}(v)$ is h-invariant, and so $h^{-1}(v) \in \mathcal{B}$.
    Now by Note 8.4.5 (ii), it follows that $v = h(h^{-1}(v)) \in \mathcal{H}$.
    The proof is complete.   ∎

Theorem 8.4.9 (Theorem 4.10 of Satyanarayana, Vijaya Kumari, Godloza, and Nagaraju, accepted for publication)

Let μ and v be fuzzy ideals of the $M\Gamma$-group G and G^1, respectively, such that Im $(\mu) = \{t_0, t_1, \ldots, t_n\}$ with $t_0 > t_1 > \ldots > t_n$, and Im $(v) = \{s_0, s_1, \ldots, s_m\}$ with $s_0 > s_1 > \ldots > s_m$.

Then the following are true:

(i) $\operatorname{Im}(h(\mu)) \subset \operatorname{Im}(\mu)$ and the chain of level ideals of $h(\mu)$ is

$$h(\mu_{t_0}) \subseteq h(\mu_{t_1}) \subseteq \cdots \subseteq h(\mu_{t_n}) = G^1.$$

(ii) $\operatorname{Im}(h^{-1}(v)) = \operatorname{Im}(v)$ and the chain of level ideals of $h^{-1}(v)$ is

$$h^{-1}(v_{s_0}) \subseteq h^{-1}(v_{s_1}) \subseteq \cdots \subseteq h^{-1}(v_{s_m}) = G.$$

Proof

(i) $h(\mu)(y^1) = \sup\limits_{h(x)=y^1} \mu(x) \in \operatorname{Im} \mu$ (since $\operatorname{Im} \mu$ is a finite set)
$\Rightarrow \operatorname{Im} h(\mu) \subseteq \operatorname{Im} \mu.$
$y^1 \in G^1, y^1 \in h(\mu_{t_i})$
$\Leftrightarrow$ there exists $x \in \mu_{t_i}$ such that $h(x) = y^1$
$\Leftrightarrow \mu(x) \geq t_i$ and $h(x) = y^1$
$\Leftrightarrow \sup\limits_{h(x)=y^1} \mu(x) \geq t_i \Leftrightarrow (h(\mu))(y^1) \geq t_i$
$\Leftrightarrow y^1 \in \left(h(\mu)\right)_{t_i}.$
Therefore, $h(\mu_{t_i}) = \left(h(\mu)\right)_{t_i}.$
Since $t_0 > t_1 > \ldots > t_n$ and $\operatorname{Im} h(\mu) \subseteq \{t_0, t_1, \ldots, t_n\}$, it follows that $\left(h(\mu)\right)_{t_0} \subseteq$ $\left(h(\mu)\right)_{t_0} \subseteq \cdots \subseteq \left(h(\mu)\right)_{t_n}$ is a sequence (of maximum length) of the level ideals of $h(\mu)$.
Since $h(\mu_{t_i}) = \left(h(\mu)\right)_{t_i}$, we can conclude that $h(\mu_{t_0}) \subseteq h(\mu_{t_1}) \subseteq \cdots \subseteq$ $h(\mu_{t_n})$ is the sequence (of maximum length) of the level ideals of the fuzzy ideal $h(\mu)$.

(ii) Since $h^{-1}(v)(x) = v(h(x))$ for all $x \in G$, and since h is onto, we have $\operatorname{Im}(h^{-1}(v)) = \operatorname{Im}(v)$.
$x \in h^{-1}(v_{s_i}) \Leftrightarrow$ there exists $y \in v_{s_i}$ such that $h^{-1}(y) = x$
$\Leftrightarrow v(y) \geq s_i$ and $y = h(x)$
$\Leftrightarrow v(h(x)) \geq s_i$
$\Leftrightarrow h^{-1}(v)(x) \geq s_i$
$\Leftrightarrow x \in \left(h^{-1}(v)\right)_{s_i}.$
Therefore, $h^{-1}(v_{s_i}) = \left(h^{-1}(v)\right)_{s_i}.$
Now, $s_0 > s_1 > \ldots > s_m$ and $\operatorname{Im}(v) = \{s_0, s_1, \ldots, s_m\}$, implying that
$\left(h^{-1}(v)\right)_{s_0} \subseteq \left(h^{-1}(v)\right)_{s_1} \subseteq \cdots \subseteq \left(h^{-1}(v)\right)_{s_m}.$
Since $h^{-1}(v_{s_i}) = \left(h^{-1}(v)\right)_{s_i}$, we conclude that
$h^{-1}(v_{s_0}) \subseteq h^{-1}(v_{s_1}) \subseteq \cdots \subseteq h^{-1}(v_{s_m}) = G$ is a chain of level ideals of $h^{-1}(\gamma)$.
The proof is complete. ∎

Example 8.4.10

Consider $G = Z$, the additive group of integers; $G^1 = Z_6$, the additive group of integers modulo 6; and $M = Z$, the Γ-near ring with $\Gamma = Z$, the set of integers. Then G and G^1 are $M\Gamma$-groups.

Let $h : Z \rightarrow Z_6$ be the canonical epimorphism of groups.

Now h is also an $M\Gamma$-homomorphism.

$2Z$ is an ideal of the $M\Gamma$-group G.

Define $\mu: G \rightarrow [0, 1]$ by $\mu(x) = \chi_{2Z}(x)$ for all $x \in G$, where χ_{2Z} is the characteristic function of $2Z$.

By Theorem 8.3.16, it follows that μ is a fuzzy ideal of G.

Now we verify that $h(\mu)$ is a fuzzy ideal of G^1.

$$h(\mu)(0) = \sup\{\mu(x) \mid x \in h^{-1}(0)\} = \sup\{\mu(x) \mid x \in 6Z\} \geq 1.$$

So, $f(\mu)(0) = 1$.

Similarly, we have $h(\mu)(2) = 1$, $h(\mu)(4) = 1$, $h(\mu)(1) = 0$, $h(\mu)(3) = 0$, $h(\mu)(5) = 0$.

The level sets of $h(\mu)$ are given by $(h(\mu))_0 = G$ and $(h(\mu))_1 = \{0, 2, 4\}$, each of which is an ideal of the $M\Gamma$-group $G^1 = Z_6$.

By Theorem 8.4.6 (i), it follows that $f(\mu)$ is a fuzzy ideal of the $M\Gamma$-group G^1.

Example 8.4.11

Consider G, G^1, Γ, and M as in Example 8.4.10.

Let $h : G \rightarrow G^1$ be the canonical epimorphism.

Define $\mu^1: G^1 \rightarrow [0, 1]$ by

$$\mu^1(x) = \begin{cases} 1 & \text{if } x \in \{0, 2, 4\} \\ 0 & \text{otherwise} \end{cases}.$$

Note that μ^1 is the characteristic function of the ideal $\{0, 2, 4\}$, and by Theorem 8.3.16, it follows that μ^1 is a fuzzy ideal of G^1. Also, we know that $h^{-1}(\mu^1)(x) = \mu^1(h(x))$ for all $x \in G$.

If x is even, then $h(x) \in \{0, 2, 4\}$, and $h^{-1}(\mu^1)(x) = \mu^1(h(x)) = 1$.

If x is odd, then $h(x) \in \{1, 3, 5\}$, and $h^{-1}(\mu^1)(x) = \mu^1(h(x)) = 0$.

Therefore, the level sets of $h^{-1}(\mu^1)$ are $(h^{-1}(\mu^1))_0 = G$ and $(h^{-1}(\mu^1))_1 = 2Z$, each of which is an ideal of the $M\Gamma$-group G.

Therefore, by Theorem 8.4.6 (ii), it follows that $h^{-1}(\mu^1)$ is a fuzzy ideal of Z.

Definition 8.4.12

Let $\mu : G \rightarrow [0, 1]$ be a given fuzzy ideal of the $M\Gamma$-group G, and $x \in G$. Then the fuzzy subset $x + \mu: G \rightarrow [0, 1]$ defined by $(x + \mu)(y) = \mu(y - x)$ is called a **fuzzy coset** of the fuzzy ideal μ.

Proposition 8.4.13 (Satyanarayana, Vijaya Kumari, and Nagaraju, accepted for publication)

Suppose that G is an $M\Gamma$-group, $\mu : G \to [0, 1]$ a fuzzy ideal, and $x, y \in G$. Then $x + \mu = y + \mu \Leftrightarrow \mu(x - y) = \mu(0)$.

Proof

Suppose that $\mu(x - y) = \mu(0)$.
   By condition (iii) of Definition 8.3.7, it follows that $\mu(y - x) = \mu(x - y) = \mu(0)$.
   To verify that $x + \mu = y + \mu$, take $z \in G$.
   Now $(x + \mu)(z) = \mu(z - x)$
$$
\begin{aligned}
&= \mu(z - y + y - x) \\
&\geq \min\{\mu(z - y), \mu(y - x)\} \\
&= \min\{\mu(z - y), \mu(x - y)\} \text{ (since } \mu(a) = \mu(-a) \text{ for all } a) \\
&= \min\{\mu(z - y), \mu(0)\} \\
&= \mu(z - y) \text{ (by Note 8.3.14 (i))} \\
&= (y + \mu)(z) \text{ (by Definition 8.4.12)}
\end{aligned}
$$
   Therefore $(x + \mu)(z) \geq (y + \mu)(z)$ for all $z \in G$.
   Similarly, we can show that $(y + \mu)(z) \geq (x + \mu)(z)$. Therefore, $x + \mu = y + \mu$.

Converse

Suppose $x + \mu = y + \mu$.
$$
\begin{aligned}
\mu(x - y) &= (y + \mu)(x) \text{ (by Definition 8.4.12)} \\
&= (x + \mu)(x) \text{ (by the given condition)} \\
&= \mu(x - x) \text{ (by Definition 8.4.12)} \\
&= \mu(0).
\end{aligned}
$$
The proof is complete. ∎

Notation 8.4.14

We write $G/\mu = \{x + \mu \mid x \in G\}$, the set of all fuzzy cosets of μ. Define
   $(x + \mu) + (y + \mu) = (x + y) + \mu$ and $m\gamma(x + \mu) = m\gamma x + \mu$ for $m \in M$, $\gamma \in \Gamma$, and $x \in G$. Then the set G/μ becomes an $M\Gamma$-group with respect to the operations defined. The $M\Gamma$-group G/μ is called the **quotient group** with respect to the fuzzy ideal μ.

Verification

Let $m_1, m_2 \in M$, $\alpha_1, \alpha_2 \in \Gamma$, and $x + \mu \in G/\mu$.

(i) $(m_1 + m_2)\alpha_1(x + \mu) = (m_1 + m_2)\alpha_1 x + \mu$
$$
\begin{aligned}
&= m_1\alpha_1 x + m_2\alpha_1 x + \mu \text{ (since } G \text{ is an } M\Gamma\text{-group)} \\
&= (m_1\alpha_1 x + \mu) + (m_2\alpha_1 x + \mu) \\
&= m_1\alpha_1(x + \mu) + m_2\alpha_1(x + \mu).
\end{aligned}
$$

(ii) $(m_1\alpha_1 m_2)\alpha_2(x + \mu) = (m_1\alpha_1 m_2)\alpha_2 x + \mu$
$$= m_1\alpha_1(m_2\alpha_2 x) + \mu \text{ (since } G \text{ is an } M\Gamma\text{-group)}$$
$$= m_1\alpha_1(m_2\alpha_2 x + \mu)$$
$$= m_1\alpha_1(m_2\alpha_2(x + \mu)).$$

Therefore, G/μ is an $M\Gamma$-group.

Definitions 8.4.15

(i) If μ is a fuzzy ideal of an $M\Gamma$-group G, then we define a *fuzzy set* θ_μ on G/μ corresponding to μ by $\theta_\mu(x + \mu) = \mu(x)$ for all $x \in G$.
(ii) If θ is a fuzzy ideal of the $M\Gamma$-group G/μ, then we define a fuzzy set σ_θ on G by $\sigma_\theta(x) = \theta(x + \mu)$ for all $x \in G$.

Theorem 8.4.16 (Satyanarayana, Vijaya Kumari, and Nagaraju, accepted for publication)

If μ is a fuzzy ideal of the $M\Gamma$-group G, then the fuzzy set $\theta_\mu: G/\mu \rightarrow [0, 1]$ given in Definition 8.4.15 (i) is a fuzzy ideal of G/μ.

Proof

Suppose μ is a fuzzy ideal of G.
 Now, we verify that θ_μ is a fuzzy ideal of the quotient $M\Gamma$-group G/μ.
Let $a + \mu, x + \mu, y + \mu \in G/\mu$, and $m \in M$.

(i) $\theta_\mu((x + \mu) - (y + \mu)) = \theta_\mu((x - y) + \mu)$ (by the definition of addition in G/μ)
$$= \mu(x - y) \qquad\qquad \text{(by the definition of } \theta_\mu)$$
$$\geq \min\{\mu(x), \mu(y)\} \qquad \text{(since } \mu \text{ is a fuzzy ideal)}$$
$$= \min\{\theta_\mu(x + \mu), \theta_\mu(y + \mu)\} \quad \text{(by the definition of } \theta_\mu).$$
(ii) $\theta_\mu((x + \mu) + (y + \mu) - (x + \mu)) = \theta_\mu((x + y - x) + \mu)$ (by the definition of addition in G/μ)
$$= \mu(x + y - x) \qquad\qquad \text{(by the definition of } \theta_\mu)$$
$$\geq \mu(y) \qquad\qquad \text{(since } \mu \text{ is a fuzzy ideal)}$$
$$= \theta_\mu(y + \mu) \qquad\qquad \text{(by the definition of } \theta_\mu).$$
(iii) $\theta_\mu(-x + \mu) = \mu(-x) \qquad\qquad \text{(by the definition of } \theta_\mu)$
$$= \mu(x) \qquad\qquad \text{(since } \mu \text{ is a fuzzy ideal)}$$
$$= \theta_\mu(x + \mu). \qquad\qquad \text{(by the definition of } \theta_\mu).$$
(iv) $\theta_\mu(m\gamma((a + \mu) + (x + \mu)) - m\gamma(a + \mu)) = \theta_\mu(m\gamma((a + x) + \mu) - (m\gamma a + \mu))$
$$\qquad\qquad \text{(by the definition of the quotient } M\Gamma\text{-group } G/\mu)$$
$$= \theta_\mu([m\gamma(a + x) - m\gamma a] + \mu) \quad \text{(by the definition of addition in } G/\mu)$$
$$= \mu(m\gamma(a + x) - m\gamma a) \qquad \text{(by the definition of } \theta_\mu)$$
$$\geq \mu(x) \qquad\qquad \text{(since } \mu \text{ is a fuzzy ideal of } G)$$
$$= \theta_\mu(x + \mu). \qquad\qquad \text{(by the definition of } \theta_\mu).$$

Therefore, θ_μ is a fuzzy ideal of G/μ.   ∎

Corollary 8.4.17

If μ and σ are two fuzzy ideals of G such that $\mu \subseteq \sigma$ and $\sigma(0) = \mu(0)$, then the mapping $h_\sigma : G/\mu \to [0, 1]$, defined by $h_\sigma(x + \mu) = \sigma(x)$ for all $x + \mu \in G/\mu$, is a fuzzy ideal. Also, $\theta_\mu \subseteq h_\sigma$ on G/μ and $\theta_\mu(0) = h_\sigma(0)$.

Proof

First, we verify that h_σ is well-defined.

Let $x + \mu, y + \mu \in G/\mu$ such that $x + \mu = y + \mu$. This implies $\mu(0) = \mu(x - y)$ (by Proposition 8.4.13). Now $\sigma(0) \geq \sigma(x - y)$ (by Note 8.3.14 (i))

$$\geq \mu(x - y) \quad \text{(since } \sigma \supseteq \mu)$$
$$= \mu(0)$$
$$= \sigma(0) \quad \text{(by the given condition in the hypothesis)}$$
$$\Rightarrow \sigma(0) = \sigma(x - y)$$
$$\Rightarrow x + \sigma = y + \sigma \text{ (by Proposition 8.4.13)}$$
$$\Rightarrow (x + \sigma)(0) = (y + \sigma)(0)$$
$$\Rightarrow \sigma(0 - x) = \sigma(0 - y) \text{ (by Definition 8.4.12)}$$
$$\Rightarrow \sigma(-x) = \sigma(-y)$$
$$\Rightarrow \sigma(x) = \sigma(y) \quad \text{(since } \sigma \text{ is a fuzzy ideal)}$$
$$\Rightarrow h_\sigma(x + \mu) = h_\sigma(y + \mu) \text{ (by the definition of } h_\sigma).$$

This shows that h_σ is well-defined.

Now we verify that h_σ is a fuzzy ideal of G/μ. For this, take $x + \mu, y + \mu, a + \mu \in G/\mu$, and $m \in M$.

(i) $h_\sigma((x + \mu) - (y + \mu)) = h_\sigma((x - y) + \mu)$
$$= \sigma(x - y) \text{ (by the definition of } h_\sigma)$$
$$\geq \min\{\sigma(x), \sigma(y)\} \text{ (since } \sigma \text{ is a fuzzy ideal)}$$
$$= \min\{h_\sigma(x + \mu), h_\sigma(y + \mu)\} \text{ (by the definition of } h_\sigma).$$

(ii) $h_\sigma((x + \mu) + (y + \mu) - (x + \mu)) = h_\sigma((x + y - x) + \mu)$ (by the definition of addition in G/μ)
$$= \sigma(x + y - x) \quad \text{(by the definition of } h_\sigma)$$
$$\geq \sigma(y) \quad \text{(since } \sigma \text{ is a fuzzy ideal)}$$
$$= h_\sigma(y + \mu) \quad \text{(by the definition of } h_\sigma).$$

(iii) $h_\sigma(-x + \mu) = \sigma(-x) = \sigma(x) = h_\sigma(x + \mu)$.

(iv) $h_\sigma(m\gamma((a + \mu) + (x + \mu)) - m\gamma(a + \mu)) = h_\sigma(m\gamma((a + x) + \mu) - (m\gamma a + \mu))$
$$\text{(by the definition of addition in } G/\mu)$$
$$= h_\sigma([m\gamma(a + x) - m\gamma a] + \mu)$$
$$= \sigma(m\gamma(a + x) - m\gamma a) \text{ (by the definition of } h_\sigma)$$
$$\geq \sigma(x) \text{ (since } \sigma \text{ is a fuzzy ideal of } G)$$
$$= h_\sigma(x + \mu) \text{ (by the definition of } h_\sigma).$$

Hence, h_σ is a fuzzy ideal of G/μ.

Let $x + \mu \in G/\mu$.

Now $h_\sigma(x + \mu) = \sigma(x) \quad$ (by the definition of h_σ)
$$\geq \mu(x) \quad \text{(since } \sigma \supseteq \mu)$$
$$= \theta_\mu(x + \mu) \quad \text{(by the definition of } \theta_\mu).$$

This is true for all $x + \mu \in G/\mu$.

This shows that $\theta_\mu \subseteq h_\sigma$.

Also, $\theta_\mu(0) = \mu(0)$ (by the definition of θ_μ)

        $= \sigma(0)$ (given by hypothesis)

        $= h_\sigma(0)$ (by the definition of h_σ).

Hence, $\theta_\mu(0) = h_\sigma(0)$. The proof is complete.   ∎

Theorem 8.4.18 (Satyanarayana, Vijaya Kumari, and Nagaraju, accepted for publication)

Let μ be a fuzzy ideal of G, and θ a fuzzy ideal of G/μ such that $\theta_\mu \subseteq \theta$ and $\theta_\mu(0) = \theta(0)$. Then $\sigma_\theta: G \rightarrow [0, 1]$, defined by $\sigma_\theta(x) = \theta(x + \mu)$, is a fuzzy ideal of G such that $\mu \subseteq \sigma_\theta$ and $\mu(0) = \sigma_\theta(0)$.

Proof

Part (a): First we show that σ_θ is a fuzzy ideal of G.

For this, take $a, x, y \in G$ and $m \in M$.

(i) $\sigma_\theta(x - y) = \theta((x - y) + \mu)$ (by the definition of σ_θ)

           $= \theta((x + \mu) - (y + \mu))$ (by the definition of addition in G/μ)

           $\geq \min\{\theta(x + \mu), \theta(y + \mu)\}$ (since θ is a fuzzy ideal)

           $= \min\{\sigma_\theta(x), \sigma_\theta(y)\}$ (by the definition of σ_θ).

      Therefore, $\sigma_\theta(x - y) \geq \min\{\sigma_\theta(x), \sigma_\theta(y)\}$.

(ii) $\sigma_\theta(x + y - x) = \theta((x + y - x) + \mu)$ (by the definition of σ_θ)

           $= \theta((x + \mu) + (y + \mu) - (x + \mu))$ (by the definition of addition in G/μ)

           $\geq \theta(y + \mu)$ (since θ is a fuzzy ideal of G/μ)

           $= \sigma_\theta(y)$ (by the definition of σ_θ).

(iii) $\sigma_\theta(-x) = \theta(-x + \mu)$ (by the definition of σ_θ)

           $= \theta(x + \mu)$ (since θ is a fuzzy ideal of G/μ)

           $= \sigma_\theta(x)$ (by the definition of σ_θ).

(iv) $\sigma_\theta(m\gamma(a + x) - m\gamma a) = \theta([m\gamma(a + x) - m\gamma a] + \mu)$ (by the definition of σ_θ)

           $= \theta(m\gamma((a + \mu) + (x + \mu)) - m\gamma(a + \mu))$

                        (by the definition of addition in G/μ)

           $\geq \theta(x + \mu)$ (since θ is a fuzzy ideal of G/μ)

           $= \sigma_\theta(x)$ (by the definition of σ_θ).

      Hence, σ_θ is a fuzzy ideal of G.

Part (b): For any $x \in G$, it follows that $\sigma_\theta(x) = \theta(x + \mu)$ (by the definition of σ_θ)

                                        $\geq \theta_\mu(x + \mu)$ (since $\theta \supseteq \theta_\mu$)

                                        $= \mu(x)$ (by the definition of θ_μ).

This shows that $\mu \subseteq \sigma_\theta$.

Also, $\sigma_\theta(0) = \theta(0 + \mu)$ (by the definition of σ_θ)

$\qquad\quad = \theta(0)$ (since $0 = 0 + \mu$ in G/μ)

$\qquad\quad = \theta_\mu(0)$ (by the hypothesis)

$\qquad\quad = \theta_\mu(0 + \mu)$ (since $0 + \mu = 0$ in G/μ)

$\qquad\quad = \mu(0)$ (by Definition 8.4.15).

Therefore, $\sigma_\theta(0) = \mu(0)$.   ■

Proposition 8.4.19 (Satyanarayana, Vijaya Kumari, and Nagaraju, accepted for publication)

Suppose $\mu : G \to [0, 1]$ is a fuzzy ideal of the $M\Gamma$-group G.

(i) The mapping $\Phi : G \to G/\mu$ defined by $\Phi(x) = x + \mu$ is an onto homomorphism with ker $\Phi = G_\mu = \{x \in G/\mu(x) = \mu(0)\}$. Hence, the $M\Gamma$-group G/μ is isomorphic to the $M\Gamma$-group G/G_μ under the mapping f: $G/G_\mu \to G/\mu$ defined by $f(x + G_\mu) = x + \mu$.

(ii) Suppose μ and σ are two fuzzy ideals of the $M\Gamma$-group G such that $G_\mu = G_\sigma$. Then the mapping $g : G/\mu \to G/\sigma$ defined by $g(x + \mu) = x + \sigma$ is an isomorphism.

(iii) If $G/\mu \cong G/\sigma$ under the isomorphism $g(x + \mu) = x + \sigma$, then $G_\mu = G_\sigma$.

Proof

(i) Define $\Phi: G \to G/\mu$ by $\Phi(x) = x + \mu$ for all $x \in G$.

Clearly, Φ is well-defined.

Let $x_1, x_2 \in G$ and $m \in M$.

Then $\Phi(x_1 + x_2) = (x_1 + x_2) + \mu$ (by the definition of Φ)

$\qquad\qquad\quad = (x_1 + \mu) + (x_2 + \mu)$ (by the definition of G/μ)

$\qquad\qquad\quad = \Phi(x_1) + \Phi(x_2)$.

$\Phi(m\gamma x_1) = m\gamma x_1 + \mu$ (by the definition of Φ)

$\qquad\quad = m\gamma(x_1 + \mu)$ (by the definition of G/μ)

$\qquad\quad = m\gamma\Phi(x_1)$ (by the definition of Φ).

Therefore, Φ is an $M\Gamma$-group homomorphism.

To verify that Φ is onto, let us consider an element $x + \mu \in G/\mu$.

Now $x \in G$, and by the definition of Φ, it follows that $\Phi(x) = x + \mu$.

This shows that Φ is onto.

Hence, Φ is an $M\Gamma$-group epimorphism.

So, by the fundamental theorem of homomorphism (Theorem 3.2.11), $G/\ker \Phi \cong \Phi(G)$, and so $G/\ker \Phi \cong G/\mu$. Now it remains to show that ker $\Phi = G_\mu$.

Now $x \in \ker \Phi \Leftrightarrow \Phi(x) = 0$

$\qquad\qquad\quad \Leftrightarrow x + \mu = 0 + \mu$ (by the definition of Φ)

$\qquad\qquad\quad \Leftrightarrow \mu(x) = \mu(0)$ (by Proposition 8.4.13)

$\qquad\qquad\quad \Leftrightarrow x \in G_\mu$.

This shows that $G/G_\mu \cong G/\mu$.

(ii) Given that $G_\mu = G_\sigma \Rightarrow G/G_\mu = G/G_\sigma$.

From Part (i), it follows that $G/\mu \cong G/G_\mu = G/G_\sigma \cong G/\sigma$ (with the association $x + \mu \leftrightarrow x + G_\mu = x + G_\sigma \leftrightarrow x + \sigma$).

So, $G/\mu \cong G/\sigma$ under the isomorphism g defined by $g(x + \mu) = x + \sigma$.

(iii) Suppose $G/\mu \cong G/\sigma$ under the isomorphism g defined by $g(x + \mu) = x + \sigma$.

Now we have to show that $G_\mu = G_\sigma$. Let $x \in G_\mu$.

Then $\mu(x) = \mu(0)$. Now $\mu(x - 0) = \mu(0)$

$$\Rightarrow x + \mu = 0 + \mu \qquad \text{(by Proposition 8.4.13)}$$
$$\Rightarrow g(x + \mu) = g(0 + \mu)$$
$$\Rightarrow x + \sigma = 0 + \sigma \qquad \text{(by the definition of } g\text{)}$$
$$\Rightarrow \sigma(x) = \sigma(0) \qquad \text{(by Proposition 8.4.13)}$$
$$\Rightarrow x \in G_\sigma \qquad \text{(by the definition of } G_\sigma\text{)}.$$

Therefore, $G_\mu \subseteq G_\sigma$. Similarly, it can be verified that $G_\sigma \subseteq G_\mu$.

Thus, we can conclude that $G_\mu = G_\sigma$.   ∎

Proposition 8.4.20 (Satyanarayana, Vijaya Kumari, and Nagaraju, accepted for publication)

Suppose that $f : G \rightarrow G^1$ is an $M\Gamma$-group homomorphism; and $\mu : G \rightarrow [0, 1]$ and $\mu^1: G^1 \rightarrow [0, 1]$ are fuzzy ideals of G and G^1, respectively. Then the following statements are true:

(i) $f(\mu)(0^1) = \mu(0)$, where 0^1 is the additive identity in G^1.

(ii) If μ is constant on ker f, then $(f(\mu))(f(x)) = \mu(x)$ for all $x \in G$.

(iii) $f^{-1}\left(G^1_{\mu^1}\right) = G_{f^{-1}(\mu^1)}$.

(iv) If f is an epimorphism, then $ff^{-1}(\mu^1) = \mu^1$.

(v) If μ is constant on ker f, then $f^{-1}f(\mu) = \mu$.

Proof

(i) Now $f(\mu)(0^1) = \sup\{\mu(x)/x \in f^{-1}(0^1)\}$

$$= \mu(0) \text{ (since } 0 \in f^{-1}(0^1) \text{ and } \mu(0) \geq \mu(x) \text{ for all } x \in G).$$

Therefore, $f(\mu)(0^1) = \mu(0)$.

(ii) Suppose that μ is constant on ker f. Now we verify that $(f(\mu))(f(x)) = \mu(x)$ for all $x \in G$.

Let $x \in G$. Write $f(x) = x^1$.

For any $z \in f^{-1}(x^1)$, we have $f(z) = x^1 = f(x)$

$$\Rightarrow f(z - x) = 0^1 \qquad \text{(since } f \text{ is a homomorphism)}$$
$$\Rightarrow z - x \in \ker f \qquad \text{(by the definition of kernel } f\text{)}$$
$$\Rightarrow \mu(z - x) = \mu(0) \qquad \text{(since } \mu \text{ is constant on kernel } f\text{)}$$
$$\Rightarrow z + \mu = x + \mu \qquad \text{(by Proposition 8.4.13)}$$
$$\Rightarrow (z + \mu)(0) = (x + \mu)(0)$$

$$\Rightarrow \mu(0 - z) = \mu(0 - x) \quad \text{(by Definition 8.4.12)}$$
$$\Rightarrow \mu(-z) = \mu(-x)$$
$$\Rightarrow \mu(z) = \mu(x) \quad \text{(since } \mu \text{ is a fuzzy ideal).}$$

Now $(f(\mu))(f(x)) = f(\mu)(x^1)$

$$= \sup\{\mu(z) \mid z \in f^{-1}(x^1)\}$$
$$= \sup\{\mu(x)\} = \mu(x).$$

This shows that $(f(\mu))(f(x)) = \mu(x)$ for all $x \in G$.

(iii) Now we show that $f^{-1}\left(G^1_{\mu^1}\right) = G_{f^{-1}(\mu^1)}$.

$x \in f^{-1}\left(G^1_{\mu^1}\right) \Leftrightarrow f(x) \in G^1_{\mu^1}$

$$\Leftrightarrow \mu^1(f(x)) = \mu^1(0^1) \text{ (by the definition of } G^1_{\mu^1})$$
$$\Leftrightarrow \mu^1(f(x)) = \mu^1(f(0)) \text{ (since } f(0) = 0^1)$$
$$\Leftrightarrow (f^{-1}\mu^1)(x) = (f^{-1}\mu^1)(0) \text{ (by the definition of } f^{-1}(\mu^1))$$
$$\Leftrightarrow x \in G_{f^{-1}(\mu^1)}.$$

Therefore, $f^{-1}\left(G^1_{\mu^1}\right) = G_{f^{-1}(\mu^1)}$.

(iv) Suppose f is an epimorphism. Now we verify that $f(f^{-1}(\mu^1)) = \mu^1$. Let $x^1 \in G^1$. Then $x^1 = f(x)$ for some $x \in G$.

$(f(f^{-1}(\mu^1)))(x^1) = (f(f^{-1}(\mu^1)))(f(x))$ (since $f(x) = x^1$)

$$= (f(\beta))(f(x)), \text{ where } \beta = f^{-1}(\mu^1), \text{ which is constant on kernel } f$$
$$\text{(by Proposition 8.4.6 (b))}$$
$$= \beta(x) \text{ (by (ii))}$$
$$= f^{-1}(\mu^1)(x) \text{ (since } \beta = f^{-1}(\mu^1))$$
$$= \mu^1(f(x)) \text{ (by Definition 1.7.19)}$$
$$= \mu^1(x^1) \text{ (since } f(x) = x^1).$$

This is true for all $x^1 \in G^1$. Hence, $f(f^{-1}(\mu^1)) = \mu^1$.

(v) Suppose μ is constant on kernel f.

Now we verify that $f^{-1}(f(\mu)) = \mu$.

Let $x \in G$. Now $(f^{-1}(f(\mu)))(x) = f(\mu)(f(x))$ (by Definition 1.7.19)

$$= \mu(x) \text{ (by (ii))}.$$

This is true for all $x \in G$.

Hence $(f^{-1}f)(\mu) = \mu$. ∎

Proposition 8.4.21 (Satyanarayana, Vijaya Kumari, and Nagaraju, accepted for publication)

If $f: G \to G^1$ is an epimorphism of $M\Gamma$-groups, then there is an order-preserving bijection between the fuzzy ideals of G^1 and the fuzzy ideals of G that are constant on ker f.

Proof

Part (i): Let $F(G) = \{\mu \mid \mu$ is a fuzzy ideal of G such that μ is constant on ker $f\}$, and $F(G^1) = \{\mu^1/\mu^1$ is a fuzzy ideal of $G^1\}$.

Define $\varphi : F(G) \to F(G^1)$ by $\varphi(\mu) = f(\mu)$ for all $\mu \in F(G)$.
Since f is well-defined, it follows that φ is well-defined.
Let $\mu_1, \mu_2 \in F(G)$ and $\varphi(\mu_1) = \varphi(\mu_2)$

$\quad\quad \Rightarrow f(\mu_1) = f(\mu_2)$ \quad\quad\quad\quad (by the definition of φ)
$\quad\quad \Rightarrow f^{-1}(f(\mu_1)) = f^{-1}(f(\mu_2))$
$\quad\quad \Rightarrow \mu_1 = \mu_2$ \quad\quad\quad\quad\quad (by Proposition 8.4.20 (v)).

Therefore, φ is one–one.
To verify that φ is onto, take $\mu^1 \in F(G^1)$. Write $\mu = f^{-1}(\mu^1)$.
By Theorem 8.4.6 (ii), it follows that μ is a fuzzy ideal of G, which is constant on kernel f.
So, $\mu \in F(G)$.
Now $\varphi(\mu) = f(\mu)$ (by the definition of φ)

$\quad\quad = f(f^{-1}(\mu^1))$ \quad\quad\quad (since $\mu = f^{-1}(\mu^1)$)
$\quad\quad = \mu^1$ \quad\quad\quad\quad\quad\quad (by Proposition 8.4.20 (iv)).

This shows that φ is onto. Hence, φ is a bijection from $F(G)$ to $F(G^1)$.

Part (ii): Now we verify that φ is an order-preserving mapping.
Let $\mu_1, \mu_2 \in F(G)$, satisfying the property $\mu_1 \subseteq \mu_2$.
Now we have to show that $\varphi(\mu_1) \subseteq \varphi(\mu_2)$.
Let $x^1 \in G^1$.
Since f is onto, there exists $x \in G$ such that $f(x) = x^1$.
Now $f(\mu_1)(x^1) = \sup\{\mu_1(t) \mid t \in f^{-1}(x^1)\}$ \quad\quad (by Definition 1.7.17)

$\quad\quad\quad\quad \le \sup\{\mu_2(t) \mid t \in f^{-1}(x^1)\}$ \quad\quad (since $\mu_1 \subseteq \mu_2$)
$\quad\quad\quad\quad = f(\mu_2)(x^1)$.

So $(f(\mu_1))(x^1) \le (f(\mu_2))(x^1)$. This is true for all $x^1 \in G^1$. This shows that $f(\mu_1) \subseteq f(\mu_2)$.
Now it is clear that $\mu_1 \subseteq \mu_2$. This implies $\varphi(\mu_1) = f(\mu_1) \subseteq f(\mu_2) = \varphi(\mu_2)$.
The proof is complete. ∎

8.5 Fuzzy Dimension in N-Groups

In this section, we consider the fuzzy ideals of an N-group G, where N is a near ring. We present the concepts of minimal elements, fuzzy linearly independent elements, and the fuzzy basis of an N-group G, and obtain fundamental related results.

Definition 8.5.1 (Saikia, 2003)

Let $\mu: G \to [0, 1]$ be a mapping. μ is said to be a **fuzzy ideal** of G if the following conditions hold:

(i) $\mu(g + g^1) \ge \min\{\mu(g), \mu(g^1)\}$
(ii) $\mu(g + x - g) = \mu(x)$

(iii) $\mu(-g) = \mu(g)$

(iv) $\mu(n(g + x) - ng) \geq \mu(x)$, for all $x, g, g^1 \in G, n \in N$.

If μ satisfies (i), (ii), and (iii), then μ is said to be a **fuzzy normal subgroup** of G.

Proposition 8.5.2 (Proposition 2.2. of Satyanarayana, Prasad, and Pradeepkumar, 2005)

Let G be an N-group with unity and $\mu : G \rightarrow [0, 1]$ be a fuzzy set with $\mu(ng) \geq \mu(g)$ for all $g \in G, n \in N$; then the following conditions are true.

(i) For all $0 \neq n \in N$, $\mu(ng) = \mu(g)$ if n is left invertible, and

(ii) $\mu(-g) = \mu(g)$

Proposition 8.5.3 (Proposition 2.4 of Satyanarayana, Prasad, and Pradeepkumar, 2005)

If μ is a fuzzy ideal of G, and $g, g^1 \in G$ with $\mu(g) > \mu(g^1)$, then $\mu(g + g^1) = \mu(g^1)$. In other words, if $\mu(g) \neq \mu(g^1)$, then $\mu(g + g^1) = \min \{\mu(g), \mu(g^1)\}$.

Proposition 8.5.4 (Proposition 2.6 of Satyanarayana, Prasad, and Pradeepkumar, 2005)

If $\mu : G \rightarrow [0, 1]$ is a fuzzy ideal, then

(i) $\mu(0) \geq \mu(g)$ for all $g \in G$,

(ii) $\mu(0) = \underset{g \in G}{\mathrm{Sup}}\, \mu(g)$.

Proposition 8.5.5 (Proposition 2.10 of Satyanarayana, Prasad, and Pradeepkumar, 2005)

Let μ be a fuzzy ideal of G and μ_t, μ_s (with $t < s$) be two level ideals of μ. Then the following two conditions are equivalent:

(i) $\mu_t = \mu_s$; and (ii) there is no $x \in G$ such that $t \leq \mu(x) < s$.

Definition 8.5.6

An element $x \in G$ is said to be a **minimal element** if $<x>$ is minimal in the set of all nonzero ideals of G.

Theorem 8.5.7 (Theorem 3.2 of Satyanarayana, Prasad, and Pradeepkumar, 2005)

If G has descending chain condition (DCC) on ideals, then every nonzero ideal of G contains a minimal element.

Proof

Let K be a nonzero ideal of G. Since G has DCC on its ideals, it follows that the set of all ideals of G contained in K has a minimal element. So, K contains a minimal ideal A (that is, A is minimal in the set of all nonzero ideals of G contained in K). Let $0 \neq a \in A$. Then $0 \neq <a> \subseteq A$, and so $<a> = A$. Since $<a>$ is a minimal ideal, it follows that a is a minimal element. ∎

Note 8.5.8

There are N-groups that do not satisfy DCC on its ideals, but contain a minimal element. For this, we observe the following example.

Example 8.5.9

Write $N = \mathbb{Z}$, $G = \mathbb{Z} \oplus \mathbb{Z}_6$. Now G is an N-group. Clearly, G has no DCC on its ideals. Consider $g = (0, 2) \in G$. Now the ideal generated by g, that is, $<g> = \mathbb{Z}g = \{(0, 0), (0, 2), (0, 4)\}$ is a minimal element in the set of all nonzero ideals of G. Hence, g is a minimal element.

Theorem 8.5.10 (Theorem 3.5 of Satyanarayana, Prasad, and Pradeepkumar, 2005)

Every minimal element is a u-element.

Proof

Let $0 \neq a \in G$ be a minimal element. Consider Na. Let $(0) \neq L$ and I be ideals of G such that $L \subseteq <a>$, $I \subseteq <a>$, and $L \cap I = (0)$. Since $L \neq (0)$, $(0) \subseteq L \subseteq <a>$, and a is minimal, it follows that $L = <a>$. Now $I = I \cap <a> = I \cap L = (0)$. This shows that L is essential in $<a>$. Hence, $<a>$ is a uniform ideal, and so a is a u-element. ∎

Note 8.5.11

The converse of Theorem 8.5.10 is not true. For this, observe the following example.

Write $G = \mathbb{Z}$, $N = \mathbb{Z}$. Since $\mathbb{Z}$ is uniform, and 1 is a generator, it follows that 1 is a u-element. But, $2\mathbb{Z}$ is a proper ideal of 1. $\mathbb{Z} = \mathbb{Z} = G$. Hence, 1 cannot be a minimal element. Thus, 1 is a u-element, but not a minimal element.   ∎

Theorem 8.5.12 (Theorem 3.7 of Satyanarayana, Prasad, and Pradeepkumar, 2005)

Suppose μ is a fuzzy ideal of G.

  (i) If $g \in G$, then for any $x \in \langle g \rangle$, we have $\mu(x) \geq \mu(g)$.
  (ii) If g is a minimal element, then for any $0 \neq x \in \langle g \rangle$ we have $\mu(x) = \mu(g)$.

Proof

  (i) By straightforward verification, we conclude that for $g \in G$, $\langle g \rangle = \cup_{i=0}^{\infty} A_i$, where

$$A_{k+1} = A_k^* \cup A_k^+ \cup A_k^0, A_0 = \{g\},$$
$$A_k^* = \{y + x - y \mid y \in G, x \in A_k\},$$
$$A_k^+ = \{n(y + x) - ny \mid n \in N, y \in G, x \in A_k\},$$
$$A_k^0 = \{x - y \mid x, y \in A_k\}.$$

We prove that $\mu(y) \geq \mu(g)$ for all $y \in A_m$ for $m \geq 1$. For this, we use induction on m. It is obvious if $m = 0$. Suppose the induction hypothesis for k, that is, $\mu(y) \geq \mu(g)$ for all $y \in A_k$. Now let $v \in A_k^* \cup A_k^+ \cup A_k^0$. Suppose $v \in A_k^*$. Then $v = z + y - z$ for some $y \in A_k$. Now $\mu(v) = \mu(z + y - z) \geq \mu(y)$ (since μ is a fuzzy ideal of G) $\geq \mu(g)$. Let $v \in A_k^0$. Then $v = y_1 - y_2$ for some $y_1, y_2 \in A_k$. Now $\mu(v) = \mu(y_1 - y_2) \geq \min \{\mu(y_1), \mu(y_2)\} \geq \mu(g)$, by induction hypothesis.
Suppose $v \in A_k^+$. Then $v = n(y + x) - ny$ for some $n \in N, y \in G, x \in A_k$. Now $\mu(v) = \mu(n(y + x) - ny) \geq \mu(x)$ (since μ is a fuzzy ideal) $\geq \mu(g)$ (by induction hypothesis). Thus, in all cases, we proved that $\mu(v) \geq \mu(g)$ for all $v \in A_{k+1}$. Hence, by the principle of mathematical induction, we conclude that $\mu(v) \geq \mu(g)$ for all $v \in A_m$ and for all positive integers m. We proved that $\mu(v) \geq \mu(g)$ for all $v \in A_m$ and for all positive integers m. Hence, $\mu(x) \geq \mu(g)$ for all $x \in \langle g \rangle$.
  (ii) Let $g \in G$ be a minimal element. Let $0 \neq x \in \langle g \rangle$. Now $(0) \neq \langle x \rangle \subseteq \langle g \rangle$. Since g is a minimal element, we have $\langle x \rangle = \langle g \rangle$. Therefore, $g \in \langle x \rangle$, and by (i), we have $\mu(g) \geq \mu(x)$. Thus, $\mu(x) = \mu(g)$.

Note 8.5.13

If G satisfies the descending chain condition on its ideals, then we say that "G has DCCI." Let K be an ideal of G. If the set $\{J \mid J$ is an ideal of $G, J \subseteq K\}$ has the descending chain condition, then we say that K has DCC on the ideals of G (we write DCCI G, in short).   ∎

Lemma 8.5.14 (Lemma 3.9 of Satyanarayana, Prasad, and Pradeepkumar, 2005)

If x is a u-element in G and G has DCCI, then there exists a minimal element $y \in \;<x>$ such that $<y> \leq_e <x>$.

Proof

Consider the ideal $<x>$. By Theorem 8.5.7, there exists a minimal element $y \in \;<x>$. Since $<y>$ is a nonzero ideal of $<x>$ and $<x>$ is a uniform ideal, it follows that $<y> \leq_e <x>$.   ∎

Now we present the concept of an inverse system that is similar to the concept of direct systems in N-groups presented in Definition 3.4.16.

Definition 8.5.15

A nonempty family $\{H_i\}_{i \in I}$ of ideals of G is said to be an **inverse system** of ideals if, for any finite number of elements $i_1, i_2, ..., i_k$ of I, there is an element i_0 in I such that $H_{i_0} \subseteq H_{i_1} \cap H_{i_2} \cap ... \cap H_{i_k}$.

Proposition 8.5.16 (Proposition 2.2 of Satyanarayana and Prasad, 2000)

The implication (i) $\Rightarrow$ (ii) $\Rightarrow$ (iii) $\Rightarrow$ (iv) is true, where

   (i) G has DCCI.
   (ii) Every inverse system of nonzero ideals of G is bounded below by a nonzero ideal of G.
   (iii) There exists a finite sum of simple ideals of G, whose sum is direct and essential in G.
   (iv) G has FGD.

Proof

(i) $\Rightarrow$ (ii): Since G has DCCI, it follows that G has FGD. By Theorem 5.1.2, there exist uniform ideals $U_1, U_2, ..., U_n$ of G such that the sum $U_1 + U_2 + ... + U_n$ is direct and essential in G.

Let $\{H_i\}_{i\in I}$ be an inverse system of nonzero ideals of G. Since $S = U_1 \oplus U_2 \oplus \ldots \oplus U_n$ is essential in G, $S \cap H_i \ne (0)$ for each $i \in I$. Write

$$S^* = \{S \cap H_i / i \in I\}.$$

Since G has DCCI, S^* contains a minimal element, say $S \cap H_{i_0}$ for some $i_0 \in I$. Let $j \in I$. Since $\{H_i\}_{i\in I}$ is an inverse system, there exists $k \in I$ such that $H_{i_k} \subseteq H_j \cap H_{i_0}$. Now $H_{i_k} \cap S \subseteq H_{i_0} \cap S \Rightarrow H_{i_k} \cap S = H_{i_0} \cap S$ (since $H_{i_0} \cap S$ is minimal in S^*). Now $H_{i_0} \cap S = H_{i_k} \cap S \subseteq H_{i_k} \subseteq H_j$. Therefore, $H_{i_0} \cap S \subseteq H_j$. This is true for all $j \in I$.

Hence, $H_{i_0} \cap S$ is a nonzero lower bound for the inverse system $\{H_i\}_{i\in I}$.

(ii) $\Rightarrow$ (iii): Let H be a nonzero ideal of G. Write $\vartheta = \{H^1/H^1 \text{ is an ideal of } G \text{ and } H^1 \subseteq H\}$. Clearly $(\vartheta, \subseteq)$ is a partially ordered set, where "$\subseteq$" is the usual set inclusion. Consider the opposite relation "$\subseteq^1$" on ϑ as $H_1 \subseteq^1 H_2$ if and only if $H_1 \supseteq H_2$. Now $(\vartheta, \subseteq^1)$ is also a partially ordered set. Let $\{H_i\}_{i\in I}$ be a chain of elements from $(\vartheta, \subseteq^1)$. Then $\{H_i\}_{i\in I}$ is also a chain in $(\vartheta, \subseteq)$. Clearly $\{H_i\}_{i\in I}$ is an inverse system of ideals with respect to "$\subseteq$". This implies $\{H_i\}_{i\in I}$ is bounded below by a nonzero ideal L of G with respect to "$\subseteq$," that is, $\{H_i\}_{i\in I}$ is bounded above by a nonzero ideal L of G with respect to "$\subseteq^1$." Therefore, by Zorn's lemma, $(\vartheta, \subseteq^1)$ has a maximal element X. This implies X is a minimal element in $(\vartheta, \subseteq)$. Hence, H contains a simple ideal X of G. So, we have $H \cap$ (sum of simple ideals of G) $\ne \{0\}$.

Hence, the sum of simple ideals is essential in G.

Now we show that $S(G) =$ sum of simple ideals, is the direct sum of finite number of simple ideals. Let us suppose $S(G)$ is the direct sum of an infinite family of simple ideals. Let $\{S_i\}_{i=1}^{\infty}$ be a countable subfamily whose sum is direct. Consider $A = \{\Sigma_{i=n}^{\infty} S_i / n = 1, 2, \ldots\}$. This is an inverse system of ideals of G. By (ii), there exists a nonzero ideal J of G such that J is a lower bound for $A \Rightarrow J \subseteq \cap_{X \in A} X = \cap_{n=1}^{\infty} \left(\Sigma_{i=n}^{\infty} S_i \right) = \{0\} \Rightarrow J = \{0\}$, a contradiction. Hence, there exists a finite collection of simple ideals whose sum is direct and essential in G.

(iii) $\Rightarrow$ (iv): Suppose $S_1, S_2, \ldots, S_k$ are simple ideals whose sum is direct and essential in G. Since each S_i is also uniform, by Theorem 5.1.2, we conclude that G has FGD. ∎

Now let us recollect the definitions of a linearly independent and essential basis given in Definition 5.2.8.

Note 8.5.17

(i) If G has FGD, then every linearly independent subset X of G is a finite set.

(ii) Suppose that dim $G = n$ and $X \subseteq G$. If X is a linearly independent set, then we have: $|X| = n \Leftrightarrow X$ is a maximal linearly independent set $\Leftrightarrow X$ is an essential basis for G.

Theorem 8.5.18 (Theorem 3.12 of Satyanarayana, Prasad, and Pradeepkumar, 2005)

If G has DCCI, then there exist linearly independent minimal elements x_1, x_2, ..., x_n in G where $n = \dim G$, and the sum $<x_1> + ... + <x_n>$ is direct and essential in G. Also, $B = \{x_1, x_2, ..., x_n\}$ forms an essential basis for G.

Proof

Since G has DCCI, by Proposition 8.5.16, G has FGD. Suppose $n = \dim G$. Then by Theorem 6.3.3, there exist u-linearly independent elements u_1, u_2, ..., u_n such that the sum $<u_1> + ... + <u_n>$ is direct and essential in G. Since G has DCCI, by Lemma 8.5.14, there exist minimal elements $x_i \in <u_i>$ such that $<x_i> \leq_e <u_i>$ for $1 \leq i \leq n$. Since $u_1, u_2, ..., u_n$ are linearly independent, it follows that $x_1, x_2, ..., x_n$ are also linearly independent.

Thus, we have linearly independent minimal elements $x_1, x_2, ..., x_n$ in G where $n = \dim G$. Since $<x_i> \leq_e <u_i>$ by a result mentioned in the introduction, it follows that $<x_1> \oplus ... \oplus <x_n> \leq_e <u_1> \oplus ... \oplus <u_n> \leq_e G$, and so $<x_1> \oplus ... \oplus <x_n> \leq_e G$.

Thus, $B = \{x_1, x_2, ... , x_n\}$ forms an essential basis for G. ∎

Now we introduce the concept of fuzzy linearly independent elements with respect to a fuzzy ideal μ of G.

Definition 8.5.19

Let G be an N-group and μ be a fuzzy ideal of G. $x_1, x_2, ..., x_n \in G$ are said to be fuzzy μ-linearly independent (or fuzzy linearly independent with respect to μ) if they satisfy the following two conditions: (i) $x_1, x_2, ..., x_n$ are linearly independent; and (ii) $\mu(y_1 + ... + y_n) = \min\{\mu(y_1), ..., \mu(y_n)\}$ for any $y_i \in <x_i>$, $1 \leq i \leq n$.

Theorem 8.5.20 (Theorem 4.2 of Satyanarayana, Prasad, and Pradeepkumar, 2005)

Let μ be a fuzzy ideal on G. If $x_1, x_2, ..., x_n$ are minimal elements in G with distinct μ-values, then $x_1, x_2, ..., x_n$ are (i) linearly independent; and (ii) fuzzy μ-linearly independent.

Proof

The proof is by induction on n. If $n = 1$, then x_1 is linearly independent and also fuzzy linearly independent. Let us assume that the statement is true for $(n - 1)$. Now suppose $x_1, x_2, ..., x_n$ are minimal elements with distinct μ values. By the induction hypothesis, $x_1, x_2, ..., x_{n-1}$ are linearly independent and fuzzy linearly independent. If $x_1, \ldots, x_n$ are not linearly independent, then the sum of $<x_1>, <x_2>, \ldots, <x_n>$ is not direct. This means $<x_i> \cap (<x_1> \oplus \ldots <x_{i-1}> \oplus <x_{i+1}> \oplus \ldots \oplus <x_n>) \neq \{0\}$. This implies $0 \neq y_i = y_1 + \ldots + y_{i-1} + y_{i+1} + \ldots + y_n$, where $y_j \in <x_j>$ for $1 \leq j \leq n$. Now $\mu(x_i) = \mu(y_i)$ (by Theorem 8.5.12) $= \mu(y_1 + \ldots + y_{i-1} + y_{i+1} + \ldots + y_n) = \min \{\mu(y_1), \ldots, \mu(y_{i-1}), \mu(y_{i+1}), \ldots, \mu(y_n)\}$ (by the induction hypothesis) $= \mu(y_k)$ for some $k \in \{1, 2, \ldots, i-1, i+1, \ldots, n\} = \mu(x_k)$ (by Theorem 8.5.12). Thus, $\mu(x_i) = \mu(x_k)$ for $i \neq k$, which is a contradiction. This shows that $x_1, x_2, \ldots, x_n$ are linearly independent.

Now we prove that $x_1, x_2, \ldots, x_n$ are fuzzy linearly independent. Suppose $y_i \in <x_i>, 1 \leq i \leq n$.

$$\mu(y_1 + y_2 + \ldots + y_{n-1}) = \min \{\mu(y_1), \ldots, \mu(y_{n-1})\} \text{ (by the induction hypothesis)}$$
$$= \mu(y_j) \text{ for some } j \text{ with } 1 \leq j \leq n-1) = \mu(x_j)$$
$$\text{(by Theorem 8.5.12)}$$

$$\text{Now } \mu(x_j) \neq \mu(x_n) \Rightarrow \mu(y_1 + y_2 + \ldots + y_{n-1}) = \mu(x_j) \neq \mu(x_n) = \mu(y_n)$$
$$\Rightarrow \mu(y_1 + y_2 + \ldots + y_{n-1} + y_n) = \min \{\mu(y_1 + \ldots + y_{n-1}), \mu(y_n)\}$$
$$\text{(by Proposition 8.5.3)}$$
$$= \min \{\min \{\mu(y_1), \ldots, \mu(y_{n-1}), \mu(y_n)\}\}$$
$$= \min \{\mu(y_1), \ldots, \mu(y_n)\}.$$

This shows that $x_1, x_2, \ldots, x_n$ are fuzzy linearly independent with respect to μ. ∎

Now we provide the definition of the concept of a "fuzzy pseudo-basis."

Definition 8.5.21

(i) Let μ be a fuzzy ideal of G. A subset B of G is said to be a **fuzzy pseudo-basis** for μ if B is a maximal subset of G such that $x_1, x_2, \ldots, x_k$ are fuzzy linearly independent for any finite subset $\{x_1, x_2, \ldots, x_k\}$ of B.

(ii) Consider the set $B = \{k \mid$ there exist a fuzzy pseudo-basis B for μ with $|B| = k\}$. If B has no upper bound, then we say that the **fuzzy dimension** of μ is infinite. We denote this fact by S-dim $(\mu) = \infty$. If B has an upper bound, then the **fuzzy dimension** of μ is sup B. We denote this fact by S-dim $(\mu) = \sup B$. If $m = $ S-dim $(\mu) = \sup B$, then a fuzzy pseudo-basis B for μ with $|B| = m$ is called a **fuzzy basis** for the fuzzy ideal μ.

Proposition 8.5.22 (Proposition 5.2 of Satyanarayana, Prasad, and Pradeepkumar, 2005)

Suppose G has FGD and μ is a fuzzy ideal of G. Then (i) $|B| \leq \dim G$ for any fuzzy pseudo-basis B for μ; and (ii) S-dim $(\mu) \leq \dim G$.

Proof

Suppose $n = \dim G$.

(i) Suppose B is a fuzzy pseudo-basis for μ. If $|B| > n$, then B contains distinct elements $x_1, x_2, \ldots, x_{n+1}$. Since B is a fuzzy pseudo-basis, the elements $x_1, x_2, \ldots, x_{n+1}$ are linearly independent; and by Theorem 6.3.3, it follows that $n + 1 \leq n$, a contradiction. Therefore, $|B| \leq n = \dim G$.

(ii) From (i) it is clear that dim M is an upper bound for the set $B = \{k \mid$ there exist a fuzzy pseudo-basis B for μ with $|B| = k\}$.

Therefore, S-dim $(\mu) = \sup B \leq \dim G$. ∎

Definition 8.5.23

An N-group G is said to have a **fuzzy basis** if there exists an essential ideal A of G and a fuzzy ideal μ of A such that S-dim $(\mu) = \dim G$. The fuzzy pseudo-basis of μ is called as **fuzzy basis** for G.

Remark 8.5.24

If G has FGD, then every fuzzy basis for G is a basis for G.

Theorem 8.5.25 (Theorem 5.5 of Satyanarayana, Prasad, and Pradeepkumar, 2005)

Suppose that G has DCCI. Then G has a fuzzy basis (in other words, there exist an essential ideal A of G and a fuzzy ideal μ of A such that S-dim $(\mu) = \dim G$).

Proof

Since G has DCCI, it has FGD. Suppose dim $G = n$. By Note 5.2.9, there exist linearly independent minimal elements $x_1, x_2, \ldots, x_n$ such that $\{x_1, x_2, \ldots, x_n\}$ forms an essential basis for G. Take $0 \leq t_1 < t_2 < \ldots < t_n \leq 1$. Define $\mu(y_i) = t_i$ for $y_i \in$ $<x_i>$, $1 \leq i \leq n$. Then μ is a fuzzy ideal on $A = <x_1> + <x_2> + \ldots + <x_n> \leq_e G$. By Theorem 8.5.20, $x_1, x_2, \ldots, x_n$ are fuzzy μ-linearly independent. So $\{x_1, x_2, \ldots, x_n\}$ is a pseudo-basis for μ. Now dim $M = n \leq \sup B \leq \dim G$ (by Proposition 8.5.22), and hence S-dim $(\mu) = \dim G$. This shows that G has a fuzzy basis. ∎

9

Fundamental Concepts of Graph Theory

9.1 Introduction

One of the most important elementary properties of a graph is that of connectedness. Intuitively, the concept of connectedness is obvious. A connected graph is in "one piece," so that we can reach any point from any other point by traveling along the lines/edges. In this section, we develop the basic properties of connected and disconnected graphs and components. We present some elementary results and examples.

Definition 9.1.1

A graph G is said to be **connected** if there is at least one path between every pair of vertices in G.

Example 9.1.2

The graph given in Figure 9.1 is a connected graph.

Definition 9.1.3

A graph G is said to be a **disconnected graph** if it is not a connected graph. A finite alternating sequence of vertices and edges (no repetition of edge allowed) beginning and ending with vertices such that each edge is incident with the vertices preceding and following it, is called a **walk**.

Example 9.1.4

The graph given in Figure 9.2 is a disconnected graph (observe that there is no path from v_4 to v_2).

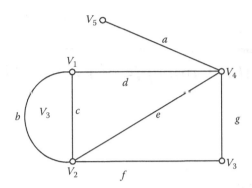

FIGURE 9.1
Connected graph.

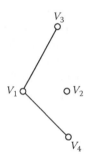

FIGURE 9.2
Disconnected graph.

Note 9.1.5

Let $G = (V, E)$ be a disconnected graph. We define a relation $\sim$ on the set of vertices as follows: $v \sim u \Leftrightarrow$ there is a walk from v to u.

Then this relation $\sim$ is an equivalence relation.

Let $\{V_i\}_{i \in \Delta}$ be the collection of all equivalence classes. Now $V = \bigcup_{i \in \Delta} V_i$.

Write $E_i = \{e \in E \mid$ an endpoint of e is in $V_i\}$ for each i.

Then (V_i, E_i) is a connected subgraph of G for every $i \in \Delta$.

This connected subgraph (V_i, E_i) of G is called a **connected component** (or **component**) of G for every $i \in \Delta$.

The collection $\{(V_i, E_i)\}_{i \in \Delta}$ of subgraphs of G is the collection of all connected components of G.

Note 9.1.6

(i) If G is a connected graph, then G is the only connected component of G.

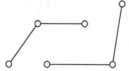

FIGURE 9.3
Graph with two components.

(ii) A disconnected graph G consists of two or more connected components.
(iii) The connected component of a graph G is a maximal connected subgraph of G.
(iv) A graph is connected if and only if it has exactly one component.
(v) Consider the graph given in Figure 9.3.

This graph is a disconnected graph with two components.

Note 9.1.7 (Formation of Components)

If G is a connected graph, then G contains only one connected component, and it is equal to G.

Now suppose that G is a disconnected graph. Consider a vertex v in G. If each vertex of G is joined by some path to v, then the graph is connected, which is a contradiction. So there exists at least one vertex that is not joined by any path to v.

The vertex v and all the vertices of G that are joined by some paths to v, together with all the edges incident on them, form a component (G_1, say).

To find another component, take a vertex u (from G) that is not in G_1. The vertex u and all the vertices of G that are joined by some paths to u, together with all the edges incident on them, form a component (G_2, say).

Continue this procedure to find the components. Since the graph is a finite graph, the procedure will stop at a finite stage. In this manner, we can find all the connected components of G. It is clear that a component itself is a graph.

Theorem 9.1.8 (Theorem 13.5 of Satyanarayana and Prasad, 2009)

A graph G is disconnected if and only if its vertex set V can be partitioned into two nonempty disjoint subsets V_1 and V_2 such that there exists no edge in G, one end vertex of which is in V_1 and the other in V_2.

Proof

Let $G = (V, E)$ be a graph.

Assume that $V = V_1 \cup V_2$, $V_1 \neq \varnothing$, $V_2 \neq \varnothing$, and $V_1 \cap V_2 = \varnothing$ such that there exists no edge in G, one end vertex of which is in V_1 and the other in V_2.

We have to show that G is disconnected.

Let us assume that G is connected.

Let $a, b \in G$ such that $a \in V_1$ and $b \in V_2$.

Since G is connected, there exists a path $a\, e_1\, a_1\, e_2\, a_2 \ldots a_{n-1}\, e_n\, b$ from a to b.

Now $b \notin V_1$. So there exists a least number j such that $a_j \notin V_1$.

Now $a_{j-1} \in V_1$, $a_j \in V_2$, and e_j is an edge between a vertex in V_1 and a vertex in V_2, which is a contradiction. Hence, G is disconnected.

Converse

Suppose G is disconnected. Let a be a vertex in G. Write

$V_1 = \{v \in V \mid$ either $a = v$, or a is joined by a walk to $v\}$.

If $V_1 = V$, then all the vertices (other than a) are joined by a path to a, and so the graph is connected, which is a contradiction.

Hence, $V_1 \subsetneq V$. Write $V_2 = V \backslash V_1$. Now V_1, V_2 form a partition for V. If a vertex $x \in V_1$ is joined to $y \in V_2$ by an edge, then y is connected to a (since $x \in V_1$) $\Rightarrow$ $y \in V_1 \Rightarrow y \in V_1 \cap V_2 = \varnothing$, which is a contradiction.

Hence, no vertex in V_1 is connected to a vertex of V_2 by an edge. The proof is complete. ∎

(Theorem 9.1.8 can also be stated as follows: A graph G is connected if and only if for any partition v of the vertex set into subsets V_1 and V_2, there is a line of G joining a point of V_1 to a point of V_2.)

Theorem 9.1.9 (Theorem 13.6 of Satyanarayana and Prasad, 2009)

If a graph (either connected or disconnected) has exactly two vertices of odd degree, then there exists a path joining these two vertices.

Proof

Let G be a graph and v_1, v_2 be the only two vertices in G whose degrees are odd.

Case (i): Suppose G is connected. Then by definition, there exists a path between any two vertices.

So, there is a path from v_1 to v_2 in G.

Case (ii): Suppose G is disconnected. Then G has two or more components.

Let G_1 be the component in which v_1 is present.

Now G_1 is a connected subgraph.

If G_1 does not contain v_2, then the number of vertices in G_1 with odd degree is 1 (an odd number), which is a contradiction (since in any graph, the number of vertices of odd degree is even).

Therefore, v_2 is also in G_1.

Hence, v_1, v_2 are in the same component.

Since every component is connected, there exists a path from v_1 to v_2 in G. ∎

Theorem 9.1.10 (Theorem 13.7 of Satyanarayana and Prasad, 2009)

A simple graph with n vertices and k components can have at most $(n - k)$ $(n - k + 1)/2$ edges (or the maximum number of edges in a simple graph with n vertices and k components is $(n - k)(n - k + 1)/2$).

Proof

Let G be a simple graph (i.e., G has no self-loops and no parallel edges) with n vertices and k components.

Suppose that $G_1, G_2, G_3, ..., G_i, ..., G_k$ are k components and $n_1, n_2, n_3, ... n_i, ..., n_k$ are the number of vertices in the components $G_1, G_2, G_3, ..., G_k$, respectively.
It is clear that

$$n = n_1 + n_2 + n_3 + ... + n_k ... \tag{9.1}$$

Now $\Sigma_{i=1}^{k}(n_i - 1) = \Sigma_{i=1}^{k} n_i - k = n - k$.

Squaring on both sides, we have $\left[\Sigma_{i=1}^{k} \right]^2 = (n - k)^2$

$$\Rightarrow \sum_{i=1}^{k}(n_i - 1)^2 + 2 \prod_{i=1}^{k} \prod_{j=1}^{k} (n_i - 1)(n_j - 1) = n^2 + k^2 - 2nk$$

$$\Rightarrow \sum_{i=1}^{k}(n_i^2 - 2n_i + 1) + \text{(some nonnegative terms)}$$

$$= n^2 + k^2 - 2nk$$

$$\Rightarrow \sum_{i=1}^{k}(n_i^2 - 2n_i) + k \le n^2 + k^2 - 2nk$$

$$\Rightarrow \sum_{i=1}^{k} n_i^2 - 2\sum_{i=1}^{k} n_i + k \le n^2 + k^2 - 2nk$$

$$\Rightarrow \sum_{i=1}^{k} n_i^2 - 2n + k \le n^2 + k^2 - 2nk \text{ (since } \sum_{i=1}^{k} n_i = n \text{ by (Equation 9.1))}$$

$$\Rightarrow \sum_{i=1}^{k} n_i^2 \le n^2 + k^2 - 2nk + 2n - k \tag{9.2}$$

Since G is a simple graph, the ith component G_i is also a simple graph. Since G_i contains n_i vertices and G_i is simple, it follows that the maximum number of edges in the ith component is $\frac{n_i(n_i-1)}{2}$.

Therefore, the maximum number of edges in all the k components is

$$\sum_{i=1}^{k} \frac{n_i(n_i-1)}{2} = \frac{1}{2} \sum_{i=1}^{k} n_i(n_i-1)$$

$$= \frac{1}{2} \sum_{i=1}^{k} (n_i^2 - n_i)$$

$$= \frac{1}{2} \left(\sum_{i=1}^{k} n_i^2 - \sum_{i=1}^{k} n_i \right)$$

$$\leq \frac{1}{2}[(n^2 + k^2 - 2nk + 2n - k) - n] \quad \text{(From Equations 9.1 and 9.2)}$$

$$= \frac{1}{2}(n^2 - nk - nk + k^2 + n - k)$$

$$= \frac{1}{2}[(n-k)n - (n-k)k + (n-k)]$$

$$= \frac{1}{2}(n-k)(n-k+1).$$

The proof is complete.   ∎

Euler formulated the concept of an Eulerian line when he solved the problem of the Konigsberg bridges. Euler lines mainly deal with the nature of connectivity in graphs. The concept of an Euler line is used to solve several puzzles and games. In the next part of this section, we discuss the relation between a local property, namely, the degree of a vertex, and global properties such as the existence of Eulerian cycles. We see that there are well–designed characterizations for Eulerian graphs.

Definition 9.1.11

Let G be a graph. A closed walk running through every edge of the graph G exactly once is called an **Euler line.**

Definition 9.1.12

A graph G that contains a Euler line is called an **Euler graph**.

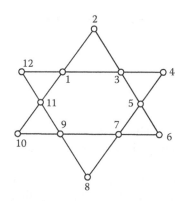

FIGURE 9.4
Euler graph.

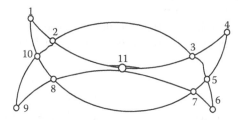

FIGURE 9.5
Euler graph.

Example 9.1.13

The graphs given in Figures 9.4 and 9.5 are Euler graphs.
(i) Consider Figure 9.4.
Here 123456789(10)(11)(12)13579(11)1 is an Euler line.
(ii) Consider Figure 9.5.
Here 23456789(10) 12(11)357(11)8(10)2 is an Euler line.

Note 9.1.14

An Euler graph may contain isolated vertices. If G is an Euler graph and it contains no isolated vertices, then it is connected.

Hereafter, we consider only those Euler graphs that do not contain isolated vertices. So the Euler graphs that we consider are connected.

Theorem 9.1.15 (Theorem 13.8 of Satyanarayana and Prasad, 2009)

A given connected graph G is an Euler graph if and only if all the vertices of G are of even degree.

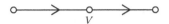

FIGURE 9.6
Euler line.

Proof

Suppose G is an Euler graph. Then G contains an Euler line. So there exists a closed walk running through all the edges of G exactly once.

Let $v \in V$ be a vertex of G. Now, in tracing the walk it goes through two incident edges on v, with one entering v and the other exiting it (Figure 9.6).

This is true not only for all the intermediate vertices of the walk, but also for the terminal vertex, because we exited and entered at the same vertex at the beginning and the end of the walk.

Therefore, if v occurs k times in the Euler line, then $d(v) = 2k$.

Thus, if G is an Euler graph, then the degree of each vertex is even.

Converse

Suppose all the vertices of G are of even degree. Now to show that G is an Euler graph, we have to construct a closed walk starting at an arbitrary vertex v and running through all the edges of G exactly once.

To find a closed walk, let us start from the vertex v. Since every vertex is of even degree, we can exit from every vertex we entered; the tracing cannot stop at any vertex other than v.

Since v is also of even degree, we eventually reach v when the tracing comes to an end.

If this closed walk (h, say) includes all the edges of G, then G is an Euler graph.

Suppose the closed walk h does not include all the edges, then the remaining edges form a subgraph h^1 of G.

Since both G and h have all their vertices of even degree, the degrees of the vertices of h^1 are also even.

Moreover, h^1 must touch h at least at one vertex a, because G is connected. Starting from a, we can again construct a new walk in graph h^1.

Since all the vertices of h^1 are of even degree, this walk in h^1 must terminate at the vertex a.

This walk in h^1 combined with h forms a new walk that starts and ends at vertex v and has more edges than those that are in h.

We repeat this process until we obtain a closed walk that travels through all the edges of G.

In this manner, one can get an Euler line.

Thus, G is an Euler graph.   ∎

Note 9.1.16 (Konigsberg Bridges Problem)

In the graph of the Konigsberg bridges problem, there exist vertices of odd degree. So, all the vertices are not of even degree.

Hence, by Theorem 9.1.15, we conclude that the graph representing the Konigsberg bridges problem is not an Euler graph.

So, we conclude that it is not possible to walk over each of the seven bridges exactly once and return to the starting point.

Note 9.1.17

The concept of an Euler graph can be used to solve a number of puzzles such as finding how a given picture can be drawn in one continuous line without retracing a line and without lifting the pencil from the paper.

To study a large graph, it is convenient to consider it as a combination of small graphs. First, we understand the properties of the small graphs involved, and then we derive the properties of the large graph.

Since graphs are defined using the concepts of set theory, we use set theoretical terminology to define operations on graphs. In defining the operations on graphs, we are more concerned about the edge sets.

Definition 9.1.18

Let $G_1 = (V_1, E_1)$ and $G_2 = (V_2, E_2)$ be any two graphs. Then the **union** of G_1 and G_2 is the graph $G = (V, E)$, where $V = V_1 \cup V_2$ and $E = E_1 \cup E_2$. We write $G = G_1 \cup G_2$.

Definition 9.1.19

Consider the graphs G_1, G_2, and G_3 (Figures 9.7 through 9.9).

It is clear that $G_1 = G_2 \cup G_3$.

Definition 9.1.20

Let $G_1 = (V_1, E_1)$ and $G_2 = (V_2, E_2)$ be any two graphs with $V_1 \cap V_2 \neq \emptyset$. Then the **intersection** of G_1 and G_2 is defined as the graph $G = (V, E)$, where $V = V_1 \cap V_2$ and $E = E_1 \cap E_2$. We write $G = G_1 \cap G_2$.

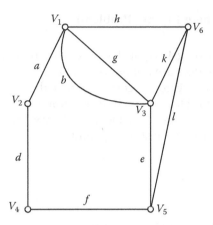

FIGURE 9.7
Graph.

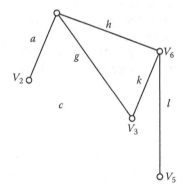

FIGURE 9.8
Subgraphs of Graph 9.7.

Example 9.1.21

The intersection graph of the graphs plotted in Figures 9.8 and 9.9 is given in Figure 9.10: $G_4 = G_2 \cap G_3$.

Definition 9.1.22

The **ring sum** of two graphs $G_1 = (V_1, E_1)$ and $G_2 = (V_2, E_2)$ is defined as the graph $G = (V, E)$, where $V = V_1 \cup V_2$ and $E = (E_1 \cup E_2)\setminus(E_1 \cap E_2)$. We write $G = G_1 \oplus G_2$.

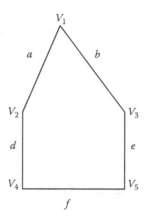

FIGURE 9.9
Subgraphs of Graph 9.7.

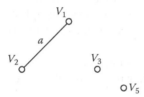

FIGURE 9.10
Intersection graph.

Example 9.1.23

Consider the graphs G_2 and G_3 given in Figures 9.8 and 9.9.
The graph $G_2 \oplus G_3$ is given by G_5 (Figure 9.11).

Note 9.1.24

(i) The three operations (union, intersection, and ring sum) on the set of all finite graphs are commutative, that is, $G_1 \cup G_2 = G_2 \cup G_1$, $G_1 \cap G_2 = G_2 \cap G_1$, and $G_1 \oplus G_2 = G_2 \oplus G_1$.

(ii) If G_1 and G_2 are edge disjoint, then $G_1 \cap G_2$ is a null graph, and $G_1 \oplus G_2 = G_1 \cup G_2$.

(iii) For any graph G, we have $G \cup G = G \cap G = G$ and $G \oplus G = \emptyset$ (a null graph).

(iv) If v is a vertex in a graph G, then $G - v$ denotes a subgraph of G obtained by deleting v from G. Deleting a vertex always implies

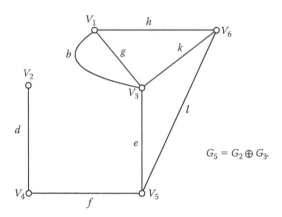

FIGURE 9.11
Ring sum of two subgraphs.

the deletion of all edges incident on that vertex. If e is an edge in a graph G, then $G - e$ denotes the subgraph of G obtained by deleting e from G. Deleting an edge does not imply the deletion of its end vertices. It is clear that $G - e = G \oplus e$.

(v) The complement of a subgraph g of G is the graph $G \oplus g$ or $G - g$, which is obtained after removing all the edges of g from G.

Definition 9.1.25

A graph G is said to be **decomposed** into two subgraphs g_1 and g_2, if (i) $g_1 \cup g_2 = G$ and (ii) $g_1 \cap g_2 = \emptyset$ (a null graph).

(Equivalently, a graph G is said to be **decomposed** into two subgraphs g_1 and g_2, if every edge of G occurs either in g_1 or in g_2, but not in both.)

However, some of the vertices may occur in both g_1 and g_2. During the decomposition, the isolated vertices are disregarded.

Definition 9.1.26

A graph G can be decomposed into more than two subgraphs. A graph G is said to have been **decomposed** into subgraphs $g_1, g_2, \ldots, g_n$, if
(i) $g_1 \cup g_2 \cup \ldots \cup g_n = G$, and (ii) $g_i \cap g_j = \emptyset$ (a null graph), for $i \neq j$.

Definition 9.1.27

A pair a, b of vertices in a graph G are said to be **fused** or **merged** or **identified** if the two vertices are replaced by a single new vertex such that every edge that is incident on either *a* or *b* or on both is incident on the new vertex. It is clear that the fusion of two vertices does not alter the number of edges, but it reduces the number of vertices by one. A closed walk in which no vertex (except the initial vertex and the final vertex) appears more than once is called a **circuit** or **cycle**.

Example 9.1.28

Observe graphs *G* and *G** given in Figures 9.12 and 9.13.
    The vertices *a* and *b* of *G* were fused to get the graph *G**.

Theorem 9.1.29 (Theorem 13.10 of Satyanarayana and Prasad, 2009)

A connected graph *G* is an Euler graph if and only if it can be decomposed into circuits.

Proof

Let *G* be a connected graph. Assume that *G* can be decomposed into circuits.

    That is, *G* is the union of edge–disjoint circuits. Since the degree of every vertex in a circuit is even, it follows that the degree of every vertex in *G* is even. By Theorem 9.1.15, it follows that *G* is an Euler graph.

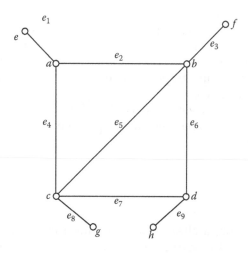

FIGURE 9.12
Graph.

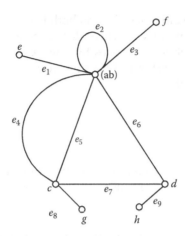

FIGURE 9.13
Fused graph of 9.12.

Converse

Suppose that G is an Euler graph.

Then all the vertices of G are of even degree.

Now consider a vertex v_1. Clearly, there are at least two edges incident at v_1.

Suppose that one of these edges is between v_1 and v_2. Since vertex v_2 is also of even degree, it must have at least one more edge (say between v_2 and v_3).

Proceeding in this way, we arrive at a vertex that has previously been traversed.

Then we get a circuit say Γ_1.

Now let use remove Γ_1 from G. (It is understood that we are removing only the edges of Γ_1 from G.) Then all the vertices in the remaining graph must also be of even degree.

Now from the remaining graph, remove another circuit Γ_2 in the same way (as we removed Γ_1 from G).

Continue this process to get circuits $\Gamma_1, \Gamma_2 \dots \Gamma_n$ (until no edges are left). Now it is clear that the graph G has been decomposed into the circuits $\Gamma_1, \Gamma_2 \dots \Gamma_n$. The proof is complete.   ∎

Definition 9.1.30

In a connected graph, a closed walk running through every vertex of G exactly once (except the starting vertex at which the walk terminates) is called a **Hamiltonian circuit**. A graph containing a Hamiltonian circuit is called a **Hamiltonian graph**.

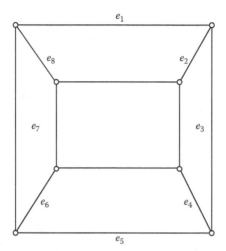

FIGURE 9.14
Hamiltonian graph.

Example 9.1.31

Observe the graph given in Figure 9.14.

In this graph, the walk $e_1, e_2, e_3, e_4, e_5, e_6, e_7, e_8$ is a closed walk running through every vertex of G exactly once. Hence, this walk is a Hamiltonian circuit.

It follows that this graph is a Hamiltonian graph.

Note 9.1.32

   (i) A circuit in a connected graph G is a Hamiltonian circuit if and only if it includes every vertex of G.
   (ii) A Hamiltonian circuit in a graph of n vertices consists of exactly n edges.

Note 9.1.33

Every connected graph may not have a Hamiltonian circuit.

Example 9.1.34

The two graphs given in Figures 9.15 and 9.16 are connected, but they do not have Hamiltonian circuits.

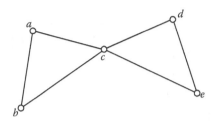

FIGURE 9.15
Graph with no Hamiltonian circuit.

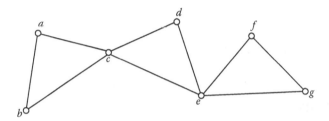

FIGURE 9.16
Graph with no Hamiltonian circuit.

The concept of a "tree" plays a vital role in the theory of graphs. First, we introduce the definition of a "tree" and then study some of its properties and applications.

Definition 9.1.35

A connected graph without circuits is called a **tree**. A tree in which one vertex (called the **root**) is distinguished from all the other vertices is called a **rooted tree**.

Example 9.1.36

Trees with one, two, three, and four vertices are shown in Figure 9.17.

Note 9.1.37

(i)   A tree contains at least one vertex.
(ii) A tree without an edge is referred to as a **null tree**.
(iii) A tree is always a simple graph.

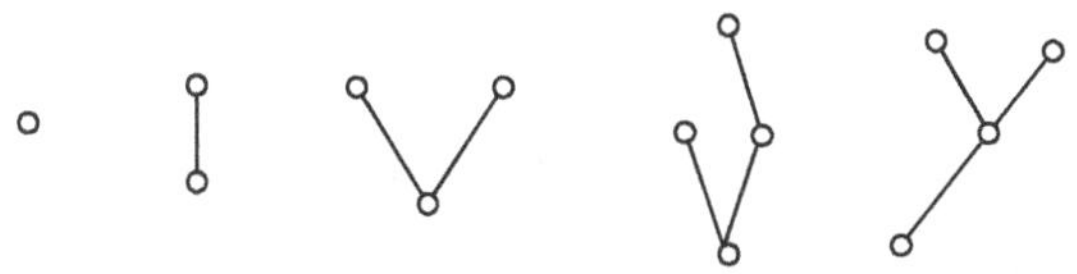

FIGURE 9.17
Trees.

Theorem 9.1.38 (Theorem 13.14 of Satyanarayana and Prasad, 2009)

In a tree T, there is one and only one path between every pair of vertices.

Proof

Suppose T is a tree. Then T is a connected graph and contains no circuits.

Since T is connected, there exists at least one path between every pair of vertices in T.

Suppose that there are two distinct paths between two vertices a and b of T.

Now, the union of these two paths will contain a circuit in T, which is a contradiction (since T contains no circuits).

This shows that there exists one and only one path between a given pair of vertices in T.

Theorem 9.1.39 (Theorem 13.15 of Satyanarayana and Prasad, 2009)

If there is one and only one path between every pair of vertices in G, then G is a tree.

Proof

Let G be a graph. Assume that there is one and only one path between every pair of vertices in G. This shows that G is connected.

If possible, suppose that G contains a circuit. Then there is at least one pair of vertices a, b such that there are two distinct paths between a and b. But this is a contradiction to our assumption. So, G contains no circuits. Thus, G is a tree. ∎

Theorem 9.1.40 (Theorem 13.16 of Satyanarayana and Prasad, 2009)

A tree G with n vertices has $(n - 1)$ edges.

Proof

We prove this theorem by induction on the number of vertices n.

If $n = 1$, then G contains only one vertex and no edge.

So, the number of edges in G is $n - 1 = 1 - 1 = 0$.

Suppose the induction hypothesis that the statement is true for all trees with less than n vertices. Now let us consider a tree with n vertices.

Let e_k be any edge in T whose end vertices are v_i and v_j.

Since T is a tree, by Theorem 9.1.38, there is no other path between v_i and v_j.

So, by removing e_k from T, we get a disconnected graph.

Furthermore, $T - e_k$ consists of exactly two components (say T_1 and T_2).

Since T is a tree, there are no circuits in T and so there are no circuits in T_1 and T_2.

Therefore, T_1 and T_2 are also trees.

It is clear that $|V(T_1)| + |V(T_2)| = |V(T)|$, where $V(T)$ denotes the set of vertices in T.

Also, $|V(T_1)|$ and $|V(T_2)|$ are less than n.

Therefore, by the induction hypothesis, we have

$|E(T_1)| = |V(T_1)| - 1$ and $|E(T_2)| = |V(T_2)| - 1$.

Now, $|E(T)| - 1 = |E(T_1)| + |E(T_2)| = |V(T_1)| - 1 + |V(T_2)| - 1$

$$\Rightarrow |E(T)| = |V(T_1)| + |V(T_2)| - 1$$
$$= |V(T)| - 1 = n - 1.$$

The proof is complete. ∎

Theorem 9.1.41 (Theorem 13.17 of Satyanarayana and Prasad, 2009)

Any connected graph with n vertices and $n - 1$ edges is a tree.

Proof

Let G be a connected graph with n vertices and $n - 1$ edges.

It is enough to show that G contains no circuits.

If possible, suppose that G contains a circuit. Let e be an edge in that circuit.

Since e is a circuit, it follows that $G - e$ is still connected.

Now $G - e$ is connected with n vertices, and so it should contain at least $n - 1$ edges, which is a contradiction (to the fact that $G - e$ contains only $(n - 2)$ edges).

So, G contains no circuits. Therefore, G is a tree. ∎

Definition 9.1.42

A connected graph is said to be **minimally connected** if the removal of any one edge from the graph provides a disconnected graph.

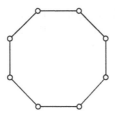

FIGURE 9.18
Graph not minimally connected.

FIGURE 9.19
Graph minimally connected.

Example 9.1.43

(i) The graph given in Figure 9.18 is not minimally connected.
(ii) The graph given in Figure 9.19 is minimally connected.
(iii) Any circuit is not minimally connected.
(iv) Every tree is minimally connected.

Theorem 9.1.44 (Theorem 13.18 of Satyanarayana and Prasad, 2009)

A graph G is a tree if and only if it is minimally connected.

Proof

Assume that G is a tree.

Now we have to show that G is minimally connected. Let us suppose that G is not minimally connected. Then there exists an edge e such that $G - e$ is connected.

That is, e is in some circuit, which implies that G is not a tree, which is a contradiction.

Hence, G minimally connected.

Converse

Suppose that G is minimally connected. Now it is enough to show that G contains no circuits. Let us suppose G contains a circuit.

Then by removing one of the edges in the circuit, we get a connected graph, which is a contradiction (to the fact that the graph is minimally connected). This shows that G contains no circuits. Thus, G is a tree.　■

Note 9.1.45

To interconnect n given distinct points, the minimum number of line segments needed is $n - 1$.

Theorem 9.1.46 (Theorem 13.19 of Satyanarayana and Prasad, 2009)

If a graph G contains n vertices, $n - 1$ edges, and no circuits, then G is a connected graph.

Proof

Let G be a graph with n vertices, $n - 1$ edges, and no circuits.
　　Let us suppose that G is disconnected.
　　G consists of two or more circuitless components (say, $g_1, g_2, \ldots, g_k$).
　　Now, $k \geq 2$. Select a vertex v_i in g_i, for $1 \leq i \leq k$.
　　Add new edges $e_1, e_2, \ldots, e_{k-1}$, where $e_i = v_i v_{i+1}$ to get a new graph G^*.
　　It is clear that G^* contains no circuits and is connected, and so G^* is a tree.
　　Now G^* contains n vertices and $(n-1)+(k-1)=(n+k-2) \geq n$ edges, which is a contradiction (since a tree contains $(n-1)$ edges). This shows that G is connected. The proof is complete.　■

Theorem 9.1.47 (Theorem 13.20 of Satyanarayana and Prasad, 2009)

For a given graph G with n vertices, the following conditions are equivalent:
　(i)　G is connected and is circuitless.
　(ii)　G is connected and has $n - 1$ edges.
　(iii)　G is circuitless and has $n - 1$ edges.
　(iv)　There is exactly one path between every pair of vertices in G.
　(v)　G is a minimally connected graph.
　(vi)　G is a tree.

Proof

　(i)　$\Leftrightarrow$　(vi) is clear.
　(vi)　$\Rightarrow$　(ii) and (iii): Theorem 9.1.40
　(ii)　$\Rightarrow$　(vi): Theorem 9.1.41
　(iii)　$\Rightarrow$　(vi): Theorem 9.1.46

(iv) ⟺ (vi): Theorems 9.1.38 and 9.1.39
(v) ⟺ (vi): Theorem 9.1.44. ∎

Theorem 9.1.48 (Theorem 13.21 of Satyanarayana and Prasad, 2009)

If T is a tree (with two or more vertices), then there exist at least two pendant vertices (a pendant vertex is a vertex of degree 1).

Proof

Let T be a tree with $|V| \geq 2$.

Let $v_0 e_1 v_1 e_2 v_2 e_3 \ldots v_{n-1} e_n v_n$ be a longest path in T. (Since T is a finite graph, it is possible to find a longest path.)

Now we wish to show that $d(v_0) = 1 = d(v_n)$.

If $d(v_0) > 1$, then there exists at least one edge e with endpoint v_0 such that $e \neq e_1$.

If $e \in \{e_1, e_2, \ldots, e_n\}$, then $e = e_i$ for some $i \neq 1$.

So, either $v_{i-1} = v_0$ or $v_i = v_0 \Rightarrow v_0$ repeated in the path, which is a contradiction.

Hence, $e \notin \{e_1, e_2, e_3, \ldots, e_n\}$.

Now, $e, e_1, e_2, e_3, \ldots e_n$ is a path of length $n + 1$, which is a contradiction.

Hence, $d(v_0) = 1$. In a similar way, we can show that $d(v_n) = 1$.

Hence, v_0, v_n are two pendant vertices. ∎

Definition 9.1.49

Let G be a connected graph. The **distance** between two vertices v and u is denoted by $d(v, u)$ and is defined as the length of the shortest path (i.e., the number of edges in the shortest path) between v and u.

Example 9.1.50

Consider the connected graph given in Figure 9.20.

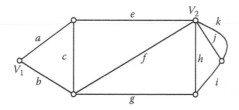

FIGURE 9.20
Connected graph.

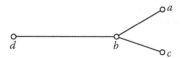

FIGURE 9.21
Tree.

Here, some of the paths between v_1 and v_2 are (a, e), (a, c, f), (b, c, e), (b, f), (b, g, h), (b, g, i, j), (b, g, i, k). Here there are two shortest paths (a, e) and (b, f), each of length 2. Hence, $d(v_1, v_2) = 2$.

Example 9.1.51

Consider the tree given in Figure 9.21.
   Here, $d(a, b) = 1$, $d(a, c) = 2$, $d(a, d) = 2$, $d(b, d) = 1$.
   In a connected graph, we can find the distance between any two given vertices.

Definition 9.1.52

Let X be a set. A real–valued function $f(x, y)$ of two variables x and y (i.e., f: $X \times X \rightarrow R$ where R is the set of all real numbers) is said to be a **metric** on X if it satisfies the following properties

   (i) Non-negativity: $f(x, y) \geq 0$ and $f(x, y) = 0$, if and only if $x = y$.
   (ii) Symmetry: $f(x, y) = f(y, x)$.
   (iii) Triangle inequality: $f(x, y) \leq f(x, z) + f(z, y)$ for x, y, z in X.

Theorem 9.1.53 (Theorem 13.22 of Satyanarayana and Prasad, 2009)

Let G be a connected graph. The distance $d(v, u)$ between two vertices v and u is a metric.

Proof

Let $v, u \in V$.
   (i) $d(v, u) = $ (the length of the shortest path between v and u)
           ≥ 0.
      Therefore, $d(v, u) \geq 0$.
      Also, $d(v, u) = 0$,
      if and only if there exists a path between v and u of length 0,
      if and only if $v = u$

(if $v \neq u$, then $d(v, u) \geq 1$, which is a contradiction).
So, $d(v, u) = 0$ if and only if $v = u$.

(ii) $d(v, u) =$ the length of the shortest path between v and u
  $=$ the length of the shortest path between u and v
  $= d(u, v)$.

(iii) Now we show that for any w in V, $d(v, u) \leq d(v, w) + d(w, u)$.
Suppose $n = d(v, w)$ and $m = d(w, u)$.
Then there exists a path of minimum length n from v to w, and there exists a path of minimum length m from w to u. Combining these two paths, we get a path from v to u of length $\leq n + m$.
So, $d(v, u) \leq n + m = d(v, w) + d(w, u)$.

Definition 9.1.54

Let G be a graph and v be any vertex in G. Then the **eccentricity** of v is denoted by $E(v)$ and is defined as the distance from v to the vertex farthest from v in G. That is, $E(v) = \max \{d(v, u) \mid u$ is a vertex in $G\}$. Sometimes, it is also referred to as the **associate number** or **separation**.

Example 9.1.55

Consider the graph given in Figure 9.22.
Here, $E(a) = 2$, because the distance from a to a vertex d farthest from a is $d(a, d) = 2$. Similarly, $E(b) = 1$, $E(c) = 2$, $E(d) = 2$.

Note 9.1.56

The eccentricities of vertices of a graph may be represented as in Figure 9.23.

Example 9.1.57

Consider the tree given in Figure 9.24.
Here $E(a) = 3$, $E(d) = 2$, $E(b) = 3$, $E(e) = 3$, $E(c) = 2$, $E(f) = 3$.

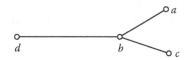

FIGURE 9.22
Tree.

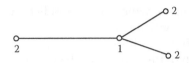

FIGURE 9.23
Tree.

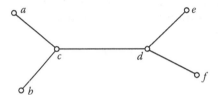

FIGURE 9.24
Tree.

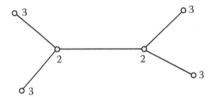

FIGURE 9.25
Tree.

The graph given in Figure 9.24 may be represented as the graph in Figure 9.25.

Definition 9.1.58

In a graph, a vertex with minimum eccentricity is called the **center** of the graph. v is the center of a graph if and only if $E(v) = \min \{E(u) \mid u \in G\}$.

Example 9.1.59

Consider Graph-1 given in Figure 9.26.
   The center of this graph is b (since b has the minimum eccentricity).

Note 9.1.60

In a circuit, every vertex has equal eccentricity.

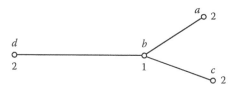

FIGURE 9.26
Tree.

Theorem 9.1.61 (Theorem 13.23 of Satyanarayana and Prasad, 2009)

Every tree T has either one or two centers.

Proof

If T contains exactly one vertex, then that vertex is the center. If T contains exactly two vertices, then these two vertices are centers. Let T be any tree with more than two vertices. Let $v \in V$. Now the maximum distance, max $d(v, v_i)$, from the given vertex v to any other vertex v_i occurs only when v_i is a pendant vertex.

Then by Theorem 9.1.48, T must have two or more pendant vertices.

Now by removing all pendant vertices from T, we get a tree T^1.

The eccentricity of a vertex v in T^1 is one less than the eccentricity of the vertex v in T. This is true for all vertices v.

Therefore, all the vertices that T had as centers will still remain centers in T^1. Now, from T^1 we again remove all pendant vertices, and get another tree T^{11}. ∎

We continue this process (as shown in Figures 9.27 through 9.30) until we are left with either a vertex (which is the center of T) or an edge (whose end-vertices are the two centers of T). Hence, every tree has either one or two centers.

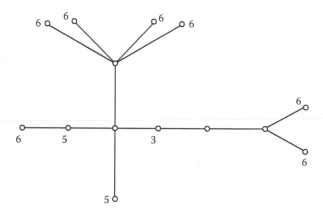

FIGURE 9.27
Tree.

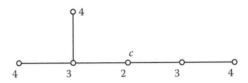

FIGURE 9.28
Tree.

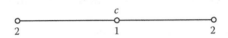

FIGURE 9.29
Tree.

FIGURE 9.30
Tree.

Corollary 9.1.62

From the preceding argument, it follows that if a tree has two centers, then the two centers must be adjacent.

Definition 9.1.63

The eccentricity of a center in a graph (or in a tree) is called the **radius** of that graph.

Example 9.1.64

Consider the graphs T, T^1, T^{11} (given in Theorem 9.1.61). The radius of T is 3; the radius of T^1 is 2, and the radius of T^{11} is 1.

Definition 9.1.65

The length of the longest path in a tree is called the **diameter** of the tree.

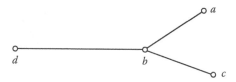

FIGURE 9.31
Tree with diameter 2.

Example 9.1.66

(i) Consider the tree given in Figure 9.31.
Clearly, the diameter of $T = 2$, radius $= 1$.
The only longest paths in T are abc, abd, cbd (and their lengths are equal to 2).

Observation 9.1.67

The diameter is not always equal to twice the radius.
For example, in the graph given in Figure 9.32, radius $= 2$ and diameter $= 3$.

Definition 9.1.68

(i) A **directed graph** $D = (V, E, h)$ consists of a finite nonempty set V, called the **vertex set**; a set E, called the **edge set**; and a mapping $h: E \rightarrow \{(u, v) \mid u, v \in V\}$. We also write $V(D)$ for V, $E(D)$ for E. We write $D = (V, E)$ instead of $D = (V, E, h)$.
If V, E are finite sets, then the directed graph is said to be **finite**. The elements of V are called **vertices** or **nodes**, and the elements of E are called **edges** or **arcs**.
(ii) If a is an arc in a directed graph D associated with an ordered pair of vertices (u, v), then a is said to **join** u to v. u is called the **origin** or the **initial vertex** or the **tail** of a, and v is called the **terminus** or the **terminal vertex** or the **head** of a.

FIGURE 9.32
Tree with diameter 3 and radius 2.

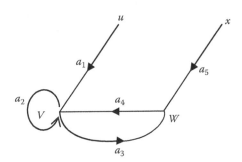

FIGURE 9.33
Directed graph.

Notation 9.1.69

In directed graphs, the arcs are represented by arrowed lines.

Example 9.1.70

Consider the directed graph given in Figure 9.33. Here the vertex set $V = \{u,$ $v, w, x\}$. The arc set $E = \{a_1, a_2, a_3, a_4, a_5\}$. a_1 joins u to v, and so a_1 is associated with the ordered pair (u, v). So, $h(a_1) = (u, v)$; similarly, $h(a_2) = (v, v)$, $h(a_3) = (v, w)$, $h(a_4) = (w, v)$, $h(a_5) = (x, w)$.

Definition 9.1.71

(i) An arc e is called a **loop** if $h(e) = (v, v)$ for some $v \in V$.
(ii) The arcs e, e^1 are said to be **multiple** arcs if there are multiple edges when the direction is ignored.
(iii) A directed graph D is said to be **simple** if it has neither multiple arcs nor loops.
(iv) Let $v_0, v_1, \ldots, v_n$ be vertices in D. If there exists a sequence of distinct arcs $v_0v_1, v_1v_2, \ldots, v_{n-1}v_n$, then the sequence $v_0v_1 \ldots, v_{n-1}v_n$ is called a **path** from v_0 to v_n of length n. If $v_0 = v_n$, then this path is called a **circuit**.
(v) Let u, v be two vertices. Then $d(u, v) = \min\{l(P) \mid P \text{ is a } v - u \text{ path}\}$ is called the **shortest distance** from u to v. The path P from u to v of length $d(u, v)$ is called the **shortest path**.
(vi) For $v \in V$, the integer $l(v)$, called the **level** of v, is defined as follows: If there is no path to v from any point of V, then $l(v) = 0$. Otherwise, $l(v) = \max \{n \mid n \text{ is the length of a path from } u \text{ to } v \text{ for some } u \in V\}$. Here, if xy is an arc, then we denote $l(x) < l(y)$.
(vii) We say that uv **starts** at level m and **ends** at level $m + 1$ if $l(u) = m$ and $l(v) = m + 1$.

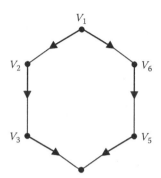

FIGURE 9.34
Directed graph with levels.

Example 9.1.72

In the directed graph given in Figure 9.34, the levels of elements are as follows: $l(v_1) = 0$; $l(v_6) = l(v_2) = 1$; $l(v_3) = l(v_5) = 2$; $l(v_4) = 3$.

Definition 9.1.73

Let $D = (V, E)$ and $D^1 = (V^1, E^1)$ be finite directed graphs. We say that D and D^1 are isomorphic if there exist bijections $f: V \to V^1$ and $g: E \to E^1$ such that $g(uv) = f(u)f(v)$.

Example 9.1.74

Consider the directed graphs given in Figures 9.35 and 9.36.

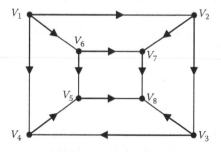

FIGURE 9.35
Isomorphic directed graphs.

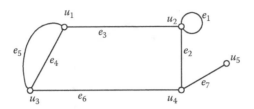

FIGURE 9.36
Isomorphic directed graphs.

Here, $V = \{v_1, v_2, v_3, v_4, v_5, v_6, v_7, v_8\}$ and $V^1 = \{u_1, u_2, u_3, u_4, u_5, u_6, u_7, u_8\}$. Define $f: V \rightarrow V^1$ by $f(v_i) = u_i$ for $1 \le i \le 8$. Clearly, f is a bijection. Now define $g: E \rightarrow E^1$ as $g(v_iv_j) = u_iu_j$. Clearly, g is a bijection and $g(v_iv_j) = u_iu_j = f(v_i)f(v_j)$.

Notation 9.1.75

Let S be any subset of the vertex set $V(G)$ of a graph G. We denote by $G - S$ the graph whose vertex set is $V(G) - S$ and whose edge set is $E(G-S) = \{xy \mid x, y\} \cap S \ne \varphi\}$.

Definition 9.1.76

(i) A **vertex cut** of G is a subset $S \subseteq V(G)$ such that $G - S$ is disconnected. If $T \subseteq E(G)$ is any subset, we denote by $G - T$ the graph whose vertex set is $V(G)$ and whose edge set is $E(G) - T$. An **edge cut** of G is a subset $T \subseteq E(G)$ such that the graph $G - T$ is disconnected.

(ii) The **(vertex) connectivity** of G is defined by $\kappa(G) = \min\{n \ge 0 \mid$ there exists a vertex cut $S \subseteq V(G)$ such that $|S| = n\}$ if G has a finite vertex cut, and $\kappa(G) = \infty$ otherwise. Similarly, the **edge connectivity** of G is defined by $\lambda(G) = \min\{n \ge 0 \mid$ there exists an edge cut $T \subseteq E(G)$ such that $|s| = n\}$ if G has a finite edge cut, and $\lambda(G) = \infty$ otherwise. We denote $\delta(G) = \min\{d(v) \mid v \in V\}$. It is clear that for a connected graph G, we have $\kappa(G) \le \lambda(G) \le \delta(G)$.

9.2 Directed Hypercubes

In this section, we present the notion of a directed hypercube of dimension n. Some fundamental definitions and results are presented. Eventually, we provide a characterization for a directed hypercube of dimension n.

We start with some preliminary definition of n-cubes.

Definition 9.2.1

For any nonempty set X, the set of all subsets of X is denoted by $\wp(X)$, and it is called the **power set** of X.

Note 9.2.2

(i) If $|X| = n$, then $|\wp(X)| = 2^n$.
(ii) The number of subsets of X containing m elements is ${}^nC_m$.
(iii) For $A \subseteq X, B \subseteq X$, we write $A \backslash B$ for the set $\{a \in A \mid a \notin B\}$.

Notation 9.2.3

We write $A \, \Delta \, B$ for the set $(A \backslash B) \cup (B \backslash A)$.
For a positive integer $n > 1$, we define an n-cube as follows.

Definition 9.2.4

Let n be a positive integer >1. The n-cube is the graph whose vertices are the ordered n-tuples of 0's and 1's; and there is an edge between two vertices u, v if and only if u and v differ in exactly one position. (Equivalently, we can define n-cube as follows: Let X be a set with $|X| = n$ and $\wp(X)$ be its power set. Then in a graph having $\wp(X)$ as its vertex set, there is an edge between two vertices A, B if and only if $|A \, \Delta \, B| = 1$, where $A \, \Delta \, B = (A \backslash B) \cup (B \backslash A)$, is called the n-**cube**.)

Example 9.2.5

(i) Take $V = \{(0, 0), (0, 1), (1, 0), (1, 1)\}$. For instance, consider $(1, 1)$ and $(1, 0)$. They differ from each other in the second coordinate.
    So, these two vertices are to be connected by an edge. Since $(0, 0)$ and $(1, 0)$ differ from each other in the first coordinate, these two vertices are to be joined by an edge. Similarly, $(1, 1)$ and $(0, 1)$ differ in exactly one position. Also $(0, 1)$ and $(0, 0)$ differ in exactly one position.
    By connecting the pairs of vertices (which differ in exactly one position) by edges, we get the graph, as shown in Figure 9.37. This is a 2–cube.
(ii) We can get a 2-cube by following the equivalent definition given earlier as follows:

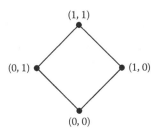

FIGURE 9.37
2-cube binary representation of vertices.

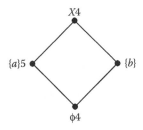

FIGURE 9.38
2-cube.

Take $X = \{a, b\}$. Then $\wp(X) = \{\emptyset, \{a\}, \{b\}, X\}$. Take $V = \wp(X)$ as the vertex set. If $A, B \in \wp(X)$ with $|A \Delta B| = 1$, then we join A and B by an edge. For instance, $|\{a\} \Delta X| = 1$. Therefore $\{a\}$, X are to be joined by an edge. $\emptyset$, $\{a\}$ are to be joined by an edge. $\emptyset$, $\{b\}$ are to be joined by an edge. $\{b\}$, X are to be joined by an edge. Then we get the graph (2-cube), as shown in Figure 9.38.

Example 9.2.6

Let $V = \{(0, 0, 0), (0, 0, 1), (0, 1, 0), (1, 0, 0), (1, 0, 1), (1, 1, 0), (0, 1, 1), (1, 1, 1)\}$. For instance, consider $(1, 1, 1)$ and $(1, 1, 0)$; these two vertices differ from each other in the third coordinate. Therefore, these two vertices are to be joined by an edge. Similarly, we get the other edges of the graph, as shown in Figure 9.39. This represents a 3-cube.

(ii) To get the 3-cube (Figure 9.40) following the power set representation of a set, take
$X = \{a, b, c\}$. Write $V = \wp(X) = \{\emptyset, \{a\}, \{b\}, \{c\}, \{a, b\}, \{b, c\}, \{a, c\}, X\}$. Since $|\{a, b\} \Delta X| = 1$, we join $\{a, b\}$ and X by an edge. Similarly, we get all edges, and finally we present the 3-cube given.

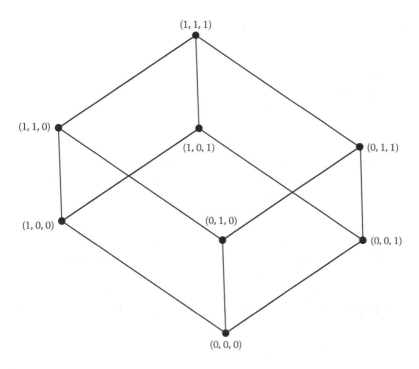

FIGURE 9.39
3-cube binary representation of vertices.

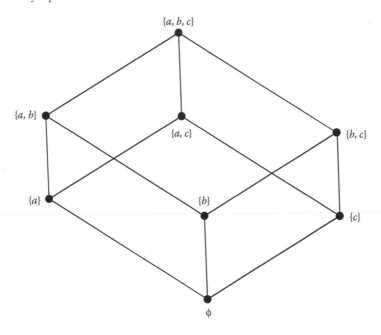

FIGURE 9.40
3-cube.

Theorem 9.2.7

(i) An n-cube is regular of degree n.
(ii) An n-cube has 2^n vertices and $n\,2^{n-1}$ edges.
(iii) The diameter of an n-cube is n.

Proof

(i) To show that an n-cube is regular of degree n, we have to show that every vertex of an n-cube is of degree n. Let $V = \wp(X)$ with $|X| = n$. Let A be any vertex. Then $A \subseteq X$.

Consider $\delta(A)$ = number of edges having an endpoint A
= number of vertices that are adjacent to A
= the number of subsets B of X with $|A \triangle B| = 1 = n$.

(ii) The vertex set $V = \wp(X)$ with $|X| = n$. Therefore, $|V| = |\wp(X)| = 2^n$. Therefore, $\Sigma\delta(v_i) = 2|E|$. Since an n-cube is regular of degree n, we have $\delta(v) = n$ for all $v \in V$.
Now $2|E| = \Sigma\delta(v_i) = \Sigma n = |V|n = 2^n n \Rightarrow |E| = n\,2^n \div 2 = n\,2^{n-1}$.
Therefore, an n-cube has $n\,2^{n-1}$ edges.

(iii) Let $D = (V, E)$ be an n-cube. Then $V = \wp(X)$ for some set X with $|X| = n$. We know that the diameter of D is
$d(D) = \max\{d(u, v) \mid d(u, v)$ is the shortest distance for any $u, v \in V\}$.
The vertices of D are subsets of X and there is an edge between the vertices $A, B \in V(D)$ if $|A \triangle B| = 1$. It follows that there is an edge between the vertex $\emptyset$ to each of the singleton subsets of X. So, the shortest distance between the vertex $\emptyset$ and each of the singleton sets is 1. Since we are joining each singleton set to a two-element set for which it is a subset, the shortest distance between the vertex $\emptyset$ and the two-element set is 2.
If we continue this process, the shortest distance between the vertex $\emptyset$ and a vertex A with $|A| = n - 1$ is $n - 1$.
Since $|X| = n$, it follows that the shortest distance between the vertex X and $\emptyset$ is n.
Also, it can be observed that for any two sets A and B, the number $d(A, B)$ is equal to

$$|A \triangle B| = |((A \backslash B) \cup (B \backslash A))|. \text{ Also, } |A \triangle B| \leq n.$$

Therefore, $d(D) = \max\{d(A, B) \mid A \subseteq X, B \subseteq X\} \leq n$ and $d(\emptyset, X) = n$. This shows that $d(D) = n$. ∎

Now we consider finite directed simple graphs. We present the concept a "directed hypercube of dimension n (or directed n-cube)."

Definition 9.2.8 (Definition 2 of Satyanarayana and Prasad, 2003)

Let X be a set with $|X| = n$. The directed graph Q_n with $V(Q_n) = \wp(X)$ (the power set of X) and $E(Q_n) = \{IJ \mid I, J \in \wp(X) \text{ and } |I| = |J| + 1, J \subseteq I\}$ is called the **directed hypercube of dimension** n (or a directed n-cube).

Example 9.2.9

Take $X = \{1, 2\}$. Then $|\wp(X)| = 2^2 = 4$, where $\wp(X) = \{\emptyset, X, \{1\}, \{2\}\}$.

Now $V = \wp(X)$, and we join two vertices A, B in V if $|A| = |B| + 1$ and $B \subseteq A$.

For instance, take X and $\{1\}$. Now $|X| = 2$ and $|\{1\}| = 1$, $|X| = 2 = 1 + 1 = |\{1\}| + 1$, and $X \supseteq \{1\}$. So, there is an arc from X to $\{1\}$. Finally, we present the graph as shown in Figure 9.41, which is a directed hypercube of dimension 2 (or a directed 2-cube).

Example 9.2.10

Take $X = \{1, 2, 3\}$. Now $|\wp(X)| = 2^3 = 8$, where
$\wp(X) = \{\emptyset, \{1\}, \{2\}, \{3\}, \{1, 2\}, \{1, 3\}, \{2, 3\}, X\}$.

Now we join two vertices A, B in $V = \wp(X)$ if $|A| = |B| + 1$ and $B \subseteq A$.

For instance, take X and $\{1, 2\}$. Now $X \supseteq \{1, 2\}$ and $|X| = 3 = 2 + 1 = |\{1, 2\}| + 1$. So, there exists an arc from X to $\{1, 2\}$. Similarly, we can get all other arcs. Finally, we get the graph, as shown in Figure 9.42, which represents a directed 3-cube.

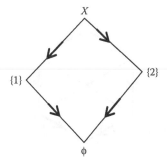

FIGURE 9.41
Directed 2-cube.

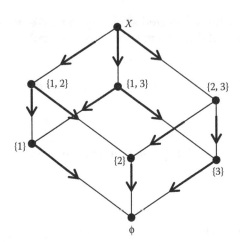

FIGURE 9.42
Directed 3-cube.

Definition 9.2.11 (Definition 1 of Satyanarayana and Prasad, 2000)

The subgraph of D generated by a vertex $v_o \in V$ is the graph $G^1 = (V^1, E^1)$ of D, where V^1 is the set of all nodes v such that there exists a path from v_o to v, that is,

$V^1 = \{v \mid$ there exists a path from v_o to $v\} \cup \{v_0\}$ and $E^1 = \{uv \in E \mid u, v \in V^1\}$.

Example 9.2.12

Consider the directed graph shown in Figure 9.43.
The subgraph generated by v_2 is shown in Figure 9.44.

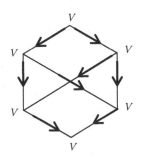

FIGURE 9.43
Graph.

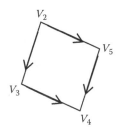

FIGURE 9.44
Subgraph generated by a vertex.

Notation 9.2.13

We denote the subgraph generated by an element $v \in V$ by $G_v = (V_v, E_v)$, where $V_v = \{u \mid \text{there exists a path from } v \text{ to } u\} \cup \{v\}$ and $E_v = \{xy \in E \mid x, y \in V_v\}$.

Example 9.2.14

Consider the directed graph shown in Figure 9.45.
    The subgraph generated by v_5 (i.e., $D_{v_5} = (V_{v_5}, E_{v_5})$ is shown in Figure 9.46.

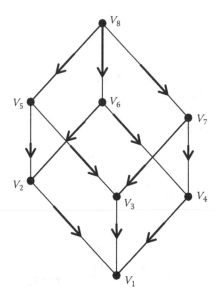

FIGURE 9.45
Graph.

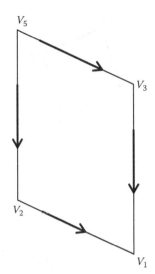

FIGURE 9.46
Subgraph generated by a vertex.

Definition 9.2.15

We write $L_i = \{v \in V \mid l(v) = n - i\}$, for $0 \le i \le n$; $F_j = \{uv \in E \mid u \in L_j\}$, for $1 \le j \le n$; $od(v)$ = the number of arcs starting at $v \in V$; and $id(v)$ = the number of arcs ending at $v \in V$. $od(v)$ is called the **out-degree** of v. $id(v)$ is called the **in-degree** of v.

Example 9.2.16

Consider the graph shown in Figure 9.47.
Here, $l(v_1) = 0$, $l(v_2) = l(v_3) = 1$,
$l(v_4) = 2$. Therefore, $L_0 = \{v_4\}$,
$L_1 = \{v_2, v_3\}$, and $L_2 = \{v_1\}$.

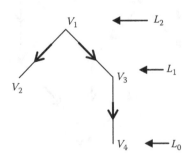

FIGURE 9.47
Directed graph: levels.

Example 9.2.17

Consider the directed 3-cube given (see Figure 9.42). Here
$L_0 = \{v \in V \mid l(v) = 3 - 0\} = \{\varnothing\};$
$L_1 = \{v \in V \mid l(v) = 3 - 1\} = \{\{1\}, \{2\}, \{3\}\};$
$L_2 = \{v \in V \mid l(v) = 3 - 2\} = \{\{1, 2\}, \{2, 3\}, \{1, 3\}\};$ and
$L_3 = \{v \in V \mid l(v) = 3 - 3\} = \{\{1, 2, 3\}\}.$

Example 9.2.18

The diagram for the directed 4-cube Q_4 is shown in Figure 9.48.
   Let $V(Q_4) = \wp(X)$, where $X = \{1, 2, 3, 4\}$.
In this directed graph,

   $L_0 = \{v \in V \mid l(v) = 4 - 0 = 4\} = \{\varnothing\};$

   $L_1 = \{v \in V \mid l(v) = 4 - 1 = 3\} = \{\{1\}, \{2\}, \{3\}, \{4\}\};$

   $L_2 = \{v \in V \mid l(v) = 4 - 2 = 2\} = \{\{1, 2\}, \{1, 3\}, \{2, 3\}, \{1, 4\}, \{2, 4\}, \{3, 4\}\};$

   $L_3 = \{v \in V \mid l(v) = 4 - 3 = 1\} = \{\{1, 2, 3\}, \{1, 2, 4\}, \{1, 3, 4\}, \{2, 3, 4\}\};$ and

   $L_4 = \{v \in V \mid l(v) = 4 - 4 = 0\} = \{\{1, 2, 3, 4\}\}.$

Definition 9.2.19 (Definition 1.1 of Satyanarayana and Prasad, 2003)

A node $v \in L_k$ is said to be **determined** by x_i, $1 \le i \le k$ if $L_1 \cap V_v = \{x_i \mid 1 \le i \le k\}$, and there is no $x \in L_k$ other than v such that $L_1 \cap V_x = \{x_i \mid 1 \le i \le k\}$. Here

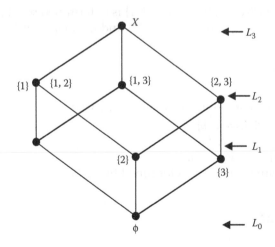

FIGURE 9.48
3-cube: levels.

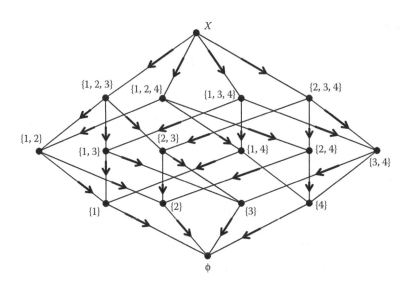

FIGURE 9.49
Directed 4-cube: levels.

we recall that $V_v = \{u \in V \mid$ there exists a path from v to $u\} \cup \{v\}$ (also, refer Notation 9.2.13).

Example 9.2.20

Consider the directed 3-cube Q_3 with $V(Q_3) = \wp(X)$, where $X = \{1, 2, 3\}$ (Figure 9.49).

In this directed 3-cube, the vertex $\{1, 2\}$ is determined by $\{\{1\}, \{2\}\}$, since $L_1 = \{\{1\}, \{2\}, \{3\}\}$, $V_{\{1, 2\}} = \{\{1, 2\}, \{1\}, \{2\}, \varnothing\}$, and $L_1 \cap V_{\{1, 2\}} = \{\{1\}, \{2\}\}$.

Example 9.2.21

Consider the directed graph shown by Figure 9.50.
$L_0 = \{e\}, L_1 = \{c, d\}, L_2 = \{a, b\},$
$V_a = \{a, b, c, e\}, V_b = \{b, c, d, e\},$
$L_1 \cap V_a = \{c, d\}, L_1 \cap V_b = \{c, d\}.$
Therefore, neither a nor b is determined by $\{c, d\}$.

Definition 9.2.22

A directed graph D is said to be **completely determined** if
(i) Every node $v \in L_k$ is determined by $L_1 \cap V_v$ and $|L_1 \cap V_v| = k$

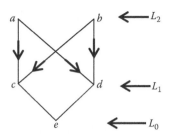

FIGURE 9.50
Graph: completely determined.

(ii) $\{x_i \mid 1 \leq i \leq k\} \subseteq L_1$ implies that there exists $v \in L_k$ such that $\{x_i \mid 1 \leq i \leq k\} = L_1 \cap V_v$.

(iii) If $x, y \in V$, $\{x_1, x_2, \ldots, x_{k-1}\} = L_1 \cap V_y$, and $\{x_1, x_2, \ldots, x_k\} = L_1 \cap V_x$, then xy.

Example 9.2.23

In Figure 9.51, $V_{v_{12}} = \{v_{12}, v_7, v_8, v_2, v_3, v_4, v_1\}$, $V_{v_6} = \{v_6, v_2, v_3, v_4, v_1\}$.

Now $L_1 \cap V_{v_{12}} = \{v_2, v_3, v_4\} \supseteq \{v_2, v_3\} = L_1 \cap V_{v_6}$ and there is no arc from v_{12} to v_6. So we may conclude that the given directed graph satisfies conditions (i) and (ii) of the definition of "completely determined" but not (iii).

For a directed graph we identify the following four properties.

(P_1): Simple and having no circuits.

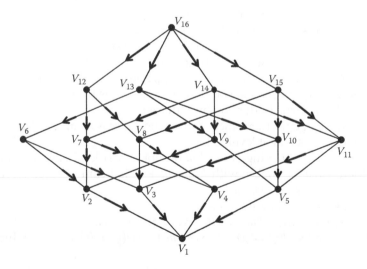

FIGURE 9.51
Graph: not completely determined.

(P_2): The number of levels is $n + 1$ and the number of nodes at level i is ^nC_i for $0 \leq i \leq n$.

(P_3): $\text{id}(v) = m$ if $l(v) = m$; and $\text{od}(v) = m$ if $l(v) = n - m$, where id and od denote in-degree and out-degree, respectively.

(P_4): Completely determined.

Lemma 9.2.24 (Satyanarayana and Prasad, 2003)

If D satisfies P_1, P_2, and P_3, then the number of arcs starting at level $i - 1$ is the same as the number of arcs ending at level i.

Proof

The number of arcs starting at level $i - 1 = {}^nC_{i-1}(n - + 1) = {}^nC_i$. $i =$ the number of arcs ending at level i.	∎

Theorem 9.2.25 (Theorem 1.5 of Satyanarayana and Prasad, 2003)

If D satisfies P_1, P_2, and P_3, then every arc starting at level $i - 1$ ends at level i. (In other words, if $x \in L_m$ and arc xy exists, then $y \in L_{m-1}$.)

Proof

We prove this by mathematical induction.

Let $P(m)$ be the statement: All the arcs starting at level k end at level $k + 1$ for all $k < m$. To show $P(1)$ is true, suppose $L_n = \{v_0\}$ and $v \in L_{n-1}$.

Since $l(v) = 1$, there exists an arc from v_0 to v.

Since $\text{od}(v_0) = n$, $\text{id}(v) = 1$ for each $v \in L_{n-1}$, and $|L_{n-1}| = n$, it follows that all the edges starting at v_0 end at some $v \in L_{n-1}$. So $P(1)$ is true.

Suppose $P(m)$ is true (induction hypothesis).

If $P(m + 1)$ is not true, then since $P(m)$ is true, there is an arc starting at level $k = m$ that does not end at level $m + 1$.

Since all the arcs starting at level m do not end at level $m + 1$, by Lemma 9.2.24, there exists a node y with $l(y) = m + 1$ that cannot receive all of its $\text{id}(y) = m + 1$ arcs from level m.

So, there exists arc xy with $l(x) \neq m$.

If $l(x) < m$, then since $P(m)$ is true, every arc starts at level $l(x)$ and ends at level $l(x) + 1$, and so $l(y) = l(x) + 1$. Now $m + 1 = l(y) = l(x) + 1 \leq m$, which is a contradiction.

If $l(x) > m$, then $m + 1 = l(y) > l(x) \geq m + 1$, which is a contradiction.

Therefore, $P(m + 1)$ is true. The proof is complete.	∎

Corollary 9.2.26 (Corollary 1.6 of Satyanarayana and Prasad, 2003)

Let $x, y \in V$ and xy. Then
(i) $L_1 \cap V_y \subseteq L_1 \cap V_x$.
(ii) If D satisfies P_1, P_2, P_3, and P_4, then $L_1 \cap V_y \subseteq L_1 \cap V_x$ and $|L_1 \cap V_y| + 1 = |L_1 \cap V_x|$.

Proof

(i) There exists arc $xy \Rightarrow V_y \subseteq V_x \Rightarrow L_1 \cap V_y \subseteq L_1 \cap V_x$.
(ii) Let $x \in L_k$. By Theorem 9.2.25, we have $y \in L_{k-1}$. Now by condition (i) of Definition 9.2.22, it follows that $|L_1 \cap V_x| = k = 1 + (k-1) = 1 + |L_1 \cap V_y|$. ∎

Note 9.2.27

From Corollary 9.2.26, we may conclude that the converse of the statement given in condition (iii) of Definition 9.2.22 is true if the graph D satisfies properties P_1 through P_4.

Theorem 9.2.28 (Theorem 2.1 of Satyanarayana and Prasad, 2000)

If $D = (V, E)$ and $D^1 = (V^1, E^1)$ are two finite directed graphs satisfying P_1, P_2, P_3, and P_4, then D is isomorphic to D^1.

Proof

Suppose $L_i^1 = \{v \in V^1 \mid l(v) = n - i\}$, $0 \leq i \leq n$, and
$F_j^1 = \{uv \in E^1 \mid u \in L_i^1\}$, $1 \leq j \leq n$. Since the sets L_i, L_i^1, F_j, F_j^1 form partitions for V, V^1, E, E^1, respectively, we can define

$$f : \bigcup_{i=0}^{m} L_i \to \bigcup_{i=0}^{m} L_i^1 \text{ and } g : \bigcup_{j=1}^{m} F_j \to \bigcup_{j=1}^{m} F_j^1$$

by induction on m.
Suppose $L_o = \{v_o\}$, $L_o^1 = \{v_o^1\}$, $L_1 = \{v_i \mid 1 \leq i \leq n\}$, and $L_1^1 = \{v_i^1 \mid 1 \leq i \leq n\}$.
Now define $f(v_o) = v_o^1$ and $f(v_i) = v_i^1$ for $1 \leq i \leq n$.
Clearly, f is a bijection from $L_o \cup L_1$ to $L_o^1 \cup L_1^1$. By Theorem 9.2.25, $F_1 = \{v_i v_o \mid 1 \leq i \leq n\}$ and $F_1^1 = \{v_i^1 v_o^1 \mid 1 \leq i \leq n\}$.
Define $g: F_1 \to F_1^1$ as $g(v_i v_o) = v_i^1 v_o^1 = f(v_i)f(v_o)$. Clearly, g is a bijection from F_1 to F_1^1.

Suppose $f : \bigcup_{i=1}^{k-1} L_i \to \bigcup_{i=1}^{k-1} L_i^1$ and $g : \bigcup_{j=1}^{k-1} F_j \to \bigcup_{j=1}^{k-1} F_j^1$ are extended bijections with $g(xy) = f(x)f(y)$.

Let $x \in L_k$. Suppose x is determined by $x_i \in L_1$, $1 \leq i \leq k$, and consider $f(x_i)$, $1 \leq i \leq k$.

By P_4, there exists a unique $x^* \in L_k^1$ such that $L_1^1 \cap V_{x^*} = \{f(x_i) \mid 1 \leq i \leq k\}$. Now define $f(x) = x^*$.

To show f is one–one, take x, y such that $x \neq y$. Suppose $L_1 \cap V_x = \{x_i \mid 1 \leq i \leq k\}$ and $L_1 \cap V_y = \{y_i \mid 1 \leq i \leq k\}$.

Since $x \neq y$, we have $\{x_i \mid 1 \leq i \leq k\} \neq \{y_i \mid 1 \leq i \leq k\}$.

Since $f: L_1 \to L_1^1$ is a bijection, we have $\{f(x_i) \mid 1 \leq i \leq k\} \neq \{f(y_i) \mid 1 \leq i \leq k\}$.

Since $f(x)$ and $f(y)$ are determined by $f(x_i)$, $1 \leq i \leq k$ and $f(y_i)$, $1 \leq i \leq k$, respectively, we have $f(x) \neq f(y)$. This shows that f is one–one.

To show f is onto, take $x^* \in L_k^1$. Since $f: L_1 \to L_1^1$ is a bijection, we can suppose that $L_1^1 \cap V_{x^*} = \{f(x_i) \mid 1 \leq i \leq k, x_i \in L_1\}$. Now $f(x) = x^*$, where $V_x \cap L_1 = \{x_i \mid 1 \leq i \leq k\}$. Therefore, f is a bijection.

Now we define $g: F_k \to F_k^1$ as follows: Let $xy \in F_k$.

Suppose $V_y \cap L_1 = \{x_i \mid 1 \leq i \leq k-1\}$.

By Corollary 9.2.26, $V_y \cap L_1 \subseteq V_x \cap L_1$, $|V_y \cap L_1| = k - 1$, and there exists $x_k \in L_1$ such that $V_x \cap L_1 = \{x_i \mid 1 \leq i \leq k\}$. Since $f(x)$ and $f(y)$ are uniquely determined by $f(x_i)$, $1 \leq i \leq k$ and $f(x_i)$, $1 \leq i \leq k-1$, respectively by P_4, we have $f(x)f(y)$.

Now we define $g(xy) = f(x)f(y)$.

To show g is onto, take $x^1 y^1 \in F_k^1$. If $V_{y1} \cap L_1^1 = \{y_i^1 \mid 1 \leq i \leq k-1\}$, by Corollary 9.2.26, there exists $y_k^1 \in L_1^1$ such that $V_x^1 \cap L_1^1 = \{y_i^1 \mid 1 \leq i \leq k\}$.

Let $x, y \in V$ such that $f(x) = x^1$, $f(y) = y^1$ and $y_i \in L_1$ such that $f(y_i) = y_i^1$. Now it is clear that $V_x \cap L_1 = \{y_i \mid 1 \leq i \leq k\} \supseteq \{y_i \mid 1 \leq i \leq k-1\} = V_y \cap L_1$ and so xy.

Hence, $g(xy) = x^1 y^1$. The fact that g is one–one directly follows from the fact that f is one–one.

Hence, by mathematical induction D and G^1 are isomorphic. ∎

Theorem 9.2.29 (Theorem 2.2 of Satyanarayana and Prasad, 2003)

Any finite directed graph D satisfying P_1, P_2, P_3, and P_4 is isomorphic to Q_n, the directed hypercube of dimension n.

Proof

If we show that Q_n satisfies P_1, P_2, P_3, and P_4, then the rest follows from Theorem 9.2.28. Suppose $V(Q_n) = \wp(X)$, where X is a set with $|X| = n$. Clearly, Q_n satisfies P_1. Since $I \in V(Q_n)$ is of level m if and only if $|I| = n - m$, it follows that the total number of levels possible is $n + 1$.

Since the number of distinct subsets I of X with $|I| = n - m$ that can be formed is ${}^nC_{n-m} = {}^nC_m$, it follows that the number of nodes at level m is ${}^nC_m$ for all $1 \leq m \leq n$. Hence, P_2 is satisfied.

Let $l(I) = m$. Then $|I| = n - m$. If the edge IJ exists, then $|J| = n - m - 1$. Since the number of distinct sets J with $|J| = n - m - 1$ that can be formed from the elements of I is $^{(n-m)}C_{n-m-1} = n - m$, it follows that there exist $n - m$ arcs, which start at I and end at subsets J of I with $|J| = |I| - 1$.

Therefore, Q_n satisfies one part of P_3.

To show the other part, suppose $J \in V(Q_n)$ with $l(J) = m$. Since IJ means $I \supseteq J$, $|I| = |J| + 1$ and $|J| = n - m$, there exist m such subsets I such that $IJ \in E(Q_n)$.

Therefore, there are m arcs that end at J. Hence, P_3 holds.

It is clear that if $I \in V(Q_n)$, then I is determined by $\{x_i\}$, $1 \le i \le k$, where $I = \{x_i \mid 1 \le i \le k\}$. Now P_4 follows from the definition of Q_n.

By combining Theorems 9.2.28 and 9.2.29, we get the following main theorem. ∎

Theorem 9.2.30 (Theorem 2.3 of Satyanarayana and Prasad, 2003)

Every directed graph satisfying P_1, P_2, P_3, and P_4 is isomorphic to Q_n (or a directed graph is a directed hypercube of dimension n if and only if it satisfies P_1, P_2, P_3, and P_4).

9.3 Vector Spaces of a Graph

The concept of a vector space is presented and illustrated in Section 1.4. In this section, we provide an illustration of a vector space that is associated with a given graph. A few fundamental results and simple examples are presented. One can refer to Sections 1.2 and 1.3 for group-theoretic and ring-theoretic notations and examples.

Theorem 9.3.1

(i) The ring sum of two circuits in a graph G is either a circuit or an edge-disjoint union of circuits.
(ii) The ring sum (given in Definition 9.1.22) of any two edge-disjoint unions of circuits is also a circuit or another edge-disjoint union of circuits.

Proof

Let Γ_1 and Γ_2 be any two circuits in a graph G.

    Case (i): If the two circuits Γ_1 and Γ_2 have no edge or no vertex in common, then the ring sum of Γ_1 and Γ_2 (i.e., $\Gamma_1 \oplus \Gamma_2$) is a disconnected subgraph of G. Clearly, it is an edge-disjoint union of circuits.

Case (ii): If the two circuits Γ_1 and Γ_2 have one or more edges or vertices in common, then we have the following situations:

(i)   Since the degree of every vertex in a circuit is 2, it follows that every vertex v in the subgraph $\Gamma_1 \oplus \Gamma_2$ has degree $d(v)$, where $d(v) = 2$, if v is in Γ_1 only or in Γ_2 only or if one of the edges formerly incident on v was in both Γ_1 and Γ_2 or $d(v) = 4$ if Γ_1 and Γ_2 just intersect at v (without a common edge). Therefore, in $\Gamma_1 \oplus \Gamma_2$ the degree of a vertex is either 2 or 4. Thus, $\Gamma_1 \oplus \Gamma_2$ is an Euler graph (by a known result). Since $\Gamma_1 \oplus \Gamma_2$ is an Euler graph, we have that $\Gamma_1 \oplus \Gamma_2$ consists of either a circuit or an edge-disjoint union of circuits (by a known result). Hence, the ring sum of two circuits is either a circuit or an edge-disjoint union of circuits. The proof of (i) is complete. ∎

(ii) Follows directly from (i).

A straightforward verification gives the following.

Theorem 9.3.2

The set consisting of all the circuits and the edge-disjoint unions of circuits (including the null set φ) in a graph G is an Abelian group under the operation ring sum $\oplus$.

Now we give the notation of a Galois field modulo m as follows.

Definition 9.3.3

(i)   Consider $\mathbb{Z}_3 = \{0, 1, 2\}$ = the ring of integers modulo 3. The addition and multiplication for $\mathbb{Z}_3$ are called **addition modulo 3** and **multiplication modulo 3**. Both of these operations put together is known as **modulo 3 arithmetic**.

Observe the addition and multiplication tables. In modulo 3 arithmetic, $1 + 1 + 2 \cdot 2 + 1 + 2 + 1 = 1 \pmod 3$ (Tables 9.1 and 9.2).

(ii) In general, we can define a **modulo m arithmetic** system consisting of elements 0, 1, 2, ..., $m - 1$ and the following relationship: for any $q > m - 1$, $q = r \pmod m$, where $q = m \cdot p + r$ and $0 \le r < m$.

TABLE 9.1

Addition Modulo 3

+3	0	1	2
0	0	1	2
1	1	2	0
2	2	0	1

TABLE 9.2

Multiplication Modulo 3

•3	0	1	2
0	0	0	0
1	0	1	2
2	0	2	1

(iii) Write $Z_m = \{0, 1, 2, ..., m-1\}$. Now Z_m is a field $\Leftrightarrow m$ is a prime number. If m is a prime number, then the field Z_m is called a **Galois field modulo** m. We denote this by GF(m).

Example 9.3.4

$Z_2 = \{0,1\}$ is a Galois field modulo 2 under the addition and multiplication modulo 2. It is denoted by GF(2), the Galois field with two elements.

Note 9.3.5

(i) In an ordinary two-dimensional (Euclidean) plane, a point is represented by an ordered pair of numbers $X = (x_1, x_2)$. The point X may be regarded as a **vector**.

(ii) In a three-dimensional Euclidean space, a point can be represented as a triplet (x_1, x_2, x_3).

Sometimes, we may represent this element as the column vector
$$\begin{pmatrix} x_1 \\ x_2 \\ x_3 \end{pmatrix}.$$

(iii) Consider GF(2).

Then every number in a triplet may be equal to either 0 or 1. Thus, there are 8 ($= 2^3$) vectors possible in a three-dimensional space. These are (0, 0, 0) (0, 0, 1), (0, 1, 0), (0, 1, 1), (1, 0, 0), (1, 0, 1), (1, 1, 0), (1, 1, 1). Extending this concept, a vector in a k-dimensional Euclidean space is an ordered k-tuple. For example, the 7-tuple (0, 1, 1, 0, 1, 0, 1) represents a vector in a seven-dimensional vector space over the field GF(2).

(iv) The numbers from a field d are called **scalars**. In GF(2), the scalars are 0 and 1.

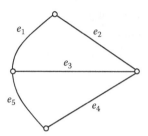

FIGURE 9.52
Graph.

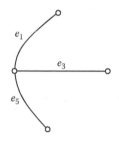

FIGURE 9.53
Subgraph (1, 0, 1, 0, 1).

Note 9.3.6

(i) Let us consider the graphs given in Figures 9.52 through 9.54. These graphs have four vertices and five edges, e_1, e_2, e_3, e_4, e_5. Any subset of these five edges (i.e., any subgraph g) of G can be represented by a 5-tuple: $X = (x_1, x_2, x_3, x_4, x_5)$, where $x_i = 1$ if the edge e_i is in g and $x_i = 0$ if e_i is not in g.

(ii) The subgraph g_1 shown in Figure 9.53 may be represented as (1, 0, 1, 0, 1). The subgraph g_2 shown in Figure 9.54 may be represented as (0, 1, 1, 1, 0). It is clear that $2^5 = 32$ such 5-tuples are possible, including the zero vector $O = (0, 0, 0, 0, 0)$ which represents a null graph and (1, 1, 1, 1, 1) which represents the graph given in Figure 9.52.

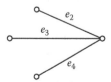

FIGURE 9.54
Subgraph (0, 1, 1, 1, 0).

Note 9.3.7

Let G be a graph with n edges. Suppose g_1, g_2 are two subgraphs and $(h_1, h_2, ..., h_n)$, $(f_1, f_2, ..., f_n)$ are the n-tuple representations for g_1, g_2, respectively. If an edge e_i is in both g_1 and g_2, then e_i is not in $g_1 \oplus g_2$. Since e_i is in both g_1 and g_2, it follows that $h_i = 1, f_i = 1$. Now $h_i + f_i = 1 + 1 = 0 \pmod 2$. So, in the binary representation for $g_1 \oplus g_2$, the ith component is 0. The ring sum operation between the two subgraphs corresponds to the modulo 2 addition between the two n-tuples $(h_1, h_2, ..., h_n)$, $(f_1, f_2, ..., f_n)$ representing the two subgraphs.

Example 9.3.8

Consider the two subgraphs g_1 and g_2 of G shown in Figures 9.52 through 9.54. The subgraph $g_1 = \{e_1, e_3, e_5\}$ is represented by $(1, 0, 1, 0, 1)$, and $g_2 = \{e_2, e_3, e_4\}$ is represented by $(0, 1, 1, 1, 0)$.

Clearly, $g_1 \oplus g_2 = \{e_1, e_2, e_3, e_4, e_5\}$ is represented by $(1, 1, 0, 1, 1)$ (9.3)

Also, the modulo 2 addition between the two 5-tuples $(1, 0, 1, 0, 1)$ and $(0, 1, 1, 1, 0)$ is given by

$$(1, 0, 1, 0, 1) + (0, 1, 1, 1, 0) = (1 + 1, 0 + 1, 1 + 1, 0 + 1, 1 + 0) = (1, 1, 0, 1, 1). \quad (9.4)$$

Observe that the results (9.3) and (9.4) are the same.

Definition 9.3.9

A **vector space** W_G **associated with a graph** G consists of
  (i) a Galois field modulo 2, that is, the set $\{0, 1\}$ with the addition operation modulo 2 (written as "+"). The addition modulo 2 is given by $0 + 0 = 0, 1 + 0 = 1, 0 + 1 = 1, 1 + 1 = 0$, and the multiplication modulo 2 is given by $0 \cdot 0 = 0 = 0 \cdot 1 = 1 \cdot 0, 1 \cdot 1 = 1$.
 (ii) 2^e vectors (e-tuples), where e is the number of edges in G.
(iii) An addition operation between two vectors X, Y in this space, defined as the vector sum $X \oplus Y = (x_1 + y_1, x_2 + y_2, ..., x_e + y_e)$, where $X = (x_1, x_2, ..., x_e)$, $Y = (y_1, y_2, ..., y_e)$, and + being the addition modulo 2.
(iv) A scalar multiplication between a scalar $c \in Z_2$ and a vector X, defined as $c \cdot X = (cx_1, cx_2, ..., cx_e)$.
      Now we recollect the definitions (given in 1.4.17, 1.4.18, 1.4.22, 1.4.27) of linear combination, linear span, linear independent, and basis of a vector space.

Example 9.3.10

(i) The **natural** or **standard basis** in a k-dimensional vector space is the

following set of k unit vectors:
$$\begin{pmatrix} 1 \\ 0 \\ : \\ : \\ : \\ 0 \end{pmatrix}, \begin{pmatrix} 0 \\ 1 \\ 0 \\ : \\ : \\ 0 \end{pmatrix}, ..., \begin{pmatrix} 0 \\ 0 \\ : \\ 0 \\ 1 \end{pmatrix}.$$

(with k elements in each column vector).

(ii) Any vector in the k-dimensional vector space (over the field of real numbers) can be expressed as a linear combination of these k vectors.

Definition 9.3.11

Consider the vector space W_G associated with a given graph G. Corresponding to each subgraph of G, there is a vector in W_G represented by an e-tuple.

The **natural basis for this vector space** W_G is a set of e linearly independent vectors, each representing a subgraph consisting of exactly one edge of G.

Example 9.3.12

Consider the graph as vertices in a 5-cube.

Here the set of five vectors $(1, 0, 0, 0, 0)$, $(0, 1, 0, 0, 0)$, $(0, 0, 1, 0, 0)$, $(0, 0, 0, 1, 0)$, $(0, 0, 0, 0, 1)$ serves as a basis for W_G. The possible subgraphs are $2^e = 2^5 = 32$. Any of these possible 32 subgraphs can be represented (uniquely) by a linear combination of these five vectors.

9.4 Prime Graph of a Ring

In this section, we consider simple graphs. We present some fundamental results from Satyanarayana, Syam Prasad and Nagaraju (2010). We present a new concept known as the "prime graph of a ring R," denoted by PG(R), where R is a ring, and exhibit some examples of PG(R) and fundamental results. This forms a new bridge between the algebraic concepts of "ring," and "graph theory."

Definition 9.4.1

Let R be a ring. A graph $G(V, E)$ is said to be a **prime graph** of R (denoted by $PG(R)$) if $V = R$ and $E = \{ xy \mid xRy = 0 \text{ or } yRx = 0, \text{ and } x \neq y \}$.

Example 9.4.2

Consider $\mathbb{Z}_n$, the ring of integers modulo n.

(i) Let us construct the graph $PG(R)$, where $R = \mathbb{Z}_2$. We know that $R = \mathbb{Z}_2 = \{0, 1\}$. So $V(PG(R)) = \{0, 1\}$. Since $0R1 = 0$, there exists an edge between 0 and 1. There are no other edges. So, $E(PG(R)) = \{01\}$. Now $PG(R)$ is shown in Figure 9.55.

(ii) Let us construct the graph $PG(R)$, where $R = \mathbb{Z}_3$. We know that $R = \mathbb{Z}_3 = \{0, 1, 2\}$. So, $V(PG(R)) = \{0, 1, 2\}$. Since $0R1 = 0$, $0R2 = 0$, there exists an edge between 0 and 1, and also an edge between 0 and 2. There are no other edges, as there are no two nonzero elements $x, y \in R$ with $xRy = 0$.

So, $E(PG(R)) = \{01, 02\}$. Now $PG(R)$ is shown in Figure 9.56.

(iii) Let us construct the graph $PG(R)$, where $R = \mathbb{Z}_4$. We know that $R = \mathbb{Z}_4 = \{0, 1, 2, 3\}$. So $V(PG(R)) = \{0, 1, 2, 3\}$. Since $0R1 = 0$, $0R2 = 0$, $0R3 = 0$, it follows that $01, 02, 03 \in E(PG(R))$. There are no other edges, as there are no two nonzero elements $x, y \in R$ such that $xRy = 0$. So $E(PG(R)) = \{01, 02, 03\}$. Now $PG(R)$ is shown in Figure 9.57.

(iv) $PG(R)$, where $R = \mathbb{Z}_5$, is shown in Figure 9.58.

(v) Note that $PG(R)$, when $R = \mathbb{Z}_n$, $1 \leq n \leq 5$, contains no triangles, and all these graphs are star graphs.

(vi) Let us construct a graph $PG(R)$, where $R = \mathbb{Z}_6$.

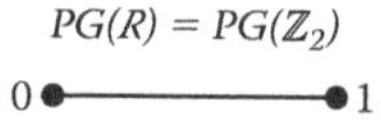

$$PG(R) = PG(\mathbb{Z}_2)$$

0 •————————————• 1

FIGURE 9.55
Prime graph on $\mathbb{Z}_2$.

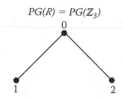

$$PG(R) = PG(\mathbb{Z}_3)$$

FIGURE 9.56
Prime graph on $\mathbb{Z}_3$.

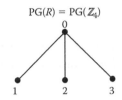

FIGURE 9.57
Prime graph on $\mathbb{Z}_4$.

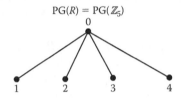

FIGURE 9.58
Prime graph on $\mathbb{Z}_5$.

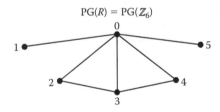

FIGURE 9.59
Prime graph on $\mathbb{Z}_6$.

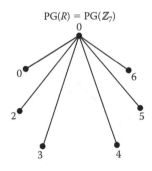

FIGURE 9.60
Prime graph on $\mathbb{Z}_7$.

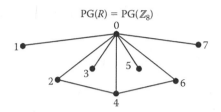

FIGURE 9.61
Prime graph on $\mathbb{Z}_8$.

We know that $R = \mathbb{Z}_6 = \{0, 1, 2, 3, 4, 5\}$. So, $V(PG(R)) = \{0, 1, 2, 3, 4, 5\}$. Since $0R1 = 0$, $0R2 = 0$, $0R3 = 0$, $0R4 = 0$, $0R5 = 0$, $2R3 = 0$, $3R4 = 0$, it follows that $01, 02, 03, 04, 05, 23, 34 \in E(PG(R))$. So, $E(PG(R)) = \{01, 02, 03, 04, 05, 23, 34\}$.

Now $PG(R)$ is shown in Figure 9.59.

(vii) $PG(R)$, where $R = \mathbb{Z}_7$, is shown in Figure 9.60.

(viii) Let us construct the graph $PG(R)$, where $R = \mathbb{Z}_8$. We know that $R = \mathbb{Z}_8 = \{0, 1, 2, 3, 4, 5, 6, 7\}$. So $V(PG(R)) = \{0, 1, 2, 3, 4, 5, 6, 7\}$. Note that $0R1 = 0$, $0R2 = 0$, $0R3 = 0$, $0R4 = 0$, $0R5 = 0$, $0R6 = 0$, $0R7 = 0$, $2R4 = 0$, $4R6 = 0$. It is clear that $E(PG(R)) = \{01, 02, 03, 04, 05, 24, 46\}$. The graph $PG(R)$ is shown in Figure 9.61.

Observation 9.4.3

Let R be a ring and $PG(R)$ be its prime graph.

(i) There are neither self-loops nor multiple edges in $PG(R)$, and so $PG(R)$ is a simple graph.

(ii) Since $0Rx = 0$ for all $0 \neq x \in R$, there is an edge from 0 to x for all $x \in V = R$. So, $d(0) = |R \setminus \{0\}| = |R| - 1$. For any two nonzero elements x, y in R, there are edges one from x to 0, and another from 0 to y. This shows that the graph $PG(R)$ is a connected graph. Moreover, $d(0, x) = 1$ and $d(x, y) \leq 2$ for any two nonzero elements $x, y \in R$.

(iii) If there are two nonzero elements x, y in R such that $xRy = 0$, then the subgraph generated by $\{0, x, y\}$ is a triangle graph. For example, refer to the graph shown in Figure 9.59. Note that $2R3 = 0$. So $\{0, 2, 3\}$ forms a triangle.

(iv) Now, it is clear that $xRy = 0$ or $yRx = 0$, if and only if $d(x, y) = 1$ or $x = 0$ or $y = 0$.

[**Verification:** Suppose $xRy = 0$ and $x \neq 0 \neq y$. Then $xy \in E(PG(R))$ and so $d(x, y) = 1$.

Conversely, suppose $d(x, y) = 1$ or $x = 0$ or $y = 0$. If $x = 0$ or $y = 0$, then $xRy = 0$ (or $yRx = 0$).

If $d(x, y) = 1$ and $x \neq 0 \neq y$, then $xy \in E(PG(R))$, which implies that $xRy = 0$ or $yRx = 0$.]

(v) $xRy \neq 0$ if and only if $d(x, y) = 2$.

[**Verification:** Suppose $xRy \neq 0$. Then there is no edge between x and y. So, $d(x, y) > 1$.

Since $xR0 = 0$, $yR0 = 0$, $x0$, $0y \in E(PG(R))$. Hence, $d(x, y) = 2$.

Converse: Suppose that $d(x, y) = 2$. Since $d(x, y) \neq 1$, there is no edge from x to y. So, $xRy \neq 0$.]

(vi) If R is a commutative ring with 1, then there exists an edge between x and y in PG(R) if and only if $xy = 0$.

[**Verification:** Given that R is a commutative ring. Suppose that there is an edge between x and y in PG(R). Then $xRy = 0 \Rightarrow xyR = 0$ (since R is commutative) $\Rightarrow xy = 0$ (since $1 \in R$).

Converse: Suppose $xy = 0$. Then $xyR = 0 \Rightarrow xRy = 0$ (since R is commutative) $\Rightarrow xy \in E(PG(R))$.]

(vii) The set $\{0\}$ is a dominating set for PG(R). Hence, the domination number of PG(R) is equal to 1.

Theorem 9.4.4 (Theorem 2.4 of Satyanarayana, Prasad, and Nagaraju, 2010)

Consider $\mathbb{Z}_n$ for some n, where $\mathbb{Z}_n$ is the ring of integers modulo n. The following conditions are equivalent:

(i) n is prime or $n = p^2$ for some prime p.
(ii) There are no triangles in PG($\mathbb{Z}_n$).
(iii) $d(x, y) = 2$ for any two distinct nonzero vertices of PG($\mathbb{Z}_n$).
(iv) PG($\mathbb{Z}_n$) is a star graph with the special vertex 0 (the additive identity).

Proof

(i) $\Rightarrow$ (ii): Suppose n is a prime number. It follows that $\mathbb{Z}_n$ is a field. So, $\mathbb{Z}_n$ is an integral domain, which means that there are no nonzero zero divisors.

Suppose that $x, y \in \mathbb{Z}_n \setminus \{0\}$ and $\{0, x, y\}$ is a triangle.

Then $0 \neq x \neq y \neq 0$ and $x\mathbb{Z}_n y = 0$, and so x is a zero divisor, which is a contradiction.

This shows that the graph PG($\mathbb{Z}_n$) contains no triangles, if n is a prime number.

Suppose $n = p^2$ for some prime p.

Then for any number $x < n$ with $x \neq p$, it follows that the g.c.d of x, p is 1.

Let us suppose that $x, y \in \mathbb{Z}_n \setminus \{0\}$ with $\{0, x, y\}$ is a triangle.

Now $x \neq y$ and $x\mathbb{Z}_n y = 0$.

So, $xy = 0$, which implies that $n = p^2$ divides xy and $x < n$, $y < n$.

This implies that p divides either x or y, which is a contradiction.

Hence, the graph $PG(\mathbb{Z}_n)$ contains no triangle, if $n = p^2$ for some prime p.

(ii) $\Rightarrow$ (i): Suppose (ii). Let us suppose that neither n is prime nor $n = p^2$ for all primes p.

Then by the prime decomposition theorem, there exist (at least two) distinct primes $p_1, p_2, \ldots, p_k$ and positive integers $s_1, s_2, \ldots, s_k$ such that $n = p_1^{s_1} p_2^{s_2} \ldots p_k^{s_k}$.

Write $x = p_1^{s_1}$ and $y = p_2^{s_2} \ldots p_k^{s_k}$. It is clear that $x \neq y$ and $x\mathbb{Z}_n y = 0$.

This implies that there is an edge between the two distinct vertices x, y in $PG(\mathbb{Z}_n)$.

Hence $\{0, x, y\}$ forms a triangle in $PG(\mathbb{Z}_n)$, which is a contradiction.

(ii) $\Rightarrow$ (iii): Let x, y be two nonzero distinct vertices.

If $d(x, y) = 1$, then $\{0, x, y\}$ forms a triangle, which is a contradiction. By Observation 9.4.3 (ii), we conclude that $d(x, y) = 2$.

(iii) $\Rightarrow$ (ii): Suppose (iii). Let us suppose that $\{0, x, y\}$ forms a triangle in $PG(\mathbb{Z}_n)$.

Then there is an edge between x, y, and so $d(x, y) = 1$, which is a contradiction.

Now we have proved (i) $\Leftrightarrow$ (ii) $\Leftrightarrow$ (iii).

(ii) $\Rightarrow$ (iv): Since $PG(\mathbb{Z}_n)$ contains no triangles, and $0x \in E(PG(\mathbb{Z}_n))$ for all $0 \neq x \in \mathbb{Z}_n$, we conclude that $PG(\mathbb{Z}_n)$ is a star graph with the special vertex 0.

(iv) $\Rightarrow$ (iii): Since there is no edge between two distinct nonzero elements $x, y \in \mathbb{Z}_n$, and $x0, 0y \in E(PG(\mathbb{Z}_n))$, we conclude that $d(x, y) = 2$. The proof is complete. ∎

Note 9.4.5

(i) A star graph with n vertices is called an *n*-**star graph**.

(ii) Let R be a ring with $n = |R|$. Consider the graph $PG(R)$. Since $0x \in E(PG(R))$ for all $0 \neq x \in R$, it follows that the n-star graph is a subgraph of $PG(R)$.

Now we compare the n-cubes and prime graphs related to the rings $R_1 = \mathbb{Z}_2 \times \mathbb{Z}_2$ and $R_2 = \mathbb{Z}_2 \times \mathbb{Z}_2 \times \mathbb{Z}_2$.

Remark 9.4.6

(i) If the ring $R_1 = \mathbb{Z}_2 \times \mathbb{Z}_2 = \{(0, 0), (1, 0), (0, 1), (1, 1)\}$, then the corresponding 2-cube is given by Figure 9.62.

(ii) The corresponding $PG(R_1)$ is given by Figure 9.63.

(iii) If the ring $R_2 = \mathbb{Z}_2 \times \mathbb{Z}_2 \times \mathbb{Z}_2 = \{(0, 0, 0), (0, 0, 1), (0, 1, 0), (0, 1, 1), (1, 0, 0), (1, 0, 1), (1, 1, 0), (1, 1, 1)\}$, then the corresponding 3-cube is given by Figure 9.64.

(iv) The corresponding $PG(R_2)$ is given by Figure 9.65.

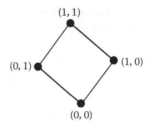

FIGURE 9.62
2-cube.

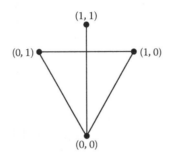

FIGURE 9.63
Prime graph on $\mathbb{Z}_2 \times \mathbb{Z}_2$.

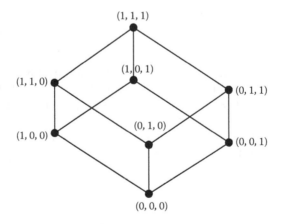

FIGURE 9.64
3-cube.

Observation 9.4.7

(i) From Remark 9.4.6 (i) and (ii), we can conclude that a 2-cube is not a prime graph.

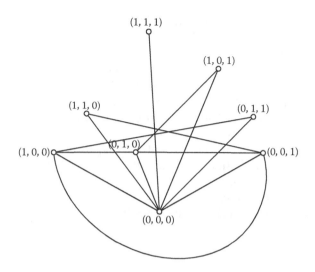

FIGURE 9.65
Prime graph on $\mathbb{Z}_2 \times \mathbb{Z}_2 \times \mathbb{Z}_2$.

(ii) From Remark 9.4.6 (iii) and (iv), we can conclude that a 3-cube is not a prime graph.

(iii) In general, every n-cube is not a prime graph.
   [Verification: Let $G_n = (V_n, E_n)$ be an n-cube related to the ring $R_n = \mathbb{Z}_2 \times \mathbb{Z}_2 \times \ldots \times \mathbb{Z}_2$ (n times).
   In the n-cube, $x = (1, 0, \ldots, 0)$ and $y = (1, 1, 0, \ldots, 0)$ are joined by an edge. If the n-cube is the same as $PG(R_n)$, then $xy \in E(PG(R_n))$, which implies that $xR_ny = 0$, and hence $xy = 0$, which is a contradiction, since $xy = (1, 0, \ldots, 0)(1, 1, 0, \ldots, 0) = (1, 0, \ldots, 0) = x \neq 0$.
   Thus, an n-cube is not the same as $PG(R_n)$.]

Theorem 9.4.8 (Theorem 3.1 of Satyanarayana, Prasad, and Nagaraju, 2010)

If R is a semiprime ring, then the following conditions are equivalent:

(i) R is a prime ring.
(ii) $PG(R)$ is a star graph.
(iii) The prime graph $PG(R)$ is a tree.

Proof

(i) $\Rightarrow$ (ii): Let $e = xy \in E(PG(R))$.
   Note that by the definition of $PG(R)$, it follows that $x \neq y$.
   Then $xRy = 0$ or $yRx = 0$. This implies that $x = 0$ or $y = 0$ (since R is a prime ring), that is, e is an edge with an endpoint 0.

This shows that every edge of PG(R) has 0 as one of its endpoints, and so PG(R) is a star graph.

(ii) $\Rightarrow$ (iii): follows since every star graph is a tree.

(iii) $\Rightarrow$ (i): Suppose that PG(R) of a ring R is a tree.
Now we have to show that R is a prime ring.
Let us suppose that R is not a prime ring. Then there exist two non-zero elements x, y in R such that $xRy = 0$. By the definition of PG(R), there is an edge between x and y.
Since $x \neq 0 \neq y \neq x$, it follows that $\{0, x, y\}$ forms a cycle in PG(R), which is a contradiction to the fact that PG(R) is a tree. This shows that R is a prime ring.
The proof is complete.   ∎
A straightforward verification shows the following.

Corollary 9.4.9 (Corollary 3.2 of Satyanarayana, Prasad, and Nagaraju, 2010)

Suppose R is a ring with $|R| \geq 2$. Then R is a prime ring if and only if the diameter is 2 and the radius is 1.

Corollary 9.4.10 (Corollary 3.3 of Satyanarayana, Prasad, and Nagaraju, 2010)

The following conditions are true

  (i) R is not a prime ring.
 (ii) The triangle is a subgraph of PG(R).
(iii) There exists a chain of length greater than 2 in PG(R).
 (iv) PG(R) is not a tree.
  (v) PG(R) is not a star graph.

Proof

  (i) $\Leftrightarrow$ (iv) $\Leftrightarrow$ (v) follows from Theorem 9.4.8.
  (v) $\Rightarrow$ (ii): Suppose that PG(R) is not a star graph. By Note 9.4.5 (ii), it follows that PG(R) has an n-star graph as subgraph with $n = |R|$.
It is clear that there is an edge e in PG(R) that is not in the n-star graph, and $e = xy$ where $x \neq 0 \neq y \neq x$. Therefore, $\{0, x, y\}$ forms a triangle in PG(R).
 (ii) $\Rightarrow$ (iii): Let $\{0, x, y\}$ be a triangle in PG(R).
Now $0x$, xy, $y0$ is a chain of length 3 in PG(R).
(iii) $\Rightarrow$ (iv): Let xy, yz, zu be a chain of length 3 in PG(N).
By the definition of PG(R), it is clear that $x \neq y$, $y \neq z$, $z \neq u$.
Case (i): If $x \neq 0 \neq y$, then $\{0, x, y\}$ forms a triangle and so PG(R) is not a tree.
Case (ii): Suppose $x = 0$. Since $x \neq y$, it follows that $y \neq 0$.

If $z = 0$, then xy, yz form a cycle, and so PG(R) is not a tree.
If $z \neq 0$, then $\{0, y, z\}$ is a triangle in PG(R), and hence PG(R) is not a tree.
In all cases, we conclude that PG(R) is not a tree. The proof is complete. ∎

Notation 9.4.11

$B(R) = \{(x, y) \mid x \neq y, x \neq 0 \neq y, xRy = 0 \text{ or } yRx = 0\} \subseteq R \times R$, where R is a ring.

Corollary 9.4.12 (Corollary 3.5 of Satyanarayana, Prasad, and Nagaraju, 2010)

(i) R is a prime ring if and only if $B(R) = \varnothing$.
(ii) The number of elements in $B(R)$ is less than or equal to the number of triangles in PG(R).
(iii) If $B(R) \neq \varnothing$, then the length of the longest walk ≥ 3.
(iv) $B(R) \neq R \times R$.
(v) If R is a prime ring, then PG(R) is not an Euler graph.

Proof

(i) R is a prime ring $\Leftrightarrow$ there are no $x, y \in R$ with $x \neq 0 \neq y \neq x$, and $xRy = 0$
 $\Leftrightarrow B(R) = \varnothing$.
(ii) Let $(x, y) \in B(R)$.
 Then $\{0, x, y\}$ forms a triangle in PG(R).
 If $(x, y), (u, v) \in B(R)$ such that $(x, y) \neq (u, v)$, then $x \neq u$ or $y \neq v$, and so the triangles $\{0, x, y\}$ and $\{0, u, v\}$ (in PG(R)) are distinct.
 This shows that the number of elements in $B(R)$ is less than the number of triangles in PG(R).
(iii) Let $(x, y) \in B(R)$. Then $0x$, xy, $y0$ is a triangle, and this walk is of length 3. Hence, the length of the longest walk is greater than or equal to 3.
(iv) Since $(x, x) \in (R \times R)\backslash B(R)$, it follows that $B(R) \subsetneq R \times R$.
(v) If R is a prime ring, then by Theorem 9.4.8, PG(R) is a star graph.
 For any $0 \neq x \in V(\text{PG}(R))$, $d(v) = 1$, an odd number.
 Now by Theorem 9.1.15, it follows that PG(R) is not an Euler graph.
 The proof is complete. ∎

Result 9.4.13 (Result 3.6 of Satyanarayana, Prasad, and Nagaraju, 2010)

Let $G = \text{PG}(R)$. Let I be an ideal of R. For any two elements $x, y \in V(\text{PG}(R))$, merge x and y in G if $x - y \in I$. Suppose the graph obtained is G^1, after removing loops and replacing the multiple edges by a single edge. G^1 is a simple graph.

There is a bijection $f: V(G^1) \rightarrow V(\text{PG}(R/I))$ such that $xy \in E(G^1)$ implies that $f(x)f(y) \in E(\text{PG}(R/I))$. Moreover, if I is a prime ideal of R, then $f(x)f(y) \in E(\text{PG}(R/I))$ implies that $xy \in E(G^1)$.

In other words, if I is a prime ideal of R, then the graphs G^1 and $PG(R/I)$ are isomorphic.

Proof

Suppose $G^1 = (V^1, E^1)$ and $x \in V^1$.

Define $f: V^1 \to V(PG(R/I))$ by $f(x) = x + I$.

Let $x = y$ in V^1. Then x and y are merged, and so $x - y \in I$.

By the definition of the coset, we get that $x + I = y + I$ and $f(x) = f(y)$.

Hence, f is well–defined.

$f(x) = f(y) \Rightarrow x + I = y + I \Rightarrow x - y \in I \Rightarrow x$ and y are merged in $G^1 \Rightarrow x = y$ in G^1.

So, f is one–one.

For any $x + I \in R/I$, we have $x \in V^1$ and $f(x) = x + I$.

This shows that f is a bijection from V^1 to $V(PG(R/I))$.

Part 1: $x - y \in I$.

Then $x + I = y + I$, and so there is no edge between $f(x) = x + I$ and $f(y) = y + I$ in $PG(R/I)$.

Since $x - y \in I$, x and y are merged, and loops are removed, it follows that there is no edge between x and y in G^1. Since $x + I = y + I$, by the definition of a prime graph, there is no edge between $x + I = f(x)$ and $y + I = f(y)$.

Part 2: Suppose $x - y \notin I$.

In this case, in the process of getting G^1 from G, x and y are not merged.

So, $xy \in E(G) \Leftrightarrow xy \in E(G^1)$.

Since $x - y \notin I$, it follows that $x + I \neq y + I$.

$xy \in E(G^1)$

$\Rightarrow xy \in E(PG(R))$

$\Rightarrow xRy = 0, x \neq 0 \neq y \neq x$

$\Rightarrow (x + I)R(y + I) = xRy + I = 0 + I$

$\Rightarrow (x + I)(R/I)(y + I) = 0 + I$ (by the definition of a product in a quotient ring)

$\Rightarrow$ there is an edge between $f(x) = x + I$ and $f(y) = y + I$ in $PG(R/I)$.

In this case, it follows that if there is an edge between x and y in G^1, then there is an edge between $f(x)$ and $f(y)$.

Part 3: Suppose that I is a prime ideal of R.

Also suppose that there is an edge between two distinct elements $f(x) = x + I$ and $f(y) = y + I$ in $PG(R/I)$. We show that either $x \in I$ or $y \in I$.

Let us suppose $x \notin I$ and $y \notin I$. Then $f(x) = x + I \neq 0$ and $f(y) = y + I \neq 0$.

Also, $f(x) \neq f(y)$.

Now $f(x)f(y) = (x + I)(y + I) \in E(PG(R/I))$

$\Rightarrow (x + I)(R/I)(y + I) = 0 + I$ (by the definition of a prime graph)

$\Rightarrow xRy + I = 0 + I$ (by the definition of a product in R/I)

$\Rightarrow xRy \subseteq I$

$\Rightarrow x \in I$ or $y \in I$ (since I is a prime ideal), which is a contradiction.

So, either $x \in I$ or $y \in I$.

Since f is a bijection and $f(x) \neq f(y)$, it is clear that $x \neq y$.

Suppose $x \in I$. If $y \in I$, then $f(x) = x + I = 0 + I$ (since $x \in I$) $= y + I$ (since $y \in I$) $= f(y)$. This implies that $f(x) = f(y)$, which is a contradiction.

Now $y \notin I \Rightarrow y - 0 \notin I \Rightarrow y$ and 0 are not merged in the process of getting G^1 from G. Also, $x \in I \Rightarrow x - 0 \in I \Rightarrow x$ and 0 are merged in the process of getting G^1 from G.

We know that the edge $0y$ is in $G = PG(R)$.

Since 0 and y are not merged, we conclude that the edge $0y$ is also in G^1.

Since 0 and x are merged in the process, it follows that $0 = x$ in G^1, and so $f(0) = f(x)$.

Now $0y = xy$ is in G^1 whenever $f(x)f(y)$ is in $PG(R/I)$.

This shows that $G^1 \cong PG(R/I)$.

The proof is complete. ∎

Remark 9.4.14 (Remark 3.7 of Satyanarayana, Prasad, and Nagaraju, 2010)

If R is a ring with $R^2 = 0$, then $PG(R)$ is a

(i) a complete graph
(ii) a Hamiltonian graph
(iii) a regular graph

Proof

(i) Suppose that R is a ring with $R^2 = 0$.
Then for any $x, y \in R$, we have $xRy = 0$, and so $xy \in E(PG(R))$ for all $x \neq y$. This shows that $PG(R)$ is a complete graph.
(ii) Since every complete graph is a Hamiltonian graph, we conclude that $PG(R)$ is a Hamiltonian graph.
(iii) By Note 1.6.19, it is clear that a complete graph is a regular graph of degree $(p - 1)$ where $p = |V|$. Hence, $PG(R)$ is a regular graph. The proof is complete.

Definition 9.4.15

Let R be a ring. A directed graph $D = (V, E)$ is said to be a **prime directed graph** of R (denoted by $PDG(R)$) if $V = R$ and $E = \{ \overrightarrow{xy} \mid xRy = 0, x \neq y \}$.

Observation 9.4.16

If R is a ring, then the following two conditions are equivalent:

(i) $a, b \in R, ab = 0 \Rightarrow ba = 0$.
(ii) $\overrightarrow{xy} \in E \Rightarrow \overrightarrow{yx} \in E$.

Observation 9.4.17

(i) For every $0 \neq x \in R$, we have $0Rx = 0 = xR0$, and so $\overrightarrow{0x}$, $\overrightarrow{x0} \in E = E(PDG(R))$. Therefore, the out-degree of $(0) = |R| - 1$; the in-degree of $(0) = |R| - 1$; and the degree of $(0) = (|R| - 1) + (|R| - 1) = 2(|R| - 1)$.

(ii) There are no self–loops in the graph PDG(R).

Definition 9.4.18

A directed graph D is said to be a **flower graph** if there exists a special/particular vertex x such that $\overrightarrow{xy}$, $\overrightarrow{yx} \in E(G)$ for all vertices y in G with $x \neq y$. A flower graph with n vertices is called an n-**flower graph**.

Note 9.4.19

(i) It is clear that an n-flower graph contains $2(n - 1)$ edges.

(ii) Figure 9.66 shows a 4-flower graph as it contains four vertices. Figure 9.67 shows a 6-flower graph.

Lemma 9.4.20

For every ring R, the graph PDG(R) contains an n-flower graph as its sub-graph where $n = |R|$.

Proof

Write $V^* = V = R$.
    Let $0 \in V^*$ be the special vertex.

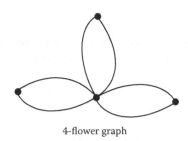

4-flower graph

FIGURE 9.66
4-flower graph.

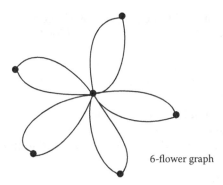

6-flower graph

FIGURE 9.67
6-flower graph.

For every $0 \neq x \in R$, by Observation 9.4.17 (i), we have $\overrightarrow{0x}, \overrightarrow{x0} \in E = E(PDG(R))$.
Write $E^* = \{\overrightarrow{0x}, \overrightarrow{x0} \mid 0 \neq x \in V^*\}$.

Now $G^* = (V^*, E^*)$ is an n-star graph, and it is also a subgraph of the graph $PDG(R)$.

Hence, an n-star graph is a subgraph of the graph $PDG(R)$ for every ring R with $|R| = n$.

The proof is complete. ■

Theorem 9.4.21

The following conditions are equivalent for a ring R.
   (i) R is a prime ring.
  (ii) The graph $PDG(R)$ is a flower graph.
 (iii) $|E| = 2|R| - 2$, where $E = E(PDG(R))$.

Proof

   (i) $\Rightarrow$ (ii): Let $\overrightarrow{xy} \in E = E(PDG(R))$.
        Then $xRy = 0 \Rightarrow x = 0$ or $y = 0$ (since R is a prime ring).
        So every arc in E has "0" as one of its endpoints.
        Consider the n-flower graph $G^* = (V^*, E^*)$ (defined in the proof of Lemma 9.4.20), which is a subgraph of the graph $PDG(R)$.
        Here we have verified that $\overrightarrow{xy} \in E \Rightarrow x = 0$ or $y = 0$. This implies that $\overrightarrow{xy} \in E^*$.
        This shows that the graph $PDG(R)$ is a subgraph of G^*.
        Hence, $G^* = PDG(R)$ is an n-flower graph, where $|R| = n$.
  (ii) $\Rightarrow$ (iii): Follows from Note 9.4.19 (i).

(iii) $\Rightarrow$ (i): By Lemma 9.4.20, the graph PDG(R) contains G^*, the n-flower graph with special vertex 0 and $n = |R|$. Now G^* is a subgraph of the graph PDG(R).

So $|E(\text{PDG}(R))| \geq |E(G^*)| = 2(|R| - 1)$.

By (iii), it follows that $|E(\text{PDG}(R))| = 2(|R| - 1)$. This shows that $G^* = \text{PDG}(R)$.

Now we verify that R is a prime ring.

Let us suppose that R is not a prime ring.

Then there exist two distinct elements $u, v \in R$ with $u \neq 0 \neq v \neq u$ such that $uRv = 0$

$\Rightarrow \overrightarrow{uv} \in E(\text{PDG}(R))$

$\Rightarrow \overrightarrow{uv} \in E(G^*)$ (since PDG(R) = G^*)

$\Rightarrow \overrightarrow{uv} \in \{\overrightarrow{0x}, \overrightarrow{x0} \mid 0 \neq x \in V^*\}$

$\Rightarrow$ either $0 = u$ or $0 = v$, which is a contradiction.

This shows that R is a prime ring. The proof is complete. ∎

Result 9.4.22

Suppose R is a ring satisfying the property $ab = 0$ which implies that $aRb = 0$ for all $a, b \in R$ (such a ring is called a ring with insertion factors property (IFP)). If R contains a nilpotent element $0 \neq x \in R$, k is the least positive integer such that $x^k = 0$, and $k \geq 3$, then PDG(R) contains at least one circuit of length 3.

Proof

Suppose $k \geq 3$ and k is the least positive integer such that $x^k = 0$.

Now $x \cdot x^{k-1} = 0$. This implies that $x \neq x^{k-1}$ and $xRx^{k-1} = 0$ (since R is a ring with IFP) $\Rightarrow xx^{k-1} \in E$.

We know that $\overrightarrow{0x}, \overrightarrow{x^{k-1}0} \in E$. Therefore, $\{\overrightarrow{0x}, \overrightarrow{xx^{k-1}}, \overrightarrow{x^{k-1}0}\}$ forms a circuit. Hence, there is a circuit of length 3 in PDG(R).

The proof is complete. ∎

Corollary 9.4.23

If R is a commutative ring and it contains a nilpotent element $0 \neq x$, and k is the least positive integer such that $x^k = 0$, $k \geq 3$, then the graph PDG(R) contains a circuit of length 3.

Proof

Let $x, y \in R$ such that $xy = 0$. Now $xy = 0 \Rightarrow xyr = 0$ for all $r \in R$

$\Rightarrow xry = 0$ for all $r \in R$ (since R is a commutative ring)

This shows that R is a ring with IFP.

By Result 9.4.22, there exists a circuit of length 3. The proof is complete.

Now we define a relation on the set of vertices of the graph PDG(R) and observe some properties of the relation.

Definition 9.4.24

Let $x, y \in R$. Define a relation $\sim$ on R by $x \sim y \Leftrightarrow \overrightarrow{xy} \in E$.

Observation 9.4.25

(i) In the graph PDG(R), every vertex is of even degree and hence by Theorem 9.1.15, it follows that PDG(R) is an Euler graph.

(ii) By the definition of the graph PDG(R), it follows that $\overrightarrow{xx} \notin E$. So $x \sim x$ is not true, and hence $\sim$ is not a reflexive relation.

(iii) Take the ring $R = \left\{ \begin{pmatrix} a & b \\ 0 & d \end{pmatrix} \middle| a, b, d \in Z, \text{ where } Z \text{ is the set of all integers} \right\}$

with respect to the matrix addition and multiplication operations.

Let $x = \begin{pmatrix} 0 & 1 \\ 0 & 0 \end{pmatrix}$ and $y = \begin{pmatrix} 1 & 0 \\ 0 & 0 \end{pmatrix}$.

For any $r = \begin{pmatrix} a & b \\ 0 & d \end{pmatrix} \in R$, $xry = \begin{pmatrix} 0 & 1 \\ 0 & 0 \end{pmatrix} \begin{pmatrix} a & b \\ 0 & d \end{pmatrix} \begin{pmatrix} 1 & 0 \\ 0 & 0 \end{pmatrix} = \begin{pmatrix} 0 & 0 \\ 0 & 0 \end{pmatrix}$.

So, $xRy = 0$, and hence $x \sim y$.

For $r = \begin{pmatrix} a & b \\ 0 & d \end{pmatrix} \in R$, we get

$$yrx = \begin{pmatrix} 1 & 0 \\ 0 & 0 \end{pmatrix} \begin{pmatrix} a & b \\ 0 & d \end{pmatrix} \begin{pmatrix} 0 & 1 \\ 0 & 0 \end{pmatrix} = \begin{pmatrix} a & b \\ 0 & 0 \end{pmatrix} \begin{pmatrix} 0 & 1 \\ 0 & 0 \end{pmatrix}$$

$$= \begin{pmatrix} 0 & a \\ 0 & 0 \end{pmatrix} \neq 0 \text{ if } a \neq 0.$$

So, $yRx \neq 0$, and hence $y \sim x$ is not true.

This shows that $\sim$ is not a symmetric relation, in general.

(iv) Consider R, x, y as in (ii). We know that $yR0 = 0, 0Rx = 0 \Rightarrow y \sim 0$, and $0 \sim x$.

By (ii), $y \sim x$ is not true. So, $y \sim 0, 0 \sim x$, and the statement $y \sim x$ is not true.

This shows that $\sim$ is not a transitive relation.

Observation 9.4.26

(i) From Observation 9.4.25 (ii), we conclude that the graph PDG(R) contains no self-loops.
(ii) From Observation 9.4.25 (i), we conclude that $\overrightarrow{xy} \in E \Rightarrow \overrightarrow{yx} \in E$ is not true, in general, where $E = \text{PDG}(R)$.
(iii) From Observation 9.4.25 (ii), we conclude that $\overrightarrow{xy}, \overrightarrow{yz} \in E \Rightarrow \overrightarrow{xz} \in E$ is not true, in general, where $E = \text{PDG}(R)$.

Note 9.4.27

(i) If v_1, v_2, v_3 are vertices of a graph $G = (V, E)$ and the maximal subgraph with vertex set $\{v_1, v_2, v_3\}$ forms a triangle, then we say that the set $\{v_1, v_2, v_3\}$ forms a triangle.
(ii) The number of triangles in the graph given by Figure 9.62 is zero. The number of triangles in the graph given by Figure 9.63 is one. The number of triangles in the graph given by Figure 9.65 is six.

Lemma 9.4.28 (Lemma 4.5 of Satyanarayana, Prasad, and Nagaraju, 2010)

Suppose I_1 and I_2 are two ideals of a ring R such that $I_1 \cap I_2 = (0)$. If $0 \neq a_1 \in I_1$ and $0 \neq a_2 \in I_2$, then the set $\{0, a_1, a_2\}$ forms a triangle.

Proof

Let $0 \neq a_1 \in I_1$ and $0 \neq a_2 \in I_2$.

Since I_1 is also a right ideal, it follows that $a_1 R a_2 \subseteq I_1$.

Since I_2 is also a left ideal, it follows that $a_1 R a_2 \subseteq I_2$.

$a_1 R a_2 \subseteq I_1 \cap I_2 = (0) \Rightarrow a_1 R a_2 = 0 \Rightarrow$ there exists an edge between a_1 and a_2. This implies that $\overline{0a_1}, \overline{a_1a_2}, \overline{a_20}$ is a triangle. That is, $\{0, a_1, a_2\}$ forms a triangle. The proof is complete. ∎

Corollary 9.4.29 (Corollary 4.6 of Satyanarayana, Prasad, and Nagaraju, 2010)

Suppose U_1 and U_2 are two uniform ideals of a ring R such that $U_1 \cap U_2 = (0)$. If $0 \neq a_1 \in U_1$ and $0 \neq a_2 \in U_2$, then the set $\{0, a_1, a_2\}$ forms a triangle.

Theorem 9.4.30 (Theorem 4.7 of Satyanarayana, Prasad, and Nagaraju, 2010)

Let R be a ring with FDI (finite dimension on ideals). Then the number of triangles in PG(R) $\geq$ (dim R)(dim $R - 1$). In other words, PG(R) contains at least (dim R)(dim $R - 1$) triangles.

Proof

Suppose $k =$ dim R.

Then by Theorem 5.1.2, there exist uniform ideals U_i, $1 \leq i \leq k$ of R such that $U_1 \oplus U_2 \oplus \ldots \oplus U_k \leq_e R$.

Let us fix $0 \neq a_i \in U_i$ for $1 \leq i \leq k$.

If $a_i = a_j$ for $i \neq j$, then $a_i = a_j \in U_i \cap U_j = (0)$, which is a contradiction.

Hence, $a_1, a_2, \ldots, a_k$ are all distinct elements.

By Corollary 9.4.29, if $i \neq j$, the set of vertices $V_{ij} = \{0, a_i, a_j\}$ forms a triangle in PG(R).

This is true for all $1 \leq i, j \leq k$.

So, the number of such distinct V_{ij}'s that can be formed using $a_1, a_2, \ldots, a_k$ is $k(k-1)$.

Hence, the number of distinct triangles in the graph is PG(R) $\geq k(k-1) =$ (dim R)(dim $R - 1$).

The proof is complete. ∎

Theorem 9.4.31 (Theorem 4.8 of Satyanarayana, Prasad, and Nagaraju, 2010)

Suppose dim $R = k$ and U_i, $1 \leq i \leq k$ is a collection of uniform ideals whose sum is direct and essential in R.

The number of triangles in PG(R) $\geq \Sigma_{j=1}^{k-1} n_j \left(\Sigma_{i=j+1}^{k} n_i \right)$, where $n_i = |U_i| - 1$ for $1 \leq i \leq k$.

In other words, PG(R) contains at least $\Sigma_{j=1}^{k-1} n_j \left(\Sigma_{i=j+1}^{k} n_i \right)$ triangles.

Proof

If $0 \neq a_1 \in U_1$ and $0 \neq a_2 \in U_2$, then by Corollary 9.4.29, it follows that $\{0, a_1, a_2\}$ forms a triangle in PG(R).

The number of such triangles (with one nonzero element x from U_1 and one nonzero element y from U_2) that can be formed is $|U_1 \backslash \{0\}| \times |U_2 \backslash \{0\}| = n_1 \cdot n_2$.

Using the elements of U_1 and U_3, the number of such triangles that can be formed is $n_1 \cdot n_3$.

Thus, the number of such triangles that can be formed in this procedure with one element from U_1 and the other element from U_2 or .... or U_k is

$$n_1 n_2 + n_1 n_3 + \ldots + n_1 n_k = n_1 (n_2 + \ldots + n_k) = n_1 \left(\sum_{i=2}^{k} n_i \right).$$

Similarly, the number of triangles that can be formed using one element from U_2 and the other element from U_3 or ... or U_k is

$$n_2 n_3 + n_2 n_4 + \ldots + n_2 n_k = n_2 \left(\sum_{i=3}^{k} n_i \right).$$

We can continue this procedure.

Eventually, we get the number of triangles that can be formed in this way as

$$n_1 \left(\sum_{i=2}^{k} n_i \right) + n_2 \left(\sum_{i=3}^{k} n_i \right) \cdots + n_{k-1} n_k = \sum_{j=1}^{k-1} n_j \left(\sum_{i=j+1}^{k} n_i \right).$$

The proof is complete.   ∎

9.5 Homomorphism of Graphs

In this section, we present the concepts of the "homomorphism" of graphs and the "kernel" of a homomorphism with respect to a vertex in the codomain, and suitable examples are provided.

Definition 9.5.1

Let $G_1 = (V_1, E_1)$, $G_2 = (V_2, E_2)$ be two graphs. A mapping $f: V_1 \to V_2$ is said to be a **homomorphism** from G_1 to G_2 if there exists a mapping $g: B \to E_2$ for some subset B of E_1 satisfying the following condition:

If $xy \in E_1$ and $f(x) \neq f(y)$, then $xy \in B$ and $g(xy) = f(x)f(y)$.

In this case, we say that f is a **homomorphism** with g (or (f, g) is (or forms) a homomorphism).

Moreover, if f is one–one, then we say that (f, g) is a **semi-monomorphism** from G_1 to G_2.

If f, g are both one–one, then we say that (f, g) is a **monomorphism** from G_1 into G_2. If there exists a monomorphism from G_1 into G_2, then we say that G_1 is **embedded** into G_2 by (f, g).

If f is onto, then we say that (f, g) is a **semi-epimorphism**.

If f, g are both onto, then we say that (f, g) is an **epimorphism** from G_1 onto G_2.

Note 9.5.2

(i) If $B = E_1$ and f, g are both bijections, then (f, g) is an isomorphism from G_1 onto G_2.
(ii) A graph $G = (V, E)$ is said to be a **single vertex null graph** if V is a singleton set and $E = \emptyset$.

Example 9.5.3

Let $G_1 = (V_1, E_1)$, $G_2 = (V_2, E_2)$ be two graphs (given in Figures 9.68 and 9.69) with $V_1 = \{v_1, v_2, v_3, v_4\}$ and $V_2 = \{a, b, c\}$.
  Define $f: V_1 \to V_2$ by $f(v_1) = a, f(v_2) = b, f(v_3) = c, f(v_4) = c$.
  Write $B = \{v_1v_2, v_2v_3, v_4v_1\} \subseteq E_1$.
  Define $g: B \to E_2$ by $g(v_1v_2) = ab, g(v_2v_3) = bc, g(v_4v_1) = ca$.
  Now, f together with g is a homomorphism from G_1 to G_2.

Definition 9.5.4

Let f be a homomorphism from G_1 to G_2 with $g: B \to E_2$. Let $x \in V_2$. Consider the set $f^{-1}(x) = \{v \in V_1 | f(v) = x\}$. If $V^* = f^{-1}(x) \neq \emptyset$, then the maximal subgraph of G_1 with vertex set V^* is called the **kernel** of (f, x).
  Note that the kernel depends on both f and x.

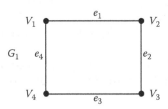

FIGURE 9.68
Graph G_1.

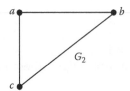

FIGURE 9.69
Graph G_2.

Example 9.5.5

Consider Example 9.5.3.

   (i) $a \in V_2$ and $f^{-1}(a) = \{v_1\}$.
      The maximal subgraph of G_1 with vertex set $\{v_1\}$ is the null graph
      $(\{v_1\}, \varnothing)$.
      Hence, the subgraph $(\{v_1\}, \varnothing)$ of G_1 is the kernel of (f, a).
  (ii) $c \in V_2$ and $f^{-1}(c) = \{v_3, v_4\}$.
      The maximal subgraph with the vertex set $\{v_3, v_4\}$ is $(\{v_3, v_4\}, \{\overline{v_3 v_4}\})$.
      So, the kernel of (f, c) is the subgraph $(\{v_3, v_4\}, \{\overline{v_3 v_4}\})$ of G_1.

Note 9.5.6

Let $G_1 = (V_1, E_1)$, $G_2 = (V_2, E_2)$, and (f, g) be a homomorphism from G_1 to G_2.
Then the following two conditions hold:

   (i) $\ker(f, x)$ is a single vertex null graph for every vertex x in G_1.
  (ii) (f, g) is a semi-monomorphism.

Procedure 9.5.7

Let $G = (V, E)$ be a graph. Let $\varnothing \neq V^* \subseteq V$.
   Step 1: Suppose $V^* = \{x_1, x_2, \ldots, x_k\}$. Merge all the vertices $x_1, x_2, \ldots, x_k$.
   Step 2: Remove the self-loops, if any.
   Step 3: If e_1, e_2 are multiple edges, then remove e_2 (i.e., replace multiple
edges by a single edge).
   Step 4: Repeat Steps 2 and 3 until the graph becomes a simple graph.
   Step 5: Denote the graph obtained from G by procedure-1 by P–1(G/V^*).

Example 9.5.8

Consider the graphs G_1 and G_2 mentioned in Example 9.5.3.

   The kernel of (f, c) is the subgraph $(\{v_3, v_4\}, \{\overline{v_3 v_4}\})$. Write $V^* = \{v_3, v_4\}$.
   Step 1: By merging v_3, v_4 we get the graph in Figure 9.70.
   Step 2: By removing the self-loop, we get the graph in Figure 9.71.
   Step 3: As there are no multiple edges, there is no question of removing
one of the parallel edges.
   Step 4: The graph obtained in Step 2 is a simple graph and P–1(G/V^*) is
given in Figure 9.71.

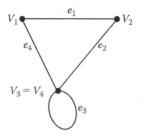

FIGURE 9.70
Graph after merging v_3, v_4.

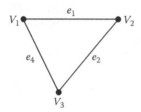

FIGURE 9.71
Graph after removing self-loop.

Example 9.5.9

Consider G_1, G_2, and $P{-}1(G_1/f^{-1}(c))$ as in Example 9.5.8.
  Define $f^*: V(P{-}1(G_1/f^{-1}(c))) \to V(G_2)$ by $f^*(v_1) = a, f^*(v_2) = b, f^*(v_3) = c$.
  Write $B^* = \{e_1, e_2, e_4\} = \{v_1v_2, v_2v_3, v_1v_3\}$.
  Define $g^*: B^* \to E(G_2)$ by $g^*(v_1v_2) = ab$, $g^*(v_2v_3) = bc$, and $g^*(v_1v_3) = ac$.
  Now (f^*, g^*) is a homomorphism from $P{-}1(G_1/f^{-1}(c))$ into G_2.
  Since both f^* and g^* are bijections, it follows that (f^*, g^*) forms an isomorphism from $P{-}1(G_1/f^{-1}(c))$ onto G_2.
  Now we explain a procedure to get a particular graph $P{-}2(PG(R)/I)$ starting with $PG(R)$ for a given ideal I of R and prove that there is a semi-onto homomorphism from $P{-}2(PG(R)/I)$ to $PG(R/I)$. We also prove that if I is a prime ideal of R, then this homomorphism is an epimorphism.

Example 9.5.10

Consider two rings $(\mathbb{Z}_6, +, \cdot)$, $(\mathbb{Z}_3, +, \cdot)$ and their prime graphs $PG(\mathbb{Z}_6)$, $PG(\mathbb{Z}_3)$ given in Figures 9.59 and 9.56.

Define a mapping f: $V(PG(\mathbb{Z}_6)) \to V(PG(\mathbb{Z}_3))$ by $f(0) = 0, f(1) = 1, f(2) = 2, f(3) = 0,$ $f(4) = 1, f(5) = 2.$

Write $B = \{\overline{10}, \overline{05}, \overline{02}, \overline{04}, \overline{23}, \overline{34}\}.$

Define a mapping g: $B \to E(PG(\mathbb{Z}_3))$ by $g(\overline{10}) = \overline{10}, g(\overline{05}) = \overline{02}, g(\overline{02}) = \overline{02},$ $g(\overline{04}) = \overline{01}, g(\overline{23}) = \overline{02},$ and $g(34) = \overline{01}.$

Now (f, g) is a homomorphism from $PG(\mathbb{Z}_6)$ onto $PG(\mathbb{Z}_3)$.

Since both f and g are onto, we conclude that (f, g) is an epimorphism.

Procedure 9.5.11

Let I be an ideal of the given ring R.

Step 1: If $x - y \in I$, then merge x and y in $PG(R)$.

Step 2: Repeat Step 1 till no such further new step occurs.

Step 3: Remove self-loops.

Step 4: Replace multiple edges (if any) by a single edge (between the same endpoints).

Step 5: The graph obtained from G through this procedure is denoted by $P\text{–}2(G/I)$.

Example 9.5.12

We apply procedure-2 to $PG(\mathbb{Z}_6)$ with the ideal $I = \{0, 3\}$.

Step 1: The vertices $1, 4 \in \mathbb{Z}_6$ with $4 - 1 \in \{0, 3\}$. So we merge 1 and 4. Then the graph obtained is given in Figure 9.72.

Step 2: Since $2, 5 \in \mathbb{Z}_6$ with $5 - 2 \in \{0, 3\}$, we merge 2 and 5, and then we get the graph given in Figure 9.73.

Step 3: Since $0, 3 \in \mathbb{Z}_6$ with $3 - 0 \in \{0, 3\}$, we merge 0 and 3. Then we get the graph given in Figure 9.74.

Conclusion

It is clear that the graph shown in Figure 9.74 is isomorphic to $PG(\mathbb{Z}_3)$.

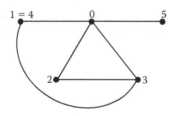

FIGURE 9.72
Graph after merging 1 and 4.

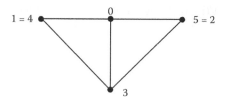

FIGURE 9.73
Graph after merging 2 and 5.

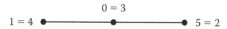

FIGURE 9.74
Graph after merging 0 and 3.

Lemma 9.5.13 (Satyanarayana, Prasad, and Nagaraju, Homomorphisms of Graphs, communicated)

Let R be a ring and I an ideal of R. Then there exists a semi-onto homomorphism f: P–2$(PG(R)/I) \rightarrow$ PG(R/I).

Proof

Suppose R is a ring and I is an ideal of R.
   Define f: $V(PG(R)) \rightarrow V(PG(R/I))$ by $f(r) = r + I$ for all $r \in R$.
   We know that R/I is the set of all equivalence classes of the form $a + I$ (which is denoted by $[a]$).
   Let us take $a_1, a_2, ..., a_n \in R$ such that $[a_1], [a_2], ..., [a_n]$ are all distinct equivalence classes.
   Write $B = \{\overline{xy} \mid \overline{xy} \in E(PG(R)), x \in [a_i], y \in [a_j]$ for some $i \neq j, 1 \leq i \leq n, 1 \leq j \leq n\}$.
   Suppose $\overline{xy} \in B$.
   Then $\overline{xy} \in E(PG(R))$.
   This means that $xRy = 0$ or $yRx = 0$.
   Suppose $x \in [a_i], y \in [a_j]$ with $i \neq j$. Now $f(x) = a_i + I$ and $f(y) = a_j + I$.
   If $xRy = 0$, then $(a_i + I)(R/I)(a_j + I) = f(x)(R/I)f(y)$
$= (x + I)(R/I)(y + I)$
$= xRy + I$
$= 0 + I = 0$,
   and so $\overline{f(x)f(y)} = \overline{(a_i + I)(a_j + I)} \in E(PG(R/I))$.
   If $yRx = 0$, then in a similar way, we can verify that
$\overline{f(y)f(x)} = \overline{(a_j + I)(a_i + I)} \in E(PG(R/I))$.
   Define g: $B \rightarrow E(PG(R/I))$ by $g(\overline{xy}) = \overline{f(x)f(y)} = \overline{(a_i + I)(a_j + I)}$.

Note that if $a, b \in V(PG(R))$ with the property that $f(a) = a + I$ and $f(b) = b + I$ are equal, then there is no edge in B between a and b.

If $f(a) = a + I$ is not equal to $f(b) = b + I$ and $\overline{ab} \in E(PG(R))$, then a, b belong to distinct equivalence classes and so $\overline{ab} \in B$.

From this discussion, we can conclude that f is a semi-onto homomorphism with $g: B \rightarrow E(PG(R/I))$.

The proof is complete.   ∎

Theorem 9.5.14 (Satyanarayana, Prasad, and Nagaraju, Homomorphisms of Graphs, communicated)

If I is a prime ideal of R, then f (defined in Lemma 9.5.13) is an epimorphism.

Proof

Suppose I is a prime ideal of R. We verify that g is an onto mapping.

Let $\overline{(a_i + I)(a_j + I)} \in E(PG(R/I))$.

Then $(a_i + I)(R/I)(a_j + I) = 0$

$\Rightarrow a_i R a_j + I = 0$

$\Rightarrow a_i R a_j \subseteq I$.

Since I is a prime ideal, it follows that $a_i = 0$ or $a_j = 0$.

If $a_i = 0$, then $\overline{0a_j} \in B$ and

$$g(\overline{0a_j}) = \overline{f(0)f(a_j)}$$

$$= \overline{(a_i + I)(a_j + I)}.$$

The other case is similar. This shows that g is onto.

Hence, $f: P\text{–}2(PG(R)/I) \rightarrow PG(R/I)$ is an epimorphism. The proof is complete. ∎

9.6 Graph of a Near Ring with Respect to an Ideal

In this section, we present some results relating to 3-prime ideals, equi-prime ideal and its corresponding graphical aspects from Babushri (2009), Satyanarayana, Prasad and Babushri (2010).

Satyanarayana and Vijaya Kumari (Prime Graph of a Nearring, communicated) presented the concept of a prime graph of a near ring. In this section, we present a method of relating a near ring N to a graph by introducing the concept of a prime graph of a near ring N. The prime graph of a near ring N

is a graph with the elements of N as the vertices, and any two vertices x, y are adjacent if and only if $xNy = 0$ or $yNx = 0$. If N is a commutative ring, then the zero–divisor graph of N is a subgraph of the prime graph of N. We present the concept called the "graph of a near ring N with respect to an ideal I" denoted by $G_1(N)$. We define a new type of symmetry (called ideal symmetry) of a near ring with respect to an ideal (denoted by $G_1(N)$). The ideal symmetry of $G_1(N)$ implies the symmetry determined by the automorphism group of $G_1(N)$.

Beck (1988) related a commutative ring R to a graph using the elements of R as vertices, and any two vertices x, y are adjacent if and only if $xy = 0$. Anderson and Livingston (1999) proposed a modified method of associating a commutative ring to a graph by introducing the concept of a zero–divisor graph of a commutative ring. The zero–divisor graph of a commutative ring R is a graph with vertices as the set of nonzero divisors of R, and there is an edge between the vertices x, y if and only if $x \neq y$ and $xy = 0$. Zero–divisor graphs are highly symmetric. We present the symmetry exhibited by $G_1(N)$. The related concepts are studied by Godsil and Royle (2001), Satyanarayana, Prasad, and Babushri (2010). For example, a human being is symmetric but only with respect to a specific vertical plane. We find that if I is a 3–prime ideal of a zero–symmetric near ring N, then I plays a role in $G_1(N)$ similar to the role of a specific vertical plane in case of a human being. Babushri (2009), Babushri, Prasad and Satyanarayana (2009, 2010), Booth and Groenewald (1991a, 1991b) studied the concepts of equiprime, 3-prime, and prime ideals of near rings. It is clear that the notions coincide in the case of rings.

Definition 9.6.1

Let I be an ideal of N. The graph of N with respect to I is a graph with each element of N as a vertex, and two distinct vertices x and y are connected by an edge if and only if $xNy \subseteq I$ or $yNx \subseteq I$. We denote the graph of N with respect to I by $G_1(N)$. An ideal I of N is called c-prime if $a, b \in N$ and $ab \in I$ implies $a \in I$ or $b \in I$. N is called c-prime near ring if $\{0\}$ is a c-prime ideal of N. An ideal I of N is called 3-prime if $a, b \in N$ and $anb \in I$ for all $n \in N$ implies $a \in I$ or $b \in I$.

Example 9.6.2

Z_2 has the ideals $I_1 = \{0\}$, $I_2 = \{0, 1\}$. Then $G_{I_1}(Z_2) = G_{I_2}(Z_2)$ is as shown in Figure 9.75.

0   1

FIGURE 9.75
Graph with respect to I_1 and I_2.

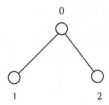

FIGURE 9.76
Graph with respect to I_1.

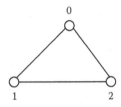

FIGURE 9.77
Graph with respect to I_2.

Example 9.6.3

Z_3 has the ideals $I_1 = \{0\}$, $I_2 = Z_3$.

    (i) $G_{I_1}(Z_3)$ (Figure 9.76)
    (ii) $G_{I_2}(Z_3)$ (Figure 9.77)

Example 9.6.4

Z_4 has the ideals $I_1 = \{0\}$, $I_2 = \{0, 2\}$, $I_3 = Z_4$.
    (i) $G_{I_1}(Z_4)$ (Figure 9.78)

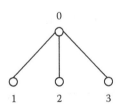

FIGURE 9.78
Graph with respect to I_1.

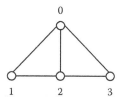

FIGURE 9.79
Graph with respect to I_2.

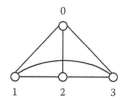

FIGURE 9.80
Graph with respect to I_3.

(ii) $G_{I_2}(\mathbb{Z}_4)$ (Figure 9.79)
(iii) $G_{I_3}(\mathbb{Z}_4)$ (Figure 9.80)

Example 9.6.5

$\mathbb{Z}_5$ has ideals $I_1 = \{0\}$, $I_2 = \mathbb{Z}_5$.
(i) $G_{I_1}(\mathbb{Z}_5)$ (Figure 9.81)
(ii) $G_{I_2}(\mathbb{Z}_5)$ (Figure 9.82)

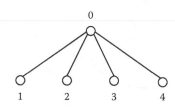

FIGURE 9.81
Graph with respect to I_1.

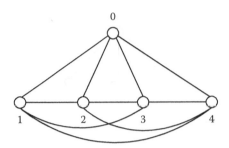

FIGURE 9.82
Graph with respect to I_2.

Example 9.6.6

$\mathbb{Z}_6$ has ideals $I_1 = \{0\}$, $I_2 = \{0, 3\}$, $I_3 = \{0, 2, 4\}$, $I_4 = \mathbb{Z}_6$.

    (i) $G_{I_1}(\mathbb{Z}_6)$ (Figure 9.83)
    (ii) $G_{I_2}(\mathbb{Z}_6)$ (Figure 9.84)
    (iii) $G_{I_3}(\mathbb{Z}_6)$ (Figure 9.85)
    (iv) $G_{I_4}(\mathbb{Z}_6)$ (Figure 9.86)

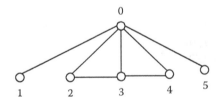

FIGURE 9.83
Graph with respect to I_1.

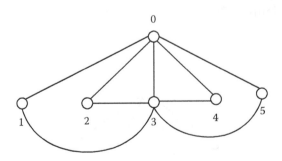

FIGURE 9.84
Graph with respect to I_2.

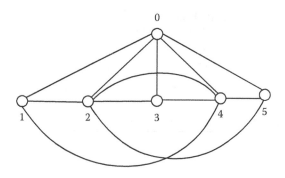

FIGURE 9.85
Graph with respect to I_3.

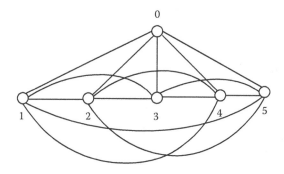

FIGURE 9.86
Graph with respect to I_4.

Example 9.6.7

Let $N = \{0, a, b, c\}$ be a set with binary operations $+$ and $\cdot$ defined as in Tables 9.3 and 9.4.

Then $(N, +, \cdot)$ is a near ring with the ideals $I_1 = \{0\}$, $I_2 = \{0, a\}$, $I_3 = \{0, b\}$, $I_4 = N$.

(i) $G_{I_1}(N)$ (Figure 9.87)
(ii) $G_{I_2}(N)$ (Figure 9.88)

TABLE 9.3

Addition Table

+	0	a	b	c
0	0	a	b	c
a	a	0	c	b
b	b	c	0	a
c	c	b	a	0

TABLE 9.4
Multiplication Table

•	0	a	b	c
0	0	0	0	0
a	0	a	0	a
b	b	b	b	b
c	b	c	b	c

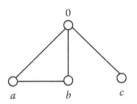

FIGURE 9.87
Graph with respect to I_1.

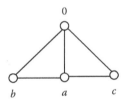

FIGURE 9.88
Graph with respect to I_2.

(iii) $G_{I_3}(N)$ (Figure 9.89)
(iv) $G_{I_4}(N)$ (Figure 9.90)

Remark 9.6.8

(i) $G_I(N)$ is a connected graph without self-loops and multiple edges.
(ii)   The maximum distance between any two vertices of $G_I(N)$ is at most 2. That is, $d(G_I(N)) \leq 2$.
      The proofs of the following proposition and remark are straight-forward.

Proposition 9.6.9 (Babushri, 2009; Proposition 3.9 of Satyanarayana, Prasad, and Babushri, 2010)

Let I and J be ideals of N such that $I \subseteq J$. Then $G_{\{0\}}(N) \subseteq G_I(N) \subseteq G_J(N) \subseteq G_N(N)$.

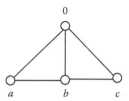

FIGURE 9.89
Graph with respect to I_3.

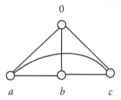

FIGURE 9.90
Graph with respect to I_4.

Remark 9.6.10

$G_{\{0\}}(N)$ is the prime graph of N defined by Satyanarayana and Vijaya Kumari (Prime Graph of a Nearring, communicated). If $|N| = n$, then $G_N(N)$ is the complete graph on n vertices K_n.

Proposition 9.6.11 (Babushri, 2009; Proposition 3.11 of Satyanarayana, Prasad, and Babushri, 2010)

Let N be integral as well as simple. Let I be an ideal of N. Then $G_I(N) = K_n$ or $G_I(N)$ is a rooted tree with root vertex 0.

Proof

If $I = N$, then $G_I(N) = K_n$. Suppose $I \neq N$. Then $I = \{0\}$. Let $0 \neq x \neq y \neq 0$. Suppose there exists an edge between x and y in $G_I(N)$. Then $xNy = 0$. As N is integral, either $xN = 0$ or $Ny = 0$. This implies that either $x^2 = 0$ or $y^2 = 0$. Again, as N is integral, either $x = 0$ or $y = 0$. This is a contradiction. Hence, $G_I(N)$ is a rooted tree with vertex 0.

Definition 9.6.12

Let G be a graph with the vertex set $V(G)$. Then a strong vertex cut of a graph G is a subset $S \subseteq V(G)$ such that $G - S$ is totally disconnected. The strong vertex connectivity of G is defined by $K(G) = \min\{n \geq 0 \mid$ there exists a strong vertex cut $S \subseteq V(G)$ such that $|S| = n\}$. The vertex connectivity (or connectivity) of a connected graph G is defined as the minimum number of vertices, whose removal from G provides the disconnected graph. We denote the vertex connectivity by $k(G)$.

Remark 9.6.13 (Babushri, 2009; Remark 3.13 of Satyanarayana, Prasad, and Babushri, 2010)

(i) $k(G) \leq K(G)$.
(ii) Consider the graph $G_{I_1}(\mathbb{Z}_6)$ given in Example 9.6.6 (i). We can observe that $\{0\}$ is a vertex cut but not a strong vertex cut of $G_{\{0\}}(N)$ (as there is an edge between vertices 2 and 3). Note that $\{0, 3\}$ is a strong vertex cut. Hence, $k(G_{\{0\}}(\mathbb{Z}_6)) = 1$ and $K(G_{\{0\}}(\mathbb{Z}_6)) = 2$.

Theorem 9.6.14 (Babushri, 2009; Theorem 3.14 of Satyanarayana, Prasad, and Babushri, 2010)

Let I be an ideal of N. If I is 3-prime, then I is a strong vertex cut of $G_I(N)$. If I is 3-semiprime and I is a strong vertex cut of $G_I(N)$, then I is 3-prime.

Proof

Let I be a 3-prime ideal of N. If $I = N$, then there is nothing to prove. Let $I \neq N$ and $x, y \in N \backslash I$ such that $x \neq y$. If possible, suppose there exists an edge between the vertices x and y in $G_I(N)$. Then $xNy \subseteq I$ or $yNx \subseteq I$. Without loss of generality, assume $xNy \subseteq I$. As I is a 3-prime ideal of N, either $x \in I$ or $y \in I$. This is a contradiction since $x \in N \backslash I$ and $y \in N \backslash I$. Hence, I is a strong vertex cut of $G_I(N)$.

Converse

Suppose I is 3-semiprime and I is a strong vertex cut of $G_I(N)$. To prove I is 3-prime, take $x, y \in N$ such that $xNy \subseteq I$. Since I is 3-semiprime, $x = y$ implies that $x \in I$. Let $x \neq y$. If possible, suppose $x \in N \backslash I$ and $y \in N \backslash I$. As I is a strong vertex cut of $G_I(N)$, there is no edge between x and y in $G_I(N)$. This implies that $xNy \nsubseteq I$ and $yNx \nsubseteq I$. This is a contradiction since $xNy \subseteq I$. Thus, I is a 3-prime ideal of N. ∎

Remark 9.6.15

(i) In Example 9.6.6 (i), we can observe that $I_1 = \{0\}$ is not a strong vertex cut.
(ii) Using Theorem 9.6.14, we can conclude from Example 9.6.7 (ii) that the ideal $I_2 = \{0, a\}$ is a 3-prime ideal of N.
(iii) The converse of Theorem 9.6.14 does not hold if the condition that I is 3-semiprime is omitted. For example, consider $G_{I_4}(\mathbb{Z}_4)$; however, I_1 is not a 3-prime ideal since $2N2 \subseteq I_1$ but $2 \notin I_1$.

Corollary 9.6.16

Let I be a minimal 3-prime ideal of N. Then $K(G_I(N)) = |I|$.

Lemma 9.6.17 (Babushri, 2009; Lemma 3.17 of Satyanarayana, Prasad, and Babushri, 2010)

(i) Let I be a 3-prime ideal of N and x be a vertex in $G_I(N)$. If $\deg(x) = \deg(0)$, then $x \in I$.
(ii) Let n be a zero-symmetric near ring and x be a vertex in $G_I(N)$. If $x \in I$, then $\deg(x) = \deg(0)$.
(iii) Let N be a zero-symmetric near ring and I be a 3-prime ideal of N. Then $x \in I$ if and only if $\deg(x) = \deg(0)$ in $G_I(N)$.

Proof

To prove (i), suppose that $\deg(x) = \deg(0)$. Then $xNy \subseteq I$ or $yNx \subseteq I$ for all $y \in N$ such that $y \neq x$. Without loss of generality, assume $xNy \subseteq I$ for all $y \in N$ such that $y \neq x$. If $I = N$, then $x \in I$. Let $I \neq N$. Choose $y \in N \backslash I$. As I is a 3-prime ideal of N and $xNy \subseteq I$, we have $x \in I$. To prove (ii), take $x \in I$. If $x = 0$, then the result is true. Let $x \neq 0$. If possible, suppose that $\deg(x) < \deg(0)$. Then there exists a vertex y such that y is not adjacent to x in $G_I(N)$. This implies that $xNy \nsubseteq I$ and $yNx \nsubseteq I$. Now as $x \in I$ and I is an ideal of N, we have $xN \subseteq I$. Hence, $xNy \subseteq Iy$. But since N is zero-symmetric, we have $Iy \subseteq I$. It follows that $xNy \subseteq I$, which is a contradiction. This proves $\deg(x) = \deg(0)$. ∎

Proposition 9.6.18 (Babushri, 2009; Proposition 3.18 of Satyanarayana, Prasad, and Babushri, 2010)

Let $|N| = n$ and I be an ideal of N. Then

(i) $I \leq k(G_I(N)) \leq \lambda(G_I(N)) \leq \delta(G_I(N)) \leq n - 1$.

(ii) Further, if N is zero-symmetric and I is a minimal 3-prime ideal of N, then $|I| \leq k(G_I(N)) \leq \delta(G_I(N))$.

Proof

For any graph G, it is well known that $k(G) \leq \lambda(G) \leq \delta(G)$. As $G_I(N)$ is a connected graph, the minimum number of vertices whose removal results in a disconnected or trivial graph is 1. Hence, $1 \leq k(G_I(N))$. As $|N| = n$, $\delta(G_I(N)) \leq \deg(0) = n - 1$. This proves (i). To prove (ii), let N be a zero-symmetric near ring and I be a minimal 3-prime ideal of N. By Corollary 9.6.16, we have $|I| = K(G_I(N))$. Let y be a vertex in $G_I(N)$ of minimum degree. By Lemma 9.6.17 (iii), y is adjacent to every element of I in $G_I(N) \geq |I| = K(G_I(N))$. ■

Definition 9.6.19

The graph $G_I(N)$ is said to be ideal symmetric if for every pair of vertices x, y in $G_I(N)$ with an edge between them, either $\deg(x) = \deg(0)$ or $\deg(y) = \deg(0)$.

Remark 9.6.20

The graphs in Examples 9.6.6 (i) and 9.6.7 (i) are not ideal symmetric, whereas all the other graphs given in Examples 9.6.2 through 9.6.7 are ideal symmetric.

Note that all the graphs given in Examples 9.6.2 through 9.6.7 are symmetric (symmetry is determined by the automorphism group) (Beineke, Wilson, and Camerong, 2004; Erdos and Renyi, 1963). The smallest nontrivial graph (apart from the one-vertex graph) that is asymmetric has six vertices.

Proposition 9.6.21 (Babushri, 2009; Proposition 3.21 of Satyanarayana, Prasad, and Babushri, 2010)

The ideal symmetry of $G_I(N)$ implies the symmetry determined by the automorphism group of $G_I(N)$.

Proof

Suppose no edges exist between any two distinct nonzero vertices of $G_I(N)$. Then $G_I(N)$ is symmetric as it is a rooted tree with root vertex 0. Suppose there is an edge between x and y in $G_I(N)$ and $0 \neq x \neq y \neq 0$. As $G_I(N)$ is ideal symmetric, either $\deg(x) = \deg(0)$ or $\deg(y) = \deg(0)$. Without loss of generality, assume $\deg(x) = \deg(0)$. This implies that x is adjacent to every other vertex in $G_I(N)$. Hence, the permutation of the vertex x with the vertex "0" preserves adjacency. Thus, the automorphism group of $G_I(N)$ is not the identity group. This proves that $G_I(N)$ is symmetric. ■

Example 9.6.22

The graph in Example 9.6.7 (i) is symmetric but not ideal symmetric. This shows that the converse of Proposition 9.6.21 does not hold in general.

Theorem 9.6.23 (Babushri, 2009; Theorem 3.23 of Satyanarayana, Prasad, and Babushri, 2010)

Let I be an ideal of N.

(a) Suppose (i) N is zero–symmetric and (ii) I is 3–prime; then $G_I(N)$ is ideal symmetric.
(b) (i) $G_I(N)$ is ideal symmetric. (ii) I is 3-semiprime. (iii) For every $x \in N$, $\deg(x) = \deg(0)$ in $G_I(N)$ implies that $x \in I$. Then I is 3-prime and I is a strong vertex cut of $G_I(N)$.

Proof

To prove (a), let x, y be distinct vertices of $G_I(N)$ with an edge between x and y. Then $xNy \subseteq I$ or $yNx \subseteq I$. Without loss of generality, assume $xNy \subseteq I$. As I is a 3-prime ideal of N, we have $x \in I$ or $y \in I$. Now as N is a zero-symmetric near ring, by Lemma 9.6.17 (ii), we get $\deg(x) = \deg(0)$ or $\deg(y) = \deg(0)$. Thus, $G_I(N)$ is ideal symmetric.

To prove (b), let $x, y \in N$ and $xNy \subseteq I$. As I is 3-semiprime, $x = y$ implies that $x \in I$. Let $x \neq y$. Now there exists an edge between x and y in $G_I(N)$. As $G_I(N)$ is ideal symmetric, $\deg(x) = \deg(0)$ or $\deg(y) = \deg(0)$. This implies that $x \in I$ or $y \in I$. Thus, I is a 3-prime ideal of N. By Theorem 9.6.14, I is a strong vertex cut of $G_I(N)$.

Theorem 9.6.24 (Babushri, 2009; Satyanarayana, Prasad, and Babushri, 2010)

Suppose N is zero-symmetric and I is a 3-semiprime ideal of N. If I is a strong vertex cut of $G_I(N)$, then $G_I(N)$ is ideal symmetric.

Proof

By Theorem 9.6.14, I is a 3-prime ideal of N. Using part (a) of Theorem 9.6.23, $G_I(N)$ is ideal symmetric.

Corollary 9.6.25 (Babushri, 2009; Corollary 3.25 of Satyanarayana, Prasad, and Babushri, 2010)

Suppose N is a commutative ring with unit element 1. Let I be a completely semiprime ideal of N. If $G_I(N)$ is ideal symmetric, then I is c-prime, and I is a strong vertex cut of $G_I(N)$. An ideal I of N is called c-prime if $a, b \in N$ and $ab \in I$ implies $a \in I$ or $b \in I$. N is called c-prime nearring if $\{0\}$ is a c-prime ideal of N.

Proof

Note that if $I \sim N$, then I is c-prime and I is a strong vertex cut of $G_I(N)$. Hence, assume $I \neq N$. Let I be completely semiprime. As N is a commutative ring, I is a 3-semiprime ideal. Suppose $G_I(N)$ is ideal symmetric. Now, let $x \in N$ such that $y \neq x$.

Case (i): Suppose $x = 1$. Then $1 \cdot 1 \cdot y \in I$ for all $y \in N$ such that $y \neq 1$. Suppose $|N| = m$. Then $|I| = m - 1$. We know that $|I| = m - 1$ must divide $|N| = m$. This happens only if $|N| = m = 2$, that is, N must have exactly two elements, namely, 0 and the unit element 1. Now, $\mathbb{Z}_2$ (Example 9.6.2) is the only ring of order with unit element 1. Note that $I = \{0\}$ is a c-prime ideal of $\mathbb{Z}_2$ and $I = \{0\}$ is a strong vertex cut of $G_I(\mathbb{Z}_2)$.

Case (ii): Suppose $x \neq 1$. Then we have $x \cdot 1 \cdot 1 \in I$. Hence, $x \in I$. Using part (b) of Theorem 9.6.23, a 3-prime ideal and c-prime ideal coincide. Thus, I is c-prime, and also a strong vertex of $G_I(N)$.

Remark 9.6.26 (Babushri, 2009; Remark 3.26 of Satyanarayana, Prasad, and Babushri, 2010)

(a) Part (a) of Theorem 9.6.23 does not hold in general if any one of the assumptions (i) and (ii) is excluded. We present the following examples.
  (i) Consider $G_{I_1}(N)$ in Example 9.6.7. I_1 is a 3-prime ideal of N. N is not a zero-symmetric near ring. Note that $G_{I_1}(N)$ is not ideal symmetric.
  (ii) Consider $G_{I_1}(N)$ in Example 9.6.6. N is a zero-symmetric near ring. I_1 is not a 3-prime ideal of N. Note that $G_{I_1}(N)$ is not ideal symmetric.
(b) The class of equiprime near rings is well known in near rings. A rigorous study of equiprime near rings can be found in Veldsman (1992) with an extensive set of examples. An equiprime near ring N is always zero-symmetric. If N is an equiprime near ring, then $I = \{0\}$ is an equiprime ideal of N. This further implies that $I = \{0\}$ is a 3-prime ideal of N. Hence, if N is an equiprime near ring, then $G_{\{0\}}(N)$ is an ideal symmetric (by part (a) of Theorem 9.6.23). Thus, if N is an equiprime near ring, then the prime graph of N is ideal symmetric. We summarize this discussion with the following corollary.

Corollary 9.6.27 (Corollary 3.27 of Satyanarayana, Prasad, and Babushri, 2010)

If N is an equiprime near ring, then $G_{\{0\}}(N)$ is ideal symmetric.

References

Abou-Zaid S. On fuzzy subnear-rings and ideals, *Fuzzy Sets Syst.* 4, 139–146 (1991).

Anderson F.W. and Fuller K.R. *Rings and Categories of Modules*, Springer-Verlag, New York, 1974.

Anderson D.F. and Livingston P.S. The zero-divisor graph of a commutative ring, *J. Algebra* 217, 434–447 (1999).

Anh P.N. and Marki L. Left orders in regular rings with minimum condition for principal one-sided ideals, *Math. Proc. Comb. Phil. Soc.* 109, 323–333 (1991).

Anh P.N. and Marki L. Orders in primitive rings with non-zero socle and posner's theorem, Commun. *Algebra* 24(1), 289–294 (1996).

Argac N. and Groenewald N.J. Weakly and strongly regular near-rings, *Algebra. Colloq.* 12, 121–130 (2005).

Atagun A.O. IFP ideals in near-rings, *Hacettepe J. Math. Stat.* 39(1), 17–21 (2010).

Babushri S.K. Fuzzy aspects of near-rings, Ph.D. Thesis, Manipal University, Manipal, India, 2009.

Babushri S.K., Prasad K.S. and Satyanarayana B. *C-Prime Fuzzy Ideals of Rn[x]* International Journal of Comtemp. Math. Sciences 8(3), 133–137 (2013).

Babushri S.K., Prasad K.S., and Satyanarayana B. Equiprime, 3-prime and c-prime fuzzy ideals of near-rings. Soft computing – a fusion of foundations, *Methodol. Appl.* 13, 933–944 (2009).

Babushri S.K., Prasad K.S. and Satyanarayana B. *Nearring Ideals, Graphs and Cliques*, International Mathematical Forum 8(2), 73–83 (2013).

Babushri S.K., Prasad K.S., and Satyanarayana B. Reference points and rough sets, *Inform. Sci.* (Elsevier, USA) 180, 3348–3361 (2010).

Babushri S.K, Satyanarayana B., and Prasad K.S. C-prime fuzzy ideals of near-rings, *Soochow J. Math.* 33(4), 891–901 (2007).

Barnes W.E. On the Γ– rings of nobusawa, *Pac. J. Math.* 18, 411–422 (1966).

Beck I. Coloring of commutative rings, *J. Algebra* 116, 208–226 (1988).

Beineke L.W., Wilson R.J., and Camerong P.J. *Topics in Algebraic Graph Theory*, Cambridge University Press, Cambridge, 2004.

Bell H.E. Near-rings in which each element is a power of itself, *Bull. Aust. Math. Soc.* 2, 363–368 (1970).

Bell H.E. Certain near-rings are rings, *J. Lond Math. Soc.* II 4, 264–270 (1971).

Birkenmeier G. and Groenewald N.J. Prime ideals in rings with involution, Dedicated to the memory of James R. Clay, *Quaestiones Math.* 20(4), 591–603 (1997).

Birkenmeier G. and Groenewald N.J. Near-rings in which each prime factor is simple, *Math. Pannon.* 10, 257–269 (1999).

Birkenmeier G. and Heatherly H. Medial near-rings, *Monatsh. Math.* 107, 89–110 (1989).

Birkenmeier G. and Heatherly H. Left self distributive near-rings, *J. Aust. Math. Soc., Ser. A* 49, 273–296 (1990).

Birkenmeier G., Heatherly H., and Lee E. Prime ideals in near-rings, *Results Math.* 24, 27–48 (1993).

Bondy J.A. and Murthy U.S.R. *Graph Theory with Applications*, The Macmillan Press Ltd, London, 1976.

Booth G.L. A note on gamma near-rings, *Stud. Sci. Math. Hung.* 23, 471–475 (1988).

Booth G.L. Radicals of gamma near-rings, *Publ. Math. Debrecen* 39, 223–230 (1990).

Booth G.L. Radicals in gamma near-rings, *Quaestiones. Math.* 14, 117–127 (1991).

Booth G.L. Fuzzy ideals in Γ-near-rings, *J. Fuzzy Math.* 11, 707–716 (2003).

Booth G.L. and Groenewald N.J. On radicals of gamma near-rings, *Math. Jpn.* 35, 417–425 (1990).

Booth G.L. and Groenewald N.J. Equiprime gamma near-rings, *Quaestiones Math.* 14, 411–417 (1991a).

Booth G.L. and Groenewald N.J. On primeness in matrix near-rings, *Arch. Math.* 56, 539–546 (1991b).

Booth G.L. and Groenewald N.J. Special radicals of near-ring modules, *Quaestiones Math.* 15(2), 127–137 (1992).

Booth G.L., Groenewald N.J., and Olivier W.A. A general type of regularity for gamma rings, *Quaestiones Math.* 14(4), 453–469 (1991).

Booth G.L., Groenewald N.J., and Veldsman S. A Kurosh-Amistur prime radical for near-rings, *Commun. Algebra* 18(9), 3111–3122 (1990).

Booth G.L., Groenewald N.J., and Veldsman S. Strongly equiprime near-rings, *Quaestiones Math.* 14(4), 483–489 (1991).

Camillo V. and Zelmanowitz J. On the dimension of a sum of modules, *Commun. Algebra* 6(4), 345–352 (1978).

Camillo V. and Zelmanowitz J. Dimension modules, *Pac. J. Math.* 91(2), 249–261 (1980).

Chatters A.W. and Hajarnavis C.R. *Rings with Chain Conditions*, Pitman Pub. Ltd., London, 1988.

Cho Y.U. A note on fuzzy ideals in gamma near-rings, *Far East J. Math. Sci.* (FJMS) 26, 513–520 (2007).

Davvaz B. Fuzzy ideals of near-rings with interval valued membership functions, *J. Sci. Islam. Repub. Iran* 12, 171–175 (2001).

Dutta T.K. and Biswas B.K. Fuzzy ideals of near-rings, *Bull. Cal. Math. Soc.* 89, 447–456 (1997).

Erdos P. and Renyi A. A symmetric graphs, *Acta Math. Hung.* 14(3–4), 295–315 (1963).

Ferrero C.C. and Ferrero G. *Nearrings: Some Developments Linked to Semigroups and Groups*, Kluwer Academic Publishers, The Netherlands, Amsterdam, 2002.

Fleury P. A note on dualizing goldie dimension, *Can. Math. Bull.* 17(4), 511–517 (1974).

Godsil C.C. and Royle G. *Algebraic Graph Theory*, Springer-Verlag, New York, 2001.

Golan J.S. Making modules fuzzy, *Fuzzy Sets Syst.* 32, 91–94 (1989).

Goldie A.W. The structure of noetherian rings, *Lectures on Rings and Modules*, Springer-Verlag, New York, 1972.

Groenewald N.J. Radical soft-rings, *Quaestiones Math.* 7(4), 337–344 (1984).

Groenewald N.J. The completely prime radical in near rings, *Acta Math. Hung.* 51(3–4), 301–305 (1988).

Groenewald N.J. Strongly prime near-rings—2, *Commun. Algebra* 17(3), 735–749 (1989).

Groenewald N.J. Different prime ideals in near-rings, *Commun. Algebra* 19, 2667–2675 (1991).

Groenewald N.J. and Olivier W.A. Generalised ideals of rings, *Quaestiones Math.* 15(4), 489–500 (1992).

Groenewald N.J. and Potgieter P.C. A generalization of prime ideals in near-rings, *Commun. Algebra* 12, 1835–1853 (1984).

Harary F. *Graph Theory*, India Naraosa Publishing House, New Delhi, 2001.

Herstein I.N. *Topics in Algebra*, Blaisdell, New York, 1964.

Hungerford T.W. *Algebra*, Rinehart and Winston. Inc., New York, 1974.

Jun Y.B. and Kim H.S. A characterization theorem for fuzzy prime ideal in near-rings, *Soochow J. Math.* 28(1), 93–99 (2002).

Jun Y.B., Kim K.H., and Ozturk M.A. On fuzzy ideals of gamma near-rings, *J. Fuzzy Math.* 9, 51–58 (2001a).

Jun Y.B., Kim K.H., and Ozturk M.A. Fuzzy maximal ideals of gamma near-rings, *Turk. J. Math.* 25, 457–463 (2001b).

Jun Y.B., Kwon Y.I., and Park J.W. Fuzzy MΓ-groups, *Kyungpook Math. J.* 35, 259–265 (1995).

Jun Y.B., Kim K.H., and Yon Y.H. Intuitionistic fuzzy ideals of near-rings, *J. Inst. Math. Comp. Sci., Math. Ser.* 12(3), 221–228 (1999).

Jun Y.B., Sapanci M., and Ozturk M.A. Fuzzy ideals in gamma near-rings, *Turk. J. Math.* 22, 449–459 (1998).

Kaarli K. and Kriis T. Prime radical near-rings, *Tartu Riikliku Ülikooli toimetised* 764, 23–29 (1987).

Kim S.D. and Kim H.S. On fuzzy ideals of near-rings, Bull. *Korean Math. Soc.* 33, 593–601 (1996).

Klir G.J. and Yuan B. *Fuzzy Sets and Fuzzy Logic, Theory and Applications*, Prentice-Hall of India Pvt. Ltd., New Delhi, 2002.

Lambek J. *Lectures on Rings and Modules*, Blaisdell Publishing Company, Waltham, MA, 1966.

Ligh S. On the commutativity of near-rings III, *Bull. Aust. Math. Soc.* 6, 459–464 (1972).

Mason G. Strongly regular near-rings, *Proc. Edinb. Math. Soc.* 23, 27–35 (1980).

Meldrum J.D.P. *Near-rings and their Links with Groups*, Pitman Advanced Publishing Program, Boston, MA, London, Melbourne, 1985.

Meldrum J.D.P. and Meyer J.H. Intermediate ideals in matrix nearrings, *Commun. Algebra* 24(5), 1601–1619 (1996).

Meldrum J.D.P. and Van der Walt A.P.J. Matrix nearrings, *Arch. Math.* 47, 312–319 (1986).

Meyer J.H. Chains of intermediate ideals in matrix nearrings, *Arch. Math.* 63, 311–315 (1994).

Murata K., Kurata Y., and Marubayashi H. A generalization of prime ideals in rings, *Osaka J. Math.* 6, 291–301 (1969).

Narsing D. *Graph Theory with Applications to Engineering and Computer Science*, Prentice-Hall of India Pvt. Ltd., New Delhi, 1997.

Nobusawa N. On a generalization of the ring theory, *Osaka J. Math.* 1, 81–89 (1964).

Pan F.-Z. Fuzzy finitely generated modules, *Fuzzy Sets Syst.* 21, 105–113 (1987).

Pan F.-Z. Fuzzy quotient modules, *Fuzzy Sets Syst.* 28, 85–90 (1988).

Pilz G. Near-Rings, Revised edition, North Holland (1983).

Ramakotaiah D. Theory of near-rings. Ph.D. Thesis, submitted to Andhra University, Waltair, A.P., India, 1967.

Reddy Y.V. and Satyanarayana B. The f-prime radical in near-rings, *Indian J. Pure Appl. Math.* 17, 327–330 (1986).

Reddy Y.V. and Satyanarayana B. A note on modules, *Proc. Jpn. Acad.* 63, Series 6, 208–211 (1987).

Reddy Y.V. and Satyanarayana B. A note on N-groups, *Indian J. Pure Appl. Math.* 19, 842–845 (1988a).

Reddy Y.V. and Satyanarayana B. Finite spanning dimension in N-groups, *Math. Stud.* 56, 75–80 (1988b).

Saikia H.K. On fuzzy N-subgroups and fuzzy ideals of near-rings and near-ring groups, *J. Fuzzy Math.* 11, 567–580 (2003).

Saikia H.K. and Barthakur L.K. On fuzzy N-groups and fuzzy ideals of near-rings and near-ring groups, *J. Fuzzy Math.* 11, 567–580 (2003).

Sambasivarao V. A characterization of semiprime ideals in nearrings, *Aust. Math. Soc., Ser. A* 32, 212–214 (1982).

Sambasivarao V. and Satyanarayana B. The prime radical in near-rings, *Indian J. Pure Appl. Math.* 15, 361–364 (1984).

Santhakumari C. On a class of near-rings, *J. Aust. Math. Soc., Ser. A* 23, 167–170 (1982).

Satyanarayana B. Tertiary decomposition in Noetherian N-groups, *Commun. Algebra* 10, 1951–1963 (1982).

Satyanarayana B. Primary decomposition in Noetherian near-rings, *Indian J. Pure Appl. Math.* 15, 127–130 (1984a).

Satyanarayana B. Contributions to near-ring theory, doctoral dissertation, Acharya Nagarjuna University, 1984b.

Satyanarayana B. On modules with finite spanning dimension, *Proc. Jpn. Acad.* 61-A, 23–25 (1985).

Satyanarayana B. A note on E-direct and S-inverse systems, *Proc. Jpn. Acad.* 64-A, 292–295 (1988).

Satyanarayana B. The injective hull of a module with FGD, *Indian J. Pure Appl. Math.* 20, 874–883 (1989).

Satyanarayana B. On modules with finite goldie dimension, *J. Ramanujan Math. Soc.* 5, 61–75 (1990).

Satyanarayana B. On finite spanning dimension in N-groups, *Indian J. Pure Appl. Math.* 22, 633–636 (1991).

Satyanarayana B. Modules with finite spanning dimension, *J. Aust. Math. Soc., Ser. A* 57, 170–178 (1994).

Satyanarayana B. On essential E-irreducible submodules, *Proc., 4th Ramanujan Symposium on Algebra and its Applications*, University of Madras Feb 1–3, 1995, pp 127–129.

Satyanarayana B. The f-prime radical in Γ-near-rings, *South East Asian Bull. Math.* 23, 507–511 (1999a).

Satyanarayana B. A Note on Γ-near-rings, Indian J. Math. (B.N. Prasad Birth Centenary commemoration volume) 41, 427–433 (1999b).

Satyanarayana B. Modules over gamma nearrings, *Acharya Nagarjuna Int. J. Math. Inform. Technol.* 1(2), 109–120 (2004).

Satyanarayana B. A note on completely semiprime ideals in nearrings, *Int. J. Comput. Math. Ideas* 1(3), 107–112 (2009).

Satyanarayana B. *Contributions to Near-ring Theory*, VDM Verlag Dr Muller, Germany, 2010, (ISBN 978-3-639-22417-7).

Satyanarayana B., Godloza L., and Vijaya Kumari A.V. Some dimension conditions in near-rings with finite dimension, *Acta Cienc. Indica* 34, 1397–1404 (2008a).

Satyanarayana B., Godloza L., and Vijaya Kumari A.V. Finite dimension in near-rings, *J. AP Soc. Math. Sci.* 1(2), 62–80 (2008b).

Satyanarayana B., Godloza L., Babu P.M., and Prasad K.S. Ideals and direct products of zero-square near-rings, *Int. J. Algebra* 4, 13–16 (2010).

Satyanarayana B. and Koteswara Rao G. On a class of modules and N-groups, *J. Indian Math. Soc.* 59, 39–44 (1993).

Satyanarayana B., Lokeswara Rao M.B.V., and Prasad K.S. A note on primeness in near-rings and matrix near-rings, *Indian J. Pure Appl. Math.* 27, 227–234 (1996).

Satyanarayana B. and Mohiddin Shaw S. *Fuzzy Dimension of Modules over Rings*, VDM Verlag Dr Muller, Germany, 2010.

Satyanarayana B. and Nagarju D. *Dimension and Graph Theoretic Aspects of Rings*, VDM Verlag Dr Muller, Germany, 2011.

Satyanarayana B. and Prasad K.S. A result on E-direct systems in N-groups, *Indian J. Pure Appl. Math.* 29, 285–287 (1998).

Satyanarayana B. and Prasad K.S. On direct and inverse systems in N-groups, *Indian J. Math.* (BN Prasad Birth Commemoration Volume) 42, 183–192 (2000).

Satyanarayana B. and Prasad K.S. An isomorphism theorem on directed hypercubes of dimension n, *Indian J. Pure Appl. Math.* 34(10), 1453–1457 (2003).

Satyanarayana B. and Prasad K.S. On fuzzy cosets of gamma near-rings, *Turk. J. Math.* 29, 11–22 (2005a).

Satyanarayana B. and Prasad K.S. Linearly independent elements in N-groups with finite goldie dimension, *Bull. Korean Math. Soc.* 42(3), 433–441 (2005b).

Satyanarayana B. and Prasad K.S. On finite goldie dimension of $M_n(N)$-group N_n, in *Near-rings and Near-fields* (Editors: H. Kiechle, A. Kreuzer, and M.J. Thomsen) (2005c), *(Proc., 18th International Conference on Near-rings and Near-fields, Universitat Bundeswar, Hamburg, Germany July 27–Aug 03, 2003)*, Springer Verlag, The Netherlands, 2005, pp 301–310.

Satyanarayana B. and Prasad K.S. *Discrete Mathematics & Graph Theory*, Prentice Hall of India, New Delhi, 2009 (ISSN:978-81-203-3842-5).

Satyanarayana B., Prasad K.S., and Babushri S.K. Graph of a near-ring with respect to an ideal, Commun. Algebra (USA) Taylor & Francis 38(5), 1957–1967 (2010).

Satyanarayana B., Prasad K.S., and Nagaraju D. A theorem on modules with finite goldie dimension, *Soochow J. Math.* 32(2), 311–315 (2006).

Satyanarayana B., Prasad K.S., and Nagaraju D. Prime graph of a ring, *J. Combin. Inform. Syst. Sci.* 35, 1–12 (2010).

Satyanarayana B., Prasad K.S., and Nagaraju D. Homomorphisms of Graphs, communicated.

Satyanarayana B., Prasad K.S., and Pradeepkumar T.V. On IFP N-groups and fuzzy IFP ideals, *Indian J. Math.* 46, 11–19 (2004).

Satyanarayana B., Prasad K.S., and Pradeepkumar T.V. On fuzzy dimension of N-groups with DCC on ideals, *East Asian Math. J.* 21(2), 205–216 (2005).

Satyanarayana B., Prasad K.S., Pradeepkumar T.V., and Srinivas T. Some results on fuzzy cosets and homomorphisms of N-groups, *East Asian Math. J.* 23(1), 23–36 (2007).

Satyanarayana B., Prasad K.S., and Ramaprasad J.L. Fuzzy cosets of prime fuzzy submodules, *J. Fuzzy Math.* (USA) 17(3), 595–603 (2009).

Satyanarayana B. and Ramaprasad J.L. *Fuzzy Prime Submodules*, VDM Verlag Dr Muller, Germany, 2010 (ISBN 978-3-639-24355-0).

Satyanarayana B. and Vijaya Kumari A.V. Prime Graph of a Nearring, communicated.

Satyanarayana B., Vijaya Kumari A.V., Godloza L., and Nagaraju D. Fuzzy ideals of modules over Γ-nearrings, *Ital. J. Pure Appl. Math.*, accepted for publication.

Satyanarayana B., Vijaya Kumari A.V., and Nagaraju D. Fuzzy cosets in modules over gamma near-rings, *IJCMI*, accepted.

Satyanarayana B. and Wiegandt R. On the f-prime radical of near-rings, in *Near-rings and Near-fields* (Editors: H. Kiechle, A. Kreuzer, and M.J. Thomsen) (2005) (*Proc., 18th International Conference on Near-rings and Near-fields, Universitat Bundeswar, Hamburg, Germany, July 27–Aug 03, 2003*), Springer Verlag, The Netherlands, 2005, pp 293–299.

Prasad K.S. Contributions to near-ring theory II, doctoral dissertation, Acharya Nagarjuna University, India, 2000.

Prasad K.S. and Satyanarayana B. A note on IFP N-groups, *Proc., 6th Ramanujan Symposium on Algebra and its Applications, Ramanujan Institute for Adv. Study in Mathematics*, University of Madras, Feb 24–26, 1999, pp 62–65.

Prasad K.S. and Satyanarayana B. Fuzzy prime ideal of a gamma nearing, *Soochow J. Math.* 31, 121–129 (2005).

Prasad K.S. and Satyanarayana B. *Finite Dimension in N-groups and Fuzzy Ideals of Gamma Near-rings*, VDM Verlag, Germany, 2011 (ISBN: 978-3-639-36838-3).

Van der Walt A.P.J. Prime ideals and nil radicals in near-rings, *Arch. Math.* 15, 408–416 (1964).

Van der Walt A.P.J. Weakly prime one-sided ideals, *J. Aust. Math. Soc.* 38(1), 84–91 (1985).

Veldsman S. On equiprime near-rings, *Commun. Algebra* 20(9), 2569–2587 (1992).

Wiegandt R. Radical theory of rings, *Math. Stud.* 51, 145–183 (1983).

Wisbauer R. *Foundations of Modules and Ring Theory*, Gordon and Breach, 1991.

Zadeh L.A. Fuzzy sets, *Inform. Control* 8, 338–353 (1965).

Zimmermann H.J. *Fuzzy Set Theory and Its Applications*, Kluwer, The Netherlands, 1985.

Index